# Analytical Laser Spectroscopy

# CHEMICAL ANALYSIS

## A SERIES OF MONOGRAPHS ON ANALYTICAL CHEMISTRY AND ITS APPLICATIONS

*Editors*

## P. J. ELVING   J. D. WINEFORDNER

*Editor Emeritus:* I. M. KOLTHOFF

VOLUME 50

**A WILEY-INTERSCIENCE PUBLICATION**

## JOHN WILEY & SONS

**New York / Chichester / Brisbane / Toronto**

# Analytical Laser Spectroscopy

*Edited by*

**NICOLÒ OMENETTO**

Institute of Inorganic and General Chemistry
University of Pavia
Pavia, Italy

**A WILEY-INTERSCIENCE PUBLICATION**

**JOHN WILEY & SONS**

**New York / Chichester / Brisbane / Toronto**

*Library of Congress Cataloging in Publication Data*:

Main entry under title:

Analytical laser spectroscopy.

(Chemical analysis; v. 50)
"A Wiley-Interscience publication."
Includes index.
1. Laser spectroscopy.  I. Omenetto, Nicolò, 1939–
II. Series.

QD96.L3A5   543'.085   78-7977
ISBN 0-471-65371-3

# AUTHORS

**S. Cova**
Istituto di Fisica
Politecnico di Milano
Milano, Italy

**W. Demtröder**
Fachbereich Physik
Universität Kaiserslautern
Kaiserslautern, Germany

**R. A. Keller**
Los Alamos Scientific Laboratory
Los Alamos, New Mexico

**K. Laqua**
Institut für Spektrochemie und Angewandte Spektroskopie
Dortmund, Germany

**A. Longoni**
Istituto di Fisica
Politecnico di Milano
Milano, Italy

**R. M. Measures**
Institute for Aerospace Studies
University of Toronto
Toronto, Canada

**N. Omenetto**
Istituto di Chimica Generale
Università di Pavia
Pavia, Italy

**E. H. Piepmeier**
Department of Chemistry
Oregon State University
Corvallis, Oregon

**C. A. Sacchi**
Centro di Studio per l'Elettronica Quantistica del CNR
Istituto di Fisica del Politecnico
Milano, Italy

**O. Svelto**
Centro di Studio per l'Elettronica Quantistica del CNR
Istituto di Fisica del Politecnico
Milano, Italy

**J. C. Travis**
Analytical Chemistry Division
United States Department of Commerce
National Bureau of Standards
Washington, D.C.

**J. D. Winefordner**
Department of Chemistry
University of Florida
Gainesville, Florida

# PREFACE

Atomic and molecular spectroscopy is one of the research fields on which the use of lasers has had a dramatic impact. New spectral information, difficult or impossible to gather by classical spectroscopy, extremely high resolution spectroscopy of atoms and molecules made possible by the overcoming of the Doppler effect, selective excitation and detection of single atomic and molecular quantum states are just a few typical examples of how laser sources have revolutionized the field, offering a unique approach to challenging problems of both fundamental and applied natures.

Several excellent books on laser spectroscopy and related topics have appeared in the past. However, these books are mainly concerned with physical applications of lasers and are generally characterized by highly specialized theoretical treatments. This book is devoted to the analytical applications of lasers in spectroscopy. Because of the great variety of applications in chemistry, biology, engineering, and related branches of science, research papers and reviews are currently published at an impressive rate. As a consequence it is practically impossible to offer an exhaustive and updated coverage of the subject. Although the analytical flavor is dominant, this book is not intended to merely provide a review of well established procedures. Very often, but especially in the field of laser spectroscopy, the analytical implementations of a new phenomenon are indeed hidden in the somewhat complicated physics behind it. Thus this work is also intended to stimulate the reader in deepening his or her understanding of the fundamentals of interaction of lasers with atoms and molecules.

The organization of the book reflects my personal evaluation of the topics to be selected for a broad coverage of the field. Chapter 1 introduces the basic physical concepts with emphasis on properties such as collimation, power, tunability, and coherence, which make lasers so peculiar for spectroscopic applications. Most types of lasers and their operational characteristics are described so as to provide the reader with a general theoretical background.

Atomic emission and absorption spectroscopy are two techniques in which the laser properties have already been, or can certainly be, exploited to the greatest advantage. The laser can play here the dual role of atomizer, that is, to produce the atomic vapor, and of primary excitation source. These two topics are treated separately in Chapters 2 and 3. Chapter 3 deals with the important subject of line profiles and broadening effects likely to be observed in most common atom reservoirs. Laser-induced atomic fluorescence is covered in Chapter 4, the discussion being confined to pulsed tunable dye lasers and flames burning at atmospheric pressure.

Combustion diagnostics by means of laser-induced fluorescence is a timely topic that will receive increasing attention in the future. The molecular counterpart of Chapters 3 and 4 is described in Chapter 5, where the reader will find a discussion on techniques such as Doppler-free, two-photon spectroscopy, laser-induced photochemistry, and optoacoustic spectroscopy. Today the importance of studying the environment is evident to most scientists. LIDAR techniques (Light Detection and Ranging) for probing atmospheric and water pollution are covered in Chapter 6.

Chapter 7 is based on a somewhat new approach to treating signals, noise, and measurements, which may not be familiar to most applied spectroscopists but should certainly be useful in covering the gap between existing literature on electronic instruments and their fundamental principles based on the theory of statistical communication, systems, probability, and random functions. Finally, Chapter 8 deals with specialized applications of analytical interest such as laser intracavity absorption, microfluorescence (of extreme importance and attractiveness in biology), isotopic analysis, and others.

During the planning of the book, I have attempted to reduce the overlap between chapters to a minimum. The repetition that remains may be helpful rather than detrimental.

Apart from some knowledge of the physical sciences, no special background is required for the reader to fully understand each subject. I hope that the effort made to collect most analytical applications of laser spectroscopy in one volume will be helpful for analytical chemists and for more physically oriented researchers, as well as for students and teachers. Due to the variety of applications reported, I also hope that this volume will be useful to the large majority of nonanalytical chemists.

The contributors to this volume are well known and active researchers in this field. Any merit of the book must certainly be attributed to them. Because this adventure started in 1974, literature coverage varies somewhat from one chapter to another and I apologize for the material which has necessarily been missed.

NICOLÒ OMENETTO

*Pavia, Italy*
*July 1978*

# CONTENTS

# Analytical Laser Spectroscopy

1

# BASIC PRINCIPLES
# OF LASERS

## C. A. SACCHI and O. SVELTO

*Centro di Studio Per l'Elettronica Quantistica del CNR*
*Istituto di Fisica del Politecnico*
*Milano, Italy*

## 1.1   INTRODUCTORY CONCEPTS

The laser utilizes three important effects that occur in the interaction between an electromagnetic wave and matter: spontaneous emission, stimulated emission, and absorption.

### 1.1.1   Spontaneous Emission, Stimulated Emission, and Absorption

Let us consider two levels (1 and 2) of an atomic system, with energy $E_1$ and $E_2$ $(E_2 > E_1)$, equal statistical weights, and population (i.e., number of atoms per unit volume) $N_1$ and $N_2$, respectively. If an atom of the system is initially on level 2, it will decay on level 1, losing the energy $E_2 - E_1$. When this energy is emitted toward the surrounding world as electromagnetic radiation, the emission is called *spontaneous emission*. The frequency $\nu$ (Hz) of the emitted radiation is given by the well-known expression

$$\nu = (E_2 - E_1)/h \tag{1}$$

where $h$ is the Planck constant. The rate of spontaneous decay (per unit volume) can be written as

$$\left(\frac{dN_2}{dt}\right)_{sp} = A_{21}N_2 \tag{2}$$

where $A_{21}$ $(\text{s}^{-1})$ is a coefficient called *the Einstein coefficient for spontaneous emission*. Its reciprocal value $\tau_{sp} = 1/A_{21}$ is the *spontaneous emission lifetime* (radiative lifetime), which is a characteristic parameter for a given transition. The decay $2 \rightarrow 1$ may also occur through nonradiative processes; in fact, corresponding energy may be given to neighboring molecules as translational energy or, more commonly, as internal energy. In the case of nonradiative decay, the rate of decays per unit volume can be written as

$$\left(\frac{dN_2}{dt}\right)_{nr} = -\frac{N_2}{\tau_{nr}} \tag{3}$$

where $\tau_{nr}$ is the *nonradiative decay time*. This parameter depends on the transition considered and on the surrounding molecules.

Let us now consider an atom initially on level 2. A plane electromagnetic (e.m.) wave with frequency $\nu$ given by Eq. (1) can interact with this atom, stimulating a transition from level 2 to level 1. The energy $E_2 - E_1$ is now emitted as electromagnetic radiation with the same frequency and phase as the incident wave. This effect is called *stimulated emission*. The rate of stimulated emissions per unit volume can be written as

$$\left(\frac{dN_2}{dt}\right)_{se} = -W_{21}N_2 \tag{4}$$

The stimulated emission probability, $W_{21}$ (s$^{-1}$), depends on the atomic system considered and on the irradiance of the incident radiation:

$$W_{21} = \sigma_{21}F \tag{5}$$

where $F$ (cm$^{-2}$ s$^{-1}$) is the photon flux, that is, the number of photons per unit surface and per unit time, and $\sigma_{21}$ (cm$^2$) is a characteristic atomic parameter with the dimensions of area, called the stimulated emission cross section.

As a final case, let us consider an atom that is initially in level 1 and interacts with an electromagnetic wave of frequency $\nu$. The atom may now undergo a transition to level 2, absorbing the required energy from the incident radiation. This is the well-known phenomenon of *absorption*. The rate of transitions per unit volume due to absorption can be written as

$$\left(\frac{dN_1}{dt}\right)_a = -W_{12}N_1 \tag{6}$$

Again, for a plane e.m. wave, the absorption probability $W_{12}$ can be written as

$$W_{12} = \sigma_{12}F \tag{7}$$

where $F$ is the photon flux and $\sigma_{12}$ is the absorption cross section.

In conclusion we may note that, as shown by Einstein, the coefficients $\sigma_{12}$ and $\sigma_{21}$ are equal for any transition. The quantity

$$\sigma = \sigma_{12} = \sigma_{21} \tag{8}$$

from now on will be called the *transition cross section*.

### 1.1.2 The Laser Concept

Let us now consider a plane electromagnetic wave propagating along the direction $z$ (Fig. 1.1) and interacting with an ensemble of atoms. Because of the simultaneous occurrence of both stimulated emission and absorption processes, the photon flux variation in a distance $dz$ is given by

$$dF = \sigma F(N_2 - N_1)\, dz \tag{9}$$

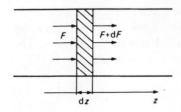

**Fig. 1.1.** Elemental change $dF$ in the photon flux $F$ for a plane electromagnetic wave in traveling a distance $dz$ through the material.

Equation (9) has been derived using Eqs. (4)–(8) and is based on a simple balance argument: each stimulated emission generates a photon, while each absorption process destroys a photon. *Equation (9) shows that the material behaves as an amplifier ($dF/dz > 0$) or an absorber ($dF/dz < 0$) according to the sign of $N_2 - N_1$.* At thermodynamic equilibrium, the population distribution is governed by the Boltzmann statistical law:

$$\frac{N_2^e}{N_1^e} = \exp[-(E_2 - E_1)/kT] \tag{10}$$

Since $N_2^e < N_1^e$, we have from Eq. (9) that any material at thermodynamic equilibrium behaves as an absorber. However, if we succeed in creating a nonequilibrium situation such that $N_2 > N_1$, the material will behave as an amplifier at frequency $\nu$. Such a situation is called *population inversion* since the population distribution is inverted with respect to equilibrium. Correspondingly, the material is called "active material." If the transition frequency $\nu$ falls in the optical frequency range (in practice from the IR to the UV) the amplifier is called LASER, which is an acronym for the expression "Light Amplification by Stimulated Emission of Radiation." In the microwave region the device is called MASER.

To obtain an oscillator from an amplifier, it is necessary to introduce a suitable positive feedback. In the laser, this is done by inserting the active material between two mirrors (for example, two plane parallel mirrors, see Fig. 1.2). The electromagnetic wave bounces back and forth along the direction normal to the mirrors, being amplified at each transit through the active material. If one of the mirrors (for example, mirror 2) is partially transparent, the light transmitted through it will constitute the output beam. The oscillation will start when the gain (per pass) equals the unavoidable losses (per pass) of the system (for example, the mirror

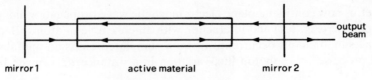

mirror 1                    active material                    mirror 2

**Fig. 1.2.** Schematic of laser design.

losses). If we call $R_1$ and $R_2$ the power reflectivities of the mirrors and ignore other losses, the threshold condition for the laser oscillation can be written, after Eq. (9), as follows:

$$R_1 R_2 \exp[2\sigma(N_2 - N_1)l] = 1 \qquad (11)$$

where $l$ is the length of the active material. Equation (11) expresses the condition that the electromagnetic radiation, after reflection at the first mirror (reflectivity $R_1$), amplification in the active material (gain $\exp[\sigma(N_2 - N_1)l]$), reflection at the second mirror (reflectivity $R_2$), and reamplification in the active material, regains its initial irradiance. The same Eq. (11) indicates that oscillation starts when the population inversion $(N_2 - N_1)$ reaches the threshold value

$$(N_2 - N_1)_{thr} = (\tfrac{1}{2}\sigma l) \ln(1/R_1 R_2)$$

When this happens, the oscillation is initiated by spontaneous emission, that is, by those photons that are spontaneously emitted normal to the mirror. Such an oscillator is called a *laser oscillator* or, more simply, *laser*.

### 1.1.3 Properties of the Laser

Laser radiation is characterized by extremely high values of the following properties: (i) directionality, (ii) monochromaticity, (iii) coherence, and (iv) radiance. We shall now describe these properties in some detail.

#### *Directionality*

This property is a direct consequence of the fact that the active material is placed in a resonant cavity formed by two plane parallel mirrors (Fig. 1.2). In fact, only a wave propagating along the direction normal to the mirrors can oscillate. From diffraction theory it is well known that a spatially coherent beam with aperture $D$ has an intrinsic divergence, due to diffraction, given by

$$\alpha_d \simeq \lambda/D \qquad (12)$$

where $\lambda$ is the radiation wavelength. If the wave is not spatially coherent, its divergence may be much higher than the minimum value given by Eq. (12). A beam the divergence of which is given by Eq. (12) is called *diffraction limited*; such beams are generated by the lasers at optimal operating conditions.

#### *Monochromaticity*

The monochromaticity of laser radiation is mainly due to the following effects: first, the resonant character of the interaction between the electromagnetic radiation and the matter, with gain in the active material, which enhances the central frequency in the interaction band; and second, the presence of a resonant cavity formed by the mirrors, which allows oscillation of characteristic frequencies only. Therefore, the linewidth of the laser is typically much narrower (by three to several orders of magnitude) than the linewidth of the passive transition as observed in spontaneous emission.

### Coherence

This can be distinguished as *spatial* coherence and *temporal* coherence. To define spatial coherence, let us consider two points $P_1$ and $P_2$ of the wave examined which, at $t = 0$, are on the same wavefront. By definition, the phases of the electric (or magnetic) fields are the same (at $t = 0$) at the two points. If the phases remain equal for $t > 0$, the two points are said to be coherent. If this happens for any two points of the wavefront, the wave is said to have perfect spatial coherence.

To define temporal coherence, let us consider the electric field of the wave at a given point and at two successive instants $t$ and $t + \tau$. If the phase difference remains constant for a given value of $\tau$ and any $t$, and if this continues to be true for any value of $\tau$, the wave is said to have a perfect temporal coherence. If this happens only for $0 < \tau < \tau_0$, we can say that the wave is partially coherent, with a coherence time $\tau_0$. Figure 1.3 gives an example of a wave coherent for time $\tau_0$; in fact, its phase changes abruptly at each time interval $\tau_0$. The concept of temporal coherence is obviously related to that of monochromaticity. Thus, a wave with bandwidth $\Delta\nu$ has a coherent time $\tau_0 \simeq 1/\Delta\nu$. The two concepts of spatial and temporal coherence are, of course, quite different: an electromagnetic wave may possess partial temporal coherence even if it is perfectly coherent in space. For example, if the wave in Fig. 1.3 represents the electric field at any point of the wavefront, the wave has perfect spatial coherence but is only partially coherent in time (over $\tau_0$).

### Radiance

For an elementary radiation source with emitting surface $dS$ (cm²), the radiant power $W$ emitted in the solid angle $d\Omega$ (sr), at an angle $\theta$ with the

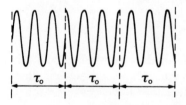

**Fig. 1.3.** Example of an electromagnetic wave with a coherence time of approximately $\tau_0$.

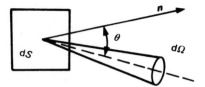

**Fig. 1.4.** Definition or radiance.

surface normal (Fig. 1.4), can be written as:

$$dP = B \cos \theta \, dS \, d\Omega \qquad (13)$$

The factor $\cos \theta$ accounts for the fact that, as far as the emission is concerned, only the projection of $dS$ on the plane normal to the emission direction is important. The coefficient $B$ is called the radiance of the source along the direction considered. If $B$ is independent of $\theta$ and $dS$, the source is isotropic (Lambert's source). The radiance of the laser is much higher (by four to 10 orders of magnitude) than the radiance of the best incoherent sources. This is a consequence of the high power and, most important, the high directionality of the laser radiation.

### 1.1.4 Pumping Schemes

We shall now describe the schemes most commonly used to obtain a population inversion. These schemes usually utilize three or more energy levels of the material system, since the interaction of an electromagnetic wave with a two-level system cannot lead to a population inversion. In fact, in this case, as the population of the upper level grows during the excitation process, the stimulated emission becomes more and more important, thus, in competition with the absorption, leading to equalization for the populations of the two levels (see Sect. 1.2.6).

The three-level laser is shown in Fig. 1.5a. In this case, the atoms are excited, in a suitable way, from level 1 to level 3. Since this excitation, because of the properties of the system, is followed by a very rapid decay to level 2, population inversion will be obtained between levels 1 and 2. The rapid decay from level 3 to level 2 is the crucial point of this scheme. In fact, not only does it generate a rapid population of level 2, but it also keeps level 3 practically empty, thus avoiding stimulated emissions that would compete with the excitation process.

The four-level laser is shown in Fig. 1.5b. Here, the atoms are excited from level 0 to level 3. If they rapidly decay to level 2, a population inversion can be obtained between levels 1 and 2. Since the laser action will bring the atoms to level 1, this level must also remain practically empty through rapid decay to level 0. It is easier to obtain a population inversion in a four-level laser than in a three-level laser. In fact, the lower

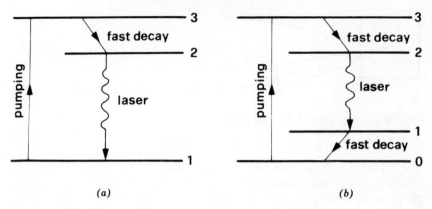

**Fig. 1.5.** Schemes for (a) three-level and (b) four-level lasers.

level of laser transition in a four-level laser (level 1) does not coincide with the ground state (as in the three-level laser) and is practically empty under equilibrium conditions provided that $E_2 - E_1$ is appreciably higher than $kT$ [see Eq. (10)].

The excitation process that brings the atoms from the ground state to excited level 3 is called "pumping" and can be achieved in several ways (see Sect. 1.3). Here, we would point out that the rate of atoms per unit volume brought to level 2 by the pumping process can be written as

$$\left(\frac{dN_2}{dt}\right)_p = W_p N_g \tag{14}$$

The constant $W_p$ is called "pumping rate"; it depends on the material and on the pumping system. $N_g$ is the ground state population.

## 1.2  RADIATION–MATTER INTERACTION

In this section we discuss the stimulated and spontaneous emission processes in more detail.

### 1.2.1  Stimulated Transitions

The formal calculation of the coefficients $W_{12}$ and $W_{21}$ defined by Eqs. (4) and (6) will not be given here. On the assumptions of (i) monochromatic electromagnetic radiation, (ii) transition mainly due to the interaction between the electric field of the wave and the electric dipole moment of the atom ("electric dipole interaction"), and (iii) a radiation wavelength much larger than the atomic dimension ("electric dipole approximation"), it can be shown (1) that the quantity $W_{12}$ is given (in mks units) by

$$W_{12} = \frac{\pi}{3\eta\epsilon_0 c_0 \hbar^2} |\mathbf{\mu}_{21}|^2 \, Ig(\omega - \omega_0) \tag{15}$$

where $\eta$ is the refractive index of the medium. $\epsilon_0$ is the vacuum dielectric constant, $c_0$ is the vacuum light velocity, $I$ (W m$^{-2}$) is the irradiance of the incident plane wave, and $|\mathbf{\mu}_{21}|$ is the absolute value of the matrix element of the electric dipole,

$$\mathbf{\mu}_{21} = \langle u_2 | e\mathbf{r} | u_1 \rangle \tag{16}$$

in which $|u_1\rangle$ and $|u_2\rangle$ are the eigenfunctions of the atomic system in levels 1 and 2, respectively; $\omega$ is the angular frequency of the radiation, $\omega_0$ is the frequency of the atomic transition $\omega_0 = (E_2 - E_1)/\hbar$, and the function $g(\omega - \omega_0)$ gives the transition lineshape and is normalized so that

$$\int g(\omega - \omega_0) \, d\omega = 1 \tag{17}$$

The quantity $W_{21}$ is obtained from Eq. (15) by exchanging the subscripts 1 and 2. Since $|\mathbf{\mu}_{12}| = |\mathbf{\mu}_{21}|$, as shown by Eq. (16), it follows that $W_{12} = W_{21}$, as already stated in Sect. 1.1. Equation (15) can be expressed in terms of the energy density $\rho$ (J m$^{-3}$) of the wave. Since for a plane wave $I = c_0\rho/\eta$, we have

$$W = \frac{\pi}{3\eta^2\epsilon_0\hbar^2} |\mathbf{\mu}|^2 \, \rho g(\Delta\omega) \tag{15a}$$

where we have put $W = W_{12} = W_{21}$, $|\mathbf{\mu}| = |\mathbf{\mu}_{12}| = |\mathbf{\mu}_{21}|$, and $\Delta\omega = \omega - \omega_0$. The transition cross section can now be obtained. From Eq. (5) we have

$$\sigma = W\frac{\hbar\omega}{I} \tag{18}$$

and from Eq. (15),

$$\sigma = \frac{\pi}{3\eta\epsilon_0 c_0 \hbar} |\mathbf{\mu}|^2 \, \omega g(\Delta\omega) \tag{19}$$

Finally, from Eq. (9), we can set

$$dI = -\alpha I \, dz \tag{20}$$

where

$$\alpha = \sigma(N_1 - N_2) \tag{21}$$

is called the *absorption coefficient* (if $N_1 > N_2$) or the *gain coefficient* (if $N_2 > N_1$). The three coefficients $W$, $\sigma$, and $\alpha$, related to each other by

Eqs. (18) and (21), make it possible to describe the stimulated transitions. Of these coefficients, only $\sigma$ depends exclusively on the considered transition, while the other two also depend either on the irradiance ($W$) or on the atomic density ($\alpha$).

## 1.2.2 Line-Broadening Mechanisms

The lineshape function $g(\omega - \omega_0)$ depends on the mechanisms that broaden the transition line. In general, these mechanisms can be divided into homogeneous and inhomogeneous ones. A broadening mechanism is homogeneous if it broadens in the same way the line of each individual atom and, therefore, of the entire system. A broadening mechanism is inhomogeneous if it spreads the resonance frequencies of the atoms over some spectral range, thereby broadening the overall system line without broadening the line of each individual atom. In a liquid or a gas, a typical homogeneous mechanism is exemplified by the atomic collisions. This appears clearly from the Heisenberg principle: since the collision time $\tau_c$ has a finite duration, the transition linewidth cannot be narrower than $1/\tau_c$. It can be shown (1) that in this case the function $g(\omega - \omega_0)$ is given by

$$g(\omega - \omega_0) = \frac{\tau_c}{\pi} \frac{1}{1 + (\omega - \omega_0)^2 \tau_c^2} \tag{22}$$

This function represents a Lorentz lineshape: its full width at half maximum, $\Delta\omega_0$, as obtained from Eq. (22) is

$$\Delta\omega_0 = \frac{2}{\tau_c} \tag{23}$$

Such a lineshape is also found with another homogeneous broadening mechanism, which is due to the spontaneous emission. Since this last phenomenon occurs for any material (solid, liquid, or gas) the corresponding broadening is called *natural*, or *intrinsic*. In this case the lineshape is again given by Eq. (22) with substitution of $\tau_{sp}/2$ for $\tau_c$, where $\tau_{sp}$ is the spontaneous emission lifetime.

For a solid material, a typical homogeneous mechanism arises from the interaction of the atom with the lattice vibrations (phonons). An important inhomogeneous mechanism typical of the gas (and liquid) state arises from the Doppler effect. Indeed, due to this effect the frequency of the electromagnetic wave is seen at a different value by each atom because of its velocity. Another way of expressing the same effect is to say that, because of the Doppler effect, each atom has a different transition frequency, according to its velocity. For example, if a given atom moves with a velocity component $v_i$ opposite to the direction of propagation of the electromagnetic wave, the interaction frequency is (nonrelativistic)

$$\nu_i = \nu_0\left(1+\frac{v_i}{c}\right)$$

This effect of frequency shift results in an overall line formed by an ensemble of homogeneously broadened Lorentzian lines centered at different transition frequencies. It can be shown (1) that the function $g(\omega - \omega_0)$ is given in this case by

$$g(\omega - \omega_0) = \frac{c}{\omega_0}\left[\left(\frac{M}{2\pi kT}\right)^{1/2}\right]\exp\left[-\frac{Mc^2}{2kT}\frac{(\omega - \omega_0)^2}{\omega_0^2}\right] \tag{24}$$

in which $M$ is the mass of the atom (or molecule) and $c$ is the velocity of light in the examined medium. This lineshape is Gaussian since it reflects the Gaussian random distribution of the physical origin of the broadening. Its full width at half-maximum $\Delta\omega_0^*$ is easily obtained by Eq. (24) as

$$\Delta\omega_0^* = 2\omega_0\left[\left(\frac{2kT}{Mc^2}\ln 2\right)^{1/2}\right] \tag{25}$$

The same type of function is found with another inhomogeneous mechanism, the inhomogeneous distribution of the local field given by the neighboring molecules. This effect gives a spread of the energy levels and therefore of the transition frequencies. The linewidth depends in this case on the spatial average fluctuation of the local field.

### 1.2.3   Spontaneous Decay

Precise calculation of the spontaneous emission coefficient $A_{21}$ requires the quantization of the electromagnetic field and is not given here. Let us only recall that, in the Wigner-Weisskopf approximation (2), population decay is described by an exponential law with a decay constant

$$\tau_{sp} = \frac{1}{A_{21}} = \frac{3\pi\hbar\epsilon_0 c_0}{\omega_0^3\eta\,|\boldsymbol{\mu}|^2} \tag{26}$$

It is possible to obtain the relationship between the spontaneous and stimulated emission coefficients without using field quantization with the help of an elegant thermodynamic argument due to Einstein. Let us consider the two-level system inside a blackbody cavity with walls at temperature $T$. In addition to the spontaneous decay processes, absorption and stimulated emission processes will also take place because of blackbody radiation. If $W'_{12}$ and $W'_{21}$ are the corresponding transition probabilities, we get

$$W'_{12} = B_{12}\rho_{\omega_0} \tag{27a}$$

$$W'_{21} = B_{21}\rho_{\omega_0} \tag{27b}$$

where $\rho_{\omega_0}$ is the spectral energy density of the blackbody radiation at the central frequency $\omega_0$ of the atomic transition and $B_{12}$ and $B_{21}$ are proportionality coefficients (Einstein $B$ coefficients). Thermodynamic equilibrium requires that the rate of transition per unit volume from the upper level to the lower (due to spontaneous and stimulated emission) be equal to the transition rate in the opposite sense (due to absorption). Therefore

$$A_{21}N_2^e + W'_{21}N_2^e = W'_{12}N_1^e \tag{28}$$

where the equilibrium populations $N_1^e$ and $N_2^e$ are related by Eq. (10). Using Eqs. (27) and (10), we get from Eq. (28) the following expression for $\rho_{\omega_0}$:

$$\rho_{\omega_0} = \frac{A_{21}}{B_{12}\exp(\hbar\omega_0/kT) - B_{21}} \tag{29}$$

The comparison of this expression with the Planck formula for the blackbody radiation (1),

$$\rho_\omega = \frac{4\nu^2}{c^3}\frac{\hbar\omega}{\exp(\hbar\omega/kT) - 1} \tag{30}$$

shows that

$$B_{12} = B_{21} \tag{31}$$

and that

$$\frac{A_{21}}{B} = \frac{\hbar\omega_0^3\eta^3}{\pi^2 c_0^3} \tag{32}$$

where we have put $B = B_{12} = B_{21}$. Equation (31) shows that the stimulated emission probability is also equal to the absorption probability for blackbody radiation. Equation (32) makes possible the calculation of $A_{21}$ once coefficient $B$ is known. This coefficient can be obtained from Eq. (15a), which is valid for monochromatic radiation. For broadband radiation such as that of a blackbody source, in fact, we call $\rho_\omega\,d\omega$ the elementary energy density in the frequency interval between $\omega$ and $\omega + d\omega$. Since the corresponding elementary wave can be assumed to be monochromatic, we obtain from Eq. (15a)

$$dW' = \frac{\pi}{3\eta^2\epsilon_0\hbar^2}|\boldsymbol{\mu}|^2 g(\omega - \omega_0)\rho_\omega\,d\omega \tag{33}$$

By integration of Eq. (33), assuming that the width of the lineshape function $g(\omega - \omega_0)$ is much narrower than that of the blackbody radiation $\rho_\omega$, we obtain

$$w' = \frac{\pi}{3\eta^2\epsilon_0\hbar^2} |\boldsymbol{\mu}|^2 \, \rho_{\omega_0} \qquad (34)$$

where Eq. (17) has been used. By comparing Eq. (34) with Eq. (22), we have

$$B = \frac{\pi}{3\eta^2\epsilon_0\hbar^2} |\boldsymbol{\mu}|^2 \qquad (35)$$

and from Eq. (32)

$$A_{21} = \frac{\eta\omega_0^3 |\boldsymbol{\mu}|^2}{3\pi\hbar\epsilon_0 c_0^3} \qquad (36)$$

Since $A_{21} = 1/\tau_{sp}$, it is easy to verify the agreement of Eq. (36) with Eq. (26).

### 1.2.4   Nonradiative Decay (3)

The calculation of the time of nonradiative decay requires a detailed knowledge of the interaction potentials and of the energy levels of the different types of molecules forming the system. In fact, the energy may often be converted to the internal energy of these molecules. Therefore, we shall not deal with a formal derivation of this time but limit ourselves to noting that it is the parameter most sensitive to the molecular environment.

With the simultaneous presence of both the radiative and nonradiative transitions, the time variation of the level 2 population can be written as

$$\frac{dN_2}{dt} = -\frac{N_2}{\tau_{sp}} - \frac{N_2}{\tau_{nr}} = -\frac{N_2}{\tau} \qquad (37)$$

where $\tau^{-1} = \tau_{sp}^{-1} + \tau_{nr}^{-1}$ is called the *lifetime* of level 2. The radiant power $P$ emitted by a volume $V$ through fluorescent emission, in the assumption of an optically thin gas, is

$$P = \frac{N_2}{\tau_{sp}} \hbar\omega_0 V \qquad (38)$$

Equation (38) shows that $P$ has the same time variation as $N_2$, namely, it decays as $\exp(-t/\tau)$. *The lifetime of the fluorescent light is therefore $\tau$ and not $\tau_{sp}$.*

Finally, we can define the fluorescence quantum efficiency $Y$ as the ratio of the total number of emitted photons to the number of atoms excited to the upper level. Therefore,

$$Y = \int (P/\hbar\omega_0) \, dt / N_2(0) \, V$$

where $N_2(0)$ is the initial population of level 2. Using Eqs. (37) and (38) and definition of $Y$, we obtain

$$Y = \frac{\tau}{\tau_{sp}} \tag{39}$$

## 1.2.5   Measurement of Decay Times and Transition Cross Section

The lifetime $\tau$ can be obtained from measurement of the temporal behavior of the fluorescent light. If the quantum yield has also been measured, the radiative time $\tau_{sp}$ can be deduced from Eq. (39). If level 1 is appreciably populated (at experimental temperature), the transition cross section $\sigma$, at a given frequency $\omega$, can easily be obtained through an absorption measurement. In fact, this measurement gives the absorption coefficient $\alpha(\omega)$ from which $\sigma(\omega)$ is readily obtained once the level populations are known, see Eq. (21). These populations can in turn be derived form Eq. (10) when the atom concentration is given. If level 1 is not appreciably populated, the absorption transition is not easily observable. In this case $\sigma$ is more conveniently obtained from the study of the fluorescent emission. Thus, from Eqs. (26) and (19) we get

$$\sigma = \left(\frac{\lambda}{2}\right)^2 \frac{(\omega - \omega_0)}{\tau_{sp}} \tag{40}$$

where $\lambda = 2\pi c_0 / \eta \omega_0$ is the transition wavelength in the medium considered. Equation (40) gives the value of $\sigma$ once $g(\omega - \omega_0)$ and $\tau_{sp}$ have been measured. Note that the lineshape $g(\omega - \omega_0)$, which is the same both in absorption and in spontaneous emission (1), is readily obtained from the spectrum of the emitted light. It is worth noting that if the quantity $\sigma$ is known (for example, from an absorption measurement), Eq. (40) can in turn be used to calculate $\tau_{sp}$. Indeed, from Eqs. (40) and (17), we obtain

$$\tau_{sp} = \frac{\left(\frac{\lambda}{2}\right)^2}{\int \sigma \, d\omega} \tag{40a}$$

Equation (40a) is particularly convenient when the lifetime $\tau$ is so short (e.g., in the picosecond range) and the quantum efficiency therefore so small ($\sim 10^{-4}$) as to make their measurement difficult.

## 1.2.6   Saturation

Let us now consider, in some detail, the interaction between a two-level system and an electromagnetic wave of irradiance $I$ and frequency $\omega$ (Fig. 1.6). Let us first consider a homogeneous broadened line. If the total number of atoms (per unit volume) of the system is constant in time, we can write the two following equations:

$$N_1 + N_2 = N \tag{41a}$$

$$\frac{dN_2}{dt} = -W(N_2 - N_1) - \frac{N_2}{\tau} \tag{41b}$$

Equation (41a) obviously follows from the assumption that has been made. Equation (41b) simply expresses the time variation of $N_2$ as the difference between the rate $WN_1$ of stimulated transitions toward level 2 and the rates of stimulated transitions $WN_2$ and spontaneous transitions $N_2/\tau$ toward level 1. In a steady-state regime the population difference $(N_1 - N_2)_{ss}$ as derived from Eqs. (41a) and (41b) is

$$(N_1 - N_2)_{ss} = \frac{N}{1 + 2W\tau} \tag{42}$$

The population difference depends therefore on the decay time of excited level 2, which is a characteristic property of the material, and on the transition probability $W$, which in turn depends linearly on the wave irradiance $I$, see Eqs. (15) and (18). The population difference decreases with increasing wave irradiance until it approaches zero as $I$ approaches infinity. In this case we should have $N_1 = N_2 = N/2$. This effect is called *saturation*.

To maintain a given population difference, a volume $V$ of the material must absorb from the incident wave a power $P_a$ given by

$$P_a = \hbar\omega W(N_1 - N_2)V = \hbar\omega \frac{NW}{1 + 2W\tau} V \tag{43}$$

and under saturation condition (i.e., $W\tau \gg 1$),

$$P_{as} = \frac{\hbar\omega W}{2\tau} V \tag{44}$$

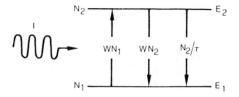

**Fig. 1.6.** Interaction between a two-level system and an electromagnetic wave of irradiance $I$.

This power is needed by the system to keep the population of the two levels equal, and it is the same as the power emitted by spontaneous emission from the upper level. Equations (43) and (44) can now be written in a more meaningful way by taking account of Eq. (18):

$$\frac{N_1 - N_2}{N} = \frac{1}{1 + \dfrac{I}{I_s}} \tag{45}$$

$$\frac{P_a}{P_{as}} = \frac{\dfrac{I}{I_s}}{1 + \dfrac{I}{I_s}} \tag{46}$$

The quantity $I_s$, defined by

$$I_s = \frac{\hbar\omega}{2\sigma\tau} \tag{47}$$

is called the *saturation intensity*.

Let us consider the modification of the absorption line with irradiance $I$ of the electromagnetic wave. Experimentally, this could be studied by measuring the absorption of a probing weak wave with frequency $\omega'$ varying around $\omega$, according to the scheme in Fig. 1.7.

If the line is homogeneously broadened, the absorption coefficient according to Eqs. (19), (21), and (45) can be written as

$$\alpha = \frac{\alpha_0}{1 + \dfrac{I}{I_s}} \tag{48}$$

where

$$\alpha_0 = \frac{\pi}{3\eta\epsilon_0 c_0 \hbar} |\boldsymbol{\mu}|^2 \, \omega' N g(\omega' - \omega_0) \tag{49}$$

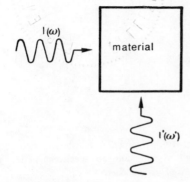

**Fig. 1.7.** Ideal scheme for measuring the absorption coefficient at frequency $\omega'$ when the material interacts with a strong wave at frequency $\omega$. The probe wave irradiance $I'(\omega')$ must be much smaller than the exciting irradiance $I(\omega)$.

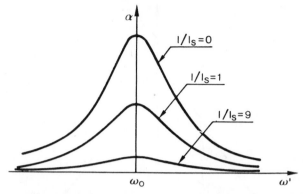

**Fig. 1.8.** Absorption coefficient of a homogeneously broadened line, at different excitation intensities. From Svelto (1b), reproduced by permission. Published by Tamburini Editore, Milano.

is the absorption coefficient at frequency $\omega'$ in the absence of the strong excitation wave at frequency $\omega$ (i.e., when $I(\omega) = 0$). Equation (48) shows that the absorption coefficient decreases with increasing excitation intensity $I$ but maintains its lineshape, which is expressed by $g(\omega' - \omega_0)$. The variation in $\alpha$ with $\omega'$ at different values of $I/I_s$ is shown in Fig. 1.8.

If the line is inhomogeneously broadened, the saturation effect is more complicated. Qualitatively, if the overall line is given by the inhomogeneous distribution of the homogeneous lines of the single atoms, the strong exciting wave $I(\omega)$ can interact only with the "packet" of atoms whose resonance frequency approaches $\omega$. Only this packet will experience the saturation effect when $I(\omega)$ becomes sufficiently high. Therefore, with the increase in excitation irradiance $I(\omega)$, a "hole" is burned in the absorption line at a frequency $\omega$. The width of this hole is approximately the width arising from the homogeneous broadening. This is shown in Fig. 1.9. It is worth noting that the measurement of the saturation intensity is a simple way of measuring, according to Eq. (47), the product $\sigma\tau$ and, if the lineshape $g(\omega - \omega_0)$ is also known, the quantum efficiency $Y$ of the transition, according to Eqs. (39) and (40). Similar considerations are also discussed in Chapters 3 and 4 of this book.

## 1.3 PUMPING PROCESSES

Liquid lasers (for example, dye lasers) and solid-state lasers (with the exception of semiconductor lasers) are usually pumped optically. The light emitted by a suitable lamp is concentrated, as described later, on the active material. Only part of this light is utilized by the active material,

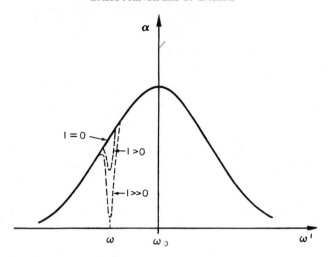

**Fig. 1.9.** Absorption coefficient of an inhomogeneously broadened line, at different excitation intensities. From Svelto (1b), reproduced by permission. Published by Tamburini Editore, Milano.

namely, the fraction that falls within the absorption band of the pump transition $1 \to 3$ of the three-level laser (Fig. 5a) or $0 \to 3$ of the four-level laser (Fig. 5b). The light emitted by the most common lamps has a broad continuous spectrum. *It is therefore preferable that the bandwidth of the pump transition be as large as possible.* For this reason, optical pumping is not used with gas lasers. In fact, the absorption lines of gases, at the pressures used, are much narrower than those of liquids of solids, and optical pumping would not be efficient. In this case electrical pumping is preferred so that population inversion is obtained through gas discharge, which is usually continuous.

Of course, other types of pumping are possible. For example, population inversion can be obtained through chemical reaction (chemical lasers) or very rapid adiabatic expansion (gas-dynamic lasers) (1).

In optical pumping, the lamp usually has a cylindrical shape, with a diameter of a few millimeters and a length of a few centimeters, and is filled with either a low-pressure (e.g., Ne or Kr) or high-pressure (e.g., Hg) gas. In a common configuration, the lamp is placed along one of the two focal axes of an elliptical cylinder (Fig. 1.10a). The active material, shaped like a rod with the same dimensions as the lamp, is placed along the other focal axis. If the inner surface of the elliptical cylinder is highly reflective, a consistent fraction of the light emitted by the lamp is concentrated on the active material, due to a well-known property of ellipses. The laser can work in a continuous fashion or in a pulsed fashion

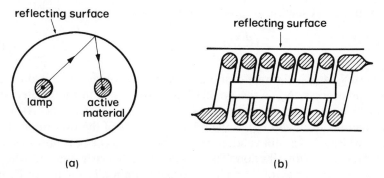

**Fig. 1.10.** Schemes for optical pumping.

depending on the lamp's operation. For pulsed operation, spiral flash-lamps are often used (Fig. 1.10$b$).

In the case of electrical pumping and with a gas of one atomic species only, the population of the upper level of the laser transition can occur only by collision of the atoms with the free electrons of the discharge. For monokinetic electrons of velocity $v$, an electronic collision cross section $\sigma_e = \sigma_e(v)$, which is effective for the given pump transition, can be defined by

$$\left(\frac{dN_2}{dt}\right)_p = N_t N_e \sigma_e v \tag{50}$$

where $(dN_2/dt)_p$ is the rate of atoms per unit volume pumped to the upper laser level by the electron–atom collision, $N_t$ is the total atomic population (per unit volume), and $N_e$ is the electronic population (per unit volume). For the free electrons of a discharge, Eq. (50) is still valid provided that the term $\sigma_e V$ is now averaged over the electron velocity distribution. Population inversion can be obtained if at least one of the following conditions is fulfilled: (i) the quantity $\langle \sigma_e v \rangle$ is higher for the transition to the upper laser level than that to the lower laser level; (ii) the lifetime of the upper laser level is much longer than that of the lower laser level. If the gas is formed by more than one atomic species (for example, atoms A and B), pumping of one species can also be achieved by the resonant transfer of the excitation (Fig. 1.11). With monokinetic

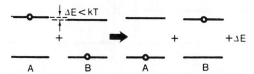

**Fig. 1.11.** Near-resonant energy transfer.

atoms the cross section $\sigma_{AB}$ effective for this process is defined by

$$\left(\frac{dN_2}{dt}\right)_p = N_A N_B \sigma_{AB} v_{AB} \tag{51}$$

where $(dN_2/dt)_p$ is the rate of atoms B (per unit volume) pumped to level 2 by this process, $N_A$ is the population of species A in the excited state (through electronic collision), $N_B$ is the population of species B at the fundamental level, and $v_{AB}$ is the relative velocity of the two atoms. For a gas atom Eq. (51) is still valid, provided the term $\sigma_{AB} v_{AB}$ is now averaged over the velocity distribution. The process is particularly efficient when (i) the excitation energy of A is almost equal (within $kT$) to that of B, thus giving a high value of $\sigma_{AB}$; (ii) the lifetime of the excited state of A is very long, thus giving a high value of $N_A$.

## 1.4  OPTICAL RESONATORS

The most common resonators utilize plane parallel or spherical mirrors (Fig. 1.12). We can define a *mode of a resonator* as *a stationary solution of Maxwell equations which meets the boundary conditions.* Actually, since laser resonators do not have side walls, perfectly stationary solutions do not exist because of diffraction losses through the lateral surface. However, since these losses are usually quite small, the previous definition of mode can also be applied to laser resonators. The electric field $E(\mathbf{r},t)$ of a stationary wave can therefore be written as

$$E(\mathbf{r},t) = u(\mathbf{r}) \sin \omega t \tag{52}$$

For a given resonator an infinite series of eigenfunctions of the form shown in Eq. (52) exist. Each of them has a well-defined spatial configuration $\mathbf{u}(t)$ and a well-defined resonance frequency $\omega$.

### 1.4.1  Plane Mirror Resonator

For a resonator formed by two plane parallel mirrors, the lowest-order mode can be taken as given by the superposition of two waves which

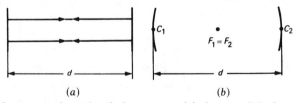

(a)                              (b)

**Fig. 1.12.**  Most commonly used optical resonators: (*a*) plane parallel mirror resonator and (*b*) confocal mirror resonator.

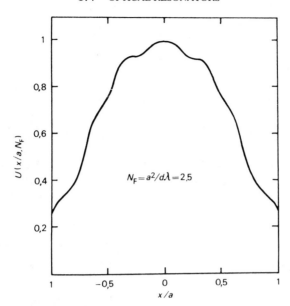

**Fig. 1.13.** Field distribution at one mirror for the lowest-order mode of a plane parallel mirror resonator and for a Fresnel number $N_F = 2.5$. $2a$ is the mirror size.

propagate in the cavity with opposite directions. The corresponding resonance frequencies are obtained very simply by making the resonator length $d$ equal to an integer number $n$ of half-wavelengths. This corresponds to the physical boundary condition of a zero electric field at the mirrors. From the condition $d = n(\lambda/2)$, the resonance frequencies $\nu$ are obtained as

$$\nu = \frac{nc}{2d} \tag{53}$$

Of course, because of the finite dimensions of the mirrors, the field distribution at the mirrors is not uniform but decreases near the edges. Thus, the effect of the diffraction losses is present mainly in the region of the mirrors. The field distribution of the lowest-order mode as given by computer calculation is shown in Fig. 1.13.

### 1.4.2   Confocal Resonator

A typical resonator employing spherical mirrors is the confocal resonator (Fig. 1.12b), where the foci ($F_1$ and $F_2$) of the two mirrors are coincident. Confocal resonators and, in general, spherical mirror resonators are more useful than plane resonators. In fact, as described later in more detail, the focusing properties of spherical mirrors give a field distribution that is more concentrated along the resonator axis ($C_1$-$C_2$) than in plane resonators. The diffraction losses are thus significantly reduced.

The modes of a confocal resonator will not be calculated here; we only present and comment on the final result. Let us assume that the field in the resonator is uniformly polarized (for example, linearly polarized), so that it can be described by one scalar function $U(\mathbf{r}, t)$ only, which represents the amplitude of the electric or magnetic field. We choose a reference system with its $z$ axis coincident with the resonator axis, the origin at the center of the resonator, and the $x$ and $y$ plane normal to the $z$ axis (Fig. 1.14). The field $U(\mathbf{r},t)$ of a confocal resonator mode is thus given by the expression

$$U'(\mathbf{r},t) = U(x,y,z,t) = \frac{w_0}{w} H_m\left(\frac{\sqrt{2x}}{w}\right) H_l\left(\frac{\sqrt{2y}}{w}\right)$$

$$\cdot \exp[-(x^2+y^2)/w^2] \sin \phi(x,y,z) \sin \omega t \qquad (54)$$

where

$$w_0 = \left(\frac{d\lambda}{2}\right)^{1/2} \qquad (55)$$

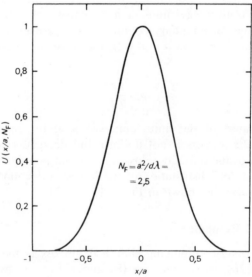

**Fig. 1.14.** Field distribution at one mirror for the lowest-order mode of a confocal resonator and for a Fresnel number $N_F = 2.5$.

with $d$ being the resonator length; $H_m$ and $H_l$ are the Hermite function of order $m$ and $l$, and the quantity $w = w(z)$ is given by

$$w^2 = w_0^2 \left[ 1 + \left( \frac{2z}{d} \right)^2 \right] \tag{56}$$

The expression for the phase $\phi$ is somewhat complicated, and we shall not give it here. We shall only remark that according to this expression, the equiphase surfaces are to a good approximation spherical, with the curvature radius $R(z)$ given by

$$R(z) = z \left[ 1 + \left( \frac{d}{2z} \right)^2 \right] \tag{57}$$

We note lastly that the resonance frequencies $v = \omega / 2\pi$ are given by the expression

$$v = \frac{c}{2d} [2n + (m + l + 1)] \tag{58}$$

Where $n$, $m$, and $l$ are positive integer numbers and where, in particular, $m$ and $l$ are the subscripts of the Hermite functions. As we can see, both the spatial configuration and the resonance frequency [i.e., the complete function $U(x,y,z,t)$] are determined by the value of the three numbers $n$, $l$, and $m$.

Let us now consider the $m = l = 0$ modes. Since the zeroth order Hermite polynomial $H_0$ is a constant, the field distribution as a function of $x$ and $y$ at a given $z$, that is, the transverse field distribution, is given by $\exp[-(x^2 + y^2)/w^2]$; it is a Gaussian function (Fig. 1.14). At $x$ (or $y$) equal to $w$, the field becomes $1/e$ times the value at $x = 0$. The quantity $w$ is therefore called the *spot size* of the beam at the plane corresponding to the $z$ coordinate. As shown by Eq. (56), the spot size is a function of $z$.

This is also shown in Fig. 1.15. We can see that the spot size is a minimum at the resonator center, where its value is $w = w_0$. It then increases with $z$ and, at the mirrors ($z = d/2$), is equal to $\sqrt{2} \, w_0$. The field amplitude distribution is thus determined. The wavefront surfaces (dashed in Fig. 1.15) are spherical surfaces, with a radius given by Eq. (57). At the resonator center the wavefront is plane, as required by

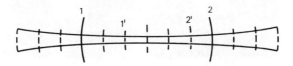

**Fig. 1.15.**   Spot size and equiphase surfaces for $\text{TEM}_{00}$ mode in a confocal resonator.

symmetry. At the mirrors the wavefronts coincide with the mirror surfaces; this is an obvious result when the mirrors are made of metallic surfaces. In this case the boundary condition of a vanishing electric field at the mirrors must be fulfilled. The resonance frequencies of the $m = l = 0$ modes are given by Eq. (58) with $m = l = 0$. Two successive modes, which differ by a unit in the value of $n$, are thus separated in frequency by $\Delta \nu = c/2d$, as for the plane mirror resonator. These modes are called $TEM_{00}$, where TEM stands for "Transverse Electric and Magnetic." Indeed, the modes of the laser resonators have electric and magnetic fields normal, to a close approximation, to the resonator axis. Since the light irradiance is proportional to $U^2$, it follows that for a $TEM_{00}$ mode the light pattern at the mirrors is circular.

Let us now consider the $m = 1$ and $l = 0$ mode. Since $H_1(x)$ is proportional to $x$, it follows from Eq. (54) that the field distribution of this mode is Gaussian along the $y$ axis, while it is a Gaussian function times $x$ along the $x$ axis (Fig. 1.17). The light pattern at the mirrors is shown in Fig. 1.16. Such a mode is called $TEM_{10}$. The $TEM_{01}$ mode $(m = 0, l = 1)$ is simply obtained from the $TEM_{10}$ by exchanging the $x$ and $y$ axis. The higher-order modes can be described with the same procedure. Figure 1.16 shows an example of the $TEM_{20}$ and $TEM_{31}$ modes. In general, the indices $m$ and $l$ represent the number of nodes of $U$ along the corresponding axes. Similarly, the number $n$ gives the nodes of $U$ along the $z$ axis.

We note finally that the quantity $\sin \phi \sin \omega t$ in Eq. (54) can be written as $[\cos(\phi - \omega t) - \cos(\phi + \omega t)]/2$; this shows that the stationary wave can be

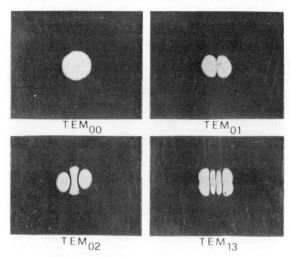

**Fig. 1.16.** Mode patterns for some low-order modes.

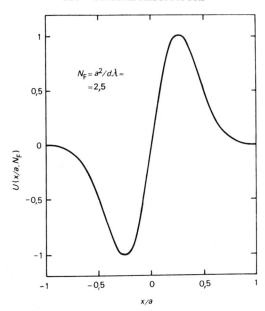

**Fig. 1.17.** Lowest-order, antisymmetric mode of a confocal resonator.

regarded as the sum of two traveling waves propagating in opposite directions in the cavity. It then follows that each of the two waves, which is a solution of the Maxwell equations with the proper boundary conditions, also represents the propagating beam out of the cavity. The field distribution of the light beam emerging from a confocal resonator is therefore obtained from Eq. (54) by replacing $\sin \phi \sin \omega t$ with $\cos(\phi - \omega t)/2$ for the wave propagating along the positive $z$ direction and with $\cos(\phi + \omega t)/2$ for the wave propagating in the opposite direction.

### 1.4.3 Nonconfocal Spherical Mirror Resonator

Nonconfocal resonators are very often used in lasers. Their properties, and in particular the calculation of the spot size and the wavefront distribution, follows quite simply from the confocal resonator theory. In fact, with reference to Fig. 1.15, if we *materialize* two equiphase surfaces such as, for example, 1' and 2' by placing there two spherical mirrors with the same radius of curvature, the overall field distribution inside and outside the resonator does not change. In fact, for given boundary conditions, the field distribution is unique. On the other hand, as seen from Eq. (57), the two new spherical surfaces are not confocal. Now we can see the problem from a different point of view: if the resonator is formed by the spherical mirrors 1' and 2', its modes can be easily found after finding the position of the corresponding confocal surfaces (i.e., surfaces 1 and 2). In other words, it is enough to determine the position of the corresponding *equivalent* confocal resonator. This can be obtained from Eq. (57). In fact, from the values of the curvature radii $R_1$ and $R_2$ of mirrors 1' and 2', the length of the equivalent confocal resonator $d$ can be found. Once $d$ is known, the electric field inside and outside the resonator is obtained from Eq. (54) with the help of Eqs. (55), (56), and (57).

## 1.5   OPERATION MODES OF LASERS

Lasers can operate in four different modes: (i) continous wave (cw), (ii) pulsed, (iii) $Q$-switched (repetitive or single pulse), and (iv) mode-locked (repetitive or pulsed).

*Continuous wave operation* can be achieved only if the lifetime of the upper laser level is longer than that of the lower level. If it is not, the laser is called *self-terminating*: it can work only with a pump pulse shorter than the upper level lifetime. For practical reasons, lasers that could, in principle, work in a cw mode are sometimes operated in a pulse mode: this happens, for example, when the cw pump intensity is so high that it imposes severe limitations on the life of the pump system.

An interesting working mode is the so called *Q-switch* or *giant pulse* mode; this makes possible the generation of light pulses with very high power (~1 MW) and very short duration (5–20 ns). To describe this mode, we may note first that in normal laser operation (both cw and pulsed), population inversion cannot greatly exceed the corresponding threshold value. In fact, laser action starts as soon as the inversion reaches threshold value, thus utilizing the inversion and preventing it from growing further. It follows that in normal laser operation the single-pass gain cannot be much higher than the single-pass losses. Now suppose we insert an opaque screen in the cavity. The effect of this screen is to prevent the laser action, thus letting the population inversion grow well beyond the normal threshold value. If now the screen is suddenly removed, the laser will experience a population inversion much higher than the threshold, that is, a gain much higher than the losses. The technique is called *Q-switching operation*, since the quality factor $Q$ of the cavity is switched from a negligible value (when the screen is inserted) to the very high value corresponding to the cavity without the screen.

For a detailed description of $Q$-switching methods we refer to the specialized literature (4,5). Here we give only a qualitative description of a very common, simple technique that makes use of saturable absorbers. Several substances, usually dyes in liquid solution, possess the property of becoming transparent when illuminated by a beam of high irradiance. Roughly speaking, these substances can be represented as a two-level system with a resonance frequency $\omega$ and with a very low saturation intensity $I_s$; see Eq. (47). Now suppose a cell containing a saturable absorber is inserted in a laser cavity and also its absorption frequency coincides with the laser frequency and its absorption is, for example, 50%. The laser oscillation can start only when the gain in the active material overcomes the high losses due to the absorber (in addition, of course, to all the other losses); this results in very high values of population inversion. However, as soon as laser oscillation starts and, more precisely, as soon as the irradiance of the laser radiation in the cavity becomes comparable to $I_s$, the absorber will become rapidly saturated (or bleached), that is, it will become transparent. If $I_s$ has a low value, the population inversion will still be very high at the time when saturation occurs, and a very intense light pulse will thus be produced. In the most common absorbers, the low $I_s$ value is due to the very high value of the absorption cross section, which is about $10^4$ times higher than that of the active material.

Another very interesting working regime is the so-called *mode locking*. This allows the generation of light pulses of high power (more than 100 MW) and ultrashort duration (from 1 ps to 1 ns). To give a simple description of this regime, we begin by recalling that the modes of a plane mirror resonator and the $TEM_{00}$ modes of a spherical mirror resonator differ in frequency by $\Delta \nu = c/2d$, see Eqs. (53) and (58). Since this difference is generally much smaller than the linewidth of the atomic transition, several laser modes usually oscillate simultaneously. Let us assume now that the phases of these modes are forced to be correlated according to, for example,

$$\phi_k - \phi_{k-1} = \alpha \tag{59}$$

where $\alpha$ is a constant. We assume for the sake of simplicity that the laser oscillates on $2n+1$ modes all having the same amplitude $E_0$. The total electric field $E(t)$, both inside and outside the resonator, can therefore be written (except for a constant phase factor) as

$$E(t) = \sum_{-n}^{n} {}_l E_0 \exp\{i[(\omega_0 + l \,\Delta\omega)t + l\alpha]\} \tag{60}$$

where $\omega_0$ is the angular frequency of the central mode and $\Delta\omega = 2\pi \,\Delta\nu = \pi c/d$. The sum of the geometric series in Eq. (60) can easily be calculated. As a result the field $E(t)$ can be written as

$$E(t) = A(t) \exp i\omega_0 t \tag{61}$$

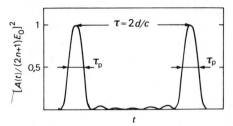

**Fig. 1.18.** Time behavior of the squared amplitude of the electric field for $2n + 1 = 7$ oscillating modes.

where

$$A(t) = E_0 \frac{\sin[(2n + 1)(\Delta\omega t + \alpha)/2]}{\sin[(\Delta\omega t + \alpha)/2]} \qquad (62)$$

Equations (61) and (62) show that the electric field $E(t)$ is expressed by a sinusoidal wave with frequency $\omega_0$ and an amplitude modulated by $A(t)$. As a representative example, the temporal behavior of $A^2(t)$ (i.e., of the output wave irradiance) for $2n + 1 = 7$ oscillating modes is shown in Fig. 1.18. It can be seen that the duration (full width at half maximum) of the light pulses $\tau_p$ is

$$\tau_p \simeq \frac{1}{\Delta\nu_{osc}} \qquad (63)$$

where $\Delta\nu_{osc} = (2n + 1) \Delta\nu$ is the oscillating bandwidth. *Since $\Delta\nu_{osc}$ may be as large as the gain bandwidth* (which spans from 1 GHz for a gas laser to more than 300 GHz for solid- or liquid-state lasers), *it is possible in this way to generate ultrashort light pulses.* By studying the function $A(t)$, we can also see that two consecutive pulses are separated in time by $\tau = 2\pi/\Delta\omega = 2d/c$, that is, by twice the cavity transit time $(d/c)$. The laser oscillation can thus be interpreted as being due to a single ultrashort pulse propagating back and forth in the laser cavity. A sequence of two pulses is shown in Fig. 1.18. Finally, note that the peak power of the pulses is greatly increased in the mode-locking operation. Indeed, it can be deduced from Eq. (62) that the peak power is proportional to $(2n + 1)^2 E_0^2$. If the phases of the modes, rather than being locked, were all random, the total power would be given by the sum of the powers of each mode, that is, proportional to $(2n + 1)E_0^2$. The relative increase in the peak power is therefore equal to the total number of oscillating modes. For a ruby or a neodimium laser, this number may be a few thousand.

For a detailed description of the techniques employed to obtain the mode-locking operation, we refer to the specialized literature (6). Here, we give a qualitative description of the technique based on mode locking by a saturable absorber (passive locking). This mode of operation can easily be obtained by inserting in a laser cavity a suitable absorber formed

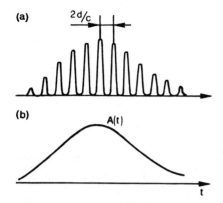

**(a)**  $2d/c$

**(b)**  $A(t)$

t

**Fig. 1.19.** (*a*) Typical temporal behavior of light emitted by a laser operating simultaneously in *Q*-switched and mode-locked modes. (*b*) Temporal behavior of the oscillating modes.

by a two-level system resonant with the laser frequency. From Eq. (45) it can easily be deduced that when $I \ll I_s$, the population of the upper level in the absorber contains, to first order, a term proportional to the total irradiance $I$. If several modes oscillate, the total irradiance, and therefore also the upper level population, will contain terms oscillating at $\Delta\omega$, which is the beat note frequency of two adjacent modes. The cavity loss due to the absorber thus contains a term oscillating at $\Delta\omega$, which will produce phase locking of the modes. Saturable absorbers are very suitable for this purpose because of their high absorption cross section (i.e., low saturation intensity $I_s$). Another requirement must be met, namely, the upper level decay time must be short compared to $1/\Delta\omega$; otherwise the amplitudes of the beat notes becomes too small. Finally, it is worth noting that, since a saturable absorber produces giant pulses (as described in Sect. 1.5), the temporal behavior of the output light pulse will be as represented in Fig. 1.19*a*. For comparison, the temporal behavior of each oscillating mode is shown in Fig. 1.19*b*. In conclusion, the saturable absorber has in this case a twofold effect: (i) it produces giant pulses; (ii) it causes locking among the phases of the modes.

## 1.6  TYPES OF LASERS

### 1.6.1  Ionic Crystal Lasers

These lasers utilize, as the active material, ions which form ionic crystals. Usually these ions belong to one of the transition series of the Mendeleieff table (e.g., $Cr^{3+}$ or rare earth ions).

#### 1.6.1.1  *The Ruby Laser*

The ruby crystal is formed by sapphire ($Al_2O_3$) with chromium ions ($Cr^{3+}$) substituting for aluminum ions ($Al^{3+}$) in some lattice sites with a

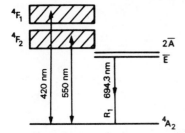

Fig. 1.20. Energy levels of a ruby laser.

typical weight concentration of 0.05%. The energy levels of the $Cr^{3+}$ ion, which is the active atom, in its host lattice are shown in Fig. 1.20. The laser transition usually occurs between levels $\bar{E}$ and $^4A_2$, at $\lambda = 694.3$ nm ($R_1$ line), while the pump transitions are $^4A_2 \rightarrow {}^4F_2$ and $^4A_2 \rightarrow {}^4F_1$. The atoms decay from the pump levels to levels $2\bar{A}$ and $\bar{E}$ through a nonradiative transition in a rather short time ($10^{-7}$ s). Since the thermodynamic equilibrium is very quickly reached, level $2\bar{A}$ is slightly less populated than level $\bar{E}$. The laser action takes place, therefore, on the $R_1$ line, and this further prevents level $2\bar{A}$ from acquiring the critical population. *The ruby represents a typical three-level system* (7).

The active material, formed by a single ruby crystal, is shaped as a cylindrical rod with a diameter of a few millimeters and a length of a few centimeters. Since the bandwidths of the pump transitions are quite large ($\sim 50$ nm), optical pumping can be used conveniently. The ruby laser normally operates in a pulsed regime, and the pumping is accomplished by flash lamps (usually low-pressure Xe lamps), although it can also operate in a cw regime (when pumped by high-pressure Hg capillary lamps). The pulsed ruby laser was the first laser to be operated in 1960 and is still widely used.

### 1.6.1.2   *The Neodymium Laser*

Another very common ion for active laser material is $Nd^{3+}$ in different host crystals such as yttrium aluminum garnet (YAG) or suitable glasses (8). A simplified scheme of the energy levels of the Nd: YAG crystal is shown in Fig. 1.21. The pump transitions $^4I_{9/2} \rightarrow ({}^4F_{5/2}, {}^2H_{9/2})$ and

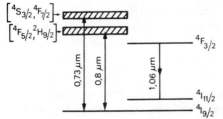

Fig. 1.21. Energy levels of a Nd:YAG laser.

$^4I_{9/2} \rightarrow (^4S_{3/2},^4F_{7/2})$ are in the red region of the spectrum, while the laser transition $^4F_{3/1} \rightarrow {}^4I_{11/1}$ occurs in the near infrared ($\lambda = 1.06\ \mu$m). Since the lower laser level ($^4I_{11/2}$) lies well above the fundamental level $^4I_{9/2}$, *this laser operates according to the four-level scheme.*

As for the ruby laser, the active material is shaped as a cylindrical rod and is optically pumped. The Nd : YAG laser can operate both in a pulsed regime (with low-pressure Xe flashlamps as the excitation source) or in the cw mode (with W or Kr pump lamps). A Nd : glass laser, on the other hand, can work only in a pulsed regime because of the low thermal conductivity of glass. An important property of this laser is the very high value of the bandwidth of the laser transition (30–40 nm), which is due to the crystal field inhomogeneity of the glass. Since the oscillating bandwidth may be as large as 10 nm, very short laser pulses can be obtained in the mode-locking operation. Indeed, if all the modes contained in the 10-nm band were phase locked, pulses as short as $10^{-13}$ s would be obtained. Actually, the shortest pulse duration normally obtained is $10^{-12}$ s. Due to the relatively low cost of the Nd : glass rods, this laser is particularly useful for obtaining high-energy, high-power pulses. Peak powers of $10^{13}$ W in a single pulse of duration of $10^{-11}$ s have been obtained with a mode-locked Nd : glass oscillator, followed by a few amplification stages with the same active material.

### 1.6.2 Gas Lasers

The broadening of the energy levels in gases is rather small (a few GHz or even less) since the line-broadening mechanisms here are less effective than in solids. At the low pressure commonly used with some gas lasers, collision broadening can be ignored so that the line broadening is essentially due to the Doppler effect. *Optical pumping with the kind of lamps used for crystal lasers would be quite inefficient in this case because of the absence of broad absorption bands.* The pumping process is therefore obtained electrically, through alternating or direct current of suitable strength. The discharge produces both free electrons and ions in the gas. The excitation process then occurs through one of the following mechanisms: (i) If the gas is formed by one constituent only, the excitation is due to electron–atom collision of the first kind:

$$e + X \rightarrow X^* + e$$

where X and $X^*$ represent the atom in the fundamental and in the excited state. (ii) When the gas is formed by more than one constituent, the excitation of a given atomic species (e.g., B) can also be produced by collision with another atomic species (A) in the excited state through an *energy transfer* mechanisms, namely,

$$A^* + B \rightarrow A + B^*$$

This process is particularly effective for pumping the atomic species B when the excited states of the two atoms A and B have approximately the same energy (with a difference $\Delta E < kT$) and when the excited state of A is metastable.

Once excited, the de-excitation process of an atom can then occur through several processes such as collision with electrons, atoms of other species, or the walls of the container or by spontaneous emission. Thus, the process of creating population inversion in a gas is much more complicated than in an ionic crystal.

Gas lasers can be divided into (i) neutral-atom gas lasers, (ii) ionized gas lasers, (iii) molecular lasers utilizing rotational–vibrational transitions, and (iv) molecular lasers using vibronic transitions. Typical examples are described below.

### 1.6.2.1 The He–Ne Laser

This is the most common laser (9). The energy levels of both He and Ne are shown in Fig. 1.22. The laser action occurs between the energy levels of Ne, while He is added to increase the pumping process. In fact, levels $2^1S$ and $2^3S$ of He are in resonance with levels 3s and 2s of Ne. Furthermore, the S levels of He are metastable (transitions are forbidden in the electric dipole approximation). It follows that levels 2s and 3s of Ne are quite efficiently pumped by the He levels through the process of

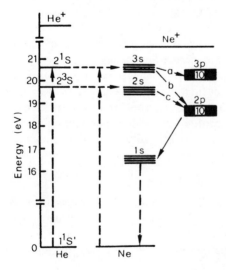

**Fig. 1.22.** Energy levels of a He–Ne laser.

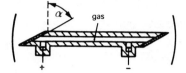

**Fig. 1.23.** General scheme for a gas laser.

energy transfer. In Ne, the permitted transitions occur from the s levels to the p levels. Since the lifetime of the s levels (~100 ns) is longer than that of the p levels by approximately one order of magnitude, all the three transitions, $a$ ($\lambda = 3.39 \mu$m), $b$ ($\lambda = 0.6328 \mu$m), and $c$ ($\lambda = 1.15 \mu$m) of Fig. 1.22, can produce cw laser action according to the four-level scheme. The operating wavelength is selected by mirror reflectivity. Of the three transitions, the $\lambda = 3.39 \mu$m one shows the highest gain. The transition most used for commercial He–Ne lasers, however, is the $b$ transition, since it corresponds to a visible (red) radiation. The $c$ transition is mainly of historical importance since it was the first transition to be operated in a gas laser in 1961.

The construction scheme of a He–Ne laser, which is common to most of the gas lasers, is shown in Fig. 1.23. The gas is contained in a glass or quartz capillary tube closed at the ends by glass windows tilted at Brewster angles. At this angle, a polarized beam with the electric field vector in the plane of the figure is transmitted at the window without reflection. The resulting reduction in cavity loss is crucial for the achievement of laser action. For this reason, the output beam of a He–Ne laser with Brewster windows is already linearly polarized. The He–Ne laser is simple to construct and to use. The power per unit length of the active material is however limited (~5 mW m$^{-1}$).

### 1.6.2.2 *The Argon Laser*

The laser action occurs between the energy levels of the Ar$^+$ ion (10). Ionized gas lasers such as Ar$^+$ lasers need a pump current density much higher than that of neutral gas lasers such as He–Ne lasers. In an ion laser, in fact, the pumping process requires two successive electron collisions: the first collision is to ionize the neutral atom while the second collision then raises the ion to an excited state. The pumping power is therefore proportional to the square of the current density of the discharge. The spacing of the energy levels for an ionized atom is larger than that of a neutral atom. This is why ionized gas lasers oscillate in the visible while neutral atom lasers operate in the infrared. In particular, the Ar$^+$ laser can oscillate on several lines the most intense of which are at $\lambda = 488.8$ nm (blue) and $\lambda = 514.6$ nm (green).

The construction scheme of the Ar$^+$ laser differs from the He–Ne

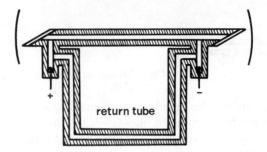

**Fig. 1.24.** Schematic diagram of an $Ar^+$ laser.

scheme by the presence of a return tube (Fig. 1.24). The electric field of the discharge causes a migration of the $Ar^+$ ions with accumulation at the cathode (cathaphoresis). The return tube then allows the diffusion of the Ar atoms back to the anode. The temperature of $Ar^+$ ions is very high ($\sim 3000$ K) due to the acceleration produced by the electric field of the discharge. The collisions of the ions with the tube walls can produce severe damage, which may be overcome by constructing the tube of graphite or beryllium oxide.

### 1.6.2.3 The Carbon Dioxide Laser

This laser utilizes rotovibrational transitions of the ground electronic state of the $CO_2$ molecule (11). This molecule, which is triatomic and linear, possesses three nondegenerate vibration modes (Fig. 1.25): (i) the longitudinal symmetric mode ($\nu_1$), (ii) the transverse (or bending) mode ($\nu_2$), (iii) the longitudinal antisymmetric mode ($\nu_3$). Its state of vibration can therefore be represented by three integers, $n_1$, $n_2$, and $n_3$, which give respectively the number of quanta in each of the three vibration modes. The lowest energy levels are shown in Fig. 1.26. The laser action usually

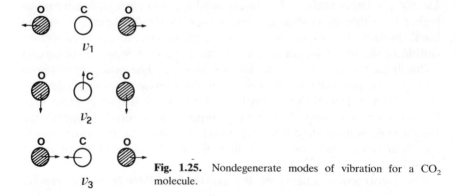

**Fig. 1.25.** Nondegenerate modes of vibration for a $CO_2$ molecule.

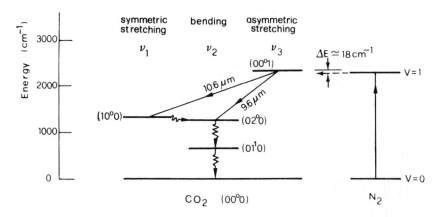

**Fig. 1.26.** Energy levels of a $CO_2$ laser.

occurs on a series of rotational lines of the $(00^\circ1) \to (10^\circ0)$ transition, with $\lambda = 10.6 \ \mu$m. It is also possible, however, to get laser action on some rotational lines of the $(001) \to (020)$ transition ($\lambda = 9.6 \ \mu$m). The upper laser level $(001)$ can be pumped by collision of the $CO_2$ molecules with electrons or, more efficiently, by resonant energy transfer from $N_2$ molecules which are purposely added to the gas. The gas mixture also contains He, which reduces the lifetime of the lower laser levels by collisions and also contributes to the cooling of the gas.

The $CO_2$ laser is at the present time the most powerful in continuous wave operation: powers of several hundred kilowatts have already been obtained. It is also the most efficient after the semiconductor laser: in fact, the conversion efficiency between the optical power and the electrical discharge power is 25%.

### 1.6.2.4 The Nitrogen Laser and the Hydrogen Laser

These lasers utilize vibronic transitions of the molecules of nitrogen or hydrogen (12). The emission wavelengths fall within the ultraviolet ($\lambda = 337.1$ nm) for $N_2$ and within the vacuum ultraviolet ($\sim 161$ and $\sim 116$ nm) for $H_2$. *Both these lasers are self-terminating* since the lifetime of the upper level is shorter than that of the lower level; *therefore, they can only be operated in a pulsed regime.*

In the $N_2$ laser, the emission occurs between the lowest vibrational levels of the electronic levels $\Pi_u$ and $\Pi_g$ (Fig. 1.27). In the $H_2$ laser (Fig. 1.28), the emission is due to a series of vibronic transitions of either the $\Sigma_u \to \Sigma_g$ transition (Lyman band $\lambda \sim 161$ nm) or the $\Pi_u \to \Sigma_g$ transition (Werner band, $\lambda \sim 116$ nm). The relative line intensities of these transitions are given by the relative transition probabilities.

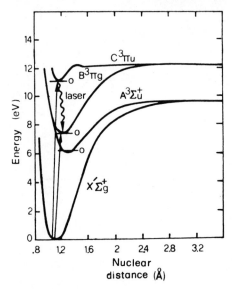

Nuclear
distance (Å)

**Fig. 1.27.** Energy levels of the $N_2$ molecule.

Since the lifetime of the upper laser level is very short ($\sim$40 ns for $N_2$ and $\sim$1 ns for $H_2$), these lasers require a very short, powerful pump current pulse. This can easily be obtained with a transverse discharge and fast electronic circuit such as the Blumlein circuit (Fig. 1.29). In this way, a current pulse of duration as short as 1 ns and a peak value of up to $10^5$ A can be achieved. Since the gain is very high, the inverted population is completely depleted in only one or two passages of the light in the gas. The use of the two mirrors is therefore unnecessary, and only one mirror, mounted opposite the output window, is used. Nanosecond laser pulses with a power of several hundred kilowatts are in this way currently obtained.

The $N_2$ laser is especially important as the pump source of dye lasers, while the $H_2$ laser is important since its wavelength is the shortest laser wavelength obtained up to the present time.

### 1.6.3  Dye Lasers

These lasers utilize as active material solutions of suitable dyes in liquid solvents such as water or alcohol (13). Dye molecules form a large family of organic compounds with conjugated double bonds. The $\pi$ electrons of these bonds form a charge cloud along the conjugated chain, along which they are more or less free to propagate. This electron delocalization is somewhat similar to that occurring in solid-state material, for example, in semiconductors. It is therefore not surprising that these delocalized electrons can develop a large dipole moment $\mu$, see Eq. (16). For this

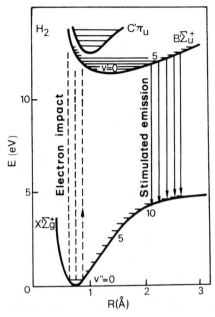

**Fig. 1.28.** Energy levels of the $H_2$ molecule.

reason dye molecules often have large absorption cross sections ($\sigma \sim 10^{-16}$ cm$^2$). If the absorption occurs in the visible, the dye solution, even with a small dye concentration, will be strongly colored, complementary to the absorbed wavelength. Because of the high value of $|\mu|$, the spontaneous emission lifetime, Eq. (26), is short (5–10 ns) and the quantum yield, Eq. (39), is often very high. Therefore, the dyes are often fluorescent. Fluorescent dyes are very good candidates as active materials for lasers.

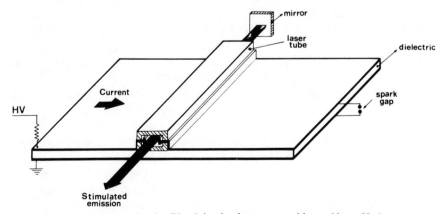

**Fig. 1.29.** Schematic of a Blumlein circuit to pump either a $N_2$ or $H_2$ laser.

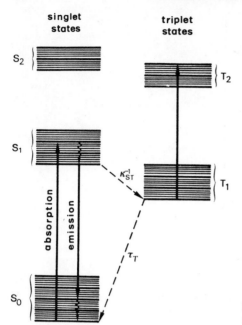

**Fig. 1.30.** Energy levels of a dye molecule.

A typical scheme of the energy levels for a dye molecule in solution is shown in Fig. 1.30. Each electronic state is actually formed by several vibrational and rotational lines. The vibrational lines are typically separated by 1400–1700 cm$^{-1}$, while the separation between rotational lines is about a hundred times smaller. Actually, since the line-broadening mechanisms are very effective in liquids, the rotational lines are not resolved and form a continuum over the vibrational levels. Since the number of $\pi$ electrons involved is usually even (compounds with unpaired electrons are called free radicals and are very reactive), the total spin $S$ is an integer and the molecule possesses singlet states ($S = 0$), triplet states ($S = 1$), and so on. The laser action occurs between the first and second singlet states. The triplet states, on the other hand, may produce losses since the population of the triplet level $T_1$ gives an absorption (on transition $T_1 \rightarrow T_2$) at the very laser wavelength. This absorption loss can in some cases be so critical as to prevent laser action.

The transitions $S_1 \rightarrow T_1$ (intersystem crossing) cannot occur by radiative process because it would require a spin change, but it may occur by collision since the levels $S_1$ and $T_1$ are almost resonant. The transition $T_1 \rightarrow S_0$ is also radiatively forbidden and, furthermore, it is much less affected by collision. A population accumulation may therefore occur on level $T_1$. A cw laser action can be obtained only by adding some special *triplet quenchers* (e.g., oxygen) to the dye solution in order to strongly reduce the lifetime of the $T_1 \rightarrow S_0$ transition. Otherwise, only the pulsed operation is permitted, with a pulse risetime shorter than the value typical for the particular material.

Dye lasers are pumped optically using (i) a rapid flashlamp (with pulse risetime shorter than $1 \, \mu s$) and (ii) another laser. The $N_2$ laser, in particular, is very useful to this end since its UV wavelength is strongly absorbed by several dyes which emit in the visible. This pumping system is very convenient; very high gain and efficiency (between visible and UV light), as high as 30–40%, can be obtained. Continuous laser action throughout the visible range can also be achieved with some of these dyes when pumped by a cw argon laser.

All dye lasers oscillate normally within a rather broad band ($\Delta\lambda \sim$ 10 nm). This makes them suitable for the generation of very short pulses when working in the mode-locking regime. Thus, pulses of $\sim 1 \, ps$ have been obtained, even in a steady-state regime, with the cw-pumped Rhodamine 6G laser. However, *the most important consequence of the broadband emission is the possibility of tuning the laser wavelength.* This can be obtained with the schemes shown in Figs. 1.31*a* and 1.31*b* over several tens of nanometers. With more sophisticated schemes (employing intracavity Fabry-Perot etalons, narrow-band filters, dispersive elements, etc.), very narrow emission lines can also be obtained. By changing the dye, it is then possible to get laser emission over any wavelength from the near UV ($\sim 350$ nm) to the near IR ($\sim 1 \, \mu m$). All these properties make the dye laser very useful in a variety of applications, particularly in spectroscopy.

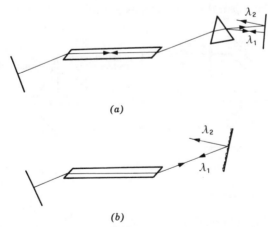

(a)

(b)

**Fig. 1.31.**   Schemes of dispersive resonators for tuning the laser wavelength at $\lambda = \lambda_1$. The dispersive element is a prism (*a*) or a diffraction grating in Littrow mounting (*b*).

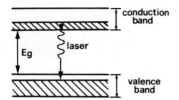

Fig. 1.32. Principle of semiconductor laser.

## 1.6.4 Semiconductor Lasers

Population inversion is here achieved between the bottom of the conduction band and the top of the valence band. This is illustrated in Fig. 1.32, where the bottom of the conduction band (hatched region) is assumed to be filled with electrons, while the top of the valence band (not hatched) is assumed to be empty of electrons; namely, filled with holes. Laser action takes place due to the stimulated transitions of the electrons from the conduction band back to the valence band. Stimulated emission thus arises from electron–hole recombination. From Fig. 1.32 two peculiar properties of semiconductor lasers are immediately apparent: (i) the frequency $\nu$ of the emitted radiation is directly related to the semiconductor bandgap $E_g$, namely, $\nu \simeq E_g/h$. (ii) Since the stimulated transition occurs between bands rather than between localized energy levels, the gain bandwidth, that is, the region over which a single laser can be frequency tuned, can be considerably large (e.g., $\Delta\lambda \sim 10$ nm with GaAs).

Population inversion is usually achieved through the use of a given semiconductor in the form of a $p$–$n$ junction that is forward biased (Fig. 1.33). The active region is localized around the plane of the junction and may typically be 1 $\mu$m thick. Electrons are injected from the $n$ side of the

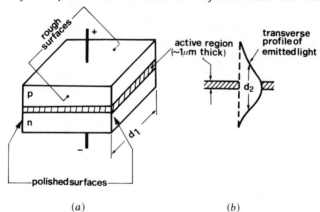

(a)                                    (b)

**Fig. 1.33.** (a) Schematic of $p-n$ junction semiconductor laser. (b) Transverse distribution of light intensity.

junction, and holes are injected from the $p$ side of the junction. Since free electrons and holes are localized respectively in the conduction band and in the valence band, these two bands of the active region are partially filled with electrons and holes, respectively. Since furthermore electrons tend to fill the bottom while holes tend to fill the top of a band (intraband relaxation is in fact very rapid), the condition depicted in Fig. 1.32 can thus be achieved.

If the threshold condition is reached, laser action will occur in the plane of the junction. The two mirrors are in this case two faces of the semiconductor that are made flat and parallel (Fig. 1.33). Usually, no coating on these surfaces is used because, due to the high refractive index of semiconductors (e.g., $\eta = 3.6$ for GaAs), the Fresnel reflectivity at the air–semiconductor interface is already sufficient to permit laser action. The laser beam has a rectangular cross section with a length equal to the diode length $d_1$ ($d_1 = 0.1$–1 mm) and height $d_2$, which is appreciably larger than the height of the active region (typically $d_2 \sim 40$ $\mu$m). The output beam is thus highly divergent, with a divergence approximately equal to $\lambda/d_1$ and $\lambda/d_2$ in the plane of the junction and in the plane perpendicular to the junction, respectively. The beam can of course be collimated through a suitable lens system.

The most interesting semiconductor lasers are the following: (i) In the visible or near infrared, we have either the GaAs laser ($\lambda = 0.84$ $\mu$m) or the ternary compound GaAs$_{1-x}$P$_x$ (with $0 < x < 0.4$). By varying the index of composition $x$, the bandgap changes, and thus the wavelength of the emitted radiation spans the region $0.84$ $\mu$m $\gtrsim \lambda \gtrsim 0.6$ $\mu$m. Lasers made of GaAs in the double heterojunction configuration can show continuous laser action at room temperature. This laser is therefore particularly attractive as transmitter of an optical communication link using optical fibers as the transmitting medium. (ii) In the middle-far infrared the lead salt semiconductor lasers can be used, particularly the ternary compounds PbS$_{1-x}$Se$_x$ (4 $\mu$m $< \lambda < 8.5 \mu$m), Pb$_{1-x}$Sn$_x$Te (6.5 $\mu$m $< \lambda < 32$ $\mu$m), and Pb$_{1-x}$Sn$_x$Se (9 $\mu$m $< \lambda < 30$ $\mu$m). Laser operation in these cases requires cryogenic temperatures ($T < 77$ K for cw operation). For a given index of composition $x$, the wavelength of the emitted radiation can be tuned either by a magnetic field or hydrostatic pressure, or by changing the diode current. This last way is particularly simple, and the change in wavelength arises from the heating of the diode due to nonradiative and ohmic losses. This heating causes a change in the refractive index of the semiconductor that changes the effective cavity length. (The effect of frequency tuning due to a change in cavity length from thermal expansion is negligible by comparison.) Continuous tuning of a single mode from this effect can be as much as 2 cm$^{-1}$ before a mode jump occurs. The jumping of modes results from a shift in peak frequency of the gain spectrum. Typical applications of lead salt lasers are found in the field of infrared spectroscopy, particularly in high-resolution spectroscopy (linewidths as narrow as 54 kHz, with a frequency stability of at least one order of magnitude better than this, have been obtained in PbSnTe diodes).

## 1.7  PHYSICAL APPLICATIONS OF LASERS: NONLINEAR OPTICAL EFFECTS

The very high monochromaticity and radiance of laser have made possible the observation of a large variety of new effects in radiation–matter interaction. Since they depend nonlinearly on the electric field amplitude of the light beam, $E$, these effects are called *nonlinear optical effects* (14). Among the most significant are optical harmonic generation effects. Suppose a laser beam with frequency $\omega$ interacts with a material far from any of its absorption lines. In this case, the material response can be described by its polarization $P$. At high irradiance, instead of the usual relation $P = \chi E$ (where $\chi$ is the optical susceptibility), we can write

$$P = \chi_1 E + \chi_2 E^2 + \chi_3 E^3 \cdots \tag{64}$$

If the material is centrosymmetric, it must be $P(-E) = -P(E)$, and hence the coefficients of the odd terms in Eq. (64) must be zero. Let us consider therefore a noncentrosymmetric material and study the effect of the first nonlinear term in Eq. (64), namely, $P^{(2)} = \chi_2 E^2$. Putting $E = E_0 \sin \omega t$, we have

$$P^{(2)} = \frac{\chi_2 E_0^2}{2} (1 - \cos 2\omega t) \tag{65}$$

Equation (65) shows that $P^{(2)}$ is the sum of a constant term and a term oscillating at frequency $2\omega$. While the former gives a dc voltage across the crystal, the latter radiates an electromagnetic wave at frequency $2\omega$ (second harmonic) *with monochromaticity and directionality properties similar to those of the incident laser beam*. In a similar way, we can consider the effect of the term $P^{(3)} = \chi_3 E^3$. With the sinusoidal field $E = E_0 \sin \omega t$ we have

$$P^{(3)} = \frac{\chi_3 E_0^3}{4} (\sin 3\omega t + 3 \sin \omega t) \tag{66}$$

The first term in parentheses generates a wave at a frequency $3\omega$ (third harmonic) which is as monochromatic and directional as the incident laser beam. The second term, being at the same frequency as that of the input beam, modifies its propagating properties. In this respect this term is

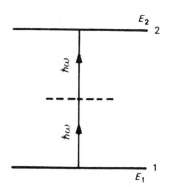

**Fig. 1.34.** Two-photon absorption.

similar to the linear term $\chi_1 E$, which is also at frequency $\omega$. While however this last term simply contributes to the refractive index of the material, as known, thus only modifying the phase velocity of the beam, the second term of Eq. (66) is responsible for a variety of nonlinear phenomena that can deeply modify the propagating beam. These effects are called *self-focusing, self-phase-modulation,* and *self-steepening* (15). Other effects occur obviously when more than one light beam is present. For example, the presence of two beams with frequencies $\omega_1$ and $\omega_2$, respectively, allows the generation of waves at the sum $\omega_1 + \omega_2$ and at the difference $\omega_1 - \omega_2$ of the frequencies (16).

If the frequency $\omega$ of the laser beam has a suitable value, a two-photon absorption process can occur. This requires a transition with a frequency $\omega_0$ such that $h\omega_0 = 2h\omega$, as shown in Fig. 1.34. While the probability of the absorption of one photon is proportional to the beam irradiance, two-photon absorption probability is proportional *to the square* of this irradiance. Since the selection rules are also different for the two processes, the two-photon absorption can be used for a new type of spectroscopy. This is just one example of nonlinear effects belonging to a new branch of spectroscopy that is based on the use of the laser, called *laser spectroscopy* (17). This new field includes not only two-photon spectroscopy but also *saturation spectroscopy, quantum beat spectroscopy,* and *fluorescence line-narrowing spectroscopy*.

## 1.8  CONCLUSIONS

In this chapter the basic principles of lasers are briefly discussed. Thus, after an introductory section (Sect. 1.1), the basic elements forming a laser are described; namely, the active material (whose characteristics are discussed in Sect. 1.2 devoted to radiation–matter interaction), the pumping system (Sect. 1.3, pumping processes), and the resonator (Sect.1.4).

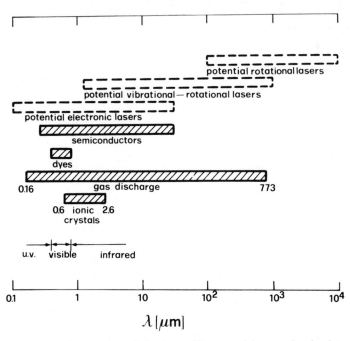

**Fig. 1.35.** Wavelength range for existing lasers. The potential ranges for the three types of transition used in lasers are also shown.

Section 1.5 is then devoted to a description of the various regimes of laser operation, while Sect. 1.6 describes a few of the most interesting types of laser. This last section constitutes the central part of the work. To sum up some of the results discussed, Fig. 1.35 shows the wavelength ranges over which the various types of laser have actually been operated. The same figure also shows the potential ranges for the three different types of laser transition that can be used: (i) transitions between electronic states, (ii) vibrational–rotational transitions, and (iii) rotational transitions. However, these ranges cannot, in general, be continuously covered by existing lasers. Dye lasers are an exception to this rule. It should be noted that the wavelengths of existing lasers already span four orders of magnitude. This wide variability is by no means characteristic of only the wavelength of a laser. The output power of a continuous wave laser can range from a few milliwatts for a He–Ne laser to perhaps a few megawatts for $CO_2$ lasers (peak powers in the terawatt range are now achievable!). Pulse duration is also very variable, as it can range from milliseconds (in pulsed solid-state lasers) to picoseconds (in mode-locked lasers). This wide variability in the parameters of the output beam, coupled with the peculiar properties of the beam (high directionality, monochromaticity, and radiance) make lasers unique for many applications. We can thus say that, after 15 years from the discovery of the laser, a great deal of progress has been made. Indeed, in the early days after the discovery, some people used to call the laser "a bright solution in search of a problem." Perhaps the laser can now be considered *the* solution to many problems where photons are involved, notably in the field of photophysics, photochemistry, and photobiology.

## References

1a. O. Svelto, *Principles of Lasers*, Plenum Press, New York (1976); O. Svelto, in *Tecniche e Dispositivi Speciali: Laser e Maser, Enciclopedia della Fisica*, Vol. II, R. Fieschi, Ed., ISEDI, Milan (1976) pp. 13.1–13.33.

1b. O. Svelto, *Principi dei Laser*, Tamburini Editore, Milano, 1972.

2. W. Louisell, *Radiation and Noise in Quantum Electronics*, McGraw-Hill, New York (1964), Chap. 5.

3. J. B. Birks, *Photophysics of Aromatic Molecules*, Wiley-Interscience, London (1970), Chap. 5.

4. O. Svelto, in *Laser Handbook*, Vol. 1, F. T. Arecchi and E. O. Schulz-Dubois, Eds., North-Holland, Amsterdam (1972), pp. 529–559.

5. R. W. Hellwarth in *Lasers*, Vol. 1, A. K. Levine, Ed., Marcel Dekker, New York (1966), p. 253.

6. P. W. Smith, M. A. Duguay, and E. P. Ippen, "Mode-locking of Lasers," in *Progress in Quantum Electronics*, Vol. 3, Part 2, J. H. Sanders and S. S. Stenholm, Eds., Pergamon Press, London (1974).

7. V. Evtuhov and J. K. Neeland, in *Lasers*, Vol. 1, A. K. Levine, Ed., Marcel Dekker, New York (1966), Chap. 1.

8. D. Findlay and D. W. Goodwin, in *Advances in Quantum Electronics*, Vol. 1, D. W. Goodwin, Ed., Academic Press, New York (1970), pp. 77–128.

9. C. K. Rhodes and A. Szöke, in *Laser Handbook*, Vol. 1, F. T. Arecchi and E. O. Schulz-DuBois, Eds., North-Holland, Amsterdam (1972), pp. 265–324.

10. W. B. Bridges, A. N. Chester, A. S. Halsted, and J. V. Parker, *Proc. IEEE*, **59**, 724 (1971).

11. C. K. N. Patel, in *Lasers*, Vol. 2, A. K. Levine, Ed., Marcel Dekker, New York (1968), Chap. 1.

12. C. K. Rhodes, *IEEE J. Quant. Electron* **QE-10**, 153 (1974).

13. F. P. Schäfer, Ed., *Dye Lasers*, Springer-Verlag, Berlin (1973).

14. R. W. Terhune and P. D. Maker, in *Lasers*, Vol. 2, A. K. Levine, Ed., Marcel Dekker, New York (1968), pp. 295–370.

15. O. Svelto, in *Progress in Optics*, Vol. XII, E. Wolf, Ed., North-Holland, Amsterdam (1974), pp. 1–51.

16. A. Yariv, *Introduction to Optical Electronics*, Holt, Rinehart and Winston, New York (1971), Chap. 8.

17. A. Yariv and J. E. Pearson, in *Progress in Quantum Electronics*, Vol. 1, J. H. Sanders and K. W. H. Stevens, Eds., Pergamon Press, London (1971), pp. 1–46.

18. R. G. Brewer and A. Mooradian, Eds., *Laser Spectroscopy*, Plenum Press, New York, (1974).

# ANALYTICAL SPECTROSCOPY
# USING LASER ATOMIZERS

## K. LAQUA

*Institut für Spektrochemie und Angewandte Spektroskopie*
*Dortmund, Germany*

## 2.1  INTRODUCTION

In order to better understand the particular analytical properties of laser atomizers, the general role of a spectrochemical excitation source is summarized here briefly. In optical spectrochemical analysis, chemical elements are identified by their optical line spectra. Therefore, part of a solid or liquid sample has to be converted into atomic vapor in which free atoms can then either be detected with the help of their absorption spectra or, if suitably excited, with the help of their emission spectra. Ionized atoms could also be subjected to mass spectroscopic analysis. This subject, however, is not treated in this chapter.

Vaporization, atomization, and excitation are often, but not necessarily, produced by one and the same excitation source. With sources of electrical excitation, for example, electrical arc sparks or discharges at low pressure, electrically conducting samples can be analyzed directly, whereas nonconducting samples have to be pretreated, usually by pulverizing and mixing them with conducting material. Other excitation sources, for example, flames and flamelike plasmas, are mostly used for the analysis of liquids in the form of aerosols. Solid materials are first dissolved; their direct conversion to an aerosol is possible in special cases, as has been reported recently.

The in situ analysis of a small spot of a surface, that is, a *local* analysis, may yield valuable information. Many special techniques have been developed for this purpose. As regards spectrochemical analysis, some of the electrical excitation sources, for example, spark discharges, have been adapted but with limited success. A spatial resolution of not much better than about 1 mm seems to be typical. In addition, *such methods are restricted to the analysis of electrically conducting surfaces.* With nonconducting surfaces, local analysis can only be performed indirectly by first mechanically removing part of a sample and then analyzing it by a standard method. It is understandable that because of the lack of fast, powerful methods for local spectrochemical analysis, the first application of lasers was in this field. The inherent possibilities of powerful lasers as spectrochemical excitation sources were explored within a short time after the first construction of a ruby laser in 1960 by Maiman (1). In 1962, a "laser microprobe" was demonstrated by Brech and Schuch (2) of the Jarrell Ash Company, which shortly afterward was marketed and put into analytical routine. Their latest model is shown in Fig. 2.1.

The principle of operation can best be understood with the help of Fig. 2.2, which is a semischematic representation of a similar instrument. The main parts are the laser including the $Q$-switch, a microscope for aligning the sample and focusing the laser radiation on the surface, and the electrode system for auxiliary spark cross excitation. Power supplies for the laser and the spark are omitted. Instruments of this type are primarily designed for emission spectroscopy using laser radiation as the thermal source of heat for vaporizing and exciting samples for local and microanalysis. In this case the latter type of analysis is reduced to local analysis because a microsample can only be analyzed if it is either already concentrated or can be concentrated locally.

The inherent possibilities of laser atomizers in analytical spectroscopy have not yet been fully exploited to afford more facilities than are normally incorporated in present-day commercial instruments. Therefore, additional components or modules may be necessary to solve a specific analytical problem. In Fig. 2.3, an attempt is made to compile the most important combinations relevant to spectrochemical analysis.

Finally, another use of lasers in analytical spectroscopy, namely, as a source of (intense) monochromatic radiation, which is the subject of Chapter 3, should nevertheless be mentioned in this context because combined procedures with lasers for both purposes, for example, in

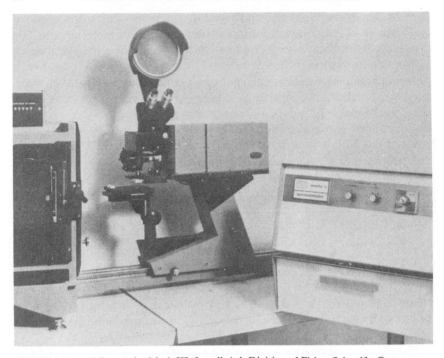

**Fig. 2.1.**  Laser Microprobe Mark III, Jarrell-Ash Division of Fisher Scientific Company.

atomic absorption or atomic fluorescence spectroscopy, seem promising (see also Chapter 4).

## 2.2  CLASSES OF LASERS

The most important properties of lasers for the purpose in question are sufficient power, a small angle of divergence of the beam, compactness, ease of operation and maintenance, long troublefree operation time, and low costs. The wavelength can be of some importance as is discussed later, whereas *spectral purity is not necessary.* Therefore, multimode operation is feasible and many kinds of lasers may be suitable, those optically pumped as well as electrically excited, and solid-state, liquid, and gas lasers. In practice, however, only a few have found application. In Table 2.1, a selection is listed together with their pertinent properties.

### 2.2.1  Solid-State Lasers

A typical laser setup using a solid-state laser is shown in Fig. 2.2. With a solid-state laser a very wide range of output energy and power is

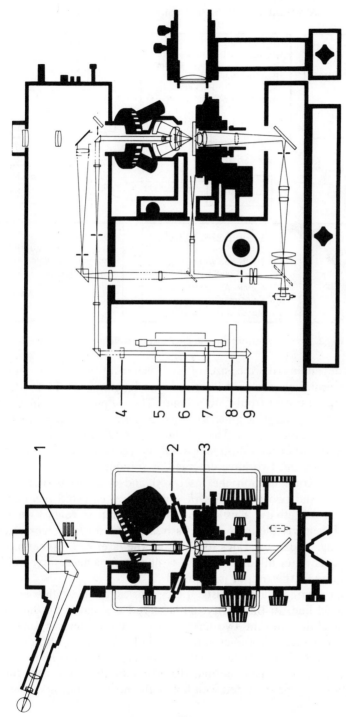

**Fig. 2.2.** Laser Micro Analysator LMA 10. Left, front view; right, side view; 1, observation microscope; 2, electrodes for cross discharge; 3, microscope stage with sample holder; 4, semitransparent resonator mirror; 5, laser cavity; 6, laser rod; 7, flash-lamp for optical pumping; 8, liquid dye Q-switch; 9, resonator prism. VEB Carl Zeiss, Jena.

51

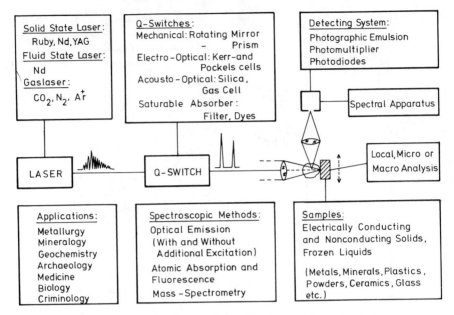

**Fig. 2.3.** Analytical facilities in combination with laser atomizers.

available depending on the size of the laser rod and of the pumping system. Typical dimensions of laser rods are lengths from 15 to 100 mm and diameters ranging from 5 to 15 mm. The output energy may range between some tenths of a joule to about 5 joules, but applications with considerably higher energy have been reported. The rod is usually pumped by one or several flash lamps. Of the geometric configurations of the pumping system, two have emerged as suitable.

In the first configuration the laser rod and a single flash lamp are positioned on the two focal lines of a cylindrical ellipsoid having inner walls of high reflectivity. This is the optimum setup as regards pumping efficiency. It is advisable to cool the flash lamp and for high reproducibility and/or high repetition rates to also cool the laser rod. Sometimes the whole cavity is water filled. Control of temperature is necessary, and high reproducibility of the output is required.

The other configuration consists of a cylindrical double ellipsoid with a common focal line. In this cavity the laser rod is situated on the middle common focal line and the two flash lamps on the respective outer focal lines. Theoretically, this configuration should have a lower efficiency than the first one, but in practice this is not so because the symmetric arrangement leads to less thermal distortion of the laser rod. Such a configuration may be operated well below the nominal rating of the flash

**TABLE 2.1**
**Some Properties of Pulsable Lasers Suitable for Atomization of Solid Samples**

| Type of laser | Wavelength, $\mu$m | Excitation | Output energy of free-running laser, J | Range of power per pulse, MW | Angle of divergence, mrad | Repetition rate, pps |
|---|---|---|---|---|---|---|
| Solid state |
| Ruby | 0.6943 ⎫ | optically pumped | 10 | 0.1–10 | 3–5 | low |
| Nd:glass | 1.06 ⎬ |  | 10 | 0.01–10 | 5 | low |
| Nd:YAG | 1.06 ⎭ |  | 10 | 0.01–0.1 | 3–5 | $10^3$ |
| Fluid state |
| Nd:POCl$_3$ | 1.06 ⎫ | optically pumped | 5 | ~10 | ~10 | 10 |
| Nd:SeOCl$_2$ | 1.06 ⎬ |  |  |  |  |  |
| Dyes | 0.22–0.74 ⎭ |  | 1 | ~0.1 | 2 | 100 |
| Gaseous state |
| CO$_2$–TEA | 10.6 ⎫ | electrically excited | 100 | ~1 | 5 | 100 |
| N$_2$ | 0.337 ⎭ |  | 0.01 | ~0.1 | ~10 | 100 |

lamps, with consequent long life expectancy of the lamps. They may then be expected to last for several 100,000 shots, which considerably adds to the reliability of the whole system. The flash lamps are typically operated with a power supply capable of delivering from several hundred up to several thousand joules from a condenser battery charged to about 2 kV. The duration and shape of the discharge current are adjusted with the help of additional impedances. A unipolar pulse of from several hundred microseconds to several milliseconds is wanted. The discharge is initiated by a high-voltage pulse which can be introduced into the circuit either in series or in parallel.

The laser resonator consists of two mirrors of interferometer quality, with high reflectivity and low loss. Dielectric coating is common. A mirror may also be formed by a suitable prism. The plane-parallel configuration is almost exclusively used. Other configurations with spherical mirrors are also possible. One of the mirrors is semitransparent in order to extract the laser radiation, and its optimum transmittance depends on the properties of the active material (laser rod) and on the geometric configuration and should be chosen accordingly.

As material for the laser rod, a ruby ($Al_2O_3$ doped with 0.05% chromium) or neodymium in glass is commonly used. Of the other materials, none has found widespread use. The advantages of ruby are the following: it radiates in the visible region (6943 Å), it is thermally robust, and it has good heat conductivity. Its main disadvantage lies in its rather high threshold energy resulting in lower efficiency. The quality of ruby rods may affect the divergence of the beam. The neodymium glass laser has a low threshold, a high $Q$ factor, and high optical quality. Its main disadvantages are its thermal sensitivity and its radiation at 1.06 $\mu$m in the infrared. Because of the invisibility of the radiation, it may be more difficult to adjust, and protection against radiation hazard may require more attention. Laser with yttrium aluminum garnet (YAG) doped with neodymium as active element would also be suitable. With these, very low threshold energies are obtainable. Therefore, they may also be pumped by continuous light sources leading to continuous laser action. For several reasons, YAG lasers should find a wide field of application.

### 2.2.2 Liquid Lasers

There seem to be several types of liquid lasers of some potential interest, for example, the neodymium-doped lasers with either selenium oxychloride or phosphorus oxychloride as host solutions. The former was developed mainly by Heller (3), the latter by Schimitschek (4). The liquid acting as the active element is usually circulated through a cuvette. The performance of these lasers is apparently similar to that of the corresponding glass laser. A very high power output is achieved, and the threshold energy is very low, of the order of that of a YAG laser. The very reactive nature of the liquids, however, especially that of $SeOCl_2$, complicates the operation and maintenance of such a system considerably.

In principle, flash lamp-pumped dye lasers of sufficient power and energy could also be used as atomizers. As atomization can be better achieved with other types of lasers, their application in this field can *only be recommended in cases where such a system is already available* and only an occasional analysis has to be made.

### 2.2.3 Gas Lasers

Of the gas lasers, nitrogen as well as argon or $CO_2$ lasers, if properly $Q$-switched, are capable of delivering outputs suitable for sample vaporization. It should be emphasized that continuous laser action is not desirable because of the comparatively low temperatures achievable on the sample surface (boiling temperature of the material). The most promising is definitely the $CO_2$ laser, but its usefulness may be hampered by its rather big physical size and by its wavelength which, at 10.6 $\mu$m, is rather far in the infrared and therefore not optimal for sampling from metal surfaces, all of which have a high reflectivity in this region. On the other hand, $CO_2$ lasers can easily be operated at higher repetition rates than either ruby or Nd glass lasers. This property could be quite useful in the field of macroanalysis. It should further be mentioned that all gas lasers in question are electrically pumped, in contrast to those discussed previously. Consequently, their energy conversion efficiency may seem high.

## 2.3 OPERATION MODES OF LASERS

### 2.3.1 Continuous Wave Operation

Continuous wave (cw) operation may be obtained from electrically pumped gas lasers and from solid-state and liquid lasers optically pumped by tungsten halogen or similar high-power lamps. In this field, the $CO_2$ laser and the Nd:YAG laser have gained special importance both of which are widely used for drilling, cutting, and welding in material processing. For atomization, even in a two-step process with additional excitation, they are not very well suited for the reasons already stated in Sect. 2.2.3.

### 2.3.2 Free-Running Operation

The output of flash lamp-pumped, solid-state lasers usually consists of a number of irregular and incoherent radiation pulses called "spikes" of less than 1 $\mu$s in duration. This laser action starts some hundred microseconds after the firing of the pumping lamp and lasts for a few hundred microseconds. Figure 2.4 shows such a laser output. In Table 2.2, column 3, some typical properties for a Nd glass laser are listed.

### 2.3.3 Q-Switched Operation

This technique is also widely used in connection with laser atomizers. First proposed by Hellwarth (5), the principle is as follows. A shutter in the laser cavity is initially closed and obstructs the path of light during part of the pumping period. Therefore, no laser action will occur under certain circumstances, and more energy will be stored in the active medium as compared to the free-running case. This energy will be

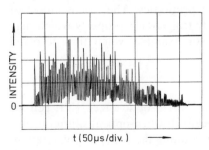

**Fig. 2.4.** Radiant power output of free-running solid-state laser. From Quentmeier (24), reproduced by permission.

**TABLE 2.2**
**Laser Properties Pertaining to Different Operating Modes Based on the Properties of the Free-Running Laser of the Second Column**

| Property | Free running mode | Q-Switched mode | | | |
|---|---|---|---|---|---|
| | | Rotating mirror or prism | Kerr or Pockels cells | Acousto-optical switch | Saturable dye |
| Maximum energy, J | 1.4 | 0.6 | 0.1 | 1 | 0.6 |
| Spike duration, $\mu$s | 0.1–0.3 | 0.1–0.3 | 0.02 | 0.2 | 0.2 |
| Total duration, $\mu$s | 270–300 | 10–15 | — | 230–250 | 230–250 |
| Energy per spike, mJ | 5 | 20 | 100 | 35 | 30–50 |
| Number of spikes | 280 | 20–25 | 1 | 30 | 10–12 |
| Power per spike, kW | 16 | 80 | $5 \times 10^3$ | 175 | 250 |
| Diameter of irradiated area, arbitrary units | 1 | 1.42 | 0.7 | 1. | 0.7 |
| Mean power, kW | 5 | 50 | — | 4 | 2.5 |

released in a short high-power radiation pulse a few nanoseconds long if the shutter is opened sufficiently fast. In slowly switched systems several pulses may occur, each of intermediate power and time duration. As both types of shutters are of interest, some examples are listed in Table 2.2.

### 2.3.3.1   Saturable Dyes

The simplest type of Q-switch is a bleachable substance which may either be a solid or a liquid, for example, the Schott filters RG 8 and RG 10. The former is mostly used in connection with ruby lasers. This switch enables the production of one to several power spikes of up to several megawatts per pumping cycle. The number and spacing of the spikes, the latter of which is of a statistical nature, depend on the properties of the whole system. They may be changed by suitably adjusting the optical thickness of the filter for the laser radiation. With neodymium lasers, only liquid filters (Q-switch solution) are feasible which however, permit easy adjustment of threshold and switching properties. The disadvantage of this switch is that it cannot be externally triggered, a property which may be desirable for certain applications, for example, when time-resolved measurements have to be made.

### 2.3.3.2   Electro-optical Shutters

They make use of either the quadratic electro-optic effect (Kerr cell) or the linear electro-optic effect (Pockels cell). Electro-optical shutters combine a polarizer with the cell and act by rotating the plane of polarization.

The shutter is usually opaque when biased by a high voltage and opens when this voltage is switched off. Since good optical crystals exhibiting a strong linear electro-optic effect have become available, for example, ADP (ammonium dihydrogen phosphate), KDP (potassium dihydrogen phosphate) and KD$^+$P (the deuterated form of KDP), Pockels cells are now extensively used because they need a switching voltage of only a few kilovolts.

In analytical practice, electro-optic shutters have been almost completely replaced by saturable dye shutters because of their ease of operation. Nevertheless, it should be emphasized that the former can be accurately triggered and synchronized with other equipment thereby allowing special but very useful techniques as time-resolved spectroscopy to be used.

### 2.3.3.3 *Mechanical Shutters*

For this purpose, rotating discs, rotating mirrors, and rotating prisms are being employed. If high-power single giant pulses have to be generated, a high speed of rotation is necessary, typically 30,000 rpm. At speeds below 10,000 rpm, several spikes of medium power are produced. In spectrochemical application speeds of 3000 rpm are customary. A typical example is listed in Table 2.2. These switches also have the advantage of offering trigger and synchronizing facilities.

### 2.3.3.4 *Acousto-optical Shutters*

There are two basically different types of systems. The first is based on periodic spatial variations in the index of refraction caused by an elastic wave in a suitable crystal. In effect, a diffraction grating is produced by which the laser beam is deviated periodically. With operating frequencies in the order of several hundred megahertz, very fast switches can be obtained. Until now, no use in our field of application has been reported.

The other type of system, which offers some properties of special importance for certain applications in spectrochemical analysis, is the acousto-optical $Q$-switch of Boissel et al. (6). Figure 2.5 shows its operating principle. The system consists of a ceramic cylindrical transducer in a chamber filled with $SF_6$ at high pressure (typical about 20 bars). The transducer is driven by an audio generator at its resonance frequency which in a typical setup may be between 50 and 100 kHz. Due to the radial movement of the transducer, the gas is periodically compressed and released causing a time-dependent density gradient. The corresponding time-dependent gradient of the index of refraction of the gas produces an oscillating dispersion of the laser beam if used outside the

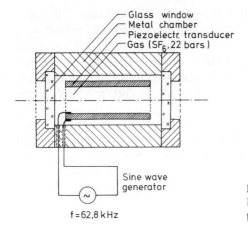

Glass window
Metal chamber
Piezoelectr. transducer
Gas ($SF_6$,22 bars)

Sine wave generator

f = 62,8 kHz

**Fig. 2.5.** Acousto-optical $Q$-switch. From Boissel et al. (6), reproduced by permission.

laser cavity. If used inside the cavity, the $Q$ value will change periodically. As the gradient disappears twice during each cycle, optimum $Q$ is obtained at twice the frequency of the oscillator. This $Q$-switch produces a series of equidistant spikes as shown in Fig. 2.6. The spikes have powers of several hundred kilowatts. The main advantage seems to be the high efficiency of such a system because it produces spikes of sufficient power for a considerable fraction of the pumping time. The performance values given in Table 2.2 are typical for an analytical setup.

It should be noted that the power of the spikes show a random distribution similar to that of the free-running laser. It should be possible to produce spikes of equal power with a spherical mirror configuration. It is doubtful, however, whether improved performance as regards reproducibility or limit of detection of a laser spectrochemical procedure would be the result. According to preliminary investigations, averaging the signals obtained with a large number of spikes of varying amplitude would give better results than relying on fewer spikes of constant amplitude.

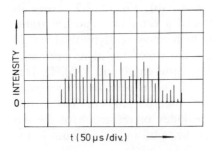

INTENSITY

0

t (50 μs /div.)

**Fig. 2.6.** Radiant power output of acousto-optically $Q$-switched solid-state laser. (Scale of ordinate reduced by 10 as compared to Fig. 2.4.) From Boissel et al. (6), reproduced by permission.

### 2.3.3.5 *Mode Locking*

In a free-running laser the irregular emission of spikes is due to multimode oscillation. If the modes are forced to oscillate in phase and with similar amplitudes, this so-called mode-locked operation enables ultrashort pulses of very high power in the picosecond range. Several methods can be used to obtain mode locking by relying on active as well as passive elements. Applications of mode-locked Nd glass or Nd-YAG lasers in laser spectrochemical analysis are conceivable. The interested reader is advised to consult the excellent *Laser Handbook* edited by Arecchi and Schulz-Dubois for relevant information (7).

### 2.3.3.6 *Frequency Doubling*

Of the lasers presently employed for the vaporization and atomization of materials for chemical analysis, most emit in the red or infrared spectral region. This is favorable for dielectrics but not for metals. In the latter case the choice of shorter wavelengths should be considered. They are accessible through frequency doubling of the original radiation, which can easily be obtained with optical crystals having nonlinear optical properties. Quartz, $LiNbO_3$, ADP, KDP, and many more are suited for this purpose.

In the preceding sections, only those properties of laser and laser setups have been reviewed that are or at least should be relevant to laser atomization. This necessarily makes the survey somewhat patchy. However, more general and fundamental information is readily obtainable (see Chapter 1).

## 2.4 INTERACTION OF LASER RADIATION WITH SOLIDS

Radiation from a laser source is emitted with a very small angle of divergence. Its radiant intensity, that is, its radiant power per solid angle, can be very high, exceeding that of all other known sources. In the power range considered here, an unfocused laser beam will produce an irradiance on a target suspended in its way that usually will vaporize very little material, if any. It is therefore necessary to focus the laser radiation onto the target with the help of a lens or a mirror. Because the laser beam is highly parallel, this focusing can be very effective in producing small focal spots with corresponding high irradiance. The gain in irradiance is typically more than about three orders of magnitude.

The ultimate diameter of the focal spot is given by diffraction of the laser beam at that aperture which determines its diameter, ultimately the effective diameter of the active element. The diffraction-limited diameter

$d$ of the focal spot is determined by the relation

$$d = 1.2 \frac{f}{D} \lambda \tag{1}$$

where $f$ is the focal length of the lens, $D$ is the diameter of the beam, and $\lambda$ is the wavelength of the laser radiation. In practice, due to imperfections of the laser rod or to production of transversal laser modes, the angle of divergence, $u$, and consequently the obtainable diameter of the focal spot is larger and may be given as

$$d = uf \tag{2}$$

In a good system, $u$ may be about 2 to 4 mrad.

The type of lens most commonly used is a microscope objective lens. The focal length should be chosen according to the kind of application and may range from about 5 to 50 mm. The lower limit is set by the minimum working distance between the front of the lens and the sample which may be of the order of several millimeters. The lens should be able to stand the high radiation density for long times of operation. Therefore, *uncemented elements are advisable.* The lens should be optically corrected as usual for such a purpose, except for chromatic aberration. Whereas for longer focal lengths lenses are common, mirror objectives permitting a comparable large working distance in relation to the focal length have also been described in the literature (8). In commercial instruments the microscope objective lens is part of the observation microscope. (Fig. 2.2).

The divergence of the laser beam can be decreased by mode selection. Figure 2.7 shows the system used by Klocke (9), which eliminated most of the transversal modes. Care must be taken with such a system to prevent the diaphragm from being damaged by the focused radiation. This is accomplished by using long focal length lenses. No selection of longitudinal modes needs to be made.

The focal spot can be further decreased if the laser beam enters the microscope through the eyepiece. In this case the microscope should be adjusted in such a way that the beam is first concentrated in the focal

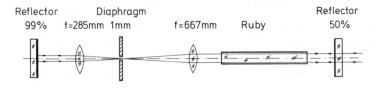

Reflector    Diaphragm                                    Reflector
99%    f=285mm 1mm    f=667mm    Ruby              50%

**Fig. 2.7.** Intercavity mode selection for suppressing transversal modes.

point of the eyepiece lens, whereas the objective lens then images this point onto the surface.

### 2.4.1 The Crater

The laser radiation focused onto a target may cause *irreversible* changes depending on the irradiance and the nature of the target. Panzer (10) has listed the following figures for irreversible changes: red blood corpuscles, at about $10^5$ W cm$^{-2}$; the human retina, at about $3 \times 10^4$ W cm$^{-2}$, metals, at about $10^6$ W cm$^{-2}$; and glass, above $10^9$ W cm$^{-2}$. These figures may serve as a very rough guide only since it is not merely the irradiance which matters but also the duration of interaction or, in other words, the radiant exposure measured in J cm$^{-2}$. If the irradiance is high enough—which can easily be obtained—material is vaporized irrespective of its physical properties.

The amount of sample material vaporized by a single laser shot and the dimensions of the crater, namely, diameter, depth, and shape, strongly depend on the properties of the sample and its surface, as well as on the properties of the laser system. One of the relevant factors is, of course, the area of the focal spot as determined by the divergence of the laser beam and the focal length of the lens according to Eq. (2). However, *the diameter of the crater does not coincide with the diameter of the focal area*. The two diameters must clearly be distinguished. The former is usually considerably larger than the latter because a substantial fraction of the energy absorbed in the focal spot is spread into the near environment by heat conduction. Consequently, material is evaporated from a larger area. Therefore, *the amount of energy more than anything else determines the size of the crater*. By decreasing this energy sufficiently, the crater diameter may approach that of the focal spot. In Fig. 2.8, some typical examples illustrate the above concepts.

It should be stressed that for such comparisons the other properties of the laser system should not be changed, particularly the mode of operation. A series of craters produced by a free-running laser of 0.6 J output energy with increasing attenuation by neutral filters showing a large range of diameters is shown in Figs. 2.9 to 2.11.

The shape of the crater also depends on the energy, which can either be changed by varying the number of spikes but keeping their energy constant, or vice versa. In each case the shape of the crater depends strongly on the mode of operation. A free-running laser emitting a large number of spikes of low energy and of low power produces a rather deep crater in relation to its diameter because of the consecutive action of a large number of spikes at intervals too short for the material to be

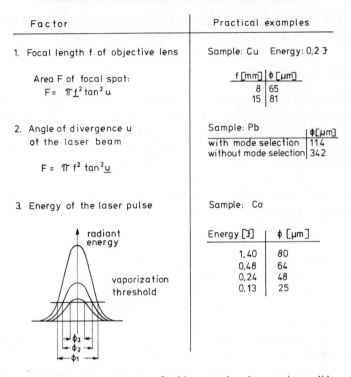

| Factor | Practical examples |
|---|---|

1. Focal length f of objective lens — Sample: Cu    Energy: 0,2 J

Area F of focal spot:
$F = \pi \underline{f}^2 \tan^2 u$

| f [mm] | Φ [μm] |
|---|---|
| 8 | 65 |
| 15 | 81 |

2. Angle of divergence u
   of the laser beam

$F = \pi f^2 \tan^2 \underline{u}$

Sample: Pb

| | Φ [μm] |
|---|---|
| with mode selection | 114 |
| without mode selection | 342 |

3. Energy of the laser pulse

Sample: Co

| Energy [J] | Φ [μm] |
|---|---|
| 1,40 | 80 |
| 0,48 | 64 |
| 0,24 | 48 |
| 0,13 | 25 |

**Fig. 2.8.**    Factors affecting diameter Φ of laser-produced crater, in a solid sample.

resolidified. It can further be said that *with decreasing energy the depth of the crater decreases more rapidly than its diameter.* This is illustrated in Figs. 2.12 and 2.13, respectively.

Photographs of craters produced by free-running lasers usually show a wall of molten material surrounding the crater. From this it can be concluded that the temperature applied has been rather low and that a considerable part of the material has been ejected as liquid or in a recondensed form. Although this mode of operation seems not optimal for atomization, it is nevertheless capable of producing very small craters.

Much better vaporization can be obtained if the power per spike is significantly increased via a Q-switch. This can be seen clearly by comparing Fig. 2.14 with Fig. 2.15.

Dielectric substances possessing high transmission and low absorption factors, for example, glass and some polymers, are best vaporized with Q-switched lasers at long wavelengths. In the case of glass there is a pronounced improvement by changing from a ruby to a Nd laser. Examples for such substances are shown in Figs. 2.16 and 2.17. Hard and

**Fig. 2.9.** Scanning electron microscope (SEM) photograph of a crater in a steel sample as produced by a free-running laser of 0.6 J radiant energy. Crater diameter 100 μm. From Klockenkämper and Laqua (34), reproduced by permission.

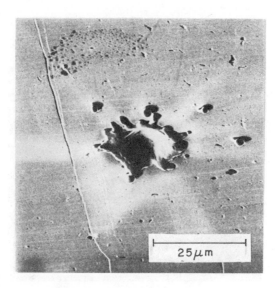

**Fig. 2.10.** Same as Fig. 2.9, except that radiant energy of laser output is reduced by filter. Crater diameter 25 μm. From Klockenkämper and Laqua (34), reproduced by permission.

63

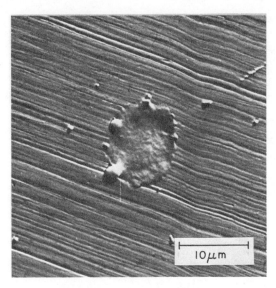

**Fig. 2.11.** Same as Figs. 2.9 and 2.10, except that laser energy output is reduced to threshold of vaporization by filter. Crater diameter 10 μm. From Klockenkämper and Laqua (34), reproduced by permission.

**Fig. 2.12.** SEM photograph of crater cross section. Values as in Fig. 2.9. Crater depth 800 μm. From Klockenkämper and Laqua (34), reproduced by permission.

64

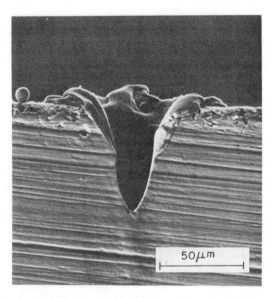

**Fig. 2.13.** SEM photograph of crater cross section. Values as in Fig. 2.10. Crater depth 60 μm. From Klockenkämper and Laqua (34), reproduced by permission.

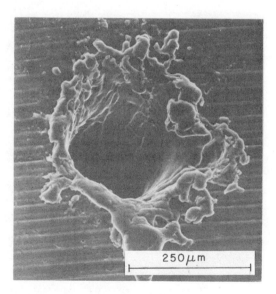

**Fig. 2.14.** SEM photograph of crater in aluminum produced by free-running solid-state laser of 0.6 J radiant energy. From Quentmeier (24), reproduced by permission.

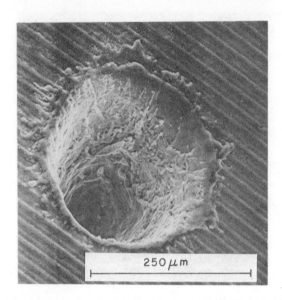

**Fig. 2.15.** Same as in Fig. 2.14 except that laser is $Q$-switched by an acousto-optical $Q$-switch. From Boissel et al. (6), reproduced by permission.

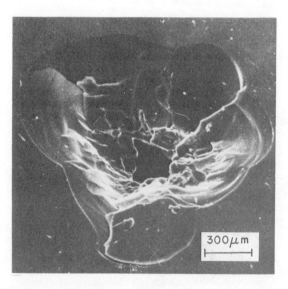

**Fig. 2.16.** SEM photograph of a crater in a thin glass plate produced with laser $Q$-switched with a mirror rotating at 1500 rpm. Crater diameter 270 $\mu$m. From Felske et al. (40$b$), reproduced by permission.

66

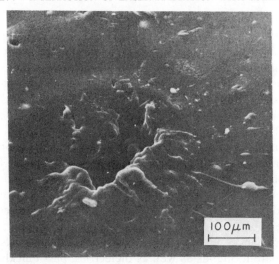

**Fig. 2.17.** SEM photograph of a crater in polythene. Laser as in Fig. 2.16. Crater diameter 80 μm. From Felske et al. (40b), reproduced by permission.

brittle substances such as glass or tooth enamel tend to develop cracks, as seen in Fig. 2.16. Sometimes this can be avoided if the laser radiation is focused slightly above instead of onto the surface.

The action of a *single giant laser pulse* is rather different. The very high irradiance on the surface results in a very shallow crater with little molten residue. The diameter of the crater may be rather large and cannot yet be reduced to that obtainable with other modes of operation. On the other hand, it is possible to further reduce the depth of the crater by decreasing the energy of the spike—values as low as 1 mJ have been reported by Allemand (11)—or by defocusing the laser radiation. The fraction of the removed material which is vaporized depends on the strength of the irradiance and can be comparably high. The range of true giant pulse operation is considered to be between $10^9$ W cm$^{-2}$ and the limit set by the dielectric breakdown of the air (laser spark) which may occur at irradiances in excess of $10^{12}$ W cm$^{-2}$.

The interaction of the laser radiation with the target is complex. A phenomenological explanation is possible, as follows: At least part of the laser radiation is absorbed by the surface within a depth of a few hundredths of a millimeter. Consequently, the temperature of this layer is raised to the boiling point and evaporation begins. At the same time heat conduction into the interior of the material occurs. As the velocity of heat transport is rather low, the layer will move into the sample. Two cases must be distinguished:

1.  At an irradiance of less than $10^8$ W cm$^{-2}$, a stream of vapor leaves the surface at a velocity of the order of $10^4$ cm s$^{-1}$. Subsequent action has been studied and described in detail by Scott and Strasheim (12). In the case of the multispike free-running laser, the process can be summarized as follows: As more energy is delivered to the surface, the initial vaporization process changes to the melting–flushing mechanism by which a significant amount of material is removed from the crater. A sheath of molten metal is flushed out well above the sample surface. The sheath however is not stable. It collapses, and additional vaporization and fragmentation is caused by subsequent incident radiation. This process may occur several times during the same laser shot. The mechanism prevails with all metals. With dielectric materials, vaporization is the only process, continuing throughout the interaction. This has been observed with graphite.

2.  For irradiances higher than $10^9$ W cm$^{-2}$, all materials, opaque as well as transparent, metals as well as dielectrics, pass quickly into a high absorbing state. The temperature of the vapor leaving the surface is higher than the boiling temperature and therefore partly ionized. A stream of vapor leaves the surface with velocities in excess of $10^6$ cm s$^{-1}$. Above the surface a plasma is formed which, because of its high conductivity, strongly absorbs the later part of the incoming laser spike. The cooling plasma resembles a nebula suspended above the sample surface. It should be noted, however, that part of the material is also ejected in liquid form, but no flushing or formation of a sheath of molten metal has been observed.

The semi-Q-switched operation (slow shutters) is somewhat intermediate in its action. More material is vaporized, but a substantial fraction is also removed by the flushing process. Because most of the shutters used for this kind of operation produce a number of spikes at rather large intervals (a few microseconds), the surface resolidifies between the spikes.

It would be of interest to predict the crater dimensions from the operation data of the laser and the physical properties of the sample. Until now, this is possible only to a limited degree and can be explained by the complexity of the processes involved. Some relationships, however, have been established. With free running or slowly Q-switched lasers, the amount of sample material removed from the surface is proportional to the radiant exposure. At low irradiances the influence of the reflectivity of the material cannot be neglected. In Table 2.3 the results of some measurements by Klocke (13) are given. Table 2.4, also taken from Klocke, reports a comparison of calculated and observed energies necessary to vaporize various metals. The measured energies are higher than

TABLE 2.3
Reflectance Before ($\rho_p$) and During ($\rho_d$) Evaporation
for a Number of Metals Vaporized by Free-Running
Laser[a]

| Element | $\rho_p$, % | $\rho_d$, % |
|---------|-------------|-------------|
| Ag | 80 | 41 |
| Cu | 74 | 38 |
| Mg | 66 | 24 |
| Mo | 36 | 25 |
| Fe | 37 | 17 |
| Cd | 59 | 17 |
| Pb | 50 | 7 |
| Bi | 51 | 8 |
| Sn | 62 | 12 |

[a] From Klocke (13).

the calculated energies for substances of low boiling temperature and are smaller for substances having a higher reflectivity than the calculated energies. For the above-mentioned mode of operation, a satisfactory formula for calculating the depths of the crater, in cm, can be written as

$$s = \frac{Et(1-r)}{\rho[H + C(T_b - T_0)]} - \frac{\lambda(T_b - T_0)}{E(1-r)} \tag{3}$$

TABLE 2.4
Energy Necessary to Vaporize a Certain Weight of a
Number of Metallic Samples with Free-Running Solid-
State Laser[a]

| Element | Energy, J | |
|---------|-----------|------------|
| | Observed | Calculated |
| Ag | 0.82 | 0.47 |
| Cu | 0.87 | 0.43 |
| Mo | 1.05 | 0.58 |
| Mg | 1.07 | 0.57 |
| Fe | 1.16 | 1.03 |
| Cd | 1.16 | 1.28 |
| Sn | 1.23 | 3.80 |
| Bi | 1.29 | 1.76 |
| Pb | 1.30 | 1.80 |

[a] From Klocke (13).

where $E$ (W cm$^{-2}$) is the beam irradiance, $t$ (s) is the pulse duration, $r$ is the reflection factor, $\rho$ (g cm$^{-3}$) is the density, $C$ (J$^{-1}$ g$^{-1}$ K$^{-1}$) is the (thermal) heat capacity, $T_0$ and $T_b$ (K) are the initial and boiling temperature, respectively, $H$ (J g$^{-1}$) is the heat of vaporization, and $\lambda$ (W cm$^{-1}$ K$^{-1}$) is the thermal conductivity. It is a modification of a formula first applied by Ready (14) and is based on the equation of thermodiffusion, theoretically treated by Landau (15). In Table 2.5 some results of measurements of crater depths are compared to calculated values. A better agreement can hardly be expected because of the many simplifications of the theoretical treatment.

In the case of giant pulse operation the influence of the reflectivity can be neglected, as shown by Krokhin (16) (Fig. 2.18), who in his survey gives a thorough description of the pertinent theories on the generation of vapors and plasmas by laser. Ready (14) has used a semiempirical approach to calculate the depth of giant pulse craters by assuming that, because of the recoiling effect of the initial vapor stream, the underlying material is superheated until the heat of vaporization falls to zero. The built-up high pressure is then released by an ejection of material. The agreement between calculated and observed depth for a number of

### TABLE 2.5
**Comparison of Observed and Calculated Crater Depths Produced by Free-Running Laser[a]**

| | | Calculated depth, $\mu$m Reflection | |
| | | --- | --- |
| Element | Observed depth, $\mu$m | Taken into account | Neglected |
| --- | --- | --- | --- |
| Sn | 1586 | 2100 | 2400 |
| Bi | 1516 | 4700 | 5100 |
| Cd | 1648 | 4240 | 5100 |
| Pb | 1554 | 3810 | 4100 |
| Mg | 1676 | 3150 | 4150 |
| Ag | 960 | 925 | 1570 |
| Cu | 830 | 615 | 995 |
| Ti | 758 | — | 970 |
| Ni | 660 | — | 741 |
| Fe | 538 | 675 | 814 |
| Mo | 570 | 550 | 730 |

[a] From Klocke (13). For the calculations the theory of Landau (15) was used.

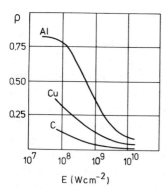

**Fig. 2.18.** Change of reflectivity $\rho$ of several materials with irradiance produced by laser radiation. From Krokhin (16), reproduced by permission.

materials is correct to within a factor of 2, typical values being about $3 \mu m$. As a further point of interest it should be mentioned that due to the absorption of radiation in the hot plasma in front of the sample, the amount of vaporized sample does not linearly increase with radiant exposure. Instead, saturation may occur.

Finally, attention should be drawn to the fact that in the case of free-running and semi-$Q$-switched lasers the observed dimensions of the craters depend on the geometry of the irradiance. A laser beam of high radiant flux in combination with a lens of long focal length is capable to penetrate deeper into an already deep crater than one which is produced the other way around. In the latter case a substantial part of the radiation flux may be impinging at a large angle of aperture and may consequently be masked by the walls of the crater. The crater formation and its dimensions are also influenced by the interaction of the plume with the surface and by the interaction of an auxiliary cross discharge (see Sect. 2.5.2) with the surface. The former has been described in detail by Yamane et al. (17) for giant pulse operation. The hot and dense plasma produced in front of the surface is further heated by absorbing a substantial fraction of the incoming radiation. Therefore, direct interaction with the surface is replaced by interaction of the plasma with the surface, which apparently lasts much longer than a single giant laser pulse. As a result, the crater dimensions are rather independent of the energy of the spike once a certain threshold has been exceeded. The latter action consists of additional heating of the surface by the cross discharge and of chemical reactions of components originating in the auxiliary electrodes with the surface (18).

The influence of the atmosphere and its pressure has also been studied, for example, by Piepmeier and Osten (19), Petrakiev et al. (20), and Treytl et al. (21).

## 2.4.2 The Plume

The properties of the plume, such as composition, chemical as well as physical, temperature, electron pressure, and rate of expansion, are strongly time and space dependent and are primarily governed by the process of generation. Ideally, the plume should be quasistationary for a length of time such that good spectroscopic signals can be obtained for chemical analysis. One could, therefore, produce a plume like that produced by a dc arc, for example, with the help of a cw-operated laser or of a slowly pulsed laser. This has been done by König and Neumann (22) with considerable success. An Ar gas laser for the production of free atoms for atomic absorption analysis was used. It should, however, be remembered that in quasistationary operation, the sample material is removed via a slow melting and boiling process and that consequently the temperature of the vapor will only be of the same order as the boiling temperature of the sample material. Therefore, strong fractional evaporation is most likely to occur.

As laser methods are suitable for a very wide range of materials, the temperature and other thermodynamic properties of the vapor vary much more than with old-established dc arc methods, which are very different in two aspects: (i) For nonconducting substances graphite is used for the supporting electrode. Therefore, it is the high sublimation temperature of this material that controls the properties of the arc more than anything else. (ii) Energy is continually supplied not only to the sample, in this case the electrode, but also to the plasma. With lasers it is highly desirable to generate a vapor at a sufficiently high temperature possibly independent of the boiling point of the sample material. Pulsed operation at elevated power is therefore compulsory. A quasistationary vapor cloud produced by intermittent high-power action of the laser is a good compromise. With some experience, the properties of the cloud can be deduced from the appearance of the crater (see, e.g., Figs. 2.14 and 2.15).

## 2.5    THE SPECTRA

### 2.5.1    Atomization by Laser Only

#### 2.5.1.1    *Atomic Absorption Spectra*

The immediate appearance of self-absorption and self-reversal in emission line spectra demonstrates the presence of a *considerable proportion of neutral ground state atoms in the vapor cloud.* Therefore, atomic absorption methods can be used to identify the elements producing these atoms. Such methods were first used for this purpose by Mossotti et al. (23).

Optimal conditions for atomic absorption spectrometry (AAS), require that the plume contains a large number of atoms capable of absorbing the primary radiation passing through it. In addition, the resulting absorption lines should be narrow and not shifted in wavelength in respect to the exciting light. It is well known that such conditions can be realized in the stationary case of flame atomic absorption with a long slit burner or with a heated graphite furnace or a tantalum boat. On the other hand, duration, temperatures, and size and shape of the laser-produced plume are in itself not very favorable for atomic absorption, but conditions and procedures can be found that permit analytical use with properties otherwise not obtainable. Some of these are discussed in this section.

The analytical signal must be separated from interfering emission signals which are usually also generated in the vapor cloud and should be corrected for nonspecific absorption. The latter is often done in a standard manner with continuum radiation from a deuterium lamp. The former can be accomplished by choosing the optimum observation distance from the sample. This situation is illustrated in Fig. 2.19 taken from Quentmeier (24). One can see how emission and absorption signals change with distance and time.

Basically, two different approaches for designing atomic absorption methods can be distinguished, namely, those in which the atomic absorption is directly observed and those in which prior to the absorption measurement, the vapor cloud is subjected to additional heating in a flame or furnace. In these cases a better geometric configuration of the

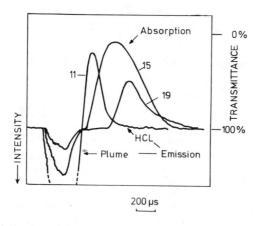

**Fig. 2.19.** Typical atomic emissions and absorption signals of spectral line Cu 3247 Å at heights of 11, 15, and 19 mm above sample surface from a laser-produced vapor. Laser operated with acousto-optical Q-switch. From Quentmeier (private communication), reproduced by permission.

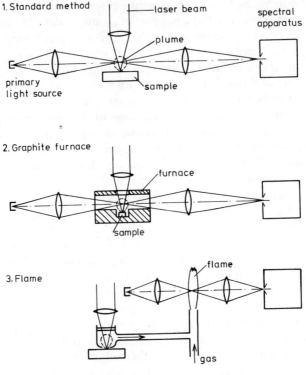

**Fig. 2.20.** Methods for atomic absorption spectroscopy on laser-atomized samples.

final absorption volume results. In Fig. 2.20, some of these methods are shown.

In all these methods, special measuring techniques to cope with the very transient nature of the optical signal must be employed. Some of the approaches may not be self-evident from Fig. 2.20. For example, in the method used by Mossotti et al. (23), a continuous primary source of high radiance produced by a flash lamp and having a similar time duration as the vapor cloud is employed for measuring atomic absorption of different species. Analytical results have been reported which, in favorable cases of a high melting point sample, allowed the determination of concentrations with limits of detection down to the ppm level or to the nanogram level. Continuum radiation as a primary source, however, necessitates the use of a high-resolution spectrometer which ideally should be able to clearly resolve the absorption linewidths.

A moderately Q-switched laser system was employed together with a rotating prism and a suitably pulsed photomultiplier power supply to reduce the dark current, and a multipass mirror system was used to increase the absorption signal. The spectral apparatus should have a practical resolution of 200,000 and a power of detection similar to that obtainable with conventional hollow-cathode lamps as primary radiation sources. For such an instrument the spectral radiance of the primary source, however, may not be high enough. Due to the advent of gratings of high resolving power and improvements in electronic instrumentation, this approach seems to offer better possibilities than hitherto recognized.

Quentmeier (24) tried to obtain a better signal by changing the laser operation conditions; namely, a simple acousto-optical $Q$-switch (see Fig. 2.5) produced a long train of equidistant spikes at a repetition rate of $10^5 s^{-1}$ of relatively high power resulting in efficient quasi-continuous vaporization. Additional experimental conditions were as follows: hollow-cathode lamp as primary source, monochromator, photomultiplier, oscilloscope with long persistence screen, $D_2$ lamp for compensation of nonspecific absorption, laser output energy of 0.6 J, sampled material about 20 $\mu g$, and height of observation 18 mm above sample. Single-shot analysis was possible with a relative standard deviation of about 0.3. The determination of Cu, Mg, and Zn in Al samples served as analytical example (24).

Limits of detection in the ppm range are possible for elements that do not form stable oxides. Due to their fast oxydization, refractory elements, for example, Mg, Al, and Si, cannot be satisfactorily determined with such a procedure if atomization is accomplished in air. An inert atmosphere is therefore necessary, as investigated by Osten and Piepmeier (25). A rather small concentration range can usually be covered by one measurement if the dependence of the absorbance on the concentration is described by the Lambert-Beer law. This is common to most absorption methods. In the case of direct AAS with laser atomizers, however, *plotting log c versus absorbance yields straight calibration curves* (Lambert-Beer law not applicable!). This results in a wider concentration range which can further be extended by simultaneous multichannel measurements at different heights of observation.

These procedures have another feature that makes them still more suited to the analysis of solid samples with widely varying concentrations. The transient nature of the signal permits measurement not only of the *peak absorption* as analytical signal but of the *half-width* of the absorbance-versus-time curve or the *area* under this curve as well. The transition of the analytical curve from one parameter to the other is smooth, and hence a *wide-range, single analytical curve* is possible. Figure 2.21 shows results of measurements by Matousek and Orr (26) supplemented by Quentmeier.

If a spectral line source, for example, a hollow-cathode lamp, is used as primary source, equipment similar to that customary in AAS will be sufficient. This is then, of course, a method for single-element analysis, whereas with a continuum source in connection with a polychromator multielement analysis could be performed.

To sum up, these two methods can be used for direct AAS analysis of materials of almost any kind with minimum sample preparation and no danger of contamination and with a wider range of concentrations than with other atomic absorption methods.

According to Osten and Piepmeier (25), an effective atomization can be achieved with the help of their Q-switched laser and a pulsed hollow-cathode lamp as primary source. Unspecific absorption due to nonatomized material was not observed. Theirs is again a single-element procedure employing a simple spectral apparatus but needing a more complicated electronic system. Some of their results are listed in Table 2.10.

Karyakin and co-workers, in a series of investigations, explored different combinations and techniques for atomic absorption analysis, for example, use of the hot, radiating walls of the crater as primary source (27), a free-running laser with a pulsed hollow-cathode lamp, and a special sample preparation technique to increase vaporization efficiency (28), and several methods for additionally atomizing the laser-produced vapor (29) in a two-step process.

A rather unusual way has been proposed by Kaporskii and Musatowa

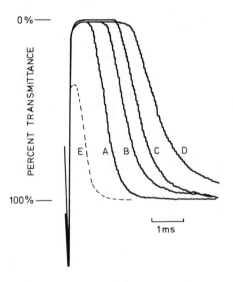

**Fig. 2.21.** Time-dependent atomic absorption signals obtained from atomizing silver–copper alloys. Ag content ranging from 1% (A) to 50% (D). Redrawn from Matusek and Orr (26), by permission. Curve E added showing nonsaturated absorption at small concentrations.

(30). In an in-line setup (Fig. 2.22), a powerful $Q$-switched laser generates a "laser spark" (see Sect. 2.10.6) emitting a strong continuous spectrum serving as the primary source, while the nonabsorbed part of the laser radiation also vaporizes the sample which is deposited on a narrow-slit plate in front of the spectrograph slit. As the continuous spectrum extends well into the ultraviolet region, this region can also be used for analysis if an inert atmosphere is provided.

Of the approaches using a two-step method, two will be described briefly. Matousek and Orr (26) vaporized the sample with the help of a pulsed $CO_2$ TEA laser at a repetition rate of 5 pps, emitting at 10.6 $\mu$m in the infrared. Peak power was 0.5 MW, pulse half-width 0.3 $\mu$s. The sample was positioned in a specially designed graphite furnace (see Fig. 2.20) which was heated to provide additional atomization for an extended time of up to 10 ms. The apparatus was completed by conventional atomic absorption instrumentation. The graphite furnace considerably prolongs the presence of free atoms in the optical path; in addition, nonspecific absorption is strongly reduced.

A laser and flame combination has been described by Kantor et al. (31). A free-running Nd glass laser of 1 J maximum output energy was used for vaporization with a standard atomic absorption spectrometer for the final atomization and measurement (see Fig. 2.20). Some analytical results are listed in Table 2.10. As the transit time of the aerosol through the transporting and mixing system is about 1 s, this system does not need any alterations of the spectrometer in existing AAS equipment.

It is noteworthy that YAG lasers have not yet been used as atomizers for AAS. YAG systems are capable of delivering laser pulses in the power range of 10–100 kW at repetition rates of about 10 kHz. They can be continuously pumped, and a quasistationary existence of useful atomic vapor in a suitable distance from the sample surface could be provided, facilitating the atomic absorption measurement. In such a mode of almost

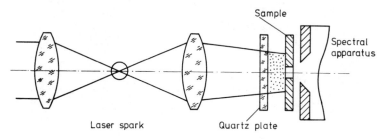

**Fig. 2.22.** Experimental setup as used by Kaporski (30) for AAS with laser spark. Reproduced by permission.

continuous operation, the laser could be turned on for a few seconds during which time a transport mechanism would have to move the target in order to sample a sufficiently large part of the surface. Similar operation should be possible with suitable $CO_2$ lasers.

Because the expansion of the vapor cloud, and hence the time of residence of the atoms within the observation zone, is pressure dependent, the sensitivity of AAS methods with laser atomizers could be improved by increasing the pressure of the surrounding atmosphere. No use of this important factor has apparently been reported.

A final evaluation of the advantages and shortcomings of AAS methods is presented in Sect. 2.9.

### 2.5.1.2   Optical Emission Spectra

Provided the laser-produced plume is hot enough, it will radiate in the optical region of the spectrum. Figure 2.23 shows a photograph of the vapor cloud as it appears to the eye. This appearance depends on the sample material and on the operating conditions of the laser. Time- and space-resolved studies of the laser plume have been carried out by a number of research groups, for example, Scott and Strasheim (12). From such investigations, conclusions can be drawn for the optimum choice of laser conditions.

The free-running laser, because of the low power of its spikes, usually produces a plume of such a low spectral radiance that direct spectroscopy

**Fig. 2.23.** Vaporization from poly(vinyl chloride) sample irradiated by laser.

is difficult if not impossible. Therefore, in this mode of operation the laser is used *almost exclusively* for the vaporization in a two-step process.

With $Q$-switched or semi-$Q$-switched operations, the plume radiates strongly owing to the high temperature of the plume. The spectra can then more easily be recorded and evaluated. The character of the spectrum differs considerably from that of conventional spark or arc discharges. It depends strongly on the power and kind of $Q$-switching as well as on the atmosphere. Giant pulses produce an intense background as well as emission lines of the sample and the surrounding gas that are broadened and in many cases shifted in wavelength. Self-reversal may also be prominent. Many workers have studied these phenomena, for example, Allemand (11), Scott and Strasheim (12), and Rasberry et al. (32). Figure 2.24 shows some typical line profiles. In Fig. 2.25, a typical example of such a spectrum recorded time integrated but spatially resolved along the y-axis is shown. As expected, the continuum is preferentially excited near the surface, whereas the spectral lines extend much

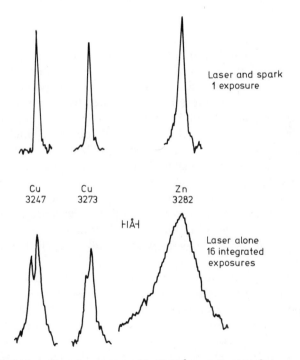

**Fig. 2.24.** Typical profiles of resonance (Cu 3247 Å and Cu 3274 Å) and nonresonance lines (Zn 3282 Å) for $Q$-switched laser alone (bottom) and for laser cross spark excitation (top). From Rasberry et al. (32), reproduced by permission.

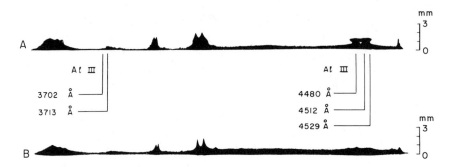

**Fig. 2.25.** Space-resolved time-integrated spectra of aluminum atomized by $Q$-switched solid-state laser. Single giant pulse, spectrum B; double giant pulse; spectrum A. From Scott (12c), reproduced by permission.

farther. This immediately shows the feasibility of improving the intensity ratio of line to background by excluding from measurement those parts of the spectrum near the surface. From an analytical point of view, the quality of the spectrum is inferior to that excited by other spectrochemical excitation sources. This should be taken into account when spectrum lines for analytical use are selected. It may well be that, although recommended, some analytical lines are not suitable because of self-absorption and self-reversal effects, which may seriously degrade their concentrational sensitivity.

The spectrum produced by slowly switching devices, because of the lower temperatures and pressures involved, yield a more favorable line-to-background intensity ratio although the resonance lines of the major constituents of the sample may show more self-reversal than in the case of fast switches.

The spectrum produced by a train of several giant pulses during the same pumping cycle depends on the repetition rate of these pulses, which in turn depends on the kind of $Q$-switching. A saturable dye switch can be adjusted to produce several giant pulses distributed over a large part of the pumping time with average intervals ranging from 1 to 100 $\mu$s. In this case the actions of the spikes, and consequently the respective spectra, are independent of each other because of the high velocity of the expanding vapor (about $10^6$ cm s$^{-1}$). The net result is a larger crater and a more intense integrated spectrum. The spectra produced by slow Q-switches, for example, by an acousto-optical switch (Fig. 2.5), depend in the same manner on the total number of spikes.

With a fast-rotating mirror or prism Q-switch, however, two or more giant pulses may follow each other at intervals of a fraction of a $\mu s$. In this case, the radiation of the following pulse may hit the hot plume still in front of the surface. It becomes partly absorbed, vaporizing new material as well as reheating the vapor produced by the previous pulse. Such a behavior has been recorded in Fig. 2.25 (spectrum A), which shows the increase in intensity at some distance from the surface caused by the action of a second pulse.

## 2.5.2 Additional Excitation in Spark Discharge

For applications affording *high spatial resolution* and *high concentrational power of detection*, low-power operation of the laser is necessary. Hence, a *two-step process is recommended*; that is, the laser only generates the vapor plume, whereas excitation is produced by an additional source. In this case the primary radiation emerging from the vapor cloud can be neglected in comparison with the one produced by the additional excitation. The spectral character is therefore mainly determined by the parameters of the latter source. Mode of operation of the laser, power, and energy again determine the crater dimensions; but of additional importance now is the transport of vaporized sample material into the excitation zone and its time of residence in the discharge.

The oldest and most common method of additional excitation is cross excitation in the spark discharge as shown in Figs. 2.26 and 2.27. Two-pointed electrodes, usually of spectrographically pure graphite, are fed from a charged condenser bank via suitable impedances (resistors and inductors). The charge stored on the condenser may be in the order of some joules, the voltage being lower than the dielectric breakdown voltage of the electrode gap. The discharge is then initiated by the charged particles of the vapor cloud entering the gap. The cross discharge *should be matched to the laser action and should last as long as the laser-produced vapor remains in the gap.* This problem has been studied in detail by Moenke et al. (33b) and by Klockenkämper and Laqua (34). In Fig. 2.28, a practical example for such close matching is given. The width of the electrode gap and the distance of the electrode tips from the surface of the sample must also be carefully selected for optimum results. An accidental flashover from an electrode to the surface must be avoided; but, on the other hand, a close distance may be very good for high power of detection.

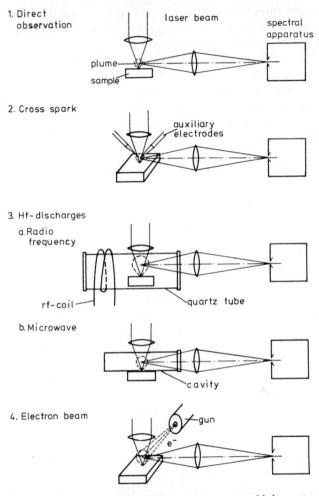

**Fig. 2.26.**  Possibilities of optical emission spectroscopy with laser atomizers.

Automatic triggering of the discharge by the vapor cloud itself may not lead to optimum results as regards line intensity and intensity ratio of line to background. This may be understood with the help of Fig. 2.29, which shows a time-resolved sequence of pictures taken with a framing camera of the firing of a cross discharge. One can distinguish the spot of the ruby laser-produced irradiance, the production of vapor, its movement toward the electrode gap, and the firing of the cross discharge with subsequent production of vapor from the auxiliary electrodes which in this case repels part of the sample vapor. One can therefore conclude that the action of the laser pulse and that of the spark

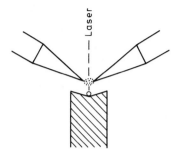

Fig. 2.27. Microanalysis with laser vaporization spark cross excitation. From Ohls et al. (70), reproduced by permission of Springer Verlag.

discharge are not well enough synchronized. Free choice of the onset of the discharge in respect to the laser action, therefore, may lead to improved performance. The cross discharge itself is very similar in its electrical parameters to customary spectrochemical medium voltage spark sources. Therefore, the spectra produced in combination with laser vaporization are also similar to conventional spark spectra, with the result that the spectral lines are much narrower and the spectral background

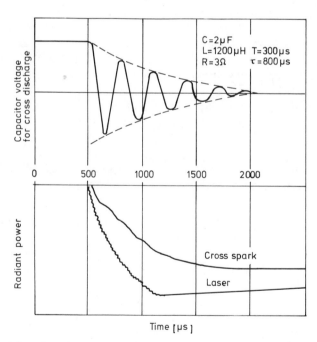

Fig. 2.28. Adaptation of cross spark discharge to vapor production of free-running laser. From Klockenkämper and Laqua (34), reproduced by permission.

**Fig. 2.29.** Sequence of time-resolved photographs of firing of a cross discharge by laser-produced plume entering the electrode gap. Exposure 10 μs per frame. From Felske et al. (40a), (40b), reproduced by permission.

lower than those observed in the primary spectrum of the laser-produced vapor plume. The intensity ratio of line to background and consequently the power of detection are improved by orders of magnitude in favorable cases (see Table 2.6). The gross intensity of the emerging radiation may also be higher by one or two orders of magnitude so that the requirements of the spectral apparatus as regards optical conductance and radiation detectors are less stringent.

The cross spark method, especially with externally triggered discharge, has been extensively studied by Quillfeldt (33a) and Treytl et al. (35), and also by Moenke et al. (33b) and Klockenkämper and Laqua (34). There are two basic disadvantages of the cross spark method, however, which must be mentioned, namely, *dilution* and *danger of contamination*. The analytical signal is not only derived from the laser-produced vapor cloud but also from the auxiliary electrodes.

As the energy of the cross spark is usually higher than the laser output energy, more vapor may originate from the auxiliary electrodes than from the sample. It is further known that, if residual impurities are present in high-purity graphite, they may not be uniformly distributed in the material but rather be contained in small clusters. It is therefore sometimes

**TABLE 2.6**
**Limits of Detection (%) for Some Elements in Optical Emission Spectrochemical Analysis Pertaining to Several Excitation Methods but All Using Lasers for Atomization**

| Analytical line, Å[a] | Laser alone[a] | Cross excitation[a] | 85 MHz discharge[a] | Microwave discharge[b] | Analytical line, Å[b] |
|---|---|---|---|---|---|
| Mg 2852 | $1 \times 10^{-3}$ | $9 \times 10^{-5}$ | $2 \times 10^{-5}$ | $4 \times 10^{-5}$ | Mg 2796 |
| Al 3944 | $2 \times 10^{-3}$ | $6 \times 10^{-4}$ | $4 \times 10^{-4}$ | $2 \times 10^{-4}$ | Al 3093 |
| Cu 3274 | — | $1 \times 10^{-4}$ | — | — | |
| Pb 4057 | — | $6 \times 10^{-4}$ | $3 \times 10^{-4}$ | $2 \times 10^{-4}$ | Pb 2833 |
| Mn 4030 | $2 \times 10^{-3}$ | $1 \times 10^{-4}$ | $1 \times 10^{-4}$ | — | |

[a] Values taken from Möde (36).
[b] Values taken from Leis (37).

difficult to distinguish the origin of certain elements if detected by their spectra, in particular, if only one shot can be made for the analysis.

### 2.5.3 Additional Excitation in Electrodeless High-Frequency Discharge

Several attempts have been made to overcome the above-mentioned difficulties. The first was the excitation of the vapor in a high-frequency electrodeless discharge. In the method of Möde (36) (see also Fig. 2.26), the vapor is excited in an induction-coupled rf discharge at a frequency of 80 MHz. The vapor cloud initiates the discharge in a similar way as in a spark cross discharge. A "clean" plasma can thus be produced having a very low spectral background and sharp spectral lines enabling low limits of detection (Table 2.6). The main disadvantage, however, is the low optical thickness of the plasma and its rather irregular and irreproducible shape. In addition, as the sample has to be introduced into the electromagnetic field, its size and electrical properties affect the discharge as well.

In an investigation by Leis (37), additional excitation in a microwave cavity was chosen at a frequency of 2.45 GHz and with a maximum power of 2.5 kW. This system evades most of the difficulties just mentioned. The cavity is shown in Fig. 2.30. It should be noted that the sample is outside the cavity and therefore does not affect the discharge. The cavity is divided into two parts in order to separate the discharge chamber from the tuning and feeding system. Such a system can be operated easily and reproducibly. The discharge atmosphere is argon at approximately normal pressure. However, the low optical density of the plasma, and consequently its low radiance, again prevents the photographic recording of the

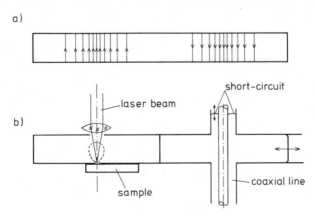

**Fig. 2.30.** Excitation of laser-produced plume in microwave discharge: (*a*) distribution of electrical field; (*b*) microwave cavity. From Leis (37), reproduced by permission.

spectral background with a single shot, even with high-speed spectrographs. This restricts the use in local analysis to the determination of the main constituents only.

The situation is much better with photoelectric radiation measurements. These problems are discussed in more detail in Sect. 2.8.2. Some analytical results are given in Table 2.6. The method as a whole seems to be quite promising, the more so as readily available microwave generators such as those from microwave kitchen stoves could be adapted with little effort. On the other hand, the method could possibly also be adapted to the incorporation of low-power microwave generators. The optically thin plasma together with its fairly low temperature of about 4300°K have the distinctive advantage of low spectral background and narrow spectral lines resulting in straight analytical curves over large concentration ranges. This is desirable for the analysis of samples of completely unknown composition. Whereas the microwave discharge, until now, has been used in combination with saturable dye $Q$-switched laser yielding a useful spectral signal for about 10 ms, it seems to be equally well suited for the acousto-optical $Q$-switched system (Fig. 2.5).

## 2.6  MEANS OF IMPROVING THE ANALYTICAL SIGNAL

For improving the analytical signal, a number of possibilities should be considered.

### 2.6.1  Artificial Gas Atmospheres

This factor influences the production of the plume as well as its excitation. As discussed by Piepmeier and Osten (19), the presence of an atmosphere affects the spectra, the crater size, and the amount of sample vaporized with $Q$-switched giant pulse lasers. The absorption of laser radiation by the laser-produced plasma is governed to a large extent by the pressure of the surrounding atmosphere. As a result, the crater diameter and the amount of sample vaporized increase with increasing energy of the laser pulse if the ambient pressure is low. Treytl et al. (21) in their studies included the gases helium, nitrogen, oxygen, and argon. Their results are not easy to interpret. However, it seems that, with otherwise constant operating conditions (i.e., pulse energy about 1–8 mJ), the highest spectral line intensities are obtainable with argon at normal pressure (this is important for the identification of major constituents in local analysis or microanalysis). For the best line-to-background ratio (see Sect. 2.9.1), vacuum was established as optimum (this is important for high power of detection as far as concentrations are concerned), but at the expense of the absolute intensity of the spectrum, which is poor.

In a number of papers, Dimitrov and Petrakiev (38) and Dimitrov and Gagov (112) described their experiments with different ambient atmospheres for two-step methods employing spark cross excitation. Apart from the redistribution of the intensities of spectral lines with respect to the different atmospheres, they generally observed up to an order-of-magnitude improvement in the absolute intensities of spectral lines, background, as well as intensity ratios. For details, their papers are well worth consulting.

Argon seems to offer overall improvement. The role of the surrounding atmosphere in atomic absorption spectrometry has been already discussed in Sect. 2.5.1.1.

Pressures higher than normal are also worthwhile to investigate. Until now, only the investigations by Stupp and Overhoff (39) have dealt with high pressures. These authors evaluated the special case of atomization and excitation of graphite at high pressures ranging from 0.1 to 4 MPa in helium, oxygen, and argon with the ultimate aim to developing an analytical method for the analysis of graphite globules in a nuclear reactor.

### 2.6.2  Use of Spatial Resolution

All workers involved in fundamental investigations on laser atomization for spectral analysis have at least to some degree discussed the spatial change in intensity of line and background.

The method of space-resolved spectroscopy can be applied with some success in all cases where the direct radiation from the laser-produced plume serves as the analytical signal. As explained before, this radiation, due to high temperature and pressure in the immediate vicinity of the laser impact area, contains a substantial amount of continuous radiation. The spectral lines are also likely to be broadened with the result that the intensity ratio of line to background and hence the power of detection may be far from optimum.

With a slow-mirror (3000 rpm) $Q$-switch with several spikes, Felske et al. (40) recommend not to use the region within 1 mm from the surface. With a single giant pulse, Treytl et al. (41) found an increase in line-to-background ratio of an order of magnitude for a combined lateral and vertical displacement of the observation zone while masking the surface-near region. Allemand (11) also studied these problems carefully.

### 2.6.3  Use of Time Resolution

The continuous background decays faster than the radiation of spectral lines of low excitation potential. Therefore, shutting off that period of time (which may be in the order of a few microseconds) will result in improved power of detection without impairing the precision of the method. Analytical examples making use of time resolution have been given, among others, by Treytl et al. (35) for biologic samples with a $Q$-switched giant pulse laser and by Felske et al. (40) with their slow $Q$-switch. In Fig. 2.31, the time dependence of the signals belonging to line and background are shown.

A combination of time as well as space resolution will give best results. In practice the latter can be accomplished with the help of a diaphragm at an intermediate image of the vapor cloud, whereas the former can be accomplished in photoelectric detection by suitably gating the electronic signal. Such a technique was originally introduced in spectroscopy by Crosswhite et al. (42) for investigations on spark discharges. Treytl et al. (35) describe a modern system for their laser work.

In photographic detection, time resolution could be obtained with the help of electromechanical, electro-optical, or magneto-optical shutters. But because of experimental difficulties it has apparently not been tried.

Finally, it should be mentioned that space and time selection leads to a better spectrum as far as linewidths are concerned. This yields a wide concentrational range of a particular analytical line and high concentrational power of detection provided a spectrograph of high resolving power is available to make use of it.

### 2.6.4 Application of a Magnetic Field

Petrakiev et al. (43), in a two-step procedure with spark cross excitation, adapted a method originally developed by Vukanović et al. (44) for

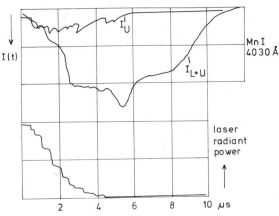

**Fig. 2.31.** Time-resolved intensities $I(t)$ of an analytical line and of spectral background. From Felske et al. (40a), reproduced by permission.

the dc arc. These investigators observed an increase in the intensities of the spectral lines as well as in the intensity ratios of line to background when the discharge took place in an inhomogeneous magnetic field of some hundred gauss. Typical enhancement factors of 2 were obtained.

## 2.7 THE SPECTRAL APPARATUS

For the purpose of spectrochemical analysis, the radiant energy of the laser-produced plume is small, even after additional excitation, especially in extremely local analysis or microanalysis. Therefore, some measures have to be taken to secure satisfactory results.

### 2.7.1 The Spectrograph Illumination System

This system has the task to conduct radiant power from the source through the spectral apparatus to the radiation detector in such a way that the desired height of the slit is illuminated and the full aperture of the instrument is filled with light. The radiant power reaching the detector is then given by

$$\phi = \tau L G \tag{4}$$

with $L$ being the radiance of the source, $\tau$ the transmittance and $G$ the optical conductance of the instrument, which can be approximated by

$$G = n^2 S_1 S_2 / a_{12}^2 \tag{5}$$

where $S_1$ and $S_2$ are the areas of the entrance slit and of the aperture, respectively, $n$ is the refractive index of the medium between $S_1$ and $S_2$, and $a_{12}$ is the distance between them.

In the case of the spectrograph it is the irradiance which matters, that is, $E = \phi/S_3$, where $S_3$ is the irradiated area of the plate. The reader is referred to reference 45 for a more detailed account of this approach.

For correct illumination it is only necessary to match the lenses or mirrors with regard to their focal length and size so as to keep the optical conductance $G$, determined by spectral apparatus, constant throughout the entire system. One can also choose the optical conductance of the source as the pertinent value and look for a spectral apparatus whose conductance is at least as good as that of the source. There are two ways of correctly illuminating the slit of the spectral apparatus. The first is to image onto the slit the source, and the second is to image onto the slit the aperture of the field lens imaging the source onto the aperture of the apparatus.

In almost all foreseeable cases, the formation of an image of the source onto the collimating optics of the spectral apparatus is *preferable* to the imaging of the source onto the slit. In the latter case the slit actually selects, according to its width, a very small width in the source from which radiation is viewed. If space selection is desired, it can be achieved with the help of a diaphragm at an intermediate image. From the point of view of speed, that is, with either irradiance of photographic emulsions or optical conductance for photomultipliers, there is no difference as long as the spectral apparatus is fully illuminated in the sense explained above. Very often, this almost trivial requirement is not met, specially if suitable short focal length lenses are not available. In low-energy local analysis with a very small source, one-step magnification of the source area to a sufficient size at the aperture of the spectral instrument is not possible because the necessary short distance from the source to the slit cannot be realized. A two-step magnification including field lenses gives better results, even when the radiation losses caused by the additional optical elements are taken into account. Apart from the gain in intensity, it is the improved resolving power and consequently the gain in power of detection which has to be considered.

In atomic absorption work, an additional limitation may be set by the rather small optical conductance of the hollow-cathode lamp which may be smaller than the optical conductance of the monochromator. Consequently, the monochromator cannot be illuminated at maximum slit height and maximum aperture area. In this case the full illumination of the aperture should be preferred to the illumination of the full slit height in order to avoid a decrease in the resolving power of the instrument.

### 2.7.2 The Spectrograph

Because of the low radiant energy of the sources in question, we have to consider the $f$-ratio of the camera of the spectrograph together with the spectral resolving power of the instrument. Since laser equipment was originally introduced into analytical practice as an addition to existing spectrographic instruments, *very often a mismatch prevented its optimum use*, with the consequence of an unavoidable disappointment. Even today, many of the available and recommended instruments are not well suited. The spectrograph should be chosen according to the kind of analytical problems that have to be solved. For the identification of main constituents in local or microanalysis, it is the $f$-ratio *only* which matters because the spectral background will be far below the line intensity.

For trace analysis in small samples, two conditions must be met; namely, the intensity of the spectral background must be measurable (i.e., the $f$-ratio should preferably be very small) and the linear dispersion small. In addition, the intensity ratio of line to background should be

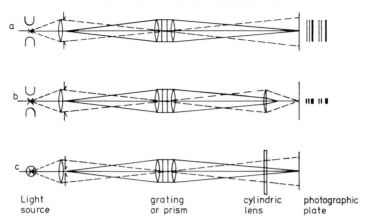

**Fig. 2.32.** Cylindric lens arrangement for a stigmatic flat field spectrograph (a) without cylindric lens, (b) with cylindric lens, shown perpendicular to dispersion, and (c) with cylindric lens, shown in plane with dispersion.

high, meaning that spectral resolution should be high. As linear dispersion and resolving power of the instrument are usually linked, these two conditions *cannot be optimized at the same time,* and a compromise has to be found. Generally, however, resolving power can always be traded in for one of the other properties. A remedy for obtaining high irradiance without sacrifice of spectral resolution is the use of a cylindric lens in front of the photographic emulsion. The principle is shown in Fig. 2.32. With such a device (which is especially useful with low-speed, long focal length spectrographs), the irradiance can be increased by an order of magnitude. There are, however, some restrictions, that is, the light should penetrate the lens as nearly perpendicular to the long axis of the lens as possible. This condition can be exactly fulfilled only if, for instance, the grating in an Ebert-Fastie spectrograph is situated in the focal point of the Ebert mirror. In addition, the focal plane in which the spectrum is recorded should be a plane, a condition which is best met if the grating is situated in the so-called $\sqrt{3}$ position, that is, at a distance $R(1 - 1/\sqrt{3})$ from the mirror. Finally, the dispersion should not be too small. With a plane spectrum, very good results can be obtained.

Last but not least, the most important property of the spectrograph should be stressed. A spectrograph, in conjunction with the photographic emulsion, permits the simultaneous multielement analysis of random samples of completely unknown composition due to the *high capacity of the photographic emulsion for storing information.* This is specially valuable for local analysis, where the actual sample is completely destroyed by one laser shot and a similar sample, spot, or inclusion may not be available. The concentrational power may also be quite good and actually not very different from that of photoelectric methods provided the radiant exposure (defined as the time integral of the irradiance) can be made large enough to blacken a sufficient number of grains of the emulsion leading to a well-measurable density. A density value of 0.1 to 0.2 will be sufficient.

There will always be a need for a spectrograph for laser spectrochemical analysis, irrespective of the kind of two-dimensional radiation detector finally used.

### 2.7.3  The Spectrometer

As stated previously, the properties relevant to the analytical problems in question are high optical conductance, good spectral resolving power (for trace analysis), and multielement capability. The first condition can either be met by a small $f$-ratio or a long slit or by both. It is this property which makes these instruments well suited for the analysis of samples restricted in mass. The simplest form of a spectrometer is a mono-chromator with a photoelectric detector behind the exit slit. This one-channel instrument is suited only for *sequential* analysis, which severely restricts its use in local or microanalysis in the above-mentioned cases.

The determination of a number of elements by photoelectric detectors has still to be satisfactorily solved, especially if random samples have to be analyzed. That means that true simultaneous multielement analysis with the help of laser vaporization is restricted to the routine determina-tion of a comparatively small number of elements—depending on the type of spectrometer available—for which the instrument is provided. Good examples of the present state of the art have been given by Meric et al. (46) for the analysis of cement raw materials and by Treytl et al. (35) for biologic samples.

All customary multislit spectrometers as used in industrial spark analysis can in principle easily be converted for laser spectrochemical analysis by some alterations of the electronic measurement system to make it suitable for fast and transient signals. With modern electronics, this no longer presents the difficulties of the past. A complete system especially developed for laser spectrochemical analysis has been described in detail by Treytl et al. (35,47).

Spectrometers of this kind are very useful for the determination of major constituents in analytical samples but may not be equally well suited for trace analysis in small samples. In the latter cases the resolving power should be high. A suitable instrument seems to be the multichan-nel Echelle spectrometer. For these instruments a resolving power of about 400,000 and an $f$-ratio of $f/10$ are typical. Unfortunately, their effective slit height is small with the consequence of a rather poor optical conductance. Nevertheless, the use of such instruments should be investigated more closely.

## 2.8 RADIATION DETECTORS

### 2.8.1 Photographic Emulsions

Some of the unique properties of this two-dimensional radiation detector have already been mentioned in the preceding section. The high inherent low noise amplification at a high quantum efficiency (10 radiation quanta may lead to a developed grain of perhaps 1 $\mu m^3$) compares well with that of a photomultiplier. The main disadvantages are the granular structure of the emulsion, which consists of many small, independent detectors, which must actually be counted for measurement, and a threshold for the radiant exposure below which no signal can be recorded. This makes photographic emulsions far inferior to photoelectric detectors for small radiant exposures as encountered in spectrochemical trace analysis with laser atomizers where the spectral background has to be measured.

It is possible to overcome to a certain degree the threshold by preexposure, but at the expense of the power of detection for concentrations. For exposures of very short duration, due to the reciprocity failure, the sensitivity of the emulsion is decreased. This has been studied in lasers by Hercher and Ruff (48), who found a strong dependence of the blackening of the emulsion on pulse duration. As to the choice of emulsion, slow but fine-grain emulsions are recommended such as Kodak SA 1 or the ORWO WU 3. If necessary, high-speed emulsions, for example, Kodak 103 or the ORWO ZU 2, should be used. The choice depends also on the wavelength range in question. In the ultraviolet below about 2800 Å, a slow, fine-grain emulsion may be more sensitive than a fast, coarse-grain emulsion.

Finally, ease of operation and costs should be considered. The processing and evaluation of a photographic emulsion is time consuming and requires skill and experience. To the costs for the photographic material, one should add the costs for darkroom and evaluation equipment. With the advent of automatic photometers and computer-assisted evaluation, the interest in photographic detectors will be revived.

### 2.8.2 Photoelectric Detectors

Low noise, high inherent amplification, and fast response are the prominent features of these detectors. The PM tube is a one-dimensional detector; therefore one tube for each spectral line or background signal must be used. This means rather large physical dimensions, and because of it background measurement, which is necessary for trace analysis, cannot be made as easily with PM tubes as with photographic emulsions. For the recording of short time phenomena with long intervals as in laser work, the dark current problem has to be considered. The dark current of selected tubes can be of the order of a few electrons per second at the photocathode, which is small compared with a typical cathode sensitivity of $50 \ \mu A \ lm^{-1}$; but because of the short exposure some time gating is necessary. Felske et al. (40) and Mossotti et al. (23) indeed have used a pulsed power supply for the PM tubes. With this method, overloading and fatigue of the last stages of a PM tube can be prevented. Allemand (11), Piepmeier and Malmstadt (49), and Treytl et al. (47) gated the signal obtained from the PM tube.

**TABLE 2.7**

**Comparison of Photographic and Photoelectric Detection of Weak Background Radiation Produced by Laser Vaporization and Excitation in Pulsed Microwave Discharge**

|  | Photographic detection | Photoelectric detection |
| --- | --- | --- |
| Spectral apparatus | quartz spectrograph | Czerny-Turner monochromator |
| Resolving power | 20,000 | 70,000 |
| Slit width, mm | 0.01 | 0.04 |
| Slit height, mm | 0.8 | 18 |
| Aperture ratio | 1:17 | 1:13 |
| Detector | photographic plate Kodak SA 1 | PM tube EMI 6256 S |
| Number of photons per $mm^2 \ s$ | 19,000 | 36,000 |
| Actual number of photons per laser shot for background measurement | 3 | 520 |
| Number of laser shots necessary to obtain blackening of $S_U = 0.2$ for background and sufficient photoelectric signal | 7,000 | 1 |
| Vaporized mass, mg | 14 | 0.002 |
| Number of photons for exposure | 21,000 | 520 |
| Quantum efficiency, % | 10 | 15 |
| Number of electrons at the photocathode of the PM tube |  | 78 |
| Relative standard deviation, % | 1 | 11 |
| Smallest measurable ratio of line-to-background | 0.04 | 0.5 |

Table 2.7 compares photoelectric with photographic radiation measurement; typical results for both methods are listed.

Other photoelectric detectors should also be mentioned. The *image dissector tube* is essentially a PM tube suited for the sequential measurement of an intensity distribution on the cathode. Because of the very small effective cathode area, the dark current can be kept very small. Combined with the spectral apparatus, part of the spectrum can be imaged onto the photocathode having a diameter between 10 and 25 mm. For this spectral range the whole system acts as a monochromator with no moving parts. In another arrangement, the image dissector tube is coupled to an Echelle spectrometer with a corresponding large spectral range. In this case, as mentioned before, the optical conductance is much smaller than that in the foregoing case. For simultaneous multielement radiation detection the *vidicon* could be useful, especially the silicon target vidicon, perhaps in combination with an image intensifier.

At present, these radiation detectors, due to their insufficient dynamic range, high cross talk, unlinearities, and blooming, are not yet fully suited for the outlined purpose. As further improvements can be expected the photographic emulsion may be replaced or supplemented in the not too far future. Higher precision and much higher speed of the analytical procedure would then be possible. Until now, however, laser spectrochemical analysis has been done with rather inexpensive equipment and therefore was also afforded by laboratories with rather small budgets. This may not be the case with expensive, computer-operated photoelectric systems with two-dimensional detectors for simultaneous multielement analysis, irrespective of their apparent virtues regarding power of detection, speed, and precision.

## 2.9  ANALYTICAL FIGURES OF MERIT

To evaluate the power of an analytical procedure, it has been found useful to classify it with the help of some figures of merit. As in the case of spectrochemical analysis with laser atomizers, many essentially different methods are compared, figures of merit which principally belong to a well defined procedure can only be taken as a rough guide.

### 2.9.1  Power of Detection

According to Kaiser (50), the limit of detection as far as concentrations are concerned is given by the following equation:

$$\underline{c} = c_U \left( \frac{I_X}{I_U} \right)^{\eta} \tag{6}$$

where $I_x/I_U$ stands for the smallest detectable intensity ratio of line to background, $c_U$ is the calibration constant, that is, the concentration belonging to that intensity of the line which equals the background intensity, and $\eta$ is a parameter which determines the curvature of the calibration curve and which in most cases is approximately unity, resulting in a straight line. The calibration constant $c_U$ can be obtained in a customary manner with the help of the calibration curve according to

$$c_U = \frac{c_1}{(I_1/I_U)^\eta} \tag{7}$$

where $c_1$ is a known concentration and $(I_1/I_U)^\eta$ is the respective intensity ratio.

The smallest intensity ratio of line to background is given by the relative standard deviation of the intensity measurement as

$$(\underline{I_x/I_U}) = 3\sqrt{2}\sigma_r(I_U) \tag{8}$$

The factor 3 stands for the level of confidence and means 99.7% confidence for a large number of measurements, typically more than 30. The $\sqrt{2}$ takes into account that two measurements, that is, line plus background and background alone, have to be made. This approach to the limit of detection, although equivalent to it, differs from the signal-to-noise concept. However, it has been found that in spectrochemical analysis the clear distinction of line-to-background ratio and noise as given by the relative standard deviation of the intensity measurement is very useful. The standard deviation, especially that of photographic measurements, may be the same for similar sets of measurements. In these cases the easily measurable intensity ratio of line to background permits an estimate of the power of detection. For the photographic plate the limit of detection can be described by the following equation, taken from Laqua et al. (51):

$$\sigma_r(I_U) = \ln 10\sigma(S)\left(1 + \frac{\kappa}{10^{S_U} - 1}\right)\bigg/ \gamma_P \tag{9}$$

where $\sigma(S)$ is the standard deviation of the blackening of the background, $S_U$ is the blackening of the background, $\kappa$ is the transformation constant of the blackening transformation according to Kaiser (52), and $\gamma_P$ is the slope of the transformed blackening curve.

For photoelectric measurements the equivalent equation according to Haisch reads as follows (53):

$$\sigma_r(I_U) = 1/\sqrt{\beta N_{q,U}} \tag{10}$$

where $N_{q,U}$ is the total number of background quanta in the wavelength interval corresponding to the apparent width of the spectral line reaching the photocathode of the PM tube and $\beta$ is the quantum efficiency of the photocathode, typically 5 to 20%.

The background in this context means continuous spectral background and not detector noise. It originates from the sample material, from material vaporized from the auxiliary electrodes, and finally from the surrounding atmosphere. With photographic recording, faint lines become visible only if there is sufficient background. With very small signals, a blackening due to the spectral background may not be obtainable with one shot. If superposition of several shots is not possible, there is only one remedy, namely, to preexpose the emulsion. This can be done by increasing the slit width, inasmuch as the irradiance of the emulsion due to background radiation is proportional to the slit width: unfortunately, this is accompanied by a decrease in the intensity ratio of line-to-background and consequently in the power of detection. However, it may make possible an analysis which otherwise could not be made at all. In cases where a wide slit is not permissible because of line interferences, a preexposure can be made with the help of a continuum source, for example, a deuterium lamp. The main difference between these two methods lies in the fact that in the former case the width of the spectral line is increased with a corresponding decrease of the noise due to the granularity of the emulsion. This method should therefore be preferred. With photoelectric detectors, especially those selected for low dark current, much lower radiation energies can be detected. In such cases the shot noise of the PM tube may determine the limit of detection. A special case, hardly encountered in other spectrochemical procedures, is the determination of the main constituents of small samples, for example, with the aim to identify a small inclusion in a large sample. In this case it is not necessary to expose the photographic emulsion to such a degree that the spectral background can be measured, because the line intensity will be many orders of magnitude higher than the background. One should then use an artificial background blackening, $S_0$, calculated according to the formula given by Siedentopf (54):

$$S_0 = \ln 10 \; F/f\sigma^2(S_0) \tag{11}$$

where $F$ is the measuring area at the photometer, $f$ is the area of a single grain of the emulsion, typically 1 $\mu m^2$, and $\sigma(S_0)$ stems from the fluctuations of photometer readings taken at a clear part within the spectrum. The formula for the limit of detection of small quantities then reads

$$\underline{q} = q_U (\underline{I_x}/I_0)^n \tag{12}$$

where $(I_x/I_0)$ is calculated with the help of Eqs. (8) and (9) by replacing $S_U$ with $\overline{S_0}$.

In practice, one can assume a minimum detectable blackening of about 0.006, which leads to a limit of detection of about $q = q_U$ if the characteristic curve of the photographic emulsion is taken into account according to Eq. (9). A main constituent can often be identified by one-shot sampling, whereas determination of a low concentration may require the superposition of many shots. Some of the different properties of photographic and photoelectric recordings of small signals are given in Table 2.8.

Because the photomultiplier is able to detect very small intensities, care should be taken that such signals are not masked by spectral background not originating from the sample, for example, from the auxiliary electrodes. With cross excitation triggered by the vapor cloud, not true limit of detection but a barrier is reached when the small vapor cloud does not contain enough charged particles to trigger the cross discharge. In this case external triggering of this discharge is compulsory to remove the barrier. It should again be emphasized that with a decrease in the amount of sampled material the danger of contamination from the auxiliary electrode increases.

All these facts may explain why for some photoelectric radiation measurements the cross spark technique does not seem to offer much improvement (55). Some typical examples of limits of detection pertaining to different methods are given in Table 2.6. Some examples for photoelectric methods using time resolution (see Sect. 2.6.3) are listed in Table 2.8.

**TABLE 2.8**
**Use of Time Resolution for Obtaining Low Limits of Detection in Analytical Procedure for Analysis of Biologic Materials, Employing Q-switched Single Laser Pulse and Electronic Gating Technique[a]**

| Element | Wavelength, Å | Delay time, $\mu$s | Integration time, $\mu$s | Detection limit, g |
|---------|---------------|--------------------|--------------------------|---------------------|
| Li | 6104 | 4 | 5 | $2 \times 10^{-13}$ |
| Mg | 2796 | 4 | 7 | $2 \times 10^{-15}$ |
| Ca | 3934 | 5 | 15 | $1 \times 10^{-14}$ |
| Fe | 3020 | 10 | 6 | $3 \times 10^{-13}$ |
| Cu | 3248 | 15 | 3 | $2 \times 10^{-15}$ |
| Zn | 2139 | 5 | 2 | $5 \times 10^{-14}$ |
| Hg | 2537 | 7 | 3 | $3 \times 10^{-13}$ |
| Pb | 4058 | 16 | 5 | $1 \times 10^{-13}$ |

[a] From Treytl et al. (35).

**TABLE 2.9**
**Limits of Detection for Analytical Procedures Using Atomic Absorption Spectrometry with Laser Atomization[a]**

| Element | Matrix | Limit of detection | | Reference no. |
|---------|--------|-----|-----|------|
| | | % | g | |
| Cu | Al alloy | $5 \times 10^{-5}$ | $10^{-11}$ | 25 |
| Mg, Cu, Mn | Al alloy | $10^{-3}$–$2 \times 10^{-4}$ | $2 \times 10^{-11}$–$7 \times 10^{-11}$ | 24 |
| Cu, Mn | steel | $10^{-2}$–$5 \times 10^{-3}$ | $2 \times 10^{-9}$–$5 \times 10^{-9}$ | 24 |
| Fe | Ni-base alloy | 0.5 | $2 \times 10^{-8}$ | 31 |
| Cu, Ag, Mn, Zn | geochem. samples | $10^{-4}$–$10^{-5}$ | $2 \times 10^{-10}$–$2 \times 10^{-11}$ | 28 |
| | (powder, pellets) | | | |
| Mn | graphite | $3 \times 10^{-3}$ | $10^{-9}$ | 25 |
| Na, Cu, Cd, Zn | Photographic emulsion | $10^{-2}$ | $4 \times 10^{-10}$ | 31 |
| Cd | oxide | — | $5 \times 10^{-10}$ | 22 |
| Cr | ruby | $5 \times 10^{-3}$ | — | 106 |
| Fe | Nd:glass | $7 \times 10^{-4}$ | — | 106 |
| Mn, Fe | biologic substance | $10^{-4}$–$10^{-5}$ | $10^{-12}$ | 110 |

[a] For a survey of the methods employed see Fig. 2.20.

For comparison some figures for atomic absorption and emission techniques are given in Table 2.9.

Mandelstam and Nedler (56) some time ago calculated the ultimated limit of detection of spectrochemical procedures using the dc arc and the hollow-cathode discharge as excitation sources. Similar calculations have been performed by Klockenkämper and Laqua (34) for procedures using laser atomization and cross excitation. Because a number of assumptions have to be made, only ranges of values are given. Figure 2.33 presents the results obtained. From these results it can be concluded that the ultimate limits are well below those experimentally obtainable. The authors ascribe this discrepancy to incomplete vaporization and the short time of residence of the particles in the excitation zone. Apparently there is ample room for improvements.

### 2.9.2 Precision

Precision shall be defined by the standard deviation derived from repeated measurements of a particular parameter. Laser instrumentation has reached such a degree of perfection that reliable operation over long times and with little wear is common nowadays. The actual analysis, however, is determined by many other factors with the result that the

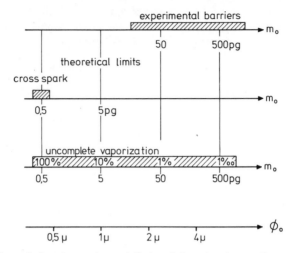

**Fig. 2.33.** Theoretical and experimental limits of detection for small quantities $m_0$ and pertinent crater diameter $\Phi$ using cross spark excitation and showing the influence of insufficient atomization. From Klockenkämper (private communication), reproduced by permission.

precision of the final analytical data is considerably worse. In Table 2.10 this is exemplified for some typical cases.

With respect to precision, atomic absorption spectrometry seems to be inferior although, for final evaluation, additional results are still warranted, in particular those pertinent to two-step methods, for example, the method of Matousek and Orr (26) or Kantor et al. (31).

### 2.9.3  Accuracy

Accuracy means freedom from systematic errors. It is the accepted and proved opinion that laser atomization results in spectrochemical analyses with a better accuracy than that obtainable by some other methods for atomization, for example, arc and spark methods. Whitehead and Heady (57) have investigated the applicability of laser atomization for the analysis of rocks and minerals and found comparatively little influence of various matrix materials on the determination of a number of elements (Table 2.11).

On the other hand, for metals, Baldwin (58) reported a definite influence of the metallurgical condition on the determination of the respective components in a two-component system. Kirchheim et al. (59) found a similar influence of the orientation of single crystals of metals on the intensity. This is, however, not quite unexpected as such effects are known from other spectrochemical excitation sources, for example, arcs

**TABLE 2.10**
**Relative Standard Deviations for Factors Related to Atomic Emission and Absorption Analysis Using Laser Atomization**

| Observed property | Relative standard deviation, % |
|---|---|
| Laser Output Energy | 4–6 |
| Dimension of laser crater[a] | |
|   Diameter | 6–15 |
|   Depth | 10–20 |
|   Volume | 15–30 |
| Emission spectroscopy | |
|   Line emission laser only | 10 |
|   Line emission with additional excitation of laser plume | |
|     by spark | 40[b] |
|     by microwave | 30[c] |
| Absorption spectroscopy | |
|   Absorbance | 30 |

[a] Strongly dependent on kind of sample.
[b] From reference 40.
[c] Taken from reference 24.

and sparks. With sparks they are usually overcome by conditioning the surface prior to measurement by presparking. In this way, a state of equilibrium is obtained in which sampling representative of the composition of the sample is effected although the composition of the actual new surface is different. It is further generally recommended to measure intensity ratios instead of intensities, because the former are much less prone to interferences. "Prespark" treatment of a surface can also be done with laser atomization sources and has been described by Felske et al. (40). In cases where this is not possible, empirically derived corrections or corrections obtained with the help of the phase diagram should be applied to ensure accurate results.

### 2.9.4 Quantitative Analysis

All spectrochemical methods are relative methods. That means that they require calibration with samples of known composition. In the case of chemical analysis with laser atomization, several types of methods according to the analytical task must be distinguished. Laser macroanalysis as treated in Sect. 2.10.4 is similar to that with conventional arc and spark methods for solid materials and therefore needs not to be specifically discussed in this respect. Special efforts have been made to work out semiquantitative methods which can be applied to a wide variety of materials with one or a few sets of calibration curves. In most

**TABLE 2.11**
**Dependence of Determination of a Number of Elements on the Matrix, Employing Laser Atomization and Excitation**[a]

|                | Intensity ratios |        |        |        | Relative standard deviation, % | Maximum percent deviation, % |
| -------------- | ------- | --------- | --------- | -------- | --------- | ----------- |
| Spectral line, Å | $BaF_2$ | $Al_2O_3$ | $Yb_2O_3$ | $WO_3$ |           |             |
| Co I   2587.2  | 0.46    | 0.61      | 0.54      | 0.56     | 11        | 33          |
| Cr II  2766.5  | 1.30    | 1.28      | 1.23      | 1.38     | 5         | 11          |
| Cr II  3124.9  | 1.49    | 1.57      | 1.49      | 1.43     | 4         | 9           |
| Fe II  2599.4  | 1.89    | 1.89      | 1.57      | 1.59     | 10        | 17          |
| Fe II  2611.8  | 1.20    | 1.13      | 0.99      | 0.99     | 10        | 18          |
| Mn II  2933.0  | 2.32    | 2.09      | 2.01      | 2.15     | 6         | 13          |
| Mn II  3441.9  | 1.44    | 1.47      | 1.61      | 1.64     | 6         | 12          |
| Fe I   3021.0  | 1.00    | 0.81      | 0.98      | 0.75     | 14        | 25          |
| Mn I   3228.0  | 1.30    | 1.05      | 1.05      |          | 13        | 19          |
| Mo I   3170.3  | 1.44    | 1.23      | 1.05      |          | 16        | 27          |
| Ni I   3437.2  | 1.85    | 1.75      | 1.58      |          | 8         | 15          |

[a] From Whitehead and Heady (57).

cases corrections to compensate for the influence of the matrix are proposed, for example, with empirically derived correction factors or with similar corrections derived from the size of the crater. The former method has been proposed by Webb and Webb (60) and the latter, by Morton et al. (61). Schroth (62,63) has described a universal method based on the arc method by Harvey (64) for the determination of 53 elements in a large variety of metallic and nonmetallic samples. In this method, it was necessary to divide the variety of samples into several groups, each with a separate calibration.

In the case of local analysis (treated in Sect. 2.10.1), true quantitative analysis can only be made if the vaporized volume is homogeneous in itself. The influence of the surrounding material of different composition must be studied and correction factors may have to be determined.

In the case of microanalysis, the weight and shape of the samples should be kept constant. Until now, little has been reported regarding the influences of size, composition, and degree of vaporization on the analytical results.

## 2.10    ANALYTICAL APPLICATIONS

### 2.10.1    Local Analysis

Laser atomizers were originally introduced for local analysis. Therefore, the name "laser microprobe analysis" in analogy to "electron

**TABLE 2.12**
**Comparison of Laser and Electron Microprobe Pertaining to Local Analysis**

| Property | Laser | Electron microprobe |
|---|---|---|
| Spatial resolution | | |
| Diameter of area, $\mu$m | 10–25 (depending on sample) | $\geq 1$ |
| Depth, $\mu$m | widely variable, depending on working conditions, $\geq 3$ | determined by matrix, 0.1–10 |
| Determinable elements | ~70; elements with persistent lines in VUV (halogens, gases, etc. not yet investigated) | all elements with atomic number $Z \geq 5$ |
| Multielement determination | simultaneous; sequential restricted because destructive | sequential; simultaneous at increased costs or at reduced performance |
| Sample preparation | no | mostly yes (grinding, polishing), surface must be electrically conductive |
| Atmosphere | any (vacuum not recommended) | vacuum |
| Destructive analysis? | yes | no |
| Concentration range | both techniques cover the same range (depending on material and spatial resolution) | |
| Quantitative analysis | yes, provided probed micro volume in itself is homogeneous | yes, same as with laser |
| Determination of elemental distribution in the sample | yes, but limited by coarse resolution and slow speed | yes, widely done |
| Precision | medium | high |
| Speed | slow with photographic, fast with photoelectric methods | fast |
| Cost | low to medium | high |

microprobe analysis" was chosen. Although this name could imply close similarity of the two methods, this is not the case. In Table 2.12 a comparison of these two methods has been attempted.

As all comparisons of this kind are unavoidably rather rough, some explanations are offered:

1.  The main difference can be described by "destructive" or "non-destructive" analysis. With lasers the actual sample, for example, a small inclusion, is destroyed during the course of the analysis, whereas with electron microprobes, the sample remains available for further investigations.

2. From the foregoing it follows that, for multielement local analysis with lasers, in most cases only a simultaneous determination of the elements in question is possible.

3. Spatial resolution with the electron microprobe is an order of magnitude higher although for some special applications, as, for example, the analysis of thin cuts, a similar resolution has been obtained with laser atomizers.

4. The possibility of directly analyzing in air at normal pressure in situ and even in vitro materials of all kinds which are not electrically conductive is a unique property of the laser atomizing methods.

5. Laser methods are less dependent on surface contamination. If the actual sample is large enough, the surface can be cleaned by removing the upper layer by a laser shot before the analysis is made.

It should further be mentioned that the sensitivity of a laser local analysis or microanalysis with spark cross excitation can be considerably improved with the help of a simple trick, namely, by removing the deposit of condensed sample material on the tips of the auxiliary electrodes by a series of additional spark discharges across the gap without further laser action and superimposing this signal onto the first produced with the actual laser atomization.

Analytical applications are reported in many different fields, especially mineralogy, metallurgy, the ceramic and glass industry, criminology, biochemistry, and medicine. It should be mentioned that biologic samples and liquids are best analyzed in a frozen condition. Figures 2.34, 2.35, and 2.36 show some typical examples.

### 2.10.2 Surface Analysis

Surface analysis represents local analysis in which the depth of erosion is kept as small as possible. Within the scope of laser operation modes, there are two possibilities to achieve this goal. The first uses a free-running laser of very low output energy and additional excitation, while the second employs a $Q$-switched low-energy, high-power giant pulse. The former may ultimately yield crater diameters of 10 $\mu$m with a depth of 3 $\mu$m; the latter, a diameter of 25 $\mu$m and a depth of 1 $\mu$m, depending also on the sample material.

An example of a surface analysis is given in Fig. 2.37, on which the spectra of clean and contaminated silicon surfaces can be compared. The differences in the spectra with respect to power of detection and general appearance can also be recognized; for example, the strong CN bands in the spectrum produced with cross excitation mask a considerable part of the spectrum.

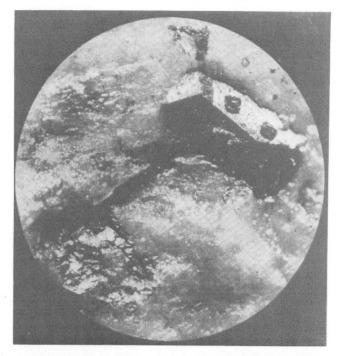

**Fig. 2.34.** Local analysis of a titanite crystal in a rock sample. Length of crystal about 1 mm. From Moenke and Moenke (66), copyright by the Institute of Physics, reproduced by permission.

### 2.10.3 Microanalysis

Laser techniques for microanalysis are closely related to those for local analysis. If a microsample can first be concentrated into a small volume and positioned onto a suitable support, it is actually a local analysis that is afterwards carried out. A typical setup for microanalysis is shown in Fig. 2.27.

For minimum matrix effects, the sample should be vaporized completely using the techniques of laser local analysis. For quantitative analysis, calibration should be done with standard samples having the same size as the analytical sample. Possibly, part of a homogeneous compact standard sample could be evaporated for this purpose and corrections applied. Such a standardization appears to give more easily accurate results with additional excitation. The amount of sample material can be as low as $1 \mu$g. Limits of detection down to 10 ppm, depending on the element and the size of the sample, can be achieved.

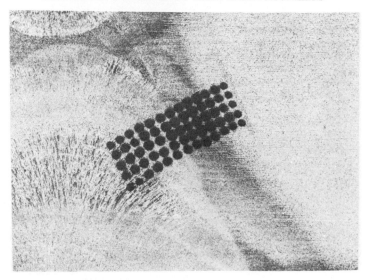

**Fig. 2.35.** Series of laser-produced craters across a welding seam in steel for investigating change in chemical composition. From Ohls et al. (65), reproduced by permission.

## 2.10.4  Macroanalysis

As has been outlined in Sect. 2.1, because of the unique property of laser atomization, homogeneous, nonconducting solid materials in addition to electrically conducting materials can for the first time be directly subjected to optical emission spectrochemical analysis. With inhomogeneous or powdered materials, sample preparation, for example, grinding

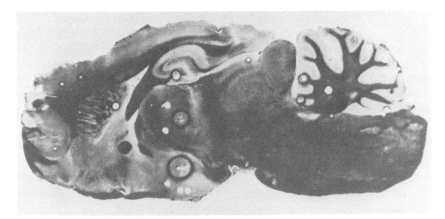

**Fig. 2.36.** Local analysis of cross section of rat brain. From (113), reproduced by permission of Jarrell-Ash Division of Fisher Scientific Company.

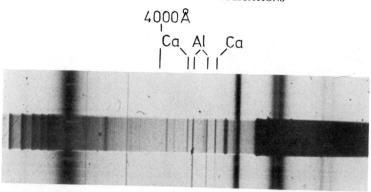

**Fig. 2.37.**   Spectra for evaluation of surface analysis on clean (top) and on contaminated (middle and bottom) silicon surface. Top and bottom, *Q*-switch laser of medium power (100 kW per spike); middle, with spark cross excitation.

and pelleting, is simple and straightforward. This should increase the power of detection while decreasing the danger of contamination. The procedures for direct analysis of solid nonconducting materials and of solid electrical conductors are identical. In the latter case, laser atomization may be preferable to analysis by using electrical discharges for several reasons:

In electrical discharges the burning spot *wanders erratically* across the surface, preferentially sticking to the grain boundaries of microcrystallites or to small inclusions. This may lead to systematic errors ascribable to the structure of the surface, because a sequence of sparks may not give a statistical means for the composition of the sparked surface. There is also the question of precision. Increasing the number of sparks for an analysis may not lead to a corresponding increase in precision because the single sparks do not act independently of each other. In laser analysis vaporization is enforced from the spot *determined solely by the focus of the laser radiation and not by sample properties.* If the precision obtainable with one-shot analysis is insufficient—it may be of the order of 10% as described by the relative standard deviation—almost any required value can be obtained by repeating the process on a hitherto unattacked part of the surface. As in this case the actions are independent of each other, counting statistics can be used for evaluating the improvement in precision.

When this kind of analysis must be performed with a comparatively large number of laser shots, automatic scanning of the surface synchronized with the repetition rate of laser shots is advisable. Although optical scanning of the laser beam with the help of mechanical, electro-optical, or acousto-optical deflection systems is possible, there are some

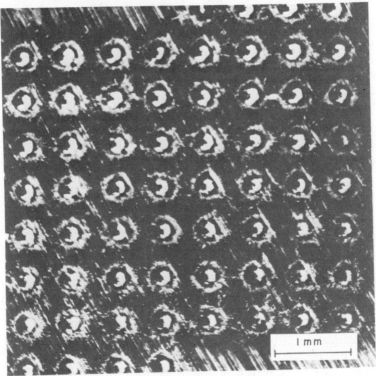

**Fig. 2.38.** Pattern of craters in steel sample showing reproducibility of vaporization. From Felske et al. (71), reproduced by permission of Springer Verlag.

difficulties which so far have prevented its use for the outlined purpose. Instead, a mechanical movement of the sample in one or two coordinates with the help of step motors can be used.

Figure 2.38 shows the typical appearance of a sample surface from which the good reproducibility of vaporization can be judged. It is advisable to *individually* measure the intensities belonging to each shot. In this case a statistical evaluation of the analysis is possible, and errors due to faults of the surface leading to different intensities can easily be detected, thus increasing the reliability of the analysis. The basic obtainable precision depends on the mode of operation. It can be as good as 10% with one-step procedure, that is, without additional excitation for a one-shot analysis and correspondingly less for a multishot analysis. If high power of detection is required, namely, with additional excitation, the basic precision may degrade to about 30%. These figures apply to laser output energies of 1 J or less than 1 J, which corresponds to vaporized sample material of 10–100 $\mu$g per shot.

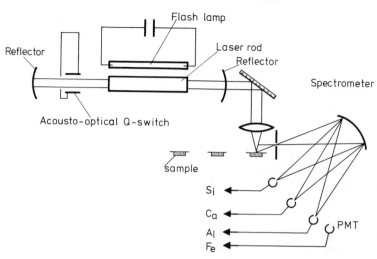

**Fig. 2.39.** Setup for laser macroanalysis. From Meric et al. (46), reproduced by permission.

It is also possible to increase the amount of sample actually used for analysis not by a multiple sampling procedure but by increasing the energy per shot. Output energies of up to 100 J have been reported. In such a case one shot may vaporize enough material for an analysis, thus representing a larger quantity. The basic precision, however, may not be improved to the same degree.

A method based on this principle has been developed by Meric et al. (46) for the analysis of cement raw materials. His experimental setup is shown in Fig. 2.39. Among the advantages of this procedure one should consider the large working distance due to the permissible long focal length of the lens and consequently the insensitivity to maladjustments of that distance. Drawbacks are the comparatively high costs of the equipment and of its maintenance.

A good compromise between the two approaches could be the use of a medium energy system with a laser output of perhaps 5 to 10 J. The maintenance costs of such a system could be low. Because less material is vaporized by each shot than in the high-energy system, a series of not too many shots should enable representative sampling while at the same time improving the precision. Experiments by Quentmeier (24) confirm the validity of this concept. It must be stated that the possibilities of using laser sources for high-precision representative analysis of homogeneous samples should definitely be further explored. Unfortunately, commercial instruments are not equipped for this kind of operation. On the other hand, most analysts will find it difficult to assemble a home-made setup or to modify existing equipment. The main modifications necessary or desirable are the following:

1. *Increase of repetition frequency of laser.* The single-shot technique for local or microanalysis and sequences of shots for representative analysis of bulk samples require different repetition rates. For the former, a repetition rate of $1\,s^{-1}$ will be sufficient. The power supply for the flash lamps can accordingly be held small. Commercial instruments are constructed for this kind of operation. For sequences of shots, a compromise must be found: high repetition rates, although desirable, not only require much larger power supplies but, in addition, better cooling of laser rod, flash lamp, and perhaps laser cavity as well. The necessary speed of the movement of the sample can always be made fast enough. A total analysis time comparable to that of spark analysis, for example, $30\,s$, seems adequate. With a repetition rate of 3 to $4\,s^{-1}$, about 100 shots per analysis would enable a relative standard deviation of 1 to 2% for a one-step method (without additional excitation) and of 3 to 4% for a two-step method (with additional excitation). A sampling by 100 shots could be quite representative even if the laser output energy is not too high. A total sample consumption of 1 to 10 mg per analysis would be comparable to that of spark analysis. The respective laser system can be realized by two flash lamps with a rating of 1000 J each (but for long life expectancy operated at less than half these values) pumping a neodymium glass laser rod of 1 cm in diameter and 15 cm in length in a double elliptical cylindrical cavity configuration. The power supply should have a 5 kVA capacity, as compared to 600 VA necessary for local or microanalysis (one 600-J flash lamp pumping a glass neodymium rod of 6 mm in diameter and 7.5 in cm length at a frequency of 1 Hz).

2. *Amendments of electronic measuring system.* This concerns the dark current problem discussed in Sect. 2.8.2.

3. *Provision for sample movement.* As sample movement is interconnected with other factors such as repetition rate, etc., it is recommended that the whole system be controlled by a small computer which then also could do the relevant calculations and the statistical evaluation.

4. *Monitoring facilities.* In order to obtain optimum results, to detect faults at an early stage, and to possibly correct for changes of operation properties, it is advisable to strictly control and monitor factors governing the laser output. It is especially useful to directly monitor this output with the help of a beam splitter and a radiation calorimeter or a silicon photodiode system. A feedback system as developed by Baues et al. (66a) could lead to an extremely constant output.

5. *Provision to adequately adjust laser output energy.* The laser should always be operated well above pumping threshold in order to ensure a

highly reproducible output. As a rule of thumb it is recommended to use a pumping energy of at least twice the threshold pumping energy. The adjustment of the radiant energy of the laser should then be made with the help of filters.

### 2.10.5 Special Methods

Lasers as atomizers in analytical spectroscopy can be used to advantage to solve very special problems. To mention a few:

1. Poisonous, radioactive, chemically unstable or reactive substances that must be kept in a closed chamber perhaps in an atmosphere other than air can be analyzed directly provided a window permits entrance of the laser radiation and another window, exit of the analytical signal.

2. As energy can be transported by a laser beam over substantial distances, material can be vaporized if a lens can be placed at an appropriate distance from the target. Experiments have been undertaken by Runge et al. (67) to analyze directly liquid steel.

3. No results are known of the exploration of vacuum spectral lines which, however, would permit the determination of several elements having their most sensitive lines in this spectral region, for example, phosphorus, sulfur, carbon, arsenic, and the halogens.

### 2.10.6 The Laser Spark

Power densities in excess of $10^{12}\,\mathrm{W\,cm^{-2}}$, which can be obtained by focusing a giant high-power laser pulse in free air or in another gas atmosphere, can cause a dielectric breakdown. This breakdown has the appearance of a spark, hence the name. It has first been observed by Meyerand and Haught (68). The initial phase of this breakdown is different from that occurring on the surface of a solid target. The production of the first free electrons, that is, the initial phase of ionization, follows a different mechanism according to the irradiance in question. With power densities below $10^{15}\,\mathrm{W\,cm^{-2}}$ it is assumed that energy transfer from the radiation field to bound electrons can take place in a similar way as from a microwave field, namely, bound electrons are forced to oscillate in the strong radiation field thereby absorbing energy until their ionization energy is exceeded. These free electrons then gain more energy from the field by the process of inverse bremsstrahlung and ionize more atoms causing a general breakdown via electron avalanche.

At power densities exceeding $10^{15}\,\mathrm{cm^{-2}}$ the prevailing process at the beginning is multiphoton ionization, that is, the ionization energy of the gas atoms is provided by the accumulated energy of several low-energy photons from the radiation field. The laser spark presents a very hot plasma. Its spectrum therefore consists of a strong continuum. In addition, spectral lines mostly belonging to higher ionic states are excited. Band spectra originating in the comparatively cold mantle of the plasma are also observed. An interesting property of the laser spark seems to be the high-speed expansion of the plasma directed mainly toward the source of the incoming laser radiation.

Mandelstam (69) has proposed the use of a laser spark for the spectrochemical analysis of gases. However, no such analytical work is known as yet. The use of a laser spark as a primary source for atomic absorption has been described in Sect. 2.5.1.

## 2.11   CONCLUSIONS

The unique property of focused laser radiation to unequivocally vaporize material at high temperatures from a predetermined spot of a sample irrespective of its chemical and physical conditions has led to the development of analytical procedures hitherto not available for optical spectrochemical analysis.

Originally introduced for local and microanalysis by optical emission spectroscopy, focused laser radiation has since found applications in surface and bulk analysis of homogeneous samples, namely, solids, liquids, metals and nonmetals, compact materials, and powders.

Procedures with laser vaporization and atomization only have supplemented the two-step approach, that is, laser atomization and additional excitation. Because of the destruction of the actual sample, laser atomization is mainly employed for simultaneous multielement analysis. This has resulted in the preference for optical emission spectroscopy because of the large range of concentrations coverable with one single measurement. Such properties are necessary for the analysis of solid random samples.

Whereas normally atomic absorption methods, apart from being practically single-element methods due to the limited concentration range of the absorption lines, are not very useful for this kind of analysis, with laser vaporization these disadvantages can partly be overcome, and consequently use can be made of the simple and straightforward AAS instrumentation.

The power of detection of procedures employing laser atomization compares quite well with that of other analytical procedures. A rough comparison is possible with the help of Table 2.13. It is in the determination of small sample weights where the method excels. Precision can be high if multiple sampling is possible. Accuracy of analytical results can be

**TABLE 2.13**
**Comparison of Power of Detection of a Number of Methods Used for the Determination of Small Concentrations and Quantities**[a]

| | | Detection limits | |
| --- | --- | --- | --- |
| Technique | Sample size | ppm | g |
| Ion microprobe | $10^{-11}$ g | $10^{-0}$–$10^{-2}$ | $10^{-17}$–$10^{-19}$ |
| | 1–5 $\mu$ diam. | | |
| Electron microprobe | $10^{-12}$ g | $10^{3}$–$10^{1}$ | $10^{-15}$–$10^{-17}$ |
| | 0.2–1 $\mu$ diam. | | |
| Laser microprobe | $10^{-8}$ g | $10^{2}$–$10^{-1}$ | $10^{-12}$–$10^{-15}$ |
| | 10–200 $\mu$ diam. | | |
| Neutron activation | mg–g | $10^{-3}$–$10^{-6}$ | $10^{-10}$–$10^{-13}$ |
| Atomic absorption | mg | $10^{-1}$–$10^{-5}$ | $10^{-9}$–$10^{-12}$ |

[a] From Treytl et al. (35).

considerably better than with other sources using also heat for vaporization.

Complete coverage of the literature has not been the objective of this chapter. The reader will find a wealth of information in the book by Moenke and Moenke (66b) and in the biannual review articles in *Analytical Chemistry* by Barnes. A number of references concerning different aspects of analytical applications are listed in Table 2.14.

**TABLE 2.14**
**Additional Literature Survey**

| Classes of samples | Reference no. |
| --- | --- |
| Metallic samples (pure metals, alloys, meteorites) | 71, 60, 43, 40, 49b, 72, 19, 58, 73, 102 |
| Nonmetallic samples (minerals, glasses. powdered materials) | 74, 75, 57, 46, 76, 43, 40, 77, 78, 79, 102, 109, 111 |
| Organic samples (biologic, medical, plastics) | 80, 81, 82, 83, 41, 35, 40, 59, 84, 85, 86, 87, 88, 102 |
| Fundamental investigations (mostly done with metallic samples) | 89, 14, 32, 90, 91, 33b, 70, 92, 11, 16, 12, 71, 62, 93, 61, 17, 72, 94, 20, 39, 19, 21, 55, 95, 96, 69, 59, 37, 97, 98, 99, 100, 101, 103, 104, 105, 106, 107, 108, 110, 112 |

Finally, it should be stressed that the use of the inherent potentialities of laser atomization in analytical chemistry strongly depends on the availability of suitable commercial instruments, possibly fully integrated ones.

## Acknowledgments

The author would like to thank his colleagues Drs. Hagenah, Leis, and Quentmeier and Mrs. Waechter for numerous discussions and suggestions during the preparation of this manuscript.

## References

1. T. H. Maiman, *Nature*, **187,** 493 (1960).
2. (*a*) F. Brech and L. Cross, *Appl. Spectrosc.*, **16,** 59 (1962); (*b*) F. Brech and K. Schuch, 10th Int. Conf. Spectroscopy, Washington, D.C. (1962).
3. A. Heller, *J. Amer. Chem. Soc.*, **90,** 3711 (1968).
4. E. J. Schimitschek, *J. Appl. Phys.*, **39,** 6120 (1968).
5. R. Hellwarth, "Advances in Quantum Electronics," J. R. Singer, Ed., Columbia University Press, New York (1961) p. 334.
6. P. Boissel, G. Hauchecorne, F. Kerhervé, and G. Mayer, *Opt. Commun.*, **4,** 44 (1971).
7. F. T. Arecchi and E. O. Schulz-Dubois, Eds., *Laser Handbook*, North-Holland, Amsterdam (1972).
8. H. Riesenberg, *Jenaer Rundsch.*, *Messesonderheft 40*, (1964).
9. H. Klocke, Ph.D. Thesis, Münster (1968).
10. S. Panzer, *Laser*, **1,** 3 (1969).
11. C. D. Allemand, *Spectrochim. Acta*, **27B,** 185 (1971).
12. (*a*) R. H. Scott and A. Strasheim, *Spectrochim. Acta*, **25B,** 311 (1970). (*b*) R. H. Scott and A. Strasheim, *Spectrochim. Acta*, **26B,** 707 (1971). (*c*) R. H. Scott, Ph.D. Thesis, University of South Africa, (1970).
13. H. Klocke, *Spectrochim. Acta*, **24B,** 263 (1969).
14. J. R. Ready, *J. Appl. Phys.*, **36,** 462 (1965).
15. H. G. Landau, *Quart. Appl. Math.*, **8,** 81 (1950).
16. O. N. Krokhin, "Generation of High-Temperature Vapors and Plasmas by Laser Radiation," in *Laser Handbook*, Vol. 2, F. T. Arecchi and E. O. Schulz-Dubois, Eds., North-Holland, Amsterdam (1972), p. 1372.
17. T. Yamane and S. Matsushita, *Spectrochim. Acta*, **27B,** 27 (1972).
18. T. Yamane, *Bunko Kenkyu*, **21,** 322 (1972).
19. E. H. Piepmeier and D. E. Osten, *Appl. Spectrosc.*, **25,** 642 (1971).
20. A. Petrakiev and G. Dimitrov, Proc. XVI. Coll. Spectr. Int., Heidelberg (1971), p. 186.

21. W. J. Treytl, K. W. Marich,. J. B. Orenberg, P. W. Carr, D. C. Miller, and D. Glick, *Anal. Chem.*, **43**, 1452 (1971).

22. (a) K. H. König and P. Neumann, *Z. Anal. Chem.*, **279**, 337 (1976); (b) P. Neumann, Ph.D. Thesis, Frankfurt/Main (1975).

23. V. G. Mossotti, K. Laqua, and W.-D. Hagenah, *Spectrochim. Acta*, **23B**, 197 (1967).

24. A. Quentmeier, Ph.D. Thesis, Düsseldorf (1977).

25. D. E. Osten and E. H. Piepmeier, *Appl. Spectrosc.*, **27**, 3, 165 (1973).

26. J. P. Matousek and B. J. Orr, *Spectrochim. Acta*, **31B**, 475 (1976).

27. A. V. Karyakin and V. A. Kaigorodov, *Zh. Anal. Khim.*, **23**, 930 (1968).

28. E. K. Wulfson, A. V. Karyakin, and A. I. Shidlovsky, *Zh. Anal. Khim.*, **28**, 1253 (1973).

29. A. V. Karyakin, Proc. Symposium on Ultrapure Materials, Dresden (1975), in press.

30. L. N. Kaporskii and G. S. Musatova, *Zh. Prikl. Spektrosk.*, **8**, 681 (1968).

31. T. Kantor, L. Polos, P. Fodor, and E. Pungor, *Talanta*, **23**, 585 (1976).

32. S. D. Rasberry, B. F. Scribner, and M. Margoshes, *Appl. Opt.*, **6**, 81, 87 (1967).

33. (a) W. Quillfeldt, *Exp. Tech. Physik*, **17**, 415 (1969); (b) H. Moenke, L. Moenke-Blankenburg, J. Mohr, and W. Quillfeld, *Mikrochim. Acta*, **6**, 1154 (1970).

34. R. Klockenkämper and K. Laqua, *Spectrochim. Acta*, **32B**, 207 (1977).

35. W. J. Treytl, J. B. Orenberg, K. W. Marich, A. J. Saffir, and D. Glick, *Anal. Chem.*, **44**, 1903 (1972).

36. U. Möde, Ph.D. Thesis, Münster (1970).

37. F. Leis, Ph.D. Thesis, Düsseldorf (1976).

38. G. Dimitrov and A. Petrakiev, *Jahrb. Univ. Sofia, Phys. Fak.*, **65**, 531 (1970).

39. H. J. Stupp and T. Overhoff, *Spectrochim. Acta*, **30B**, 77 (1975). H. J. Stupp and T. Overhoff, *Spectrochim. Acta*, **29B**, 77 (1974).

40. (a) A. Felske, W.-D. Hagenah, and K. Laqua, *Spectrochim. Acta*, **27B**, 1 (1972). (b) A. Felske, W. -D. Hagenah, and K. Laqua, *Spectrochim. Acta*, **27B**, 295 (1972).

41. W. J. Treytl, K. W. Marich, and D. Glick, *Anal. Chem.*, **47**, 1275 (1975).

42. H. M. Crosswhite, D. W. Steinhaus, and G. H. Dieke, *J. Opt. Soc. Amer.*, **41**, 299 (1951).

43. A. Petrakiev, G. Dimitrov, and L. Georgieva, *Spectrosc. Lett.*, **2**, 97 (1969).

44. V. Vukanović, V. Georgijević, D. Vukanović, and M. Todorović, *Spectrochim. Acta*, **24B**, 555 (1969).

45. International Union of Pure and Applied Chemistry, *Pure Appl. Chem.*, **30**, 653 (1972).

46. J. P. Meric, Y. LeJean, and J. F. Caron, *Rev. Mater.*, **684**, 36 (1973).

47. W. J. Treytl, J. B. Orenberg, K. W. Marich, and D. Glick, *Appl. Spectrosc.*, **25**, 376 (1971).

48. M. Hercher and B. Ruff, *J. Opt. Soc. Amer.*, **57**, 103 (1967).

49. E. H. Piepmeier and H. V. Malmstadt, *Anal. Chem.*, **41**, 700 (1969).

50. H. Kaiser, *Optik*, **21**, 309 (1964).

51. K. Laqua, W.-D. Hagenah, and H. Waechter, *Z. Anal. Chem.*, **225**, 142 (1967).

52. H. Kaiser, *Spectrochim. Acta*, **3**, 159 (1947–1948).

53. U. Haisch, *Spectrochim. Acta*, **25B**, 597 (1970).

54. H. Siedentopf, *Phys. Z.*, **38**, 454 (1937).

55. N. A. Peppers, E. J. Scribner, L. E. Alterton, R. C. Honey, E. S. Beatrice, J. Harding-Barlow, R. C. Rosan, and D. Glick, *Anal. Chem.*, **40**, 1178 (1968).

56. S. L. Mandelstam and W. W. Nedler, *Opt. i Spektroskopiya*, **10**, 390 (1961).

57. A. B. Whitehead and H. M. Heady, *Appl. Spectrosc.*, **22**, 7 (1968).

58. J. M. Baldwin, *Appl. Spectrosc.*, **24**, 429 (1970).

59. R. Kirchheim, U. Nagorny, K. Maier, and G. Tölk, *Anal. Chem.*, **48**, 1505 (1976).

60. M. S. W. Webb and R. J. Webb, *Anal. Chim. Acta*, **55**, 67 (1971).

61. K. L. Morton, J. D. Nohle, and B. S. Madson, *Appl. Spectrosc.*, **27**, 109 (1973).

62. H. Schroth, *Z. Anal. Chem.*, **253**, 7 (1971).

63. H. Schroth, *Z. Anal. Chem.*, **255**, 257 (1971).

64. C. E. Haryey, *A Method of Semi-Quantitative Spectrographic Analysis*, Applied Research Laboratories, Glendale, California (1947).

65. K. Ohls, G. Becker, and H. Grote, *Jenaer Rundsch.*, **19**, 245 (1974).

66. (a) P. Baues, U. Hundelhausen, and P. Möckel, *J. Appl. Phys.*, **44**, 4067 (1967); (b) H. Moenke and L. Moenke-Blankenburg, *Laser Micro-Spectrochemical Analysis*, Adam Hilger, London (1973).

67. E. P. Runge, S. Bonfiglio, and F. R. Bryan, *Spectrochim. Acta*, **22**, 1678 (1966).

68. R. G. Meyerand and A. F. Haught, *Phys. Rev. Lett.*, **11**, 401 (1963).

69. S. L. Mandelstam, Proc. XV. Coll. Spectr. Int., Madrid (1969), p. 115.

70. K. Ohls, K. H. Koch, and G. Becker, *Z. Anal. Chem.*, **264**, 97 (1973).

71. A. Felske, W.-D. Hagenah, and K. Laqua, *Z. Anal. Chem.*, **216**, 50 (1966).

72. A. Felske, W.-D. Hagenah, and K. Laqua, *Naturwiss.*, **57**, 428 (1970).

73. R. Ishida and M. Kubota, *Bunko Kenkyu*, **21**, 16 (1972).

74. N. Nikolov, A. P. Petkov, G. Dimitrov, and A. Dimov, *Congr. et Coll. Univ. Liège*, **59**, 267 (1970).

75. J. C. Koo, *J. Appl. Phys.*, **48,** 618 (1977).

76. D. Dimov, A. Petrakiev, G. Dimitrov, and A. Ivanova, *Bol. Geologico y Minero,* **83,** 637 (1972).

77. I. A. Fersman, L. D. Khazoo, and G. P. Tikhomirov, *Kvantovaya Elektron.,* **3,** 61 (1971).

78. J. R. Ryan, E. Ruh, and C. B. Clark, *Amer. Ceram. Soc. Bull.,* **45,** 260 (1966).

79. W. H. Blackburn, Y. J. A. Pelletier, and W. H. Dennen, *Appl. Spectrosc.,* **22,** 278 (1968).

80. G. Dimitrov and M. Marinov, *Medikon,* **5,** 4 (1976).

81. E. S. Beatrice, I. Harding-Barlow, and D. Glick, *Appl. Spectrosc.* **23,** 257 (1969).

82. E. S. Beatrice, I. Harding-Barlow, R. C. Rosan, and D. Glick, *Anal. Chem.,* **40,** 1178 (1968).

83. A. Petrakiev, G. Dimitrov, D. Doncher, and A. Ivanova, *C.R. Acad. Bulgare Sci.* **22,** 911 (1969).

84. K. W. Marich, P. W. Carr, W. J. Treytl, and D. Glick, *Anal. Chem.,* **42,** 1775 (1970).

85. S. F. Brokeshoulder and F. R. Robinson, *Appl. Spectrosc.,* **22,** 758 (1968).

86. R. C. Rosan, M. K. Healy, and W. F. McNary Jr., *Science,* **142,** 236 (1963).

87. J. R. Prine, S. F. Brokeshoulder, D. E. McVean, and F. R. Robinson, *Amer. J. Clin. Pathol.,* **45,** 448 (1966).

88. R. G. Wilson, L. Goldman, and F. Brech, *Arch. Dermatol.,* **95,** 490 (1967).

89. V. J. Vladimirov, G. M. Malyshev, G. T. Razdobarin, and V. V. Semenov, *Sov. Phys.-Tech. Phys.,* **14,** 677 (1969).

90. E. S. Beatrice and D. Glick, *Appl. Spectrosc.,* **23,** 260 (1969).

91. A. J. Saffir, K. W. Marich, J. B. Orenberg, and W. J. Treytl, *Appl. Spectrosc.,* **26,** 469 (1972).

92. J. F. Ready, E. Bernal, and L. T. Shepherd, Semiannual Report on Contract DA-18-001-AMC-1040(X), Mod. 1, Aberdeen Proving Ground, Maryland (Nov. 1967).

93. S. A. Mikhnov, V. V. Panteleev, V. S. Strizhner, and A. A. Yankovskii, *Zh. Prikl. Spektrosk.,* **17,** 394 (1972).

94. M. Berndt, H. Krause, L. Moenke-Blankenburg, and H. Moenke, *Jenaer Jahrbuch 1965,* VEB Carl Zeiss Jena (1965).

95. W.-D. Hagenah, K. Klocke, and K. Laqua, Proc. XIV. Coll. Spectr. Int., Debrecen (1967), p. 633.

96. W.-D. Hagenah, *Z. Angew. Math. Phys. (ZAMS),* **16,** 130 (1965).

97. B. Bieber and J. Drexlerova, *Hutn. Listy,* **27,** 369 (1972).

98. T. Yamane, S. Szuki, and S. Matsushita, *Bunko Kenkyu,* **19,** 147 (1970).

99. O. I. Putrenko and A. A. Yankovskii, *Zh. Prikl. Spektrosk.,* **15,** 596 (1971).

100. Yu. V. Afanas'ev, N. G. Basov, O. N. Morachevskii, and G. V. Sklizkov, *Sov. Phys.-Tech. Phys.*, **14**, 669 (1969).

101. E. H. Piepmeier, Ph.D. Thesis, University of Illinois, Urbana (1966).

102. H. Neuninger, *Arch. Kriminol.*, **144**, 121 (1969).

103. G. A. Klotzbaugh, A. L. Wolfe, G. E. Paterson, and T. A. Osial, *Def. Doc. Center AD*, **622**, 102 (1965).

104. A. V. Karyakin, A. M. Pschelnizer, A. I. Shidlovsky, E. K. Wulfson, and M. N. Zingarelli, *Zh. Prikl. Spektrosk.*, **18**, 610 (1973).

105. E. K. Wulfson, A. V. Karyakin, and A. I. Shidlovsky, *Zh. Prikl. Spektrosk.*, **12**, 14 (1975).

106. E. P. Krivchikova and V. S. Demin, *Zh. Prikl. Spektrosk.*, **14**, 592 (1971).

107. E. P. Krivchikova and K. I. Taganov, *Zh. Prikl. Spektrosk.*, **19**, 601 (1973).

108. V. I. Kasatkin and K. I. Taganov, *Zh. Prikl. Spektrosk.*, **8**, 223 (1968).

109. N. I. Eremin, *Mineral. Mag.*, **40**, 312 (1975).

110. L. T. Sukhov, G. E. Zalotukhin, and S. M. Zyabkina, *Zh. Prikl. Spektrosk.*, **25**, 199 (1976).

111. J. Debras-Guedon, N. Liodec, *C.R. Habd. Séances Acad. Sci.*, **257**, 3336 (1963).

112. G. Dimitrov and V. Gagov, *Spectrosc. Lett.*, **10**, 337 (1977).

113. Jarrell Ash, Co., *Spectrum Scanner*, **21**, 3 (1966).

# ATOMIC ABSORPTION SPECTROSCOPY WITH LASER PRIMARY SOURCES

## EDWARD H. PIEPMEIER

*Department of Chemistry*
*Oregon State University*
*Corvallis, Oregon*

## 3.1 INTRODUCTION

Narrowband tunable lasers have made feasible new applications of atomic absorption measurements that are at best difficult, if not impossible, with the commonly used hollow-cathode lamp. Indeed, this chapter would only be a paragraph or two long if the narrowband tunable laser were envisioned simply as a replacement for relatively inexpensive hollow-cathode lamps. Instead, after considering some of the unique properties of lasers, this chapter reviews how the laser has been used in atomic absorption spectroscopy and considers how it could be used, and will hopefully stimulate the reader to envision new applications.

Modern tunable dye lasers may have any of several advantages over ordinary primary light sources used in atomic absorption spectroscopy, including (a) extremely narrow linewidths, (b) continuous tunability over selected wavelength regions, (c) high intensity, (d) collimation and excellent focusing properties, and (e) extremely short pulse duration (<1 ps). Not all tunable lasers have these characteristics, however, and in some cases special care may be needed to achieve the desired performance.

Wavelength-selective devices placed in the cavity of a tunable dye laser allow the laser to be tuned to the very center of a sample cell absorption line where the absorption signal is strongest. Atomic and ionic wavelengths that are not observed in ordinary emission line sources can also be produced. Linewidths can be made narrower than natural absorption linewidths of free atoms—several decades narrower than the narrowest emission lines in hollow-cathode lamps or other common line sources. This eliminates finite spectral bandwidths as a cause of non-linearity in analytical curves, especially at high absorbances. A laser beam of high intensity minimizes the influence of shot noise, thereby allowing more precise measurements to be made of short-lived species, and of the transient absorption peaks of resistive heated atomizers, laser plumes, and sparks. High irradiances can promote a large fraction of an atomic population into an excited state allowing atomic absorption measurements to be made from the excited state. The additional wavelengths that become available might be used to eliminate spectral interferences. Even higher irradiances produce two-photon absorptions that can populate excited states where selection rules prohibit significant direct population by single-photon absorption.

In addition to continuous wave (cw) dye lasers, pulsed dye lasers have been made with pulses ranging from about 1 $\mu$s in duration for a flash lamp-pumped dye laser down to less than 1 ps for a mode-locked laser (1–5). A well-collimated (low divergence angle) dye laser beam can be focused to a submillimeter spatial region allowing very small sample cells to be used and very high irradiances to be obtained for populating selected excited states. The focal spot diameter (in cm) is given approximately by the product of the divergence angle (in rad) and the focal length of the lens (in cm). High irradiances can also be obtained with unfocused beams by using pulsed lasers with laser beam amplifiers.

Although a particular laser may have some or all of these advantages, there are other characteristics of typical lasers that are important enough to deserve special consideration: (*a*) expense, (*b*) wavelength setting accuracy and stability, (*c*) scanning linearity, (*d*) change in intensity with wavelength, and (*e*) intensity stability. Lasers that are suitable for atomic absorption measurements are presently 10 to 100 times more expensive than ordinary light sources with their power supplies. *A laser with a narrow linewidth does not also ensure the wavelength accuracy or stability that so often accompanies ordinary atomic line sources.* A dye laser having a narrow line that is tunable over a wide range may not have a wavelength dial setting that is as accurate as the line is narrow. Of course, for an atomic absorption measurement the laser could be fine tuned to the absorption peak of a standard sample. However, if there were more then one absorption peak within the error range of the laser wavelength dial setting, then it might be necessary to use a high-resolution spectrometer or spectrograph to unambiguously determine the laser wavelength.

The stability of the laser wavelength must be on the order of 1 part in 1–10 million if the laser is to remain tuned to the center of a typical absorption peak during a set of measurements. This currently requires state-of-the-art technology, implying that a *scan* across an absorption peak would produce more reliable results than the stationary wavelength technique. Although smooth wavelength scans can be obtained, care should be used in the design of the scanning system to ensure that the intensity of the beam does not fluctuate too much even across narrow wavelength intervals. When high feedback is used in a laser cavity, minor imperfections in high feedback components (mirrors) can cause large changes in intensity (including loss of lasing) even over a narrow wavelength region. These changes are unpredictable and may be altered by replacing a component. In most cases a double-beam technique can be readily used to compensate for intensity fluctuations since the relative shot noise for a high-intensity beam is negligible.

Minor changes in the average spatial position of a laser beam and in the intensity distribution throughout the beam may occur during scanning or nonscanning operation. These changes may be due to changes in the mode structure of the beam, variations in the refractive index of the lasing medium, and changes in position of the tuning elements during a wavelength scan. Spatial position and beam intensity distribution are important considerations when small sample cells are used for atomic absorption measurements. Such changes in the beam should not extend outside a sample cell region in which the absorbing atoms are homogeneously distributed. Variations in irradiance across the diameter of the beam will cause spatial variations in the fraction of excited atoms produced by the beam when a high irradiance is being used to excite a significant fraction of atoms to an excited state.

Even for ordinary sample cells, if the beam were directed to one part of a photodetector, a minor movement to another region of the detector would most likely cause a change in the intensity signal due to nonuniform sensitivity across the active surface of the detector. This type of fluctuation can be minimized by spreading the beam and placing a diffuser in front of the detector so that some light from all parts of the beam reaches each part of the active surface.

The next part of this chapter considers in detail the ways in which low-intensity laser beams with narrow linewidths may be used to advantage in analytical and other types of atomic absorption measurements. Then, the uses of high-intensity laser beams that excite a significant fraction of atoms to an excited state and that produce nonlinear effects are discussed, including techniques that minimize the spectral influence of Doppler broadening.

## 3.2  ATOMIC ABSORPTION WITH LOW-INTENSITY LASERS

Tuning a very narrow linewidth laser beam across the absorption spectrum of a sample cell will show many of the fine structural details of the absorption spectrum that are not ordinarily observed in atomic absorption measurements. The fine structural details of an atomic absorption line are reviewed below for several types of sample cells since they contain valuable analytical and diagnostic information. Later, the advantages of using a narrow linewidth laser beam are discussed, including improved spectral resolution and improved time resolution capability due to reduced relative shot noise.

### 3.2.1  Atomic Absorption Line Profiles

The general theory of atomic spectral line profiles has been covered in several books (6–10) and reviewed in several papers (11,12) for low irradiances, and new developments continue to appear in the literature (13–20), especially for cases where the population is subjected to high irradiances. However, specific experimental and theoretical information about profiles is available for only a relatively few atomic lines. Measurements using narrowband lasers are needed to provide experimental details about many other analytically useful lines. A review of the general theory will indicate the type of information available and provide a basis for discussing analytical applications of narrowband lasers.

In general, the spectral line profile that corresponds to an electron transition in an atom may exhibit several hyperfine components due to nuclear spin splitting and the presence of isotope shifts (6,7,11,21). The isotope shift is due to the influences of the nuclear mass and the nuclear volume. The isotope shift due to the nuclear mass can be estimated by

$$\Delta\lambda = \frac{-q}{1836} \frac{M_A - M_B}{M_A M_B} \lambda \tag{1}$$

where $\Delta\lambda$ is the isotope shift, expressed in wavelength units, for a line at wavelength $\lambda$ and for two isotopes with masses $M_A$ and $M_B$, in atomic mass units (amu); and the quantity $q$ is a constant that depends upon the type of mass effect and the type of electrons involved in the energy level transition. For the *normal* mass effect, $q = 1$, which shifts the transition of the heavier isotope toward shorter wavelengths. The normal mass effect is only $2 \times 10^{-4}$ nm at 400 nm for a mass difference of 1 at 33 amu and is less for heavier isotopes. A *specific* mass effect produces values of $q$ from $-20$ to $+16$ for $d$-electron transitions (11).

The nuclear volume effect causes a shift of either sign and is caused by the way in which the nuclear charge is distributed over the volume of the nucleus. The volume effect shares in importance with the mass effect for intermediate elements and dominates for heavy elements, that is, $>140$ amu (21, p. 167).

About two thirds of the stable atomic nuclei have a net nuclear spin that causes hyperfine splitting, typically on the order of $10^{-3}$ nm. The relative intensities of the nuclear spin hyperfine components can often be predicted from theoretical considerations based upon Russel-Saunders coupling, but their relative positions on the wavelength axis usually must be determined experimentally.

Figure 3.1 shows the hyperfine structure of the 404.7-nm Hg line (22). The vertical lines represent the relative positions and intensities of the unbroadened components. Two of the six isotopes have an odd number of neutrons and therefore show hyperfine splitting (up to 1.1 cm$^{-1}$, or 0.018 nm). An isotope shift of 0.03 cm$^{-1}$, or 0.0005 nm, is apparent between successive even isotopes.

In general, the magnitude of each hyperfine component of an atomic transition is characterized by an *atomic absorption coefficient*. This coefficient has a wavelength dispersion, or broadening, that can be caused by several effects. Regardless of the broadening process, however, at low irradiances where stimulated emission is negligible, the integral over all wavelengths of the spectral profile of the absorption coefficient $k(\lambda)$

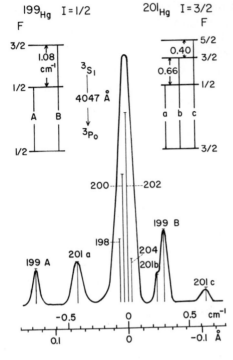

**Fig. 3.1.** Hyperfine structure of Hg 404.7-nm line emitted by low-pressure mercury lamp. Vertical lines in the spectrum indicate theoretical positions and intensities. From de Galan and Wagenaar (11), reproduced by permission.

remains a *constant* given by

$$\int_0^\infty k(\lambda)\, d\lambda = \frac{\pi e^2 \lambda_0^2 f n}{mc^2} \tag{2}$$

for a homogeneous absorption cell, where $e$ and $m$ respectively are the charge (in esu) and mass (in g) of an electron; $c$ is the speed of light (in cm s$^{-1}$); $n$ is the concentration (in cm$^{-3}$) of atoms in the lower energy level of the transition; and $\lambda_0$ is the wavelength of the line (in cm). The quantity $f$ is the *oscillator strength* of the transition and is a dimensionless constant for a given line. The oscillator strength is related to the natural lifetime $\tau_{rad}$ (in s) of the transition by

$$f\tau_{rad} = \frac{f}{A_{21}} = \frac{mc}{8\pi^2 e^2}\frac{g_2}{g_1}\lambda_0^2 = 1.51 \frac{g_2}{g_1}\lambda_0^2 \tag{3}$$

where $A_{21}$ is the Einstein transition probability for spontaneous emission, in events per second; $\lambda_0$ is in cm; and $g_2$ and $g_1$ are the statistical weights of the upper and lower energy levels, respectively. The overall absorption profile of a line is obtained by summing the individual hyperfine component profiles over all components in the line (Fig. 3.1).

## 3.2.2 Broadening Processes

An absorption coefficient is often described as *heterogeneously* or *homogeneously* broadened and shifted. Heterogeneous broadening occurs when the absorption coefficients for different subgroups of atoms in a total population are influenced to different extents by the radiation. For instance, the Doppler effect causes the absorption coefficient to be shifted to the blue for atoms traveling with a velocity component $v$ (in cm s$^{-1}$) toward the detector, while a red shift occurs for atoms traveling away from the detector. The blue shift may be obtained from the equation

$$(\lambda_0 - \lambda) = \frac{\lambda_0 v}{c} \tag{4}$$

where $\lambda_0$ is the wavelength of the stationary source, $\lambda$ is the wavelength observed at the stationary detector, and $c$ is the velocity of light (in cm s$^{-1}$). *The emission profile and absorption profile are of course shifted by the same amounts and in the same direction.*

Net Doppler shifts have been observed in both the emission and absorption profiles of lines emitted from pulsed hollow-cathode lamps (23) and may be present in pulsed atomizers. A Doppler-broadened profile is described by a Gaussian function if the atoms have a Maxwellian velocity distribution as they do when they are in thermodynamic equilibrium. The Gaussian function is of the form

$$k(\lambda) = k(\lambda_0) \exp\left( - \left( \frac{2(\lambda - \lambda_0)}{\Delta\lambda_D} \right)^2 \ln 2 \right) \tag{5}$$

where the width of the peak at half-height, hereafter called simply the linewidth, is given by

$$\Delta\lambda_D = 7.16 \times 10^{-7} \lambda_0 \left( \frac{T}{M} \right)^{1/2} \tag{6}$$

where the temperature $T$ is in °K, $M$ is in atomic mass units, $\Delta\lambda_D$ has the same units as $\lambda_0$ (e.g., nm), and $k(\lambda_0)$ is the peak value of the absorption coefficient obtained by integrating Eq. (5) over all wavelengths and substituting the results into Eq. (2):

$$k(\lambda_0) = (4\pi \ln 2)^{1/2} \frac{e^2 \lambda_0^2 fn}{mc^2 \Delta\lambda_0} \tag{7}$$

An important characteristic of the Gaussian function is that it drops off relatively rapidly on either side of its center. In fact, 99.994% of the area under the curve is contained within the bounds of ±4 linewidths.

Homogeneous broadening occurs when all atoms in a population are equally influenced by the broadening process. Examples are natural lifetime broadening and (to a first approximation) collisional broadening, which are usually described by a Lorentzian function:

$$k(\lambda) = k(\lambda_0, \delta) \cfrac{1}{1 + \left( \cfrac{2[\lambda - (\lambda_0 + \delta)]}{\Delta\lambda_L} \right)^2} \tag{8}$$

where the peak value for $k(\lambda)$ occurs at the wavelength $\lambda = (\lambda_0 + \delta)$ and is given by

$$k(\lambda_0, \delta) = \frac{2}{\Delta\lambda_L} \frac{e^2 \lambda_0^2 fn}{mc^2} \tag{9}$$

The term $\delta$ is included to indicate that the wavelength of the peak of the profile may be shifted by an amount $\delta$ from $\lambda_0$. The term $\Delta\lambda_L$ is the width of the Lorentzian profile at half-height:

$$\Delta\lambda_L = \frac{\lambda_0^2}{2\pi c} \left( \sum_m d_{i,m} + \sum_m d_{j,m} \right) = \frac{\lambda_0^2}{2\pi c} \left( \frac{1}{\tau_i} + \frac{1}{\tau_j} \right) \tag{10}$$

where $i$ and $j$ represent respectively the upper and lower energy states for the transition, $m$ is any state (including $i$ or $j$), and $d_{i,m}$ is the transition probability, per second, that an atom in the $i$th state will leave that state for the $m$th state. The summation of all $d_{i,m}$ probabilities, therefore, is the inverse of the mean lifetime, $\tau_i$, in seconds, of the $i$th state. For a resonance line, where the ground state lifetime is very long, the term in parentheses is the sum of the Einstein probability per second for spontaneous emission plus a collisional rate constant for all collisions that interrupt the coherence of the interaction between the emitted or absorbed radiation field and the atom. Such collisions may be *adiabatic* (no energy transfer) or *nonadiabatic* (quenching or exciting). The term $2\pi$ in Eq. (10) represents the conversion from radians per second (rad s$^{-1}$) to hertz (Hz), and $(\lambda_0)^2/c$ represents the conversion from a small difference in frequency in Hz to a small difference in wavelength $\Delta\lambda_L$ in nm.

A *wavelength shift* of the entire profile indicated by the quantity $\delta$ in Eq. (8) accompanies some Lorentzian broadening processes. The shift is to longer wavelengths when $\delta$ is a positive quantity. Lindholm's approximate theory for adiabatic collisions predicts not only a Lorentzian-shaped profile but also a wavelength shift caused by the continual changes in the forces (approximately proportional to $1/r^m$) that occur during a collision between two partícles separated by a varying distance $r$. Both the shift and width are proportional to the concentration of the perturbing particles. Attractive forces such as van der Waals forces ($m = 6$) cause a red shift. Behmenburg (24) has shown that blue shifts are predicted by a repulsive force (with $m = 12$) used in addition to the attractive forces. Both red and blue shifts have been observed. The interaction of atoms with others of the same element causes *resonance broadening* ($m = 3$). Interaction with charged particles, such as electrons, causes *Stark broadening*. The linear Stark effect, due to direct coulombic interaction ($m = 2$), is usually important only for hydrogen lines and some helium lines. A quadratic Stark effect ($m = 4$) results from the dipole induced in an absorbing atom by the approach of a charged collision partner.

When $m$ is even, a slight asymmetry is also predicted in the Lorentzian profile. The asymmetry is not present in Eq. (8) and is usually only important when describing the far wings of the profile.

Other processes that limit the time the absorbing atom has to interact with the laser beam will also cause broadening. *Transit broadening* occurs when a rapidly moving atom passes through a narrow laser beam. For instance, an atom with a velocity of $10^6 \, \text{cm s}^{-1}$ that passed through a focused laser beam having a diameter of $100 \, \mu\text{m}$ would be under the influence of the beam for only $10^{-8} \, \text{s}$. Since this is comparable to the natural lifetime of an excited atom, the linewidth would be comparable to the natural linewidth. *Such broadening would be negligible in common analytical cells.* A similar effect would occur if the duration of a laser pulse were very short, so that the photon packet were in contact with the atoms for only a very limited time. A picosecond mode-locked laser pulse has only $10^{-12} \, \text{s}$ in which to interact with an atom. This is considerably shorter than the time between collisions in common analytical cells, and the observed absorption profile broadening (about 0.1 nm at 500 nm) due to the short duration of the laser pulse is therefore considerably greater than collisional broadening.

*Saturation broadening* of the absorption line profile will be observed if the irradiance of a narrowband laser beam is sufficient to cause the excited atom population to deviate from the original value it has in the absence of the laser beam. When the laser is tuned to the wings of the absorption profile, relatively few atoms will be excited, and the absorption of the beam will be the same as is observed with a low-intensity light source. However, as an intense laser beam is tuned toward the center of the profile, the beam will excite more and more atoms, causing the fraction of atoms in the ground state to decrease from its usual value. The absorption of the laser beam near the center of the profile will then be less than that observed with a low-intensity light source. The observed absorption profile curve therefore is broader, and its integral over all wavelengths is less at very high irradiances than at low irradiances. This case is considered in more detail in Sect. 3.3.

Because an intense laser beam can pump a large fraction of a population of atoms to an excited state, absorptions that originate from the excited state can readily be observed. During the laser pumping of the excited state, *stimulated transitions* between the excited state and the ground state occur very rapidly because of the high irradiance by the laser beam. This shortens the effective lifetime of the excited state during the pumping time. Therefore, absorption and spontaneous emission profiles of lines that originate from the excited state will be broader if they are observed during pumping than if they are observed shortly after pumping is terminated.

### 3.2.3 The Voigt Profile

When a hyperfine component in a uniform sample cell is broadened by both Lorentzian and Gaussian (equilibrium Doppler) effects, a good approximation to the absorption profile may be obtained by assuming that every velocity subset of atoms within the wavelength band $d\lambda'$ about wavelength $\lambda'$ in the Gaussian-shaped Doppler profile centered at $\lambda_0 + \delta$ is broadened to the same extent by the Lorentzian profile centered at $\lambda'$. The absorption coefficient $k(\lambda)$ for photons with wavelength $\lambda$ is due to the absorption caused by all of the velocity subsets. To calculate the total absorption coefficient, the absorption coefficient of each velocity subset is calculated from the Lorentzian distribution, Eq. (8), with $\lambda_0 + \delta$ replaced by $\lambda'$. The result is multiplied by the fraction of all atoms within that velocity subset, which is obtained from the product of $d\lambda'$ and a normalized Gaussian distribution $g(\lambda')$:

$$g(\lambda') = \frac{2}{\Delta\lambda_D}\left(\frac{\ln 2}{\pi}\right)^{1/2} \exp(-y^2) \qquad (11)$$

where

$$y \equiv [\lambda' - (\lambda_0 + \delta)]2(\ln 2)^{1/2}/\Delta\lambda_D \qquad (12)$$

and the integral of $g(\lambda')$ over all wavelengths is unity, to account for the total atom population. The result for the total absorption coefficient for photons with wavelength $\lambda$ is

$$k(\lambda) = k(\lambda_0)\int_{-\infty}^{\infty} g(\lambda')\frac{1}{1+\left(\dfrac{2(\lambda - \lambda')}{\Delta\lambda_L}\right)^2}\, d\lambda' \qquad (13)$$

Equation (13) is usually rearranged and presented in the form

$$k(\lambda) = k(\lambda_0)\frac{a}{\pi}\int_{-\infty}^{\infty} \frac{\exp(-y^2)}{a^2 + (\omega - y)^2}\, dy = k(\lambda_0) V(a,\omega) \qquad (14)$$

where we define

$$a = \frac{\Delta\lambda_L}{\Delta\lambda_D} (\ln 2)^{1/2} \tag{15}$$

$$\omega = \frac{[\lambda - (\lambda_0 + \delta)]}{\Delta\lambda_D} 2(\ln 2)^{1/2} \tag{16}$$

$$y = \frac{[\lambda' - (\lambda_0 + \delta)]}{\Delta\lambda_D} 2(\ln 2)^{1/2} \tag{17}$$

and from Eqs. (7) and (9) we find and make use of the fact that

$$k(\lambda_0) = k(\lambda_0, \delta) a(\pi)^{1/2} \tag{18}$$

The $a$ parameter is essentially the ratio of the widths of the Lorentzian and Gaussian profiles, since $(\ln 2)^{1/2} = 0.833$. The coefficient $V(a, \omega)$ of $k(\lambda_0)$ in Eq. (14) is the Voigt integral, for which tabulated values are available (10,25). The Voigt integral for values of $a$ greater than 0.5 can be evaluated numerically to better than six significant figures by using 80 equal intervals to integrate $y$ between the limits of $-4$ to $+4$. These narrow limits can be used because the exponential term in Eq. (14) dies away very rapidly on either side of $y = 0$. For smaller values of $a$, the Lorentzian part of the profile becomes narrower, and more intervals are needed to maintain a sufficient number of intervals within the width of the Lorentzian profile to accurately describe that part of the function. For instance, 400 intervals are needed for $a = 0.1$ to maintain five-digit accuracy.

Strictly speaking, the Gaussian and Lorentzian broadening processes are not completely independent of each other, as implied by the Voigt convolution integral, because both processes depend upon atom velocity (26). However, Mizushima (27) has shown that this interdependence is negligible for the central region of the absorption coefficient profile. Caution should be exercised in trusting the Voigt profile in the wings of the line profile. It should also be emphasized that the symmetric Lorentzian function does not predict asymmetry that may be noticeable in the far wings.

### 3.2.4 Absorption Cell Line Profiles

The relative importance of each of the broadening mechanisms is now considered for several types of sample absorption cells. It is worthwhile to recall that the emission profile in the absence of self-absorption is identical in shape to the profile of the absorption coefficient observed with low-intensity (nonsaturating) light sources. Conclusions concerning the absorption profile may therefore be based upon observations of nonself-absorbed emission profiles.

Doppler broadening accompanies the motion of free atoms and is present to a significant extent in all high- and low-pressure sample cells that are ordinarily used for analytical atomic absorption measurements. It is typically one or two decades greater than natural broadening and is comparable to collisional broadening in sample cells operating near atmospheric pressure. Doppler broadening is the major source of broadening in hollow-cathode discharges and other typical low-pressure sample cells. The overall absorption line profile often exhibits several peaks due to hyperfine structure (Sect. 3.2.1.). Non-Gaussian Doppler broadening or *net* Doppler shifts of the profile may occur in sample cells where the entire absorbing population may not be in translational equilibrium. Net Doppler shifts have been observed in the self-absorption profiles of pulsed hollow-cathode lamps (22). For the study of closely spaced hyperfine components, Doppler broadening can be significantly reduced by the use of atomic beams. This technique introduces atoms into a vacuum through a series of apertures that produces a nearly unidirectional beam of atoms. The laser beam intersects the atomic beam at a single angle. In this case the main sources of broadening may be natural broadening, Doppler broadening caused by angular divergence of either the atomic beam or the laser beam, and transient broadening due to the transit time of the atoms through the laser beam.

In addition to Doppler broadening, atmospheric flame cells produce collisional broadening and shifts due to interactions with neutral particles. The linewidths of the Doppler and collisional broadenings are comparable to each other. The shifts due to van der Waals attractive forces are predicted to be about one third (0.36) of the collisional linewidth. Shifts for 14 lines of seven elements in a nitrous oxide–acetylene flame (28) have been observed to be only from 0.1 to 0.3 times the collisional linewidth, indicating the presence of repulsive forces during collisions (24,29) in addition to van der Waals attractive force. Hyperfine components broaden the absorption peak in an atmospheric flame, and in some cases the components are far enough apart to cause resolvable peaks (28,30).

Resistance-heated atomizer cells (carbon rods, graphite furnaces, tungsten filaments, etc.) operating at atmospheric pressure and at or somewhat below flame temperatures should have absorption profiles similar to those in flames. Changes in temperature during sample atomization will cause (minor) temporal variations in the profiles. In small atomizers, transport of a significant fraction of the sample atoms to cooler regions of the cell may occur. The Gaussian (Doppler) part of the profiles from each of the regions with different translational temperatures will combine together to form a non-Gaussian shape.

Stark broadening and shifts are present in atmospheric electrical discharges such as rf plasmas, sparks, and dc arcs. In addition, Doppler and other collisional broadening mechanisms are present as they are in flames.

Resonance broadening, caused by collisions between like atoms, can be ignored in most sample cells because of the low concentration of analyte atoms in the other gases present. An exception may be for a major constituent during the first few nanoseconds of a laser plume (31) when a solid sample is beginning to expand from its superheated state where concentrations approach those of the solid state and little foreign gas is initially mixed in with the analyte atoms.

### 3.2.5 Measuring Line Profiles

One of the advantages of a narrowband tunable dye laser is its ability *to scan across a line to directly observe the entire profile of the absorption coefficient in a sample cell.* Advantages of observing the entire profile during a chemical analysis measurement include the resolution of overlapping absorption lines and are discussed later. A few absorption profiles have been directly measured using an ordinary light source with Zeeman scanning (21) or a continuum source with a synchronized monochromator–interferometer scan (30). Both these techniques suffer from some experimental complexities and from limited wavelength resolution that usually requires deconvoluting the experimental results with a source or instrument function to obtain the true absorption profile. They have not been used routinely for analytical measurements. When feasible, it is easier to observe the emission profile under conditions of no self-absorption and use this profile as the profile of the absorption coefficient (28). This technique usually involves a deconvolution procedure also.

A laser with a spectral bandwidth that is wider than an atomic absorption line could be used as a continuum source to help observe the atomic absorption line profile. Preliminary results have been reported by Shimazu, Takubo, and Yoshii (32). A Fabry-Perot interferometer is used in the ordinary way (30) to select and scan the narrow wavelength region viewed by the detection system. The entire spectral bandwidth of the laser beam should be contained within one free spectral range (mode or order) of the Fabry-Perot interferometer. Using a laser in this way has several advantages over using ordinary light sources. The laser beam can be of high enough intensity to eliminate or minimize shot noise and can be tuned to almost any desired wavelength. Time resolved absorption profiles could be measured with better precision because of reduced shot noise. The collimated laser beam can be made to pass through a small selected area of the interferometer plates, allowing higher finesse (and wavelength resolution) to be achieved than is ordinarily possible when the entire aperture of the plates is used. High finesse (>40) could eliminate the need for deconvoluting the interferometer instrument function if the absorption profile were not too complex.

A narrowband tunable dye laser may be used to *directly observe* an absorption profile. Commercial dye lasers are available with total bandwidths whose short-term jitter is less than 10 MHz (3 × $10^{-5}$ nm, $10^{-3}$ cm$^{-1}$), which is 50 to 1000 times narrower than the absorption linewidths in flames (0.1–1 cm$^{-1}$) or even in low-pressure cells such as hollow-cathode lamp sputtering cells (0.05–0.2 cm$^{-1}$). *Routine and rapid measurements of absorption profiles in any sample cell should therefore be possible without the need for deconvolution.*

There is such a wide variety of laser tuning methods available that it is worthwhile to consider some of them, along with problems that may occur. Not all narrowband lasers are ideal line sources. An ideal laser cavity consists of an active lasing medium and two end reflectors that are spaced so that one round trip of the beam between them exactly equals an integer number of wavelengths. An end reflector spacing of 25 cm is equivalent to a round trip distance of $10^6$ wavelengths at a wavelength of 500 nm. If a broadband lasing medium is used, the laser may lase at 500 nm with $5 \times 10^5$ wavelengths between the end reflectors ($10^6$ wavelengths round trip), at 500.0005 nm with 499,999.5 wavelengths between the end reflectors and at any other wavelength that allows exactly an integer number of wavelengths to fit one round trip between the end reflectors. The wavelength difference between adjacent lasing lines or modes is called the longitudinal *mode* spacing (Fig. 3.2) and is usually expressed in units of frequency rather than wavelength, 0.0005 nm = 600 MHz in this case. The relationship between a small difference in optical frequency, $\Delta \nu$ (in Hz), and a small difference in optical wavelength, $\Delta \lambda$ (in nm), may be found by differentiating the relationship between frequency and wavelength, $\nu \lambda = c/n$, to obtain the very close approximation

$$\Delta \nu = \frac{\Delta \lambda}{\lambda^2} \frac{c}{n} \qquad (19)$$

where the wavelength of the line $\lambda$ is in nm, the frequency of the line $\nu$ is in Hz, $n$ is the refractive index, and the speed of light $c$ is $3.00 \times 10^{17}$ nm s$^{-1}$. Small changes in optical frequency are also expressed in wavenumbers (cm$^{-1}$) or millikaisers (1 mK = $10^{-3}$ cm$^{-1}$ = $n/(3.00 \times 10^5$ Hz).

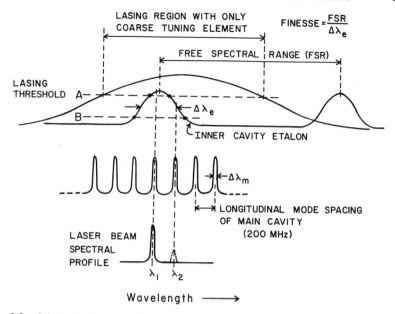

**Fig. 3.2.** Schematic diagram indicating how the spectral characteristics of a laser beam are determined by the lasing threshold and bandpass characteristics of a laser cavity containing an internal etalon and a coarse tuning element such as a birefringent filter.

Optical imperfections in the laser cavity will allow the laser to lase not only exactly at the wavelength of a mode but at wavelengths nearby. Each lasing mode will therefore have a finite spectral bandwidth that becomes the narrowest bandwidth that can be obtained from the laser. Typical imperfections include imperfect surfaces of the end reflectors, variations in the refractive index of the lasing medium and other media in the cavity (including air), and mechanical vibrations of the cavity. A stable, vibration-isolated table may be necessary. Thermal stability of the cavity is also important to minimize slow mechanical and refractive index changes.

A narrowband laser beam is usually obtained by using one or two other wavelength-selective devices within the laser cavity to allow only a *single* longitudinal mode to lase. A coarse tuning device restricts the lasing bandwidth to less than about 0.1 nm (Fig. 3.2). A birefringent filter, grating, or prism is usually used. Gratings tend to be damaged by high-power densities. The lasing region (Fig. 3.2) is controlled by the intersections of the spectral bandpass profile of the tuning element with the threshold condition. The threshold condition is controlled by the net gain of the laser cavity. To obtain a narrower spectral bandwidth, a second device, usually a Fabry-Perot interferometer (or etalon) is inserted in the cavity. This etalon has two dielectric reflecting surfaces usually spaced less than 1 cm apart. As with the laser cavity, this etalon has a longitudinal mode spacing, or *free spectral range*, that allows to pass through it only those wavelengths for which an integer number of wavelengths will fit round trip between the reflecting surfaces. Since other wavelengths are reflected rather than passed, the etalon is tilted very slightly to the axis of the laser beam to prevent reflected rays from entering the lasing medium.

The etalon has a finite spectral bandpass profile whose shape and width, $\Delta\lambda_e$ in Fig. 3.2, depend upon the reflectivity of its surfaces and imperfections similar to those of the laser cavity as discussed above. The ratio of the free spectral range (mode spacing) to the width at half-height of the spectral bandpass profile is called the *finesse*. Ideally, the finesse is a constant, but in practice it may decrease with time or during a scan as imperfections of the etalon become worse. The finesse of the etalon must be high enough so that the shape of its spectral bandpass profile allows the lasing of only one longitudinal mode of the laser cavity. Note in Fig. 3.2 that decreasing the lasing threshold from A to B allows both modes at $\lambda_1$ and $\lambda_2$ to lase. An etalon with higher finesse (narrower profile) would prevent this. To scan an atomic line profile less than a few tenths of a nanometer wide may require only the inner-cavity etalon to be scanned. Narrower lines may require scanning of both the birefringent filter and the laser cavity. Both must be scanned together, however. If only the inner-cavity etalon is scanned, the resulting laser beam output will consist of a sequence of time-varying peaks spaced at wavelength intervals equal to the longitudinal mode spacing of the laser cavity. Scanning only the laser cavity will give a sequence of peaks each of which is scanned only over the wavelength range determined by the bandpass of the inner-cavity etalon.

To synchronously scan the laser cavity and the inner-cavity etalon, one reflector of each of them may be translated along the optical axis by piezoelectric drives. Although the linearity of such drives has been only about 5%, drives with linearities better than 0.3% are now commercially available. During a wavelength scan, the wavelength of the laser beam is controlled by the position of the laser cavity reflector, while the cavity loss, and therefore gain and intensity, is controlled by how well the central wavelength of the bandpass profile of the inner-cavity etalon tracks the laser cavity wavelength. If tracking is not sufficiently accurate, a jump to an adjacent laser mode (several hundred megahertz away) may occur during a scan and give a distorted atomic absorption profile.

A piezoelectric drive may slightly tilt a reflector during a scan, causing the finesse of the laser cavity and inner-cavity etalon to change. Tilting the laser cavity etalon may cause an increase in bandwidth of the laser beam, but this should cause little change in the observed atomic absorption profile if the cavity bandwidth is narrow enough to begin with. A change in finesse of the inner-cavity etalon will not cause trouble unless the shape of the top of the profile of the spectral bandpass of the etalon becomes wide enough to let more than one longitudinal mode of the main laser cavity lase.

The inner-cavity etalon may be mechanically scanned by changing its angle of incidence. The relationship between the angle of incidence $\theta_1$ of the beam passing through a solid etalon of refractive index $n_2$ and the central wavelength $\lambda$ passed by the etalon is given by

$$\lambda = \frac{2t}{m}\left(n_2^2 - n_1^2 \sin^2 \theta_1\right)^{1/2} \tag{20}$$

where $t$ is the thickness of the etalon, $m$ is a large integer, and $n_1$ is the refractive index of the medium outside the etalon (e.g., air). The variables $t$ and $\lambda$ have the same units (e.g., nm). This type of scan is inherently nonlinear and more difficult to synchronize with a piezoelectric scan of the laser cavity reflector. The laser cavity however may also be scanned mechanically by putting an optically transparent plate in the cavity, placed at an angle to the laser beam axis. Rotating the plate scans the laser cavity by causing the optical path between the two end reflectors to change in length. Both the etalon and plate may be mounted on the same platform so that they both rotate through the same angle. Proper choice of the thickness and angle of the plate then causes both the intercavity etalon and the laser cavity to be simultaneously scanned as the platform is rotated (33).

### 3.2.6 Analytical Curves

It is tempting to compare the atomic absorption analytical curves obtained with a narrowband laser to those obtained with the commonly used hollow-cathode lamps. Dye laser linewidths have been achieved that are 100 to 1000 times narrower than the Doppler- and collision-broadened half-widths of absorption peaks in atmospheric flames. Hollow-cathode lines are broadened by equilibrium and nonequilibrium (23,28) Doppler effects and self absorption. Typical hollow-cathode lines are only one to five times narrower than flame lines. Source emission linewidths of this magnitude generally cause the observed transmittance of the sample cell to be less than its maximum value because the absorption coefficient in the flame decreases toward the wings of the lamp line profile. The slight wavelength shift accompanying the collisional broadening in the flame enhances this effect. Therefore the initial slope of the analytical curve is less than could be obtained if a very narrowband laser were tuned to the peak of the flame absorption profile. The relatively wide line profile of a hollow-cathode lamp also causes the analytical curve to bend toward the concentration axis. The shape of the curve can be described by a quadratic equation (34) up to absorbances of 2. A narrowband laser would effectively eliminate this cause of curvature.

Wagenaar, Novotny, and de Galan (34) have studied these influences using measured line profiles of both simple and complex lines of flames and hollow-cathode lamps. They have found that the line curvature caused by even self-reversed hollow-cathode lines causes deviations from a straight-line analytical curve of less than 7% at an absorbance of 1. They also calculated that an infinitely narrow line source (narrowband laser) would improve the slope of the analytical curve by less than 20%. A worse case would occur if a hollow-cathode discharge were used for both the source and cell. Then both the emission and absorption line profiles would be Doppler broadened with equal half-widths. Calculations in this case show that the initial slope of the analytical curve would be improved by less than 30% using an infinitely narrow line source.

Clearly, these improvements in the slope and shape of analytical curves as they are used today *are little justification for simply replacing relatively inexpensive hollow-cathode lamps with narrowband tunable lasers.* On the other hand, the beauty of the tunable laser and its justification in atomic absorption spectroscopy lie in its ability *to make possible new and useful types of analytical atomic absorption measurements* that are not possible, or are at best difficult, with hollow-cathode lamps.

### 3.2.7 Analytical Applications of Line Profile Measurements

The use of an intense tunable dye laser to easily and accurately measure absorption line profiles in sample cells should open the way for new analytical applications that have been difficult or impossible due to the overlap of adjacent absorption line profiles. Compensation for both spectral emission and absorption interferences can be achieved with the high wavelength resolution that is possible. Spectral line interferences, such as the Sm 492.41-nm and Nd 492.45-nm lines, the Co 341.47-nm and Ni 341.48-nm lines, and the OH band system on the Cu 324.75-nm line, could be minimized or compensated by measuring the entire absorption line profiles. Broadband interferences, such as nonspecific scattering, continuum absorption and emission in electrical discharges, the Lorentzian wings of lines from high-concentration elements (e.g., sodium), and unresolved molecular band systems, may also be readily compensated.

Broadband or line interferences may be readily compensated with ordinary narrowband light sources using a blank *if* the interference is present to the same extent for all samples. If the interference varies from sample to sample, a dual-wavelength technique may be used with ordinary light sources to compensate for broadband absorption *if* the interfering absorption at the two wavelengths has a constant ratio. On the other hand, absorption line profiles that are accurately measured with narrowband lasers *should provide the best obtainable compensation for line interferences that vary from sample to sample.*

Another use for highly resolved absorption line profiles where spectral overlap usually occurs is in *isotope analysis*. Isotope analysis by atomic absorption spectroscopy is possible when the isotope shifts are greater than the linewidths in the sample cell. The shifts must be larger than the nuclear spin hyperfine splittings, and the absorption line profiles of the different isotopes must not be overlapped into one peak by collisional broadening. Low-pressure sample cells such as hollow-cathode discharges may be used to minimize collisional broadening.

Using ordinary light sources, atomic absorption methods for the isotope analysis of helium (35), lithium (36–38), mercury (39; see also Fig. 3.1), and uranium (40, 41) have been developed. L'vov (42, p. 260) predicts that isotope analysis for about a dozen elements should be possible using ordinary light sources. The spectral bandwidth of these light sources may cause nonlinear analytical curves, while accurate line profile measurements using a tunable laser should provide linear results. When absorbance is linearly related to concentration, as it is when high-wavelength resolution is used, linear simultaneous equations can be used to obtain the concentrations from overlapping lines. This should allow isotope analysis to be extended to other elements which were thought to have isotope shifts insufficient to be resolved using ordinary light sources. Isotope analysis for even more elements should be possible using atomic absorption saturation spectroscopy, discussed in Sect. 3.3, to reduce the *influence* of Doppler broadening.

When an atomic absorption profile is completely scanned during an analysis, it may be desirable to fit the observed profile to a theoretical equation that describes its shape rather than, for instance, using the absorbances at only two appropriate wavelengths with two simultaneous equations to calculate the concentrations of two species. Curve fitting may improve the precision of the final results by using all of the available data rather than a limited amount. Also, systematic residual differences between the observed and theoretical curves would indicate the presence of an unexpected peak.

When the line positions and profile shapes of the atomic species are well known, the only unknown variables in the equations are the concentrations (peak amplitudes) of the analyte elements. This would considerably reduce the computer time needed to obtain a fit to the observed profiles.

### 3.2.8 Diagnostic Applications of Line Profile Measurements

When the sample cell is reasonably homogenous, atomic absorption line profiles can be used to obtain Doppler temperatures and information about collisional broadening processes (10–12, 28, 42) discussed in Sect. 3.2.2. Emission or fluorescence line profiles could also be used, but care is required to eliminate self-absorption. Both emission (12, 28, 43–46) and absorption (29, 30, 47, 48) line profiles using ordinary light sources have been used. These procedures have required deconvoluting the data with an instrument function whose shape may not be accurately known. A method for accurately evaluating an interferometer instrument function has recently been reported (49). Absorption profiles obtained with the help of a tunable laser should be *relatively* noise free and have negligible instrument function broadening, thereby improving the precision and accuracy of the results and eliminating bias due to an uncertain instrument function.

Even when an accurate absorption coefficient (or absorbance) line profile has been obtained, it is necessary to separate the Gaussian and Lorentzian contributions to the profile to determine the Doppler temperature. The resulting Lorentzian half-width is the linear sum of the half-widths due to all Lorentzian broadening processes. When one hyperfine component is well resolved, the tables of Posener (50) may be used to help determine the relative contributions. When several hyperfine components broaden the line profile, curve fitting techniques are required. Wagenaar, Pickford, and de Galan (30) have determined the Gaussian and Lorentzian contributions for the Cu 325-nm absorption line in an acetylene–air flame by using a relatively simple graphic technique to interpolate mathematical parameters of the experimental absorption profile between values for theoretical profiles generated with the known hyperfine structure of the line.

The high photon rates that can be achieved with tunable lasers, and the resulting low relative shot noise, make them particularly valuable for diagnostic work with transient sample cells. However, interpretation of the absorption line profiles in a transient sample cell is complicated when the temperature and collisional processes are not homogeneous throughout the absorbing region. For instance, the superposition of two pure Gaussian profiles with different bandwidths and peak heights may produce an approximately Voigt-shaped profile (51) rather than a Gaussian profile. In any case, however, the wavelength integral of the observed absorption coefficient profile at a given time (or for a short time interval) is a measure of the concentration–path length product along the optical axis at that point in time and should be useful in helping to evaluate atomic vapor atomization, transport, and loss (condensation) processes. The high cross-sectional spatial resolution that can be obtained with a collimated laser beam should also be helpful in such studies.

A laser can be tuned to wavelengths that are absorbed by ions, radicals, molecules, and excited species as well as atoms. The study of these species should be particularly valuable in helping to understand chemical equilibrium and loss processes of atoms in sample cells. Ions of several rare earths have been observed by fluorescence with a tunable dye laser (52).

### 3.2.9 Improving the Signal-to-Noise Ratio

The use of a high-intensity narrowband tunable laser may improve the signal-to-noise ratio of an atomic absorption measurement in several ways. As discussed in Sect. 3.2.6, an improvement in absorbance (but by less than a factor of 2) is expected if an infinitely narrow laser line source is used to make the absorbance measurement rather than a conventional emission source. This improvement alone would not ordinarily justify the use of a laser.

On the other hand, the high photon rate that is possible with a laser reduces the relative shot noise, particularly for transient sample cell measurements, and also considerably reduces the influences of the signal and noise caused by *emission* from a sample cell. Therefore, shot noise need not be limiting, and simultaneous double-beam compensation for fluctuation noise in a laser beam should be possible with a properly designed optical and electronic system. High-intensity laser beams should extend the usefulness of routine atomic absorption measurements to many high-temperature sample cells that are ordinarily considered as emission sources and to transient sample cells where only a limited amount of time is available to make a measurement.

Shot noise is a major source of uncertainty when ordinary light sources are used for atomic absorption measurements of transient sample cells such as high-voltage sparks and laser plumes (53). It is also an important source of noise for many elements when using transient sample cells that allow longer measurement times such as electrically heated resistance furnaces (e.g., carbon rods, graphite furnaces, and tungsten filaments and ribbons). In these cases, the relative shot noise could be made negligible by using a high-intensity narrowband laser beam as the primary source.

The influence of both the sample cell emission signal and the noise accompanying this signal can also be reduced by using a laser beam that has a much higher photon rate than the sample cell emission photon rate. The cell emission thereby becomes a negligible part of the total signal observed by the detection system. More sensitive atomic absorption measurements should therefore be possible in the visible region where sample cell emission is significant in flame cells, and in high-temperature plasma discharges where emission is very high. If a *pulsed* laser is used as the light source, the detection system should be *gated* to be open only during the laser pulse. Otherwise, additional and unwanted emission from the sample cell will unnecessarily contribute to the final signal.

When a high-intensity laser beam is used to reduce relative shot noise to a negligible value, noisy fluctuations and drift in the laser intensity can be compensated by a *double-beam technique* that measures the intensity of the beam both before and after it passes through the sample cell and produces a signal proportional to the intensity ratio of the two beams. Best results should be obtained by using a uniform beam splitter and continuous observation of both beams rather than a chopped, alternating double-beam system. The temporal response characteristics of the signal channels for both beams should be identical. The response time should be fast enough to follow the noisy fluctuations, yet long enough so that shot noise fluctuations are less than other fluctuations. The two detectors must observe identical regions of the beams with reproducibly proportional sensitivities. As already stated, since the sensitivities of detectors may vary erratically over their surfaces (54), identical diffusers should be placed in front of each detector so that each part of a beam illuminates all parts of the detector surface. This is an ideal requirement that may be difficult to achieve in practice. (Obviously, there is a need for detectors with uniform sensitivities across their surfaces.) If these conditions are met, the noise in the final signal should be mainly due to noise generated in the detection system and noise due to absorption and scattering fluctuations in the sample cell. Detection system noise is most important near the detection limit (low absorbances), while sample cell fluctuation noise becomes important at high absorbances (55).

If it is necessary to obtain a laser beam with a high enough intensity, the laser oscillator may be followed by a laser amplifier. However, an upper limit exists to the advantages to be gained by increasing the laser intensity for a sample cell having a limited cross-sectional area. If the *irradiance* (W cm$^{-2}$) at wavelengths within the absorption line profile in the sample cell is too great, the rate of transitions to the excited level of the absorption line will begin to cause a significant loss in the fraction of atoms in the lower level. When a large fraction of the atom population is in the excited state, the condition is known as *saturation* (Sect. 3.3). The initial slope of the analytical curve will then be reduced although it will remain essentially linear at low absorbances. The analytical curve will become nonlinear at high absorbance values, in part because the far side of the sample cell will experience a lower irradiance (due to beam absorption on the near side of the cell), and therefore a different fraction of atoms in the lower level, than the near side of the cell that receives the full laser irradiance. When a sample cell has a limited cross section, a compromise is obviously necessary between a reduction in noise and a loss in slope and departure from linearity of the analytical curve that accompanies high irradiance.

## 3.3  SATURATION EFFECTS

The improvement in signal-to-noise ratio that could be achieved by using high-intensity laser beams to increase the photon rate for transient atomic absorption measurements was discussed in Sect. 3.2.9. If the cross section of the sample cell is limited, the irradiance, or photon rate per cm$^2$, for wavelengths within the absorption line profile might become high enough to reduce the absorption coefficient and therefore change the *slope* and *shape* of the analytical curve. We shall now consider the influence of irradiance upon the absorption coefficient in more detail.

The absorption coefficient of a population of atoms or other absorbers is generally dependent not only upon wavelength and the oscillator strength, or the Einstein coefficients of the transition, but also upon the fraction of atoms that are in the lower (absorbing) state. The photon flux caused by conventional light sources is so low that an atom which absorbs a photon usually returns to its lower level long before another photon comes along to be absorbed by that atom. Consequently, the fraction of atoms in the lower energy level deviates very little from its equilibrium value, regardless of the magnitude of the low irradiance caused by the source. *Therefore, the absorption coefficient appears to be independent of the irradiance for low irradiances.*

As the irradiance is increased, the rate at which atoms are excited by absorbing photons increases, and the fraction of atoms that are found in the upper level at any time increases. When a laser is used, the irradiance may become so high that a large fraction of atoms will be excited. The influence of irradiance upon the absorption coefficient will then become apparent.

The absorption coefficient is not only dependent upon the fraction of atoms in the lower level but also directly upon the fraction in the upper level. This latter dependence is due to the influence of stimulated emission, which causes an excited atom to emit a photon in the direction of the incident beam. Excited atoms can thereby add photons to the beam, just as ground state atoms can remove photons from the beam. The absorption coefficient is therefore directly proportional to the *difference* between the fraction of atoms in the upper state and the fraction of atoms in the ground state. At very high irradiances, the fraction of atoms in the upper state will approach in value the fraction of atoms in the lower state, the rate of stimulated emissions will approach the rate of (stimulated) absorptions, and the absorption coefficient will become very small. The system is said to be approaching *saturation.*

We shall first use a rate equation approach to determine the spectral profile of the absorption coefficient for a group of atoms that all have exactly the same Doppler (axial) velocity with respect to a well-collimated (parallel), *monochromatic* laser beam. The result will show that the spectral profile is a Lorentzian function whose half-width increases and whose peak value and wavelength integral decrease with increasing irradiance. The influence of Doppler broadening is then considered and will be found to be more complex to deal with than for low irradiances. In particular, collisions that cause a change in the Doppler velocity of an excited atom without quenching the atom must now be taken into account. We then discuss a two-beam technique called *saturation spectroscopy* that can improve wavelength resolution by reducing the influence of Doppler broadening.

The absorption coefficient $k$ is usually defined as

$$E = E_0 \exp(-kx) \qquad (21)$$

where $x$ is the thickness of the absorbing layer, $E_0$ is the (cross-sectional) irradiance $(W\,cm^{-2})$ of the collimated incident beam, and $E$ is the irradiance of the beam after it has passed through the absorbing layer. This equation may be written in differential form as follows:

$$k = -\frac{dE/dx}{E} \qquad (22)$$

which shows how the absorption coefficient is related to the irradiance lost $-dE$ by a beam passing through a thin layer of thickness $dx$. The absorption coefficient will now be expressed in terms of intrinsic parameters of the absorbing atom that are related to $-dE$. The irradiance $E$ of a beam is obtained from the spectral irradiance $E_\lambda$ $(W\,cm^{-2}\,nm^{-1})$ by multiplying the spectral irradiance by the spectral bandwidth (nm) under consideration. The irradiance of a parallel beam of photons with wavelengths in the range of $\lambda$ to $\lambda + d\lambda$ is therefore $E_\lambda\,d\lambda$.

The rate at which this beam induces a transition from energy level $i$ to energy level $j$ is given by

$$B_{ij}(\lambda)E_\lambda\,d\lambda \qquad \text{(transitions s}^{-1}\,\text{atom}^{-1}) \qquad (23)$$

where $B_{ij}(\lambda)$ is the probability (per second) per unit irradiance that a transition will occur when the atom is stimulated by photons with frequencies in the spectral range $\lambda$ to $\lambda + d\lambda$. Quantum mechanical considerations show that $B_{ij}(\lambda)$ has a Lorentzian shape given by

$$B_{ij}(\lambda) = B_{ij}\mathscr{L}(\lambda,\lambda_0) \qquad (24)$$

where $\mathscr{L}(\lambda,\lambda_0)$ is the Lorentzian function

$$\mathscr{L}(\lambda,\lambda_0) = \frac{1/2\pi}{1 + \left[\dfrac{2(\lambda - \lambda_0)}{\Delta\lambda_L}\right]^2} \qquad (25)$$

which has an integral over all wavelengths equal to unity and a width at half-maximum of $\Delta\lambda_L$. The half-width is determined by transition rate constants as shown by Eq. (10), and $\lambda_0$ is the central wavelength of the function, that is, the wavelength of a monochromatic beam that will cause the strongest interaction between the beam and the atom.

The total probability (per second) for an induced transition can now be obtained by integrating Eq. (23) over all wavelengths. If a *continuum source* is used and the resulting spectral irradiance is constant over the wavelength range where the atom can absorb, then the total transition probability (per second) becomes

$$\int_\lambda B_{ij}(\lambda)E_\lambda\,d\lambda = E_\lambda B_{ij}\int\mathscr{L}(\lambda,\lambda_0)\,d\lambda = E_\lambda B_{ij} \qquad \text{(transitions s}^{-1}\,\text{atom}^{-1})$$

$$(26)$$

where $B_{ij}$ is the Einstein transition probability (per second) per unit irradiance for a stimulated transition.

When a line source is used and its spectral bandwidth is much narrower than the width $\Delta\lambda_L$ of the absorption line, $B_{ij}$ is effectively constant for nonzero values of $E_\lambda$ and is removed from under the integral. In this case the source may be considered to be *monochromatic* and the total transition probability (per second) becomes

$$\int_\lambda B_{ij}(\lambda)E_\lambda\,d\lambda = E(\lambda)B_{ij}\mathscr{L}(\lambda,\lambda_0) \qquad \text{(transitions s}^{-1}\,\text{atom}^{-1}) \qquad (27)$$

where $E(\lambda) = E_\lambda \, d\lambda$ is the irradiance at wavelength $\lambda$ due to the mono-chromatic beam. Equation (27) shows that the transition probability for this case depends upon the wavelength of the monochromatic source as indicated by the Lorentzian dispersion function $\mathscr{L}(\lambda,\lambda_0)$.

The net number of photons lost per second from a *monochromatic* collimated beam that produces an irradiance $E(\lambda)$ when passing through a unit volume having $n_2$ excited atoms per cm$^3$ and $n_1$ absorbing (e.g., ground state) atoms per cm$^3$ is equal to the difference between stimulated absorptions and stimulated emissions which occur each second per unit volume, or

$$n_1 E(\lambda) B_{12} \mathscr{L}(\lambda,\lambda_0) - n_2 E(\lambda) B_{21} \mathscr{L}(\lambda,\lambda_0) \qquad \text{(transitions s}^{-1}\,\text{cm}^{-3}\text{)}$$

$$(28)$$

where the rate constants by which $n_1$ and $n_2$ are multiplied have been obtained from Eq. (27). *Spontaneous emissions have been ignored because they are usually relatively infrequent and they go in all directions.* The decrease in irradiance $-dE(\lambda)$ of the beam is obtained by multiplying this expression by the energy (W s) per photon $h\nu$ (where $h$ is the Planck constant in J-s and $\nu$ is the frequency of the electromagnetic wave in Hz) and the thickness of the absorbing layer $dx$ to obtain

$$-dE(\lambda) = h\nu[n_1 E(\lambda) B_{12} \mathscr{L}(\lambda,\lambda_0) - n_2 E(\lambda) B_{21} \mathscr{L}(\lambda,\lambda_0)] \, dx \qquad \text{(W cm}^{-2}\text{)}$$

$$(29)$$

Equation (29) can be rearranged to give

$$k(\lambda) = \frac{-dE(\lambda)/dx}{E(\lambda)} = h\nu[n_1 B_{12} - n_2 B_{21}]\mathscr{L}(\lambda,\lambda_0) \qquad (30)$$

which by analogy with Eq. (22) is recognized as being equal to the absorption coefficient for photons of wavelength $\lambda$.

The $B_{ij}$ coefficients are related to $A_{21}$, the Einstein transition probability (per second) for spontaneous emission from state 2 to state 1, by

$$A_{21} = \frac{8\pi h c^2}{\lambda^5} B_{21} \qquad (31)$$

and

$$g_1 B_{12} = g_2 B_{21} \qquad (32)$$

where $c$ is the speed of light and the degeneracy factors $g_1$ and $g_2$ are the number of electrons that can occupy energy levels $E_1$ and $E_2$, respectively. It is worthwhile to emphasize that $B_{ij}$ *has been defined in the wavelength rather than frequency domain,* and in terms of irradiance rather than power density or intensity.

Equations (30), (31), and (32) can now be combined to express the absorption coefficient in terms of $A_{21}$:

$$k(\lambda) = \frac{\lambda^4 A_{21}}{8\pi c} \left( \frac{g_2}{g_1} n_1 - n_2 \right) \mathscr{L}(\lambda, \lambda_0) \tag{33}$$

The absorption coefficient therefore has a Lorentzian spectral distribution due to $\mathscr{L}(\lambda, \lambda_0)$. Another important observation is that the absorption coefficient depends upon the difference between the number of atoms in the two energy levels (with due consideration for the degeneracies).

For low irradiances and low thermal excitation, $n_2$ is negligible compared to $n_1$. At higher irradiances, a significant number of atoms may be excited, and $n_2$ becomes important in Eq. (33). When the irradiance has been on for a long period of time compared to the lifetimes of the relaxation processes of the energy levels, the system has had a chance to reach a steady state. Under steady-state conditions the number of excitations per second equals the number of de-excitations per second in a unit volume:

$$n_1[k_{12} + B_{12}E(\lambda)\mathscr{L}(\lambda, \lambda_0)] = n_2[k_{21} + A_{21} + B_{21}E(\lambda)\mathscr{L}(\lambda, \lambda_0)] \tag{34}$$

where $k_{ij}$ is the rate constant (transitions s$^{-1}$ atom$^{-1}$) for collisions that cause transitions from energy level $i$ to level $j$. When $g_1 = g_2$, Eqs. (33) and (34) can be combined to give

$$k(\lambda) = \frac{\lambda^4 A_{21}}{8\pi c}(n_1 + n_2)\left[\frac{k_{21} + A_{21} - k_{12}}{k_{21} + A_{21} + k_{12} + 2E(\lambda)B_{12}\mathscr{L}(\lambda, \lambda_0)}\right]\mathscr{L}(\lambda, \lambda_0)$$

$$\tag{35}$$

From the denominator in the brackets, the irradiance $E(\lambda_0)_{sat}$ (using photons of wavelength $\lambda_0$) that reduces the absorption coefficient to one half its original value at zero irradiance is found to be

$$E(\lambda_0)_{sat} = \frac{k_{21} + A_{21} + k_{12}}{2B_{12}} \tag{36}$$

since $\mathscr{L}(\lambda_0, \lambda_0) = 1$. This is the *saturation irradiance* defined by Greenstein and Bates (14) and by Kuhl, Neumann, and Kriese (56) for a monochromatic beam, and it is similar to the *saturation spectral irradiance* defined by Omenetto, Benetti, Hart, Winefordner, and Alkemade (57) for a continuum source.

The peak absorption coefficient, $k(\lambda_0)$, for low irradiance is found by setting $\lambda = \lambda_0$ and $E(\lambda) = 0$ in Eq. (35). Substituting the result back into Eq. (35) gives a simplified form of Eq. (35):

$$k(\lambda) = k(\lambda_0) \frac{\mathscr{L}(\lambda, \lambda_0)}{1 + [E(\lambda)/E(\lambda_0)_{sat}]\mathscr{L}(\lambda, \lambda_0)} \tag{37}$$

This equation shows how the spectral profile of the absorption coefficient changes with irradiance. By using Eq. (25) for $\mathcal{L}(\lambda,\lambda_0)$, it can be shown that the fraction on the right-hand side of Eq. (37) is also a Lorentzian function which, compared to $\mathcal{L}(\lambda,\lambda_0)$, has a maximum peak height that is reduced by the factor $[1+(E(\lambda)/E(\lambda_0)_{sat})]^{-1}$, a peak width that is increased by the factor $[1+(E(\lambda)/E(\lambda_0)_{sat})]^{1/2}$, and an area that is reduced by the factor $[1+(E(\lambda)/E(\lambda_0)_{sat})]^{-1/2}$.

For any constant irradiance, Eq. (35) shows that the absorption coefficient is still proportional to the total concentration of absorbing atoms. However, when the irradiance is significantly decreased by absorption as the beam passes through the sample, $E(\lambda)$ will decrease with $x$, and the linear relationship suggested by Eq. (35) will no longer be valid.

### 3.3.1 Absorption of Strong Monochromatic Irradiation by a Thick Sample

The transmittance of a sample that is strongly irradiated is not only dependent upon the total sample concentration (via $k(\lambda_0)$ but is also dependent upon the strength of the irradiation, as indicated in previous paragraphs. The way in which the strength of the irradiation influences the transmittance can be readily shown by considering the case of a monochromatic beam and a sample that has negligible Doppler broadening. The absorption coefficient for this case can be obtained from Eq. (37). The relative loss in irradiance for a beam that passes through an infinitely small increment $dz$ of sample thickness is given by

$$\frac{dE(\lambda,z)}{E(\lambda,z)} = -k(\lambda)\,dz \qquad (38)$$

Substituting Eq. (37) into Eq. (38), separating the variables, integrating the irradiance between the limits of the incident and transmitted irradiances $E(\lambda,0)$ and $E(\lambda,b)$, and integrating the distance $z$ traveled by the beam from $z=0$ to $z=b$ gives

$$-\ln T = [k(\lambda_0)b - (1-T)E(\lambda,0)/E(\lambda_0)_{sat}]\mathcal{L}(\lambda,\lambda_0) \qquad (39)$$

where $T = E(\lambda,b)/E(\lambda,0)$ is the transmittance of the sample. The logarithm of the transmittance is not linearly related to $k(\lambda_0)$ and therefore is not linearly related to concentration, because of the second term in brackets. However, when $E(\lambda,0) \ll E(\lambda_0)_{sat}$, the irradiance causes negligible saturation, and Eq. (39) reduces to the well-known Beer-Lambert law.

If we define $T_0$ as the transmittance of the sample for the case of

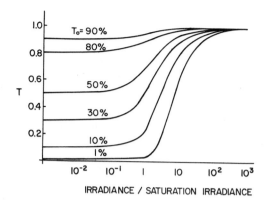

**Fig. 3.3.** Theoretical curves showing the influence of sample irradiance upon observed transmittance $T$. The transmittance at low irradiances is equal to $T_0$. From Kuhl, Neumann, and Kriese (56), reproduced by permission.

negligible irradiance, then $-\ln T = k(\lambda_0)b\mathscr{L}(\lambda,\lambda_0)$ and Eq. (39) is equivalent to

$$\ln(T/T_0) = (1-T)\frac{E(\lambda,0)}{E(\lambda_0)_{sat}}\mathscr{L}(\lambda,\lambda_0) \qquad (40)$$

Equation (40) is transcendental in $T$, and Fig. 3.3 shows how $T$ varies with the irradiance term $E(\lambda,0)\mathscr{L}(\lambda,\lambda_0)/E(\lambda_0)_{sat}$ for samples having several values of low-irradiance transmittance $T_0$. The curves in Fig. 3.3 predict that samples with any low-irradiance $T_0$ will asymptotically approach complete transparency ($T = 1$) for sufficiently high irradiances.

### 3.3.2 Doppler Broadening in Cases of Strong Irradiance

Doppler broadening is significant in all commonly used atomic absorption sample cells, but its influence on the spectral profile of the absorption coefficient does not always produce the Voigt function that is commonly used to describe the profile under equilibrium low-irradiance conditions. The fraction of excited atoms will be larger in a Doppler velocity subset that has its central wavelength $\lambda_0$ nearer the laser beam wavelength $\lambda$ than in a Doppler velocity subset that has its central wavelength far removed from the laser beam wavelength. Equation (37) could be used to help determine the absorption coefficient for each velocity subset, except that there is now the possibility that collisions will change the velocity of an excited atom, causing it to jump from one Doppler velocity subset to another. If the atom is not quenched (de-excited to the lower level) by the collision, such a collision provides an additional mechanism to increase the fraction of excited atoms in Doppler velocity subsets with resonant wavelengths that are not necessarily near the laser wavelength. The fraction of excited atoms in a far removed Doppler velocity subset is larger than it would be without such velocity-changing collisions, and the absorption coefficient predicted by Eq. (37) for that subset is too large.

Before considering line profiles for such cases, let us consider one case where Eq. (37) does apply and which may be applicable to flames commonly used in analytical atomic absorption spectrometry. This is the case where an excited atom generated in one Doppler velocity subset does not have its Doppler velocity significantly changed by collisions before it leaves the excited energy level. The change in Doppler shift must be much less than the Doppler profile width and also less than the Lorentzian profile width. This case would also occur if any collision that causes a significant change in Doppler velocity also quenched an excited atom to the lower energy level (58). This would seem to be the case for common acetylene-fueled analytical flames where the quenching rate is high. *The transfer of excited atoms between Doppler velocity subsets would be negligible in this case*, and Eq. (37) could be used to give the absorption coefficient for each Doppler velocity subset. Assuming a Maxwellian velocity distribution for the population of atoms, the relative number of atoms in each velocity subset that has been shifted to frequency $\lambda_0$ is given by a Gaussian distribution $g(\lambda_0,\lambda_D)$, where $\lambda_D$ is the center of the distribution. The absorption coefficient for photons of wavelength $\lambda$ for each of the Doppler groups is weighted by the Gaussian function $g(\lambda_0,\lambda_D)$ and integrated (added) together to obtain the total absorption coefficient $k(\lambda)$ due to absorption of photons of wavelength $\lambda$ by all Doppler subsets. The resulting absorption coefficient is expressed as

$$k(\lambda) = k(\lambda_0) \int g(\lambda_0,\lambda_D) \left[ \frac{\mathscr{L}(\lambda,\lambda_0)}{1 + (E(\lambda)/E(\lambda_0)_{sat})\mathscr{L}(\lambda,\lambda_0)} \right] d\lambda_0 \qquad (41)$$

This is a Voigt integral where the function within the brackets is the Lorentzian function, whose broader half-width and shorter peak height were compared to the Lorentzian function $\mathscr{L}(\lambda,\lambda_0)$ several paragraphs earlier. The $a$ parameter for this Voigt integral is $[1 + E(\lambda)/E(\lambda_0)_{sat}]^{1/2}$ times greater than the $a$ parameter for cases where the irradiance $E(\lambda)$ approaches zero (and therefore the Lorentzian part of the integral approaches $\mathscr{L}(\lambda,\lambda_0)$ divided by unity). The spectral profile of the observed absorption coefficient is therefore broader (and reduced in height) when the beam strongly irradiates the population than when the irradiation is weak. In the far wings of the spectral profile, however, the term in brackets in Eq. (41) approaches $\mathscr{L}(\lambda,\lambda_0)$ because $\mathscr{L}(\lambda,\lambda_0)$ in the denominator is small (providing the ratio $E(\lambda)/E(\lambda_0)_{sat}$ is not too large). Therefore, the absorption coefficients for the cases of weak and of moderately strong irradiation approach each other in value in the far wings of the spectral profiles. In the limit of zero irradiance, Eq. (41) reduces to Eq. (14).

### 3.3.3 Irradiation with an Arbitrary Spectral Profile

Irradiation that is not purely monochromatic can be described by its spectral irradiance $E(\lambda)$ (irradiance per unit wavelength interval) so that the total irradiance $E$ is

$$E = \int E_\lambda \, d\lambda \tag{42}$$

The beam consists of adjacent wavelength intervals each $d\lambda$ wide and each having an irradiance $E_\lambda \, d\lambda$. The irradiance in each wavelength interval contributes to the saturation of each Doppler velocity subset, causing the fraction of excited atoms to increase and the fraction of absorbing atoms in the subset to decrease. When there is negligible transfer of excited atoms between Doppler subsets, the total saturation due to all the wavelength intervals is obtained by integrating over all wavelengths in the beam. The denominator in Eq. (41) (the saturation term) now becomes

$$1 + (1/E(\lambda_0)_{sat}) \int \mathscr{L}(\lambda,\lambda_0) E_\lambda \, d\lambda \tag{43}$$

Following the approach of Greenstein and Bates (14), let us now define a dimensionless function

$$L(\lambda_r,\lambda_0) \equiv \frac{\int \mathscr{L}(\lambda,\lambda_0) E_\lambda \, d\lambda}{\int E_\lambda \, d\lambda} \tag{44}$$

where $\lambda_r$ is the center or characteristic frequency of the laser beam spectral irradiance distribution $E_\lambda$. The value of $L(\lambda_r,\lambda_0)$, between zero and unity, will depend upon the irradiation lineshape but will be independent of the total irradiance $E$. The saturation term, Eq. 43, can now be written

$$1 + (E/E(\lambda_0)_{sat}) L(\lambda_r,\lambda_0) \tag{45}$$

which expresses the expected fact that *the total irradiance $E$ of the polychromatic beam has the same influence on the saturation of atoms within a* *velocity subset as the total irradiance* $E(\lambda)$ *of a monochromatic beam, provided that the shape of the spectral profile of the irradiation is constant.* The absorption coefficient for a velocity subset of atoms with a resonant wavelength $\lambda_0$ absorbing a monochromatic beam of wavelength $\lambda$ can now be written as

$$k(\lambda,\lambda_0) = \frac{k(\lambda_0)\mathscr{L}(\lambda,\lambda_0)}{1+(E/E(\lambda_0)_{sat})L(\lambda_r,\lambda_0)} \tag{46}$$

The total absorption coefficient for a velocity subset of atoms for the case of a polychromatic beam can be defined as the relative decrease $dE/E$ in irradiance per increment of sample thickness $dz$, or

$$k_{poly} \equiv -\frac{dE/dz}{E} \tag{47}$$

The relationship of this absorption coefficient for a polychromatic beam to the absorption coefficient in Eq. (46) (for a monochromatic beam where the sample is irradiated by a polychromatic beam) will now be considered since the latter is related to concentration via $k(\lambda_0)$. The relationship can be substantially simplified by considering the case where we may neglect the change in $k(\lambda,\lambda_0)$ with $z$, due to the presence of $E$ in Eq. (46), as the beam propagates through a thick sample. This simplification is valid when no more than, for instance, about 10% of the beam is absorbed as it passes through the sample. (When a larger fraction of the beam is absorbed, both the irradiance $E$ and the irradiation lineshape, which influences $L(\lambda_r,\lambda_0)$, change with $z$).

The denominator in Eq. (47) is given by Eq. (42), and the numerator is obtained by integrating over all wavelengths in the beam, the decrease in irradiance $dE_\lambda \, d\lambda$ for each wavelength interval per increment of sample thickness, or

$$dE/dz = \int (dE_\lambda/dz) \, d\lambda \tag{48}$$

The decrease in irradiance $d(E_\lambda \, d\lambda)$ for each wavelength interval is equal to $k(\lambda,\lambda_0)$ for that interval, times $E_\lambda \, d\lambda$. Equation (48) then becomes

$$dE/dz = \int E_\lambda k(\lambda,\lambda_0) \, d\lambda \tag{49}$$

and the polychromatic absorption coefficient, Eq. (47), becomes

$$k_{poly} = \frac{-\int E_\lambda k(\lambda,\lambda_0) \, d\lambda}{\int E_\lambda \, d\lambda} \tag{50}$$

Substituting Eq. (46) into Eq. (50) and using an analogy with Eq. (44) gives

$$k_{poly} = \frac{k(\lambda_0)L(\lambda_r,\lambda_0)}{1+(E/E(\lambda_0)_{sat})L(\lambda_r,\lambda_0)} \tag{51}$$

for the polychromatic beam absorption coefficient for each Doppler velocity subset of atoms irradiated by the polychromatic beam. The polychromatic beam absorption coefficient due to all Doppler velocity subsets, $\bar{k}_{poly}$, is obtained by integrating over all subsets to obtain

$$\bar{k}_{poly} = k(\lambda_0)\int g(\lambda_0,\lambda_D)\left[\frac{L(\lambda_r,\lambda_0)}{1+(E/E(\lambda_0)_{sat})L(\lambda_r,\lambda_0)}\right]d\lambda_0 \tag{52}$$

Note that in the limiting case of monochromatic irradiation $E_\lambda$ behaves like a delta function in Eq. (44), and Eq. (52) reduces to Eq. (41). It should be emphasized that Eq. (52) was derived for cases where no excited atoms are interchanged between Doppler subsets and is valid for cases where only a small fraction (say, less than 0.1) of the polychromatic beam is absorbed. For these cases, the equations show that the influences of the total irradiances $E$ and $E(\lambda)$ upon the magnitude of the observed absorption coefficients are very similar.

Opposite to the case where collisions cause no excited atoms to be transferred between Doppler velocity subsets is the case where excited atoms change velocities so rapidly (without quenching) that all Doppler velocity subsets have the same fraction of excited atoms. That is, the pumping is spread over all atoms rather than being essentially limited to Doppler velocity subsets that are near the laser excitation wavelength. McIlrath and Carlsten (15) have considered this case (their case 3) with the additional requirement that (dephasing) collisions that interrupt the coherence of the interaction between the radiation field and the atom be much more frequent than velocity-changing collisions. This requirement is often met in analytical atomic absorption spectroscopy and allows correlations between the various types of collisions to be ignored. For instance, they suggest that this case is applicable for excited-state lifetimes (due to quenching collisions and spontaneous emissions) on the order of $10^{-8}$ s if a relatively heavy, nonquenching buffer gas is used at atmospheric pressure to encourage a frequent change in velocities for the excited atoms (before they decay to their lower energy level). This would seem to be the case for electrically heated furnace sample cells that used argon as a buffer gas near atmospheric pressure. At lower pressures, approaching those found in hollow-cathode discharge sample cells, this case might be applicable to relatively weak transitions, which have longer lifetimes.

The absorption coefficient for a continuous monochromatic beam for this case where velocity-changing, nonquenching collisions for excited atoms are frequent enough to spread the pumping over all atoms (and the dephasing collisions are even more frequent) is obtained (15) by replacing the Lorentzian function $\mathcal{L}(\lambda,\lambda_0)$ in both the numerator and denominator of Eq. (37) with the (low irradiance) Voigt integral $V(a,\omega)$ in Eq. (14). [The Lorentzian part of the Voigt integral is the low-irradiance Lorentzian $\mathcal{L}(\lambda,\lambda_0)$, not the bracketed term in Eq. (41)]. By analogy with Eq. (41), the spectral profile of the absorption coefficient is therefore in the limit of low irradiance, equal to the Voigt profile given by Eq. (14), but is broader at high irradiances.

When short laser pulses are used to irradiate the sample, there may not be sufficient time during the laser pulse for velocity-changing collisions to cause a significant fraction of excited atoms to change their velocity. This would be the case particularly at low pressures and when a light buffer gas such as helium were used; a collision with a light particle does not cause as great a change in velocity as a similar collision with a heavy particle. In effect, the no-velocity-changing case would be applicable, and Eq. (47) would give a more accurate representation of the spectral profile of the atomic absorption coefficient. Consequently, the spectral line profile would be narrower than for the long-pulse or continuous-wave (cw) laser beam. McIlrath and Carlsten (15) present equations that can be used to describe the temporal evolution of the absorption coefficient during short laser pulses (for cases where dephasing collisions are much more frequent than Doppler velocity-changing collisions).

The above discussion shows that strong irradiance broadens the spectral profile of the absorption coefficient and reduces its magnitude compared to low-irradiance cases. Let us now consider a two-beam technique that, under some conditions of analytical interest, can improve the wavelength resolution compared to the low-irradiance wavelength resolution.

## 3.4 MINIMIZING THE INFLUENCE OF DOPPLER BROADENING

The influence of Doppler broadening upon an absorption line profile can be minimized or essentially eliminated by a two-beam technique that primarily responds to only those atoms that are in a particular Doppler velocity subset (17,18, 59–61). This technique has been used to resolve hyperfine components of atomic lines and would be particularly useful in improving wavelength resolution in low-pressure sample cells, such as hollow-cathode discharge cells, since Doppler broadening is the main source of line broadening at low pressure. The technique, sometimes called "saturation spectroscopy," uses a strong, monochromatic beam to alternately saturate the atom population in a particular Doppler velocity subset of the sample vapor. The absorption of a weak monochromatic probe beam is then used to observe the resulting alternating change in the sample absorption coefficient caused by the alternating strong beam. The magnitude of the change in the absorption coefficient is, of course, proportional to concentration for an optically thin sample. We shall assume that there are no collisions that cause excited atoms to change velocity and thereby jump from one Doppler velocity subset to another. Such collisions broaden the observed spectral line profile and are considered later.

In order to restrict the response to only those atoms that are in a narrow Doppler velocity subset, the two beams are collimated and are positioned to intersect each other with a very small angle at the observation volume. The beams may travel essentially in the same direction or in opposite directions. (The advantages of using opposite directions is considered later.) Since the two beams are almost parallel, each atom has almost the same velocity component in the direction of each beam, and a Doppler velocity subset of atoms for one beam consists of the *same* atoms that are in the corresponding Doppler velocity subset of atoms for the other beam. Consequently, so far as these two beams are concerned, the atoms in one Doppler velocity subset (with a spectral width comparable to the Lorentzian linewidth) are the only ones being saturated by the strong beam and are therefore the only ones to produce the *changing* absorption coefficient that is observed by the weak probing beam. Atoms in the other Doppler velocity subsets, far from the wavelength (and therefore influence) of the strong beam, contribute relatively little to the *changing* absorption coefficient signal even when the wavelength of the weak probing beam is scanned to the central wavelengths of the other Doppler velocity subsets.

Scanning the wavelength of one beam while holding the other one constant would produce an absorption profile with a width comparable to and somewhat greater than the Lorentzian width of the line. Equation (41) could be used to derive the general shape of the resulting profile by setting $\lambda$ in the numerator equal to the wavelength of the probe beam and $\lambda$ in the denominator equal to the wavelength of the saturating beam. As the saturating irradiance is alternately turned off and on, the absorption coefficient alternates between values of $k(\lambda,0)$ for a weak irradiance where $E(\lambda)$ approaches zero and $k[\lambda,E(\lambda)]$ where the irradiance is $E(\lambda)$. The relative change in the transmitted photon rate $\Delta P/P_0$ of the probe beam is given by

$$\Delta P/P_0 = \exp\{-k[\lambda,E(\lambda)]\} - \exp[-k(\lambda,0)] \qquad (53)$$

When the fraction of the probe beam that is absorbed is less than, say, 0.1, the Taylor series expansion of $\exp(-\chi)$ can be truncated to $\exp(-\chi) \doteq 1 - \chi$. In this case Eq. (53) becomes

$$\Delta P/P_0 \doteq -k[\lambda, E(\lambda)] + k(\lambda, 0) \tag{54}$$

When the two beams are traveling in the same direction, Eq. (41) may be used with Eq. (53) to give

$$\Delta P/P_0 \doteq k(\lambda_0) \int g(\lambda_0, \lambda_D) \mathscr{L}(\lambda_p, \lambda_0) \left[ 1 - \frac{1}{1 + (E(\lambda_s)/E(\lambda_0)_{sat})\mathscr{L}(\lambda_s, \lambda_0)} \right] d\lambda_0 \tag{55}$$

where $\lambda_p$ is the wavelength of the probe beam and $\lambda_s$ is the wavelength of the strong beam. In Eq. (55), $g(\lambda_0, \lambda_D)$ accounts for the fact that the absorption caused by a velocity subset of atoms that is Doppler shifted to wavelength $\lambda_0$, is proportional to the relative number of atoms in that subset. The second term, $\mathscr{L}(\lambda_p, \lambda_0)$, is the Lorentzian dispersion function that accounts for a decrease in absorption when the wavelength $\lambda_p$ of the monochromatic probing beam does not coincide with the resonant wavelength $\lambda_0$ of the atoms in a velocity subset. The term in brackets is a saturation term that accounts for the fact that the net absorption by atoms within a velocity subset at $\lambda_0$ is reduced by strong irradiation (having wavelength $\lambda_s$), which increases the excited atom population while it reduces the absorbing atom population. The term in brackets reflects the fact that the signal is proportional to the *difference* between two absorption signals. When the irradiance $E(\lambda_s)$ of the alternating strong beam is zero, the alternating difference signal observed by the probing beam is zero.

When the two beams are traveling in opposite directions, they interact with the same Doppler velocity subset when their respective wavelengths are located equidistant on opposite sides of the zero-velocity Doppler wavelength $\lambda_D$. In this case we may arbitrarily assume that the probe beam is traveling in the usual direction (for which the Doppler shift was originally defined), while the strong beam is turned around. The wavelengths of the probe beam $\lambda_p$ and the Doppler velocity subsets $\lambda_0$ then remain in the same frame of reference for Eq. (55), while the wavelength of the Doppler velocity subset that is being irradiated by the strong beam of wavelength $\lambda_s$ is now given by the term $\lambda_D + (\lambda_D - \lambda_s) = (2\lambda_D - \lambda_s)$. Therefore, $\lambda_s$ in Eq. (55) is replaced by the term $(2\lambda_D - \lambda_s)$ for the case where the two beams travel in opposite directions.

Haroche and Hartman (16) have shown that Eq. (55) is only valid providing the irradiance of the strong beam is somewhat less than the saturation irradiance $E(\lambda_0)_{sat}$. For stronger irradiances, the fraction in Eq.

(55) must be multiplied by a complex coherence factor, which depends upon irradiance, to account for the fact that the atomic dipoles which absorb the probe beam are *coherently* driven by the strong beam. The coherence factor reduces the height of the observed peak and broadens the peak. In contrast to Eq. (55), the model of Haroche and Hartman also shows that in the case of large Doppler broadening the sample never becomes completely transparent to the probe beam but asympotically approaches a transmittance less than unity for high irradiances.

Resolving two closely spaced absorption lines that are heavily overlapped because of Doppler broadening requires not only a way to eliminate the effect of the Doppler *broadening* from the line profile that is scanned but also a way to ensure that only the Doppler velocity subset with *zero velocity* (and therefore no Doppler shift from $\lambda_D$) is the subset that is being observed. This is readily accomplished using the case where *the two beams always have the same wavelength and travel in opposite directions.* (An obvious advantage of this method is the need for only one tunable laser to generate the two beams.) In this case the two beams will interact strongly with the Doppler velocity subset that has no (zero) velocity in the direction parallel to the beams; that is, $\lambda_s = \lambda_p = \lambda_D$. For all other velocity subsets, the atoms that are moving toward one beam will be moving away from the other beam. The Doppler shift with respect to one beam will be equal but *opposite* to the Doppler shift for the other beam. Since the wavelengths of the two beams are equal, they will always be mainly interacting with two *different* groups of atoms (in velocity subsets on opposite sides of the zero Doppler wavelength $\lambda_D$). The alternating signal is strongest, however, when the two beams are tuned to interact strongly with the *same* group of atoms; this occurs only for the zero Doppler velocity subset, where $\lambda_0 = \lambda_D$.

Equation (55) can be used to estimate the lineshape, again by replacing $\lambda_s$ with $(2\lambda_D - \lambda_s)$ as indicated above. In this case $\lambda_s$ also equals $\lambda_p$, the strong beam and probe beam wavelengths being equal; and the observed line profile that is obtained by scanning *both* beams simultaneously and in effectively opposite directions from $\lambda_D$ will be half the width of the observed profile that is obtained by scanning only one beam.

Figure 3.4 shows the resolution of the hyperfine components of the sodium 589-nm line, obtained by Hänsch, Shahin, and Schawlow (17) in a glass cell filled with a low pressure ($3 \times 10^{-7}$ torr) of sodium vapor. The temperature of the cell was 110°C corresponding to a Doppler width of 1500 MHz (0.0017 nm). The two beams crossed each other at an angle of <10 mrad, and each beam had a spectral line width of 7 MHz ($8 \times 10^{-6}$ nm). The narrowest linewidth observed for one component was $40 \pm 4$ MHz. When 3.1 torr of argon buffer gas was introduced at room

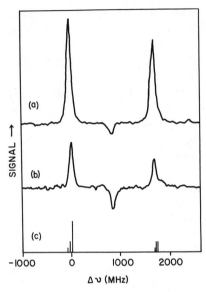

**SIGNAL →**

(a)

(b)

(c)

-1000          0          1000          2000

Δ𝜈 (MHz)

**Fig. 3.4.** (*a*) Saturation spectrum of the Na $D_2$ line without delay. (*b*) Like (*a*) but with the weak probe beam delayed 0.7 μs. (*c*) Theoretical hyperfine structure. From Hänsch, Shahin, and Schawlow (17), reproduced by permission.

temperature (corresponding to a Doppler width of 1300 MHz), the narrowest observed linewidth was 150 MHz, compared to a natural linewidth of 10 MHz (16 ns radiative lifetime) and a collisional broadening linewidth of 51 MHz (3 ns between dephasing collisions). The fact that the observed linewidth was three times greater than the expected Lorentzian linewidth appears to be due to velocity-changing collisions that caused the transfer of excited atoms between different Doppler velocity subsets during the observation time of about 30 ns. By delaying the probe beam with respect to the strong beam it was found that the observed linewidth increased and approached, but did not reach, the thermal equilibrium Doppler-broadened linewidth after about 150 ns. Consequently, in order to obtain high resolution in the presence of nonquenching velocity-changing collisions, it may be desirable to use a short-pulsed laser rather than a cw laser to ensure that the saturated absorption measurement is made before collisions can cause the transfer of excited atoms between widely separated Doppler velocity subsets. The use of a lighter buffer gas and a lower pressure would also help to reduce the observed linewidth, even when using a cw laser. Time-resolved measurements of the type indicated above could, of course, be used to study nonquenching velocity-changing collisions in analytical cells to predict conditions for optimum wavelength resolution. This example shows that the wavelength resolution was improved by almost a decade using a 30-ns laser pulse and an atomic vapor having conditions similar to those found in hollow-cathode discharge sample cells.

An anomalous dip appears midway between the two absorption lines because these two lines have the same upper energy level, and they overlap each other due to Doppler broadening in a normal absorption experiment. There is some Doppler velocity subset for which the strong laser beam appears to coincide with one transition and the probe beam appears to coincide with the other transition when the laser beams are tuned to the wavelength midway between the two transitions. Since the signal is due to a *difference* in absorptions when the strong beam is on and off, a peak in the upward direction in Fig. 3.4 occurs whenever the strong beam *reduces* the population in the lower level, causing an accompanying decrease in the absorption of the probe beam by those atoms.

The inverted direction of the anomalous dip, therefore, indicates that the population of atoms in the lower level being observed by the probe beam *increases* whenever the strong beam is turned on. This increase in population of one lower level is caused by a pumping process whereby the strong beam excites atoms from the other lower level to the common upper level, from which the atoms then decay to the former lower level being observed by the probe beam. When the two transitions share a common *lower* level, the anomalous peak is in the same, upward, direction as the two peaks that correspond to the two ordinary transitions. The anomalous peaks will only appear in the spectrum when the Doppler broadening is large enough so that there is a Doppler velocity subset that has a Doppler shift $(\lambda_0 - \lambda_D)$ equal to one half the wavelength difference between the two lines that share a common level. The anomalous lines will not appear if the transitions belong to different atoms.

## 3.5   QUENCHING RATE CONSTANT DETERMINATIONS

For any given moderate to high irradiance, the quenching rate constant $k_{21}$ influences the magnitude of the absorption coefficient and the wavelength resolution of an absorption experiment, by its influence upon the linewidth. The quenching rate constant is also important in atomic fluorescence experiments (Chapter 4). As suggested by Omenetto et al. (57), the rate constant $k_{21}$ can be found if $E(\lambda_0)_{sat}$ is determined, see Eq. (36). In the experimental determination of $E(\lambda_0)_{sat}$ it may be desirable to eliminate confusion which results from collisions that transfer excited atoms between Doppler velocity subsets, by homogeneously irradiating the atoms in all Doppler velocity subsets by a continuum source with constant spectral irradiance. A laser whose wavelength modes are spaced much more closely to each other than the Lorentzian width of the atomic transition would satisfy the continuum requirement if the modes had equal intensities.

For a spectral irradiance $E_\lambda$ that is the same for all wavelengths, Eq.

(43) becomes $1+[E_\lambda/E(\lambda_0)_{sat}]$, since $E_\lambda$ is now a constant and the integral of the Lorentzian function that remains under the integral sign is unity. Equation (41) then shows that for the population of atoms irradiated with the continuum source, the absorption coefficient for photons of wavelength $\lambda$ is now

$$k(\lambda) = \frac{k(\lambda_0)}{1+[E_\lambda/E(\lambda_0)_{sat}]} \, V(a,\omega) \tag{56}$$

which is the same as Eq. (14) except for the denominator term. The denominator term accounts for the reduction in the absorption coefficient that occurs because the moderately high continuum irradiance reduces the fraction of atoms in the lower energy level. Since the wavelength integral of the Voigt function $V(a,\omega)$ is unity, integrating both sides of Eq. (56) gives

$$\int k(\lambda) \, d\lambda = \frac{k(\lambda_0)}{1+[E_\lambda/E(\lambda_0)_{sat}]} \tag{57a}$$

which upon rearrangement gives

$$\frac{k(\lambda_0)}{\int k(\lambda) \, d\lambda} = \frac{E_\lambda}{E(\lambda_0)_{sat}} + 1 \tag{57b}$$

Equation (57a) shows that $k(\lambda_0)$ is simply the wavelength integral of the absorption coefficient for low irradiances $E_\lambda$ approaching zero. Equation (57b) shows that a plot of the ratio of wavelength integral of the absorption coefficient $k(\lambda)$ at low irradiances to the same integral at high irradiances versus the spectral irradiance $E_\lambda$ will have a slope of $1/E(\lambda_0)_{sat}$. Combining Eq. (31), (32), and (36) gives

$$E(\lambda_0)_{sat} = \frac{k_{21}+A_{21}+k_{12}}{A_{21}} \left[ \frac{g_1 8\pi hc^2}{g_2 2\lambda^5} \right] \tag{58}$$

The value of $k_{12}$ can be found (62) from the detailed-balance relationship

$$k_{12}/k_{21} = (g_1/g_2) \exp(-\Delta E/kT) \tag{59}$$

where the right-hand side of the equation is the well-known Boltzmann equilibrium ratio. Usually this ratio is much less than unity so that $k_{12}$ can be ignored. Then Eq. (58) can be used to obtain the value of the Stern-Volmer (63) fluorescence quantum efficiency $A_{21}/(A_{21}+k_{21})$; and if $A_{21}$ is known, the value of $k_{21}$ can be determined.

Experimentally, the values of $k(\lambda)$ (the absorption coefficient at each wavelength $\lambda$) to be integrated in Eq. (57) could be determined by placing a high-resolution spectrometer (e.g., a Fabry-Perot interferometer) between the sample and the detection system, that is, in the transmitted continuum beam, and scanning the wavelength. Alternatively, a low-intensity monochromatic laser beam could be used to determine $k(\lambda)$ at each $\lambda$.

## 3.6 ABSORPTION FROM EXCITED STATES

When a large fraction of a population is excited into a selected upper energy level by the high irradiance of a laser beam, it becomes feasible to observe absorption lines that *originate* from the selected excited state. Consequently a completely different series of energy levels becomes accessible for sensitive spectral measurements. This two-step absorption technique would be useful in analytical work where the first absorption step happened to be in a wavelength region where spectral interferences existed, while the second absorption (from the excited state) was relatively free from interference. Also, the influence of Doppler broadening could be essentially eliminated if the excited state were populated by a *two-photon absorption* step, where each photon contributes one half of the energy necessary to excite the atom. Elements that have been studied using absorption from laser-excited levels include Ba (63, 64), Ca (65–67), Rb (68), and Mg (63,69).

Two-step absorption measurements provide other diagnostic information that is valuable in chemical analysis. For instance, the identification of spectral lines is made easier when the excited-state population is increased by high irradiance, since lines originating from the excited state increase in strength while absorption lines that originate from the ground-state decrease in strength. Lines not belonging to the selected atom would remain essentially unchanged. Lifetime measurements of the excited state are of course possible by using the absorption signal originating from the excited state to monitor the decay of its population. Intermediate energy levels, and processes by which they are populated from the selected excited state, may also be studied.

Surveys of absorption lines originating from laser-excited upper energy levels have been made for a few elements using continuum sources. A short-pulsed flash lamp (65, 66) or broadband-pulsed laser continuum (68) provide both a spectral continuum and a means of delaying the observation for a desired time following an excitation pulse from the laser. Photographic recording of the spectra has been frequently used because it provides information about many lines at once. Carlsten (64) used a long-pulsed flash lamp with a photomultiplier and oscilloscope to observe the temporal behavior of the relatively long-lived absorption originating from the $^3D_2$ metastable level of barium. Broadband emission from a fluorescing dye solution has also been used as the absorption source. In these cases (63,67) the fluorescent dye source was pumped by the same laser that was used to pump the tunable dye laser which excited the upper level. Nanosecond observation times, delayed with respect to or coincident with the pumping time, can be achieved by properly adjusting the optical paths of the laser beams.

Line profiles of rather broad (several nanometers wide) autoionization transitions of magnesium have been determined using photographic films (63). Narrower line profiles of absorption lines could be observed using a second tunable laser to probe the spectral profile. Saturation broadening of the excited level caused by the high irradiance of the pumping pulse would of course be observed unless the probing laser pulse were delayed with respect to the pumping pulse.

The best sensitivity for analytical purposes occurs when a large fraction of the total population is pumped into the excited state. This condition occurs very shortly after the pumping pulse is turned on, on a time scale that is short compared to the mean times between collisions and spontaneous emissions that cause other energy levels to become populated from the originally pumped level. If the laser pump is left on long enough, all energy levels in the system will reach a new steady-state condition where a large fraction of atoms may be in energy levels other than the originally pumped level. Consequently, analytical use of absorption lines from selectively excited energy levels may best be made using *pulsed* laser systems, including mode-locked cw lasers, rather than continuous-mode lasers.

Excitation of an upper energy level by the simultaneous absorption of two photons traveling in opposite directions to each other provides a way to minimize or eliminate Doppler broadening (70–75). Absorption occurs when the sum of the energies of the two photons equals the energy difference between the two energy levels. Because the photons are traveling in opposite directions, any Doppler wavelength shift of the atom with respect to one photon is compensated by the opposite Doppler shift of the atom with respect to the other photon. Consequently, *all* atoms in the total population are equally influenced, irrespective of any Doppler velocity component they may have. This is in contrast to saturation spectroscopy (Sect. 3.4), where only atoms with (close to) a zero Doppler velocity component are observed by the weak probing beam. Consequently, the saturation spectroscopy signal depends upon the Doppler broadening distribution function, while the two-photon signal does not.

In general, when the photons all have the same energy, the narrow resonance line resulting from two-photon absorption from beams traveling in opposite directions is superimposed on a Doppler-broadened background caused by two-photon absorption from a beam moving in one direction. The background could be eliminated if the two beams had photons of different energies, $E_1$ and $E_2$, and the energies $2E_1$ and $2E_2$ fell outside the Doppler profile. The sum of the two energies, $E_1 + E_2$, must of course still be equal to the energy of the transition.

The two-photon irradiances necessary to cause a significant fraction of the population to reach the upper energy level are much higher than if a one-photon absorption step were used. Even so, the required irradiances are well within the reach of pulsed dye lasers, and adequate irradiances have been achieved using focused cw laser beams. The required irradiance is reduced, or the excitation rate increased, when intermediate energy levels are positioned so that their energies above the lower energy level are close to the energy of one of the photons. For instance, Hänsch, Harvey, Meisel, and Schawlow (70) estimated that they obtained on the order of $10^{10}$ excitations per second with a power density of about $3000 \text{ W cm}^{-2}$ for a Na $3S–4D$ transition that had an intermediate $3P$ level present. The excitation rate would have been reduced by a factor of 3000 had the $3P$ level not been about halfway between the $3S$ and $4D$ energy levels.

Experimentally, the counterpropagating laser beams can be generated with two lasers, or with one laser by carefully reflecting the laser beam back upon itself. In practice, some means is used to minimize feedback of the reflected beam into the laser cavity. For instance, a solution of $\alpha$-bromonaphtalene in a 10,000-gauss magnetic field has been positioned between the laser and the cell to rotate the polarization vector of the beam by 45° each of the two times it passes through the cell. By the time the reflected beam returns to the laser cavity, it has been rotated a total of 90° with respect to the original beam, and it is relatively ineffective in a laser cavity that only supports lasing in the other direction of polarization (71). In a similar manner, quarter-wave plates have been used on both sides of the sample cell to generate circularly polarized beams to isolate the reflected beam from the laser cavity (70). This method has the additional advantage of eliminating the Doppler-broadened background caused by two-photon absorption from one beam for energy level transitions where selection rules forbid the absorption of two photons of the same circular polarization (73). Other techniques have been used, such as vibrating the reflecting mirror with a loudspeaker coil to rapidly modulate the phase of the backreflected light for a cw laser, or to simply provide sufficient optical delay between the time a pulse is generated by a laser and the time its reflection returns to the laser cavity (73).

The analytical advantages to be gained by eliminating Doppler broadening have already been discussed in Sect. 3.2.7. A disadvantage of the two-photon technique is the dependence of the analytical signal upon the power densities of *both* beams, which in the case of only one laser beam reflected back upon itself becomes proportional to the *square* of the power density of the laser, at least until the upper energy level approaches saturation.

## 3.7   ENHANCED ABSORPTION USING AN INTRACAVITY CELL

A particularly promising technique to improve atomic absorption measurements is to place the absorption cell in the cavity of a laser. A very large enhancement in the absorption signal occurs as discussed in Chapter 8. The detection limits for atomic species have been limited by the use of spectrographs that have insufficient resolution for the narrow absorption lines and by the use of photographic detection with its modest noise level. The type of laser used is also an important experimental variable (Chapter 8). Although this technique improves the absorption enhancement factor, it is not clear to what extent it improves the signal-to-noise ratio of the absorbance determination. Further experimental and theoretical studies are needed to clarify the significance of this potentially important measurement technique.

## 3.8   CONCLUSIONS

One beautiful aspect of the tunable laser with a narrow spectral bandwidth is its ability to make possible new types of analytical atomic absorption measurements that are difficult or impossible to achieve with the common and reliable hollow-cathode lamp. Most analytical atomic absorption experiments using laser sources, up to now, have been of a preliminary nature, using available instrumentation rather than optimized systems, demonstrating new possibilities rather than achieving high performance on complex chemical samples. Improvements in laser characteristics and in entire measurement systems should occur as the need for particular improvements are identified and demonstrated. In this way preliminary experiments are as important to laser atomic absorption as they were to the development of the highly reliable hollow-cathode atomic absorption instruments in such wide use today. In contrast to the hollow-cathode lamp, the tunable laser makes an entire spectrum of wavelengths available for atomic absorption measurements of atomic, ionic, and molecular species, in excited states as well as ground states. It may soon be possible to obtain a *total* analysis of an element in all of the forms in which it exists in a sample cell. This would help to compensate for chemical and physical changes that occur in the sample cell when the matrix changes from one sample to the next.

Accompanying the continuum of wavelengths that is available, however, is an uncertainty in the exact value of the wavelength that is lasing for a particular laser instrument setting. Although this seems to be a disadvantage when compared to the excellent wavelength stability of the hollow-cathode lamp, when a narrowband tunable laser is scanned across the spectral profile of the absorbing line, the center of the profile becomes a reference point for the measurement and automatically compensates for minor pressure-broadening wavelength shifts. Scanning the line profiles with a narrow bandwidth laser has the advantage of providing the best possible compensation for spectral interferences, especially for those that vary from sample to sample. The high wavelength resolution available with a narrow bandwidth laser should extend and improve the use of atomic absorption for the determination of isotopes.

The high intensity of a laser provides the best possible reduction in relative shot noise, which has previously limited the usefulness of atomic absorption measurements for short-lived species in transient sample cells or dynamic chemical systems. For steady-state sample cells, measurements may be made more quickly for situations where analyses are needed in a hurry. High intensity also reduces the influence of emission noise from high-temperature sample cells. Unfortunately, due to the saturation effect, the sample cannot be exposed to too high an irradiance without significantly decreasing the absorption signal.

Accompanying the fundamental limitation of the saturation effect, however, is the ability to pump the analyte species into an excited energy level which can, in turn, be used as the lower energy level for absorption measurements, producing a whole new series of absorption lines that might be used to eliminate specific spectral interferences. Of greater importance may be the use of the saturation effect with two counter-propagating laser beams to effectively eliminate or minimize Doppler broadening, particularly in low-pressure sample cells where Doppler broadening has been the major source of line broadening. This technique would seem to hold great promise for isotope analysis where lines are very close together. Two-photon excitation of an upper energy level also offers the possibility of minimizing Doppler broadening.

Another unique application of the laser involves placing the sample inside the laser cavity to obtain an enhanced absorption signal. This technique offers the promise of very low detection limits.

As these new uses for lasers are explored, boundaries will be imposed by fundamental limitations, such as the saturation effect, or by present technology. Each boundary, however, is simply a mirror to help point our dreams in other, often unexplored directions, opening our thoughts to other exciting and perhaps useful possibilities.

## References

1. J. M. Harris, R. W. Chrisman, and F. E. Lytle, *Appl. Phys. Lett.* **26,** 16 (1975).

2. Z. A. Yasa and O. Teschke, *Appl. Phys. Lett.*, **27,** 446 (1975).

3. A. Dienes, E. P. Ippen, and C. V. Shank, *Appl. Phys. Lett.*, **19,** 258 (1971).

4. D. J. Kuizenga, *Appl. Phys. Lett.*, **19,** 260 (1971).

5. E. P. Ippen, C. V. Shank, and A. Dienes, *Appl. Phys. Lett.*, **21,** 348 (1972).

6. I. I. Sobel'man, *Introduction to the Theory of Atomic Spectra,* Pergamon Press, New York (1972).

7. H. G. Kuhn, *Atomic Spectra,* 2nd ed., Academic Press, New York (1969).

8. C. Candler, *Atomic Spectra and the Vector Model,* Hilger and Watts, London (1964).

9. R. G. Breene, Jr., *The Shift and Shape of Spectral Lines,* Pergamon Press, Oxford (1961).

10. A. C. G. Mitchell and M. W. Zemansky, *Resonance Radiation and Excited Atoms,* Cambridge University Press. Cambridge (1961).

11. L. de Galan and H. C. Wagenaar, *Methodes Phys. Anal.,* **3,** 10 (1971).

12. C. F. Bruce and P. Hannaford, *Spectrochim. Acta,* **26B,** 207 (1971).

13. R. P. Srivastava and H. R. Zaidi, *Can. J. Phys.,* **53,** 84 (1975).

14. H. Greenstein and C. W. Bates, Jr., *J. Opt. Soc. Amer.,* **65,** 33 (1975).

15. T. J. McIlrath and J. L. Carlsten, *Phys. Rev. A,* **6,** 1091 (1972).

16. S. Haroche and F. Hartmann, *Phys. Rev. A,* **6,** 1280 (1972).

17. T. W. Hänsch, I. S. Shahin, and A. L. Schawlow, *Phys. Rev. Lett.* **27,** 707 (1971).

18. T. W. Hänsch and P. Toschek, *IEEE J. Quant. Electron.,* **QE-4,** 467 (1968).

19. H. R. Schlossberg and A. Javan, *Phys. Rev.,* **150,** 267 (1966).

20. P. A. Apanasevich, *Opt. Spectrosc.,* **16,** 387 (1964).

21. H. Kopfermann, *Nuclear Moments* (English translation by E. E. Schneider), Academic Press, New York (1958).

22. D. H. Rank, G. Skorino, D. P. Fastman, G. D. Saksena, T. K. McCubbin, and T. A. Wiggins, *J. Opt. Soc. Amer.,* **50,** 1045 (1960).

23. G. J. DeJong and E. H. Piepmeier, *Spectrochim. Acta,* **29B,** 159 (1974); E. H. Piepmeier and L. de Galan, *Spectrochim. Acta,* **30B,** 263 (1975).

24. W. Behmenburg, *J. Quant. Spectrosc. Radiat. Transfer,* **4,** 177 (1964).

25. C. Young, *Tables for Calculating the Voigt Profile,* University of Michigan, Rept. 058637-T, Ann Arbor, Michigan (1965); J. T. Davies and J. M. Vaughan, *Astrophys. J.,* **137,** 1302 (1963).

26. E. W. Smith, J. Cooper, W. R. Chappell, and T. Dillon, *J. Quant. Spectrosc. Radiat. Transfer,* **11,** 1567 (1971).

27. M. Mizushima, *J. Quant. Spectrosc. Radiat. Transfer,* **7,** 505 (1967).

28. H. C. Wagenaar and L. de Galan, Spectrochim. Acta, **28B,** 157 (1973).

29. T. Hollander, B. J. Jansen, T. J. Plaat, C. Th. J. Alkemade, *J. Quant. Spectrosc. Radiat. Transfer,* **10,** 1301 (1970).

30. H. C. Wagenaar, C. J. Pickford, and L. de Galan, *Spectrochim. Acta,* **29B,** 211 (1974).

31. E. H. Piepmeier and D. E. Osten, *Appl. Spectros.,* **25,** 642 (1971).

32. M. Shimazu, Y. Takubo, M. Yoshii, *Oyo Butsuri,* **42,** 1234 (1973) (in Japanese).

33. F. Schuda, M. Hercher, and C. R. Stroud, Jr., *Appl. Phys. Lett.,* **22,** 360 (1973).

34. H. C. Wagenaar, I. Novotný, and L. de Galan, *Spectrochim. Acta,* **29B,** 301 (1974).

35. B. V. L'vov and V. I. Mosichev, *Zh. Prikl. Spektrosk.* **4,** 491 (1966).

36. A. N. Zaidel' and E. P. Korennoi, *Zavodskaya Lab.,* **29,** 1449 (1963).

37. D. C. Manning and W. Slavin, *Atomic Absorption Newsletter,* **1,** 39, (1962).

38. J. A. Goleb and Y. Yokoyama, *Anal. Chim. Acta,* **30,** 213 (1964).

39. J. V. Sullivan and A. Walsh, *Spectrochim. Acta,* **21,** 727 (1965).

40. J. A. Goleb, *Anal. Chim. Acta,* **34,** 135 (1966).

41. J. A. Goleb, *Anal. Chem.,* **35,** 1978 (1963).

42. B. V. L'vov, *Atomic Absorption Spectrochemical Analysis,* American Elsevier, New York (1970).

43. J. B. Willis, *Spectrochim. Acta,* **25B,** 487 (1970).

44. K. Yasuda, *Anal. Chem.,* **38,** 592 (1966).

45. D. K. Davies and M. A. Biondi, *J. Appl. Phys.,* **37,** 2969 (1966).

46. K. Bergstedt, *Z. Phys.,* **155,** 23 (1959).

47. H. F. van Heek, *Spectrochim. Acta,* **25B,** 107 (1970).

48. T. Hollander and H. P. Broida, *Combust. Flame,* **13,** 63 (1969).

49. D. N. Stacey, V. Stacey, and A. R. Malvern, *J. Phys. E: Sci. Instru.,* **7,** 405 (1974).

50. D. W. Posener, *Aust. J. Phys.,* **12,** 184 (1959).

51. W. C. Kreye and F. L. Roesler, *J. Opt. Soc. Amer.,* **60,** 1100 (1970).

52. N. Omenetto, N. N. Hatch, L. M. Fraser, and J. D. Winefordner, *Anal. Chem.,* **45,** 195 (1973).

53. D. E. Osten and E. H. Piepmeier, *Appl. Spectrosc.,* **27,** 165 (1973).

54. E. W. Peterson and D. H. Holland, *J. Opt. Soc. Amer.,* **40,** 253 (1950).

55. J. D. Ingle, Jr., *Anal. Chem.,* **46,** 2161 (1974).

56. J. Kuhl, S. Neumann, and M. Kriese, *Z. Naturforsch.,* **28a,** 273 (1973).

57. N. Omenetto, P. Benetti, L. P. Hart, J. D. Winefordner, and C. Th. J. Alkemade, *Spectrochim. Acta,* **28B,** 289 (1973).

58. E. H. Piepmeier, *Spectrochim. Acta,* **27B,** 431 (1972).

59. T. W. Hänsch, M. D. Levenson, and A. L. Schawlow, *Phys. Rev. Lett.,* **26,** 946 (1971).

60. P. W. Smith and T. W. Hänsch, *Phys. Rev. Lett.,* **26,** 946 (1971).

61. C. Bordé, *C.R. Acad. Sci. Paris,* **271,** 371 (1970).

62. C. Th. J. Alkemade and P. J. Th. Zeegers, "Excitation and De-Excitation Processes in Flames," in *Spectrochemical Methods of Analysis* (Advances in Analytical Chemistry and Instrumentation, Vol. 9), J. D. Winefordner, Ed., Wiley, New York (1971).

63. D. J. Bradley, P. Ewart, J. V. Nicholas, and J. R. D. Shaw, *J. Phys. B: Atom. Mol. Phys.,* **6,** 1594 (1973).

64. J. L. Carlsten, *J. Phys. B: Atom. Mol. Phys.,* **7,** 1620 (1974).

65. T. J. McIlrath, *Appl. Phys. Lett.* **15,** 41 (1969).

66. T. J. McIlrath, *J. Phys. B: Atom. Mol. Phys.,* **7,** 393 (1974).

67. T. J. McIlrath and J. L. Carlsten, *J. Phys. B: Atom. Mol. Phys.,* **6,** 697 (1972).

68. D. J. Bradley, G. M. Gale, and P. D. Smith, *J. Phys. B: Atom. Mol. Phys.,* **3,** L-11 (1970).

69. J. J. Wright, J. F. Dawson, and L. C. Balling, *Phys. Rev. A,* **9,** 83 (1974).

70. T. W. Hänsch, K. C. Harvey, G. Meisel, and A. L. Schawlow, *Opt. Commun.,* **11,** 50 (1974).

71. F. Biraben, B. Cagnac, and G. Grynberg, *Phys. Lett.,* **49A,** 71 (1974).

72. F. Biraben, B. Cagnac, and G. Grynberg, *Phys. Rev. Lett.,* **32,** 643 (1974).

73. M. D. Levenson and N. Bloembergen, *Phys. Rev. Lett.,* **32,** 645 (1974).

74. B. Cagnac, G. Drynberg, and F. Biraben, *J. Phys.* (Paris), **34,** 845 (1973).

75. L. S. Vasilenko, V. P. Chebotaev, and A. V. Shishaev, *Pisma Zh. Eksp. Teor. Fiz,* **12,** 161 (1970); *JETP Lett.,* **12,** 113 (1970).

# 4

# ATOMIC FLUORESCENCE SPECTROSCOPY WITH LASER EXCITATION

### N. OMENETTO

*Istituto di Chimica Generale*
*Università di Pavia*
*Pavia, Italy*

and

### J. D. WINEFORDNER

*Department of Chemistry*
*University of Florida*
*Gainesville, Florida*

## 4.1  INTRODUCTION

Atomic fluorescence emitted by several atomic species has been largely studied by physicists in the early twentieth century (1). However, its *analytical* birth, foreseen by Alkemade in 1962 (2), dates back to 1964, when Winefordner and Vickers (3) described their first experiments carried out with a conventional atomic absorption atomizer—a flame burning at atmospheric pressure. Since the fluorescence radiance is linearly proportional to the source radiance over the absorption linewidth of the atoms, it soon became apparent that atomic fluorescence, like all types of luminescence, was a source intensity-starved method. Indeed, except in a few favorable cases, the excitation provided by the available atomic absorption line sources, such as hollow-cathode discharge lamps, has proved to be completely inadequate for obtaining good signals and signal-to-noise ratios.

Therefore, the major field of technologic research in atomic fluores-cence spectroscopy has been the investigation of a great variety of light sources possessing, among other desirable features such as stability, long life, low cost, and high versatility, the most critical of all, namely, high intensity. Several types of sources gave satisfactory results (4). These include high-intensity hollow-cathode lamps provided with a boosting auxiliary discharge, demountable hollow-cathode lamps, and spectral vapor discharge lamps. A common feature of these sources is their lack of versatility, since a rather limited number of elements can be successfully excited. Continuum sources such as high-pressure xenon arcs should overcome this problem, despite their lack of intensity in the ultraviolet region. The most promising source has been the microwave-excited elec-trodeless discharge lamp because of its greater spectral output over the absorption bandwidth compared to the other sources. A great deal of research has been devoted to optimize the output performances of such lamps and to eliminate operating inconveniences as well as a somewhat poor reliability. Both disadvantages seem to be overcome by careful thermostating of the lamp (5).

In addition to the problem of source intensity, it is generally recognized that atomizer noise is of primary importance for flames and electrother-mal atomizers in the spectral region above 300 nm. Of all noises present in a fluorescence setup, assuming negligible scattering, the total photo-detector shot noise is probably the dominating noise in the system. If this is the case, best results would be achieved by pulsing the excitation source with a low duty cycle (therefore achieving high peak intensities) and gating the detector so that the noise is measured only during the short "on" time of the detector.

When all these considerations are put together, one can easily understand why the pulsed, tunable dye laser (6–8) seemed to be the problem-solving tool for atomic fluorescence analysis. In fact, the dye laser has the following unique features: (i) it is continuously tunable over the visible region and, with frequency doubling, down to 250 nm; (ii) it provides extremely high peak powers, of several tens of kilowatts; (iii) it has a great degree of coherence, both spatial as well as temporal, thereby leading to very high power densities (small spot size) and narrow linewidths (monochromaticity); and (iv) it is pulsed with a very low duty cycle so that the maximum benefit of signal-to-noise ratio should be obtained for background noise-limited systems by gated operation. As a consequence of laser excitation, the expected results are therefore better detection sensitivities (because of the source intensity) and better information power and selectivity (because of the source resolution and bandwidth).

This chapter deals essentially with the theory and analytical results obtained when pulsed dye lasers are used in combination with conventional flames and electrothermal atomizers in atomic fluorescence spectroscopy. Other fluorescence results obtained by excitation via a cw laser are reported elsewhere in this book (Chapter 8). The features of the laser-excited fluorescence technique are discussed in terms of excitation and deexcitation processes (Sects. 4.2 and 4.3), saturation effects for two- and three-level systems (Sect. 4.4), the influence of source irradiance upon the shape of the calibration curves (Sect. 4.5), and the possibility of local sensing of physical parameters such as temperature, quantum efficiency, and concentrations (Sect. 4.6). The general instrumental setups used by different workers are discussed in Sect. 4.7, and the analytical results, mostly given in terms of detection limits referred to aqueous solutions, are reported in Sect. 4.8. The reader is also referred to two review articles (7,8) concerned with the general applicability of tunable dye lasers to analytical spectroscopy, thereby including the fluorescence technique.

## 4.2 TYPES OF FLUORESCENCE TRANSITIONS

During the rapid development of atomic fluorescence, several authors have described the fluorescence process under study by different names so that some confusion and ambiguity has resulted as far as terminology is concerned. With the advent of tunable dye lasers, this possibility of confusion increased since many new fluorescence transitions were observed. Because of this, an effort was made by the authors of this chapter (9) to propose a consistent nomenclature with an attempt to utilize,

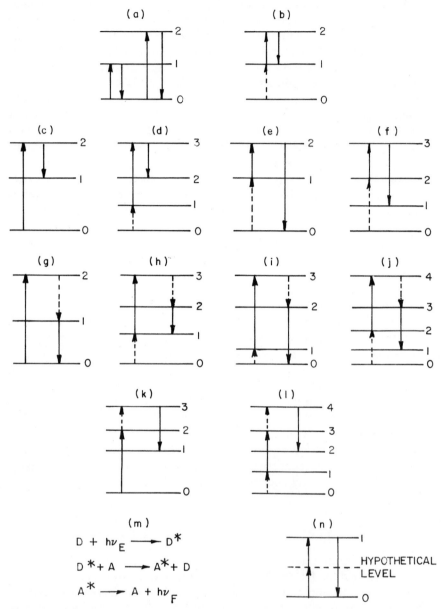

**Fig. 4.1.** Types of atomic fluorescence transitions (the spacings between atomic levels is not indicative of any specific atom): (a) resonance fluorescence (either process); (b) excited-state resonance fluorescence; (c) Stokes direct-line fluorescence; (d) excited-state Stokes direct-line fluorescence; (e) anti-Stokes direct-line fluorescence; (f) excited-state anti-Stokes direct-line fluorescence; (g) Stokes stepwise-line fluorescence; (h) excited-state

170

whenever possible, the terminology previously reported by the different authors in the field.

Basically, there are five types of atomic fluorescence transitions, which are reported, together with possible variations, in Fig. 4.1. *Resonance fluorescence*[1] results when the same lower and upper levels are involved in the excitation–deexcitation processes. Because of the usually high value of the fundamental constants characterizing the resonance fluorescence transitions, their radiance is significantly greater than for other transitions; and as a result most analytical measurements have involved these lines. *Direct-line fluorescence* results when the *same* upper level is involved in the radiational excitation and deexcitation processes. If *different* upper levels are involved in the radiational excitation and deexcitation processes, then *stepwise-line fluorescence* results. *Sensitized fluorescence* results when one species, called the donor, is excited and transfers excitation energy to an atom of the same or another species, called the acceptor, either of which deexcites radiationally. Finally, *multiphoton fluorescence*[2] results when two or more photons excite an atomic species which then radiationally deexcites. It should be pointed out that these two last processes are likely to be unimportant for analytical studies in flames and other common atomizers.

If the excitation energy is *greater* than the fluorescence energy, this type of fluorescence is termed *Stokes*; if the fluorescence energy is greater than the excitation energy, the process is called *anti-Stokes*. If the radiational excitation *and* fluorescence processes involve only excited states, the fluorescence process is said to be *excited*. If the excitation

---

[1] Strictly speaking, resonance fluorescence implies that the lower level of the transition is the ground state. Although this process is the most common in atomic fluorescence, the definition given above is more general than this and includes transitions not involving the ground state as the lower state in the transition.

[2] Excitation can be via two photons of same frequency with virtual level involved or via two photons of different frequency involving real levels.

---

Stokes stepwise-line fluorescence; ($i$) anti-Stokes stepwise-line fluorescence; ($j$) excited-state anti-Stokes stepwise-line fluorescence; ($k$) thermally assisted Stokes or anti-Stokes stepwise-line fluorescence (depending upon whether the absorbed radiation has shorter or longer wavelengths than the fluorescent radiation); ($l$) excited-state thermally assisted Stokes or anti-Stokes stepwise-line fluorescence (depending upon whether the absorbed radiation has shorter or longer wavelengths) than the fluorescence radiation); ($m$) senstized fluorescence ($D$ = donor; $D^*$ = excited donor; $A$ = acceptor; $A^*$ = excited acceptor; $h\nu_E$ = exciting radiation; $h\nu_F$ = fluorescence radiation); ($n$) two-photon excitation fluorescence (multiphoton processes involving more than two photons are even less probable than the two-photon process). From N. Omenetto and J. D. Winefordner, *Appl. Spectrosc.* **26,** 555 (1972). Reproduced by permission.

process involves a collisional excitation *following* the radiational excitation, the process is called *thermally assisted.*

The identification of the excitation and deexcitation processes involved in a fluorescence transition is important for several reasons. First of all, it helps elucidating the relative contributions of the different exciting lines as well as those of thermal collisions in populating the fluorescent level, thereby permitting the correct nomenclature to be used for the process. For example, there are several cases (Ga, In, etc.) in which the "resonance fluorescence" signal observed is due to a mixture of "resonance fluorescence" and "anti-Stokes direct-line fluorescence" ((*a*) and (*e*) in Fig. 4.1) *unless* one of the transitions is filtered from the source. Secondly, the evaluation of the quantum efficiency requires a clear knowledge of the overall process. Finally, one can take advantage of the difference existing between the excitation and fluorescence wavelengths to eliminate scattered light, which is known to interfere to variable extents (depending upon the atomizer and instrumental setup) in all resonance fluorescence measurements.[3] To accomplish this task with conventional sources is rarely easy, occasionally feasible, and most of the time impossible because of the stringent requirements on filters to isolate the line(s) of interest and the consequent decrease of the source intensity to unacceptably low levels. Moreover, the intensity of the different potentially exciting lines emitted by the source is governed and fixed by the discharge parameters. From this point of view it is clear that the laser represents an ideal source because of its tunability, with similar output intensities, over any of the lines of interest and because no filters are required to isolate the transitions. Also, the excited level population is greatly enhanced, thereby favoring efficient mixing between adjacent levels resulting in strong fluorescence emission.

Several fluorescence processes, depicted in Fig. 4.1, have been observed with laser excitation of some transition elements in a nitrous oxide–acetylene flame (10). As examples, the partial Grotrian diagrams for the elements vanadium and scandium are given in Figs. 4.2 and 4.3; the corresponding (partial) fluorescence spectra are reported in Figs. 4.4 and 4.5. Table 4.1 explains the different excitation and deexcitation processes for scandium according to the nomenclature described. Figure 4.6 shows the very minute signals obtained in an air–hydrogen flame and attributed to a two-photon excitation process for the elements zinc and cadmium (11).

To our best knowledge, no other example of multiphoton excitation in flames at atmospheric pressure has been reported so far. Steinfeld (8)

---

[3] When a continuum excitation source with no disperser is used, scatter occurs at *all* fluorescence wavelengths, that is, at resonance and nonresonance lines.

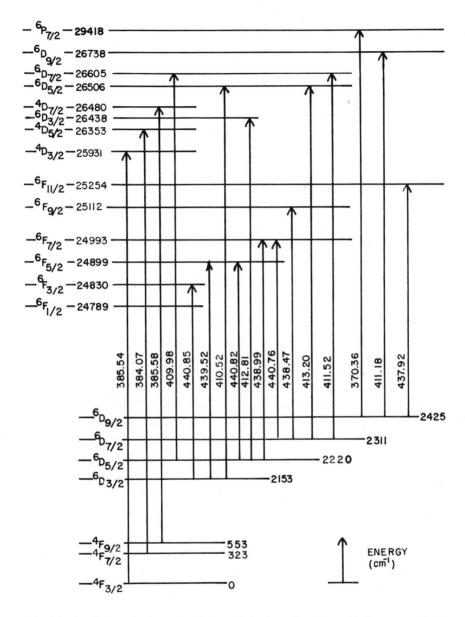

**Fig. 4.2.** Partial atomic energy level diagram for vanadium: From N. Omenetto, N. N. Hatch, L. M. Fraser, and J. D. Winefordner, *Spectrochim. Acta*, **28B**, 65 (1973). Reproduced by permission.

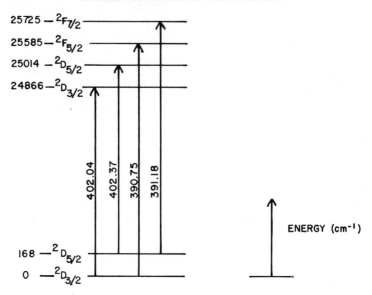

**Fig. 4.3.** Partial atomic energy level diagram for scandium. From N. Omenetto, N. N. Hatch, L. M. Fraser, and J. D. Winefordner, *Spectrochim. Acta,* **28B,** 65(1973). Reproduced by permission.

indicates that the energy levels involved in the so-called two-photon atomic fluorescence observed for both Cd (228.8 nm) and Zn (213.9 nm) are difficult to understand because the same upper and lower levels cannot be connected by both two-photon and one-photon transitions (parity forbidden). However, due to the poor detection limits obtained (~10 ppm) for Zn and Cd (11), no further work was performed in this area, and the results still remain a mystery.

## 4.3  KINETIC CONSIDERATIONS

In all resonance fluorescence measurements the signal has been stated to be linearly dependent upon the value of the quantum efficiency $Y_{21}$ of the transition involving a lower state 1 and an upper state 2. By definition, $Y_{21}$ is generally given by the ratio between the number of quanta per unit time emitted as fluorescence and the number of quanta per unit time absorbed by the atoms in the lower level of the transition. This ratio results in the so-called *Stern-Volmer relationship* and represents the ratio of the transition probability per unit time for spontaneous emission to the total probability of deexcitation per unit time (12,13). Clearly, in the atomizer, there are several processes other than the

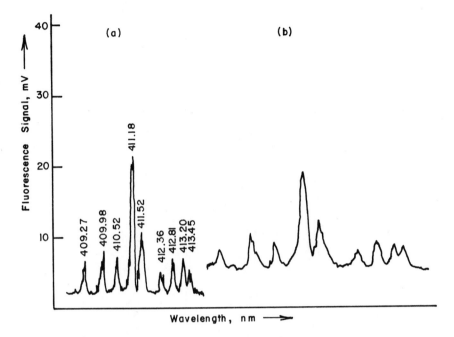

**Fig. 4.4.** Partial atomic fluorescence spectrum of vanadium-wavelength scanning; 50 $\mu$m slit (0.1 nm spectral bandwidth); 1000 ppm vanadium. ($a$) Laser excitation set at 411.18 nm; ($b$) laser excitation set at 370.35 nm, full-scale sensitivity increased 2.5 times as compared to ($a$), scan speed decreased to half as compared to ($a$). All these fluorescence transitions are due to excited Stokes stepwise-line fluorescence. From N. Omenetto, N. N. Hatch, L. M. Fraser, and J. D. Winefordner, *Spectrochim. Acta,* **28B,** 65 (1973). Reproduced by permission.

radiative ones, such as inelastic collisions and chemical reactions, concurring to the formation and/or destruction of an excited state. If a metal atom M in a gas of molecules Z is considered, the following inelastic process can be written

$$M_1 + Z \rightleftharpoons M_2 + Z \qquad (1)$$

proceeding in both directions because of the principle of microscopic reversibility (assuming the system is in thermodynamic equilibrium). The *overall rate* or velocity of the forward process is indicated as

$$v_{12} = k'_{12}[M_1][Z] \qquad (2a)$$

where $[M_1]$ and $[Z]$ are the numbers of atoms and molecules per unit volume. A similar relation holds for the rate of the reverse process (quenching). The coefficient $k'_{12}$ is the bimolecular *rate constant* for the reaction. Usually, the concentration $[Z]$ is included in the rate constant by

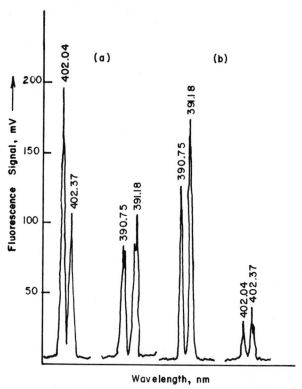

**Fig. 4.5.** Atomic fluorescence transitions of scandium-wavelength scanning; 50 $\mu$m slit (0.1 nm spectral bandwidth); 1000 ppm scandium. (*a*) Laser excitation set at 402.14 nm; (*b*) laser excitation set at 391.0 nm. From N. Omenetto, N. N. Hatch, L. M. Fraser, and J. D. Winefordner, *Spectrochim. Acta*, **28B**, 65 (1973). Reproduced by permission.

defining a unimolecular rate constant $k_{12} \equiv k'_{12}[Z]$ (pseudo-first-order kinetics). The *overall rate* of the reverse process shown in Eq. (1) is

$$v_{21} = k'_{21}[M_2][Z] \tag{2b}$$

The definition of quantum efficiency is given by

$$Y_{21} = \frac{A_{21}}{A_{21} + k_{21}} \tag{3}$$

where $A_{21}$ is the Einstein transition probability, in s$^{-1}$, for spontaneous emission. In terms of *lifetimes*,

$$Y_{21} = \tau/\tau_{sp} \tag{4}$$

where $\tau [\equiv (A_{21} + k_{21})^{-1}]$ represents the mean lifetime of level 2 and $\tau_{sp} [\equiv (A_{21})^{-1}]$ gives the radiative lifetime.

## TABLE 4.1
### Observed Fluorescence Transitions for Atomic Scandium[a]

| Laser excitation, nm | Fluorescence wavelength, nm | Type of fluorescence transitions |
|---|---|---|
| | 402.04 | resonance fluorescence and anti-Stokes stepwise-line fluorescence |
| 402.04 | 402.37 | excited resonance fluorescence and thermally assisted Stokes stepwise-line fluorescence |
| 402.37 | | |
| | 390.75 | thermally assisted anti-Stokes stepwise-line fluorescence (2 transitions) |
| | 391.18 | thermally assisted anti-Stokes stepwise-line fluorescence and excited thermally assisted anti-Stokes stepwise-line fluorescence |
| | 390.75 | resonance fluorescence and anti-Stokes stepwise-line fluorescence |
| 390.75 | 391.18 | thermally assisted Stokes stepwise-line fluorescence and excited resonance fluorescence |
| 391.18 | | |
| | 402.04 | Stokes stepwise-line fluorescence (2 transitions) |
| | 402.37 | Stokes stepwise-line fluorescence and excited Stokes stepwise-line fluorescence |

[a] From N. Omenetto, N. N. Hatch, L. M. Fraser, and J. D. Winefordner, *Spectrochim. Acta*, **28B**, 65 (1973). Reproduced by permission.

If we take a system characterized only by a ground level 1 and an excited level 2, then by definition

$$Y_{21} = \frac{(n_2 - n_2^{th})A_{21}}{B_{12}\rho_{\nu_{12}}n_1} \tag{i}$$

where $n_i$ are the actual populations of the levels, $n_i^{th}$ are the thermal population of the levels, $A_{21}$ is the spontaneous transition probability, $B_{12}$ is the Einstein coefficient for induced absorption, and $\rho_{\nu_{12}}$ is the spectral radiation density of a continuum-exciting source. Neglecting stimulated emission, the balance equations (see also Sect. 4.4) are

$$n_1(B_{12}\rho_{\nu_{12}} + k_{12}) = n_2(A_{21} + k_{21}) \tag{ii}$$

$$n_1^{th}k_{12} = n_2^{th}(A_{21} + k_{21}) \tag{iii}$$

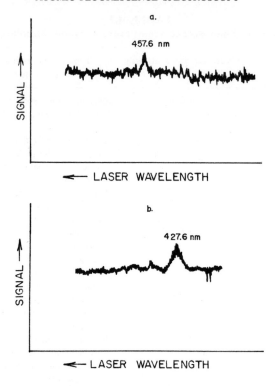

**Fig. 4.6.** Two-photon atomic fluorescence of cadmium and zinc. (*a*) Cadmium-monochromator at 228.8 nm, excitation scanned over 457.6 nm; (*b*) zinc-monochromator at 213.8 nm, excitation scanned over 427.6 nm. Reprinted with permission from L. M. Fraser and J. D. Winefordner, *Anal. Chem.*, **44**, 1444 (1972). Copyright by the American Chemical Society.

Subtracting (ii) from (iii) and remembering that $(n_1 - n_1^{th}) = -(n_2 - n_2^{th})$, then after rearranging

$$B_{12}\rho_{\nu_{12}}n_1 = (n_2 - n_2^{th})(A_{21} + k_{21} + k_{12}) \tag{iv}$$

and making use of (i),

$$Y_{21} = A_{21}/\{(A_{21} + k_{21})[1 + k_{12}/(k_{21} + A_{21})]\} \tag{v}$$

It can be easily seen from Eq. (iii) that the ratio between the square brackets in Eq. (v) is in most cases negligible compared to unity because of the Boltzmann factor. This derivation demonstrates that the formula given by Eq. (3) can be safely applied in most cases. See also reference 14.

In an atomizer where thermodynamic equilibrium prevails, the rate at which each particular process proceeds exactly equals the rate of the reverse process because of the *principle of detailed balance*. Therefore, $(v_1)_e = (v_2)_e$, and

$$(k_{12})_e/(k_{21})_e = [M_2]_e/[M_1]_e = (g_2/g_1)\exp(-E/kT) \tag{5}$$

where the Boltzmann distribution has been introduced. When various types of molecules exist, such as in a flame at atmospheric pressure, the overall rate constant $k$ is a summation of the specific rate constants corresponding to the various kind of particles. The general expression for the rate constant $k$ is given by the following relation:

$$k_{21} = \sum_j n_j \sigma_j \bar{u}_j \tag{6}$$

where $n_j$ is the density of quenching particles of kind $j$, in $cm^{-3}$; $\sigma_j$ is the effective cross section (at a given temperature $T$), in $cm^2$; and $\bar{u}_j$ is the mean relative velocity, in $cm\,s^{-1}$, of atoms and quenching particles. Remembering that

$$n_j = P_j/kT \quad \text{and} \quad \bar{u}_j = (8kT/\pi\mu_j)^{1/2} \tag{7}$$

where $P_j$ is the partial pressure of particles of kind $j$ and $\mu_j$ is the reduced mass of the atom and the particle of kind $j$, and introducing typical values for $n_j$ $(\sim 10^{18}\,cm^{-3})$, $\sigma_j$ $(\sim 10^{-15}\,cm^2)$, and $\bar{u}_j$ $(\sim 10^5\,cm\,s^{-1})$, an order-of-magnitude estimate for the quenching rate constant in a flame of $k_{21} \sim 10^8\,s^{-1}$ is obtained.

As will clearly emerge later, the quantum efficiency plays an essential role in the theoretical approach to saturation of an atomic transition in the presence of the strong radiation field provided by the laser.

## 4.4   BASIC FLUORESCENCE SIGNAL EXPRESSIONS

The dependence of the fluorescence signal upon the type and intensity of the excitation source, the quantum efficiency, and the concentration of absorbing atoms has been treated very extensively in the literature (4,13), and it can be generally stated that, with some unavoidable simplifying assumptions, the theoretical background of the technique has been firmly established and in most cases demonstrated experimentally.

The use of lasers as excitation source has the important consequence that the conventional photon transport equation (Beer's law) is strictly valid in the limit of zero incident light flux and is therefore an accurate approximation only for low-intensity sources. The high radiation density of a laser focused into an atomic vapor is able to completely redistribute the populations of the levels involved in the absorption process to the point where no more absorption occurs, that is, where the absorption coefficient decreases to zero as stimulated emission balances absorption. This effect has been commonly referred to as saturation of the optical transition (16–21). The fluorescence signal will then no longer be proportional to the source irradiance and will reach a limiting value depending upon the properties of the atomic system.

We present here a general derivation of the fluorescence radiance expression based upon simple kinetic considerations. For the sake of simplicity, the atomic system is characterized as an ensemble of atoms having only two energy levels 1 and 2 without multiplet splitting, with equal statistical weights $g_1 = g_2$, separated by the energy difference $E = h\nu_0 \geq 2$ eV, and with populations (number of atoms per unit volume) $n_1$ and $n_2$, with $n_1 + n_2 = n_T$, where $n_T$ is the total density of atoms.

Several important assumptions are made: (i) atoms are present as a trace component in a gas of molecules at thermodynamic temperature $T$ (2000–5000 K), that is, no self-absorption effects are considered; (ii) the pulsed laser radiation does not affect the energy distribution of the gas molecules, the velocity distribution of the atoms, or the temperature $T$ of the system; (iii) any coherence effects between absorbed and emitted photons are neglected; (iv) polarization effects are neglected; (v) the atomic system is homogeneous with regard to concentration and temperature; and finally (vi) the radiation density of the source is constant.

### 4.4.1 Temporal Behavior of Fluorescence Excited by a Pulsed Continuum Source

The temporal dependence of the concentration of the excited state can be represented by the following equation:

$$dn_2/dt = B\rho_{\nu_0}(n_1 - n_2) + k_{12}n_1 - n_2(A_{21} + k_{21}) \tag{8}$$

where $\rho_{\nu_0}$ stands for the uniform spectral volume energy density at $\nu = \nu_0$ (erg $Hz^{-1}$ $cm^{-3}$) of the laser, assumed to act as a spectral continuum, that is, with a frequency bandwidth much larger than the absorption bandwidth; $B$ (erg$^{-1}$ cm$^3$ s$^{-1}$ Hz) represents the Einstein coefficient for absorption and stimulated emission (in this case, $B_{12} = B_{21} = B$ because of our assumption of equal statistical weights); and all other terms have been previously defined (Sect. 4.3). Remembering that $\tau \equiv (A_{21} + k_{21})^{-1}$ and neglecting $k_{12}$ in comparison with $k_{21}$ because of the negligible Boltzmann factor for the transition and temperature values considered, Eq. (8) can be rearranged as follows:

$$dn_2/dt + \frac{n_2}{\tau}(2B\rho_{\nu_0}\tau + 1) = B\rho_{\nu_0}n_T \tag{9}$$

From the theory of radiation (1), the following relationship occurs:

$$B = (c/h\nu_0)\{1/(n_1 - n_2)\} \int k(\nu) \, d\nu$$

or

$$B = (c/h\nu_0) \int \sigma(\nu) \, d\nu \tag{10}$$

where $k(\nu)$ is the frequency-dependent absorption coefficient of the atomic vapor and $\sigma(\nu)$ is the frequency-dependent absorption cross section. Therefore, by substitution of Eq. (10) into Eq. (9),

$$\frac{dn_2}{dt} + \frac{n_2}{\tau} \left\{ \frac{2\tau}{h\nu_0} E_{\nu_0} \left[ \int \sigma(\nu) \, d\nu \right] + 1 \right\} = \frac{E_{\nu_0}}{h\nu_0} \left[ \int \sigma(\nu) \, d\nu \right] n_T \tag{11}$$

where use has been made of the relation $\rho_{\nu_0} = E_{\nu_0}/c$, where $c$ is the velocity of light and $E_{\nu_0}$ (erg s$^{-1}$ cm$^{-2}$ Hz$^{-1}$) represents the spectral irradiance of the source. Equation (11) can be simplified by defining (16) the saturation parameter $E_{\nu_0}^s$ as

$$E_{\nu_0}^s / h\nu_0 \equiv 1 \Big/ \left\{ 2\tau \left[ \int \sigma(\nu) \, d\nu \right] \right\} \tag{12}$$

Making use of Eq. (12), Eq. (11) becomes

$$\frac{dn_2}{dt} + \frac{n_2}{\tau} \left( 1 + \frac{E_{\nu_0}}{E_{\nu_0}^s} \right) = \frac{E_{\nu_0}}{h\nu_0} \, n_T \left[ \int \sigma(\nu) \, d\nu \right] \tag{13}$$

With the assumptions made at the beginning of this section, the proportional dependence of the fluorescence signal upon $n_2$ is given by

$$B_F = n_2 h\nu_0 \frac{l}{4\pi} (\tau_{sp})^{-1} \tag{14}$$

where $B_F$ is the absolute fluorescence radiance (erg s$^{-1}$ cm$^{-2}$ sr$^{-1}$) and $l$ (cm) denotes the depth of the homogeneous fluorescing volume in the direction of observation. By introducing this relation into Eq. (13),

$$\frac{dB_F}{dt} + \frac{B_F}{\tau} \left( 1 + \frac{E_{\nu_0}}{E_{\nu_0}^s} \right) = C E_{\nu_0} \tag{15}$$

where $C \equiv (l/4\pi)(n_T/\tau_{sp})[\int \sigma(\nu) \, d\nu]$. Equation (15) describes the temporal behavior of the fluorescence radiance during the excitation when the atomic vapor is irradiated with a pulsed spectral continuum source. We may apply Eq. (15) to describe several analytically useful limiting cases.

#### 4.4.1.1  Low Irradiance

In this case, $E_{\nu_0} \ll E^s_{\nu_0}$, and therefore Eq. (15) reads

$$\frac{dB_F}{dt} + \frac{B_F}{\tau} = CE_{\nu_0} \tag{16}$$

If we consider an idealized flat-topped pulse of light, of constant irradiance after time zero, that is, $\rho_{\nu_0}(t) = \rho_{\nu_0}$ at $t \geq 0$ and $\rho_{\nu_0}(t) = 0$ for $t < 0$, and with the boundary condition that $B_F$ equals zero at time zero, we have

$$B_F = CE_{\nu_0} \tau (1 - e^{-t/\tau}) \tag{17}$$

From this equation one can see that the approaching of a steady-state condition is governed by the ratio $(t/\tau)$. If steady-state conditions are attained, which is the usual case of atomic fluorescence excited with conventional sources and conventional modulation techniques, then $dB_F/dt = 0$, and the fluorescence radiance expression becomes

$$B_F = CE_{\nu_0} \tau = (l/4\pi)(\tau/\tau_{sp}) \left[ \int \sigma(\nu)\, d\nu \right] E_{\nu_0} n_T \tag{18a}$$

or

$$B_F = (l/4\pi) Y_{21} E_{\nu_0} \left[ \int k(\nu)\, d\nu \right] \tag{18b}$$

Equation (18b) shows the well-known *linear* dependence of the fluorescence radiance upon the source spectral irradiance and the quantum efficiency of the transition. The linear dependence upon the total atomic density holds here because of the assumption of dilute atomic vapor.

#### 4.4.1.2  High Irradiance

When $E_{\nu_0}/E^s_{\nu_0}$ is no longer negligible with respect to unity, then, proceeding as in the low-intensity case, the solution of Eq. (15) gives

$$B_F = CE_{\nu_0} \tau [E^s_{\nu_0}/(E_{\nu_0} + E^s_{\nu_0})] \left[ 1 - \exp\left\{ -\frac{t}{\tau}\left( \frac{E_{\nu_0} + E^s_{\nu_0}}{E^s_{\nu_0}} \right) \right\} \right] \tag{19a}$$

or

$$B_F = CE_{\nu_0} t_r [1 - \exp(-t/t_r)] \tag{19b}$$

where the *response time* of the system, $t_r$, has been introduced. This quantity is by definition given by

$$t_r \equiv \tau [E^s_{\nu_0}/(E_{\nu_0} + E^s_{\nu_0})] \tag{20}$$

One can easily see now that the attainment of steady-state conditions is governed by the ratio $t/t_r$ and therefore also by the spectral irradiance of the source. If the spectral irradiance is much less than the saturation irradiance, then $t_r$ is simply the effective lifetime. However, if $E_{\nu_0} \gg E_{\nu_0}^s$, then steady-state conditions might be warranted even if the width of the excitation pulse is narrow. For steady-state conditions, we have

$$B_F = CE_{\nu_0}\tau[E_{\nu_0}^s/(E_{\nu_0} + E_{\nu_0}^s)]$$

$$= (l/4\pi)E_{\nu_0}[E_{\nu_0}^s/(E_{\nu_0} + E_{\nu_0}^s)](\tau/\tau_{sp})n_T\left[\int \sigma(\nu)\,d\nu\right] \quad (21a)$$

or

$$B_F = (l/4\pi)Y_{21}\left[\int k(\nu)\,d\nu\right]E_{\nu_0}[E_{\nu_0}^s/(E_{\nu_0} + E_{\nu_0}^s)] \quad (21b)$$

or

$$B_F = (l/4\pi)Y_{21}\left[\int k^*(\nu)\,d\nu\right]E_{\nu_0} \quad (21c)$$

where

$$k^*(\nu) = k(\nu)[E_{\nu_0}^s/(E_{\nu_0} + E_{\nu_0}^s)] \quad (21d)$$

It is easy to see that when $E_{\nu_0} \ll E_{\nu_0}^s$, that is, for conventional excitation sources, Eq. (21) reduces to Eq. (18). However, for laser excitation, $E_{\nu_0}$ can approach $E_{\nu_0}^s$ and/or be much greater than $E_{\nu_0}^s$. If $E_{\nu_0} \gg E_{\nu_0}^s$, Eq. (21) becomes

$$(B_F)_{max} = (l/4\pi)(\tau/\tau_{sp})E_{\nu_0}^s n_T\left[\int \sigma(\nu)\,d\nu\right] \quad (22)$$

and remembering Eq. (12) and that $(\tau_{sp})^{-1} \equiv A_{21}$,

$$(B_F)_{max} = (l/4\pi)h\nu_0 A_{21}(n_T/2) \quad (23)$$

This equation gives the *maximal* fluorescence radiance obtainable for a given $n_T$. An *absolute* measurement of $(B_F)_{max}$ yields the product $n_T A_{21}$, all other parameters being known. (In principle, densities of species important in diagnosis of plasmas can be measured.) The advantage of determining this product through Eq. (23) is that neither the source parameters nor the temperature of the atomic system or the value of the quantum efficiency of the transition need be known (20).

The physical significance of the saturation parameters $E_{\nu_0}^s$ clearly results from Eq. (21). When $E_{\nu_0} = E_{\nu_0}^s$, $B_F$ becomes equal to $(B_F)_{max}/2$. In other words, $E_{\nu_0}^s$ represents the source spectral irradiance for which the maximum attainable value of the fluorescence radiance is decreased by a factor of 2. The experimental measurement of $E^s_{\nu_0}$ should offer a possibility of measuring the quantum efficiency of the transition (see Sect. 4.6.2).

### 4.4.2 Temporal Behavior of Fluorescence Excited by a Pulsed Monochromatic Source

When the laser output bandwidth is narrowed by some means to a value markedly smaller than the spectral absorption linewidth, then the first term in the right-hand side of Eq. (8), accounting for the radiatively induced processes, has to be replaced by the term $[\int B\rho(\nu)g(\nu - \nu_0)\,d\nu]\times$ $(n_1 - n_2)$. The factor $g(\nu - \nu_0)$ represents the *absorption lineshape function.* The function $g(\nu - \nu_0)\,d\nu$ can be defined as the probability that a given transition will result in the absorption (emission) of a photon whose energy lies between $h\nu$ and $h(\nu + d\nu)$. The curve $g(\nu - \nu_0)$ versus $\nu$ is normalized so that the total area under it is always unity. It is well known (1, 22) that the shape of this curve depends upon the physical processes responsible for it. See also Chapter 3.

If the absorbing atoms contribute *equally* to the absorption at any frequency within the absorption profile, then the function $g(\nu - \nu_0)$ is described by a Lorentzian profile and the broadening process is called *homogeneous.* If, on the other hand, a *particular frequency interval* in the absorption profile curve can be associated with a *particular set* of absorbing atoms, then the function $g(\nu - \nu_0)$ is described by a Gaussian profile and the broadening process responsible for it is called *inhomogeneous.* Typically, homogeneous broadening is due to collisions while inhomogeneous broadening is due to the Doppler effect. When both broadening processes are present in the atomic system, the lineshape function will be given by their convolution. With the assumption that both mechanisms act independently, the result is a convolution integral, known as the Voigt function. The profile of the absorption coefficient is therefore entirely determined by the Voigt function (1, 22).

Clearly, when the radiation density of the laser is such that the saturation regime will be approached, it is very important to know whether the laser can strongly interact with the entire atomic population. From the above definitions it follows that this will happen only if the laser bandwidth is greater than the inhomogeneous absorption line profile. On the contrary, if the laser bandwidth is much smaller than the inhomogeneously broadened absorption profile, then the atomic population may be thought as grouped in a number of Doppler intervals shifted in frequency relative to the laser frequency. In this case, the population numbers $n_1$ and $n_2$ in the derivation given for the continuum case now become $n_1(\nu)$ and $n_2(\nu)$ and refer *only* to a particular Doppler interval (18, 19).

Let the inhomogeneously broadened Doppler absorption profile be described as a superposition of many narrow subprofiles, each having a homogeneous width $\delta \nu_h$ due to collisions and spontaneous processes. The homogeneous width characterizes the bandwidth of interaction of the atoms and the monochromatic laser. The laser beam, at frequency $\nu$, traversing the atomic system in the $x$ direction can only be absorbed by atoms having a velocity component within the interval $v_x \pm (\delta v_x/2)$, where $v_x = (\nu - \nu_0)c/\nu_0$ and $\delta v_x = \delta \nu_h c/\nu_0$. If saturation occurs, the velocity-selective absorption process results in a "hole burning" in the population distribution of ground state atoms. The width of the hole will be given by the homogeneous width. By using two beams traversing the atomic system in opposite directions and tuning the laser across the absorption frequency of the atoms, the Doppler effect is nearly eliminated (23); this process is known as *saturation spectroscopy*.

For analytical purposes, we may formally evaluate the difference between continuum and line excitation by remembering that the definition of the absorption coefficient, for $g_1 = g_2$, is given by

$$k(\nu) = B(h\nu_0/c)(n_1 - n_2)g(\nu - \nu_0) \tag{24}$$

The lineshape function, with the help of Eq. (10), therefore becomes

$$g(\nu - \nu_0) = k(\nu) \Big/ \left[ \int k(\nu)\, d\nu \right] \tag{25a}$$

or, at the peak frequency $\nu = \nu_0$,

$$g(\nu_0) = k(\nu_0) \Big/ \left[ \int k(\nu)\, d\nu \right] \tag{25b}$$

where the integration extends over all frequencies for which the absorption coefficient differs noticeably from zero.

For *pure* Lorentzian and Gaussian shapes it is known that

$$[g(\nu_0)]_L = 2/\pi\delta\nu_L = 0.637/\delta\nu_L \tag{26a}$$

and

$$[g(\nu_0)]_D = (2/\delta\nu_D)(\sqrt{\ln 2}/\sqrt{\pi}) = 0.939/\delta\nu_D \tag{26b}$$

where $\delta\nu_L$ and $\delta\nu_D$ are the Lorentzian and Doppler half-widths, respectively. Therefore, the excitation rate evaluated at the peak frequency $\nu = \nu_0$ becomes

$$\rho B g(\nu_0) \equiv \rho B \left\{ k(\nu_0) \Big/ \left[ \int k(\nu)\, d\nu \right] \right\} \tag{27a}$$

where $\rho \equiv \int \rho(\nu)\, d\nu$ represents now the integrated spectral density of the source, where the integration extends over the whole spectral width of the exciting line. By introducing now the quantity

$$\delta\nu_{\text{eff}} \equiv \left[ \int k(\nu)\, d\nu \right] \bigg/ k(\nu_0) \qquad (27b)$$

as the effective width of the absorption line (12,15), we have

$$\delta\nu_{\text{eff}} = 1.571\delta\nu_L \qquad (28a)$$

in the case of a purely broadened Lorentzian line and

$$\delta\nu_{\text{eff}} = 1.064\delta\nu_D \qquad (28b)$$

in the case of a purely broadened Doppler line. In the case of combined Doppler and Lorentzian absorption profiles, $\delta\nu_{\text{eff}}$ will range somewhere between these two limiting values.

In the case of a line excitation, we may therefore replace $\rho_{\nu_0}$, the spectral energy density of the continuum source, by $(\rho/\delta\nu_{\text{eff}})$. This leads to the conclusion that a continuum source, whose irradiance within the limiting spectral bandpass (i.e., $\rho_{\nu_0}\delta\nu_s$, where $\delta\nu_s$ is the bandpass of the quasi-continuum laser or any spectrometric device used between the source and the flame, depending upon which bandpass is the smaller) equals the integral irradiance of a line source (i.e., $\rho = \int \rho_\nu\, d\nu \sim \rho'_{\nu_0}\delta\nu_{\text{eff}}$, where $\rho'_{\nu_0}$ is the peak line source spectral density) will be *less* effective in exciting the atomic fluorescence by a factor equal to $\delta\nu_{\text{eff}}/\delta\nu_s$. While this is well known for conventional low-irradiance sources, it does not hold for a laser if saturation is already achieved with a continuum laser source. However, it is clear that, for a given atomic system, the laser power needed to attain a certain degree of saturation decreases by a factor $\delta\nu_{\text{eff}}/\delta\nu_s$ when the laser bandwidth is very narrow.

If we can disregard frequency fluctuations of the monochromatic output, a narrow band laser is to be preferred over a broad band one in the case of scattering problems, which are definitely more severe with continuum excitation.

### 4.4.3  Numerical Evaluation of Nonlinear Fluorescence Behavior

It is now useful to give a quantitative estimation of the experimental conditions necessary to approach saturation in atomic fluorescence spectrometry (20). When Eq. (8) is solved for the ratio of the atomic populations, $n_2/n_1$, with the assumption of steady-state conditions, the following expression is obtained:

$$\frac{n_2 g_1}{n_1 g_2} = \frac{k_{12} + B_{12}\rho_{\nu_0}}{k_{12}\exp(E/kT) + (D + \rho_{\nu_0})B_{12}} \qquad (29)$$

where the statistical weights of the levels have now been considered and use has been made of the detailed balance relationship between the collisional rate constants and of the relations

$$A_{21} = B_{21}(8\pi h\nu_0^3/c^3) \equiv B_{21}D \tag{30}$$

and

$$B_{21}g_2 = B_{12}g_1 \tag{30a}$$

From the Stern-Volmer equation for the fluorescence quantum efficiency $Y_{21}$,

$$k_{21} = A_{21}(1 - Y_{21})/Y_{21} \tag{31}$$

Combining the above equations, the following value for $k_{12}$ is obtained:

$$k_{12} = B_{12}D[(1 - Y_{21})/Y_{21}]\exp(-E/kT) \tag{32}$$

which, after substitution into Eq. (29), gives the following general expression:

$$\frac{n_2 g_1}{n_1 g_2} = \frac{D[(1 - Y_{21})/Y_{21}]\exp(-E/kT) + \rho_{\nu_0}}{(D/Y_{21}) + \rho_{\nu_0}} \tag{33}$$

The general validity of Eq. (33) can be realized as follows: (i) When $\rho_{\nu_0} = 0$, i.e., in the absence of external radiation, the ratio of the populations deviates from the Boltzmann equilibrium ratio by the factor $1 - Y_{21}$; this factor is well known to account for radiative disequilibrium. (ii) The Boltzmann equilibrium ratio is obtained when $\rho_{\nu_0}$ is given by $D[\exp(E/kT) - 1]^{-1}$, i.e., when $\rho_{\nu_0}$ represents the Planck radiation density at the temperature $T$. (iii) When $\rho_{\nu_0}$ grows very large, $n_2$ is no more linearly related to the source density $\rho_{\nu_0}$; and when $\rho_{\nu_0}$ approaches infinity, $n_2/n_1$ approaches unity; with the assumption of equal statistical weights, $n_2$ approaches half of the total population density.

Let us now approximate the laser to a spectral continuum source traversing the atomic vapor as homogeneous beam of cross section $S$ (in $cm^2$) and characterized by a radiant flux $\Phi$ (in $erg\,s^{-1}$) and a spectral width $\delta\nu_s$ (in Hz) much larger than the absorption spectral linewidth. Therefore,

$$\rho_{\nu_0} = \frac{E_{\nu_0}}{c} = \frac{\Phi}{\delta\nu_s Sc} \tag{34}$$

where $c$ ($cm\,s^{-1}$) is the speed of light. Using this equation, expressing $D$ numerically and converting frequencies into wavelengths (in nm), Eq. (33) now reads

$$\frac{n_2 g_1}{n_1 g_2} = \frac{(1 - Y_{21})\exp(-E/kT) + z}{1 + z} \tag{35}$$

where $z = 6.6 \times 10^{-25} Y_{21} \Phi \lambda_0^5 / \delta \lambda_s S$ or $z \equiv 6.6 \times 10^{-25} Y_{21} \lambda_0^5 E_{\lambda_0}$, where the units are $\text{erg s}^{-1}$ for $\Phi$, nm for $\lambda_0$, nm for $\delta \lambda_s$, $\text{cm}^2$ for $S$, and $\text{erg s}^{-1} \text{cm}^{-2} \text{nm}^{-1}$ for $E_{\lambda_0}$. If the laser bandwidth is narrow, $E_{\lambda_0}$ must be replaced by $E/\delta \lambda_{\text{eff}}$, where $E$ is the line integral irradiance (in $\text{W cm}^{-2}$) of the laser and $\delta \lambda_{\text{eff}}$ will be an intermediate value between those given by Eqs. (28a) and (28b).

The parameter $z$ clearly defines the nonlinear optical behavior of the fluorescence signal. If $z$ is much less than unity, the linear case applies, while if $z$ is much greater than unity, saturation is approached (if $z > 10$, saturation is reached for all practical purposes). It is important to stress here that the saturation behavior is governed by the product $\Phi \cdot Y_{21}$ and therefore depends upon the medium in which atoms are present. If strong quenching occurs, $Y_{21}$ will be very low and more source power will be needed to saturate the optical transition. On the other hand, the outcome of Eq. (23), that is, the removal of the $Y_{21}$ dependence when saturation is reached and the maximal fluorescence radiance is attained, clearly results also from Eq. (35) when $z$ is very large because of $\Phi$ being *above* a certain threshold value.

As an example of calculation in practical situations, the following parameters were measured (20) for a tunable dye laser: $\Phi = 7.0 \pm 0.7 \text{ kW}$, $\delta \lambda_s = 3.5 \pm 0.3 \text{ Å}$, $S = 0.4 \pm 0.1 \text{ mm}^2$. Thus, for the thallium transition at 377.6 nm in an air–acetylene flame ($Y = 0.03$), $z = 7.64$, that is, saturation was approached within 12%.

As already stated, the saturation spectral irradiance $E_{\nu_0}^s$ is defined as the spectral irradiance for which the fluorescence radiance is within a factor of 2 of the maximum possible fluorescence. This parameter is related to the atomic system via Eq. (12). From the classical theory of radiation,

$$\int \sigma(\nu) \, d\nu = (\pi e^2 / mc) f_{12} \tag{36}$$

and

$$(\tau_{sp})^{-1} = (8 \pi^2 \nu_0^2 e^2 / mc^3) f_{12} \tag{37}$$

where $e$ and $m$ are the charge and mass of the electron respectively, and $f_{12}$ is the absorption oscillator strength. Therefore,

$$E_{\nu_0}^s = (4 \pi h \nu_0^3 / c^2)(\tau_{sp}/\tau) = (4 \pi h \nu_0^3 / c^2)(Y_{21})^{-1} \tag{38}$$

or, converting again frequencies to wavelengths (in nm),

$$E_{\lambda_0}^s = 7.6 \times 10^{23} (\lambda_0)^{-5} (Y_{21})^{-1} \tag{39}$$

If line excitation is used, $\delta \lambda_{\text{eff}}$ has to be taken into account, as shown before.

It is finally instructive to evaluate the radiative transition rates as compared to the collisional (quenching) rates estimated in Sect. 4.3. For a continuum source, the radiational rate constant will be given by the product $B\rho_{\nu_0}$. Expressing $\rho_{\nu_0}$ in terms of $E_{\nu_0}$, see Eq. (34), and using the relationship between $B$ and $\tau_{sp}$, Eq. (30), as well as that between $\tau_{sp}$ and $f_{12}$, Eq. (37), the following expression is obtained:

$$B\rho_{\nu_0} = 4.4 \times 10^3 \lambda^3 E_{\lambda_0} f_{12} \qquad (40)$$

Again, $\delta\lambda_{\text{eff}}$ has to be considered for a line source.

Now if we take the experimental laser parameters used for the calculation of $z$ (20), we estimate that, for wavelengths in the visible near-ultraviolet and for strong lines ($f \approx 1$), the excitation rate constant is of the order of $10^{11}\,\text{s}^{-1}$, that is, at least three orders of magnitude larger than the quenching rate constant in a flame at atmospheric pressure.

### 4.4.4 Extension to Multilevel Atomic Systems

The theory developed in the previous sections was aimed at emphasizing as simply as possible the fundamental outcome of the influence of the laser radiation upon a two-level atomic system. Clearly, a more general approach is needed in most experimental situations in atomic fluorescence spectroscopy, because very few, if any, systems can be approximated as two-level systems. *The presence of one or more additional levels modifies the equations given, with the important consequence that the value of the laser radiation density necessary to achieve a certain degree of saturation and the limiting value of the excited atom fraction in respect to the total atomic population are changed.* We assume here that the laser excitation still couples two levels, which we indicate as level $p$ (ground state) and level $s$ (one of the possible excited states); but we also consider the presence of a level manifold of energy greater than $s$, designated as $q$, and of a level manifold, designated as $r$, whose energy is intermediate between levels $p$ and $s$. By considering all possible processes of excitation and deexcitation of level $s$, we may write for the rate of change of the population $n_s$, in the *absence* of laser excitation, the following general equation:

$$dn_s/dt = \left[ n_p k_{ps} + \sum_{q>s} n_q (k_{qs} + A_{qs}) + \sum_{r<s} n_r k_{rs} \right] +$$
$$- n_s \left[ (k_{sp} + A_{sp}) + \sum_{s>r} (k_{sr} + A_{sr}) + \sum_{s<q} k_{sq} \right] \qquad (41)$$

The first term on the right-hand side gives the total (collisional plus spontaneous radiative) rate of population of level $s$, while the second term takes into account all possible channels of depopulation for level $s$. When the laser-induced radiation rates are considered, we have to add an extra term given by

$$B_{ps}\rho_{\nu_{ps}}(n_p - n_s)$$

in the case of spectral continuum excitation, or

$$B_{ps}\left[\int \rho(\nu)\, g(\nu - \nu_0)\, d\nu\right](n_p - n_s)$$

for narrowband excitation. Correspondingly, we can write a similar equation for the temporal dependence of the population of the ground state $p$ as well as the intermediate state $s$.

In order to present the results in a reasonably simple way, we restrict ourselves to the case of a three-level system (see Fig. 4.7). Here, levels 1 and 3 are coupled by the laser excitation while level 2 is an intermediate level. For the sake of generality, level 2 can be radiatively coupled with level 1 only, with level 3 only, or with both. Again, we consider the laser radiation as a spectral continuum at frequency $\nu_{13}$.

According to Eq. (41),

$$dn_3/dt = R_{13}n_1 + R_{23}n_2 - n_3(R_{32} + R_{31}) \tag{42}$$

and

$$dn_2/dt = R_{12}n_1 + R_{32}n_3 - n_2(R_{23} + R_{21}) \tag{43}$$

where, in the case of radiative coupling of level 2 with both levels 3 and 1, the following expressions hold:

$$\begin{aligned}
R_{13} &\equiv k_{13} + B_{13}\rho_{\nu_0} \\
R_{23} &\equiv k_{23} \\
R_{32} &\equiv k_{32} + A_{32} \\
R_{31} &\equiv k_{31} + A_{31} + B_{31}\rho_{\nu_0} \\
R_{12} &\equiv k_{12} \\
R_{21} &\equiv k_{21} + A_{21}
\end{aligned} \tag{44}$$

As mentioned before, level 2 can be coupled radiatively with level 1 only

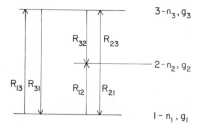

Fig. 4.7. Oversimplified sketch of a three level system, where levels can be coupled both radiationally and collisionally.

$(A_{32} = 0)$ or with level 3 only $(A_{21} = 0)$, and the $R$ values change accordingly. Solving for $n_1$, $n_2$, and $n_3$, assuming photostationary conditions,

$$n_1 = R_{21}R_{32} + R_{31}(R_{21} + R_{23})$$
$$n_2 = R_{31}R_{12} + R_{32}(R_{13} + R_{12}) \tag{45}$$
$$n_3 = R_{12}R_{23} + R_{13}(R_{21} + R_{23})$$

Since we are interested in the behavior of the system under the influence of a powerful laser beam, we can further simplify the equations by safely assuming that the collisional excitation rates to level 3 from levels 1 and 2 (i.e., $k_{13}$ and $k_{23}$) are negligible with respect to the radiatively induced rates. With these simplifications, the use of Eqs. (45) and (44) gives the following ratio for the population densities:

$$\frac{n_3}{n_1} = \frac{B_{13}\rho_{\nu 0}}{B_{31}\rho_{\nu 0} + (k_{32} + A_{32} + k_{31} + A_{31})} \tag{46}$$

If we define the quantum efficiency $Y_{31}$ as

$$Y_{31} = \frac{A_{31}}{(A_{31} + k_{31} + A_{32} + k_{32})} \tag{47}$$

then, with the help of Eq. (30),

$$\frac{n_3 g_1}{n_1 g_3} = \frac{\rho_{\nu_0} Y_{31}}{D + \rho_{\nu_0} Y_{31}} \tag{48}$$

At the limit of infinite radiation density, the laser is therefore again able to completely redistribute the populations of levels 1 and 3. However, because of the *lower* value of $Y_{31}$ as compared with that of a two-level system, *the power needed to achieve a certain degree of saturation is increased.* This is in accordance with the definition of saturation irradiance given by Eq. (39). Therefore, *in this particular case*, neglecting an intermediate level would lead to underestimate the laser power necessary to approach saturation.

Another consequence can be derived if we consider the fraction of excited atoms in level 3 with regard to the *total* atomic density (18). From Eqs. (45) and (44), we have

$$\frac{n_3}{n_T} \equiv \frac{n_3}{n_1 + n_2 + n_3}$$

$$= \frac{B_{13}\rho_{\nu_0}(k_{21} + A_{21})}{\begin{aligned}&B_{13}\rho_{\nu_0}(k_{21} + A_{21}) + (k_{31} + A_{31} + B_{31}\rho_{\nu_0})k_{12} + (k_{32} + A_{32}) \times \\ &\times (B_{13}\rho_{\nu_0} + k_{12}) + (k_{21} + A_{21})(k_{32} + A_{32} + k_{31} + A_{31} + B_{31}\rho_{\nu_0})\end{aligned}} \tag{49}$$

Equation (49), after rearranging, becomes

$$\frac{n_3}{n_T} = \frac{(B_{13}\rho_{v_0}) \Big/ \left(1 + \dfrac{k_{12}}{A_{21} + k_{21}}\right)}{B_{31}\rho_{v_0} + (A_{31} + k_{31} + A_{32} + k_{32}) + \\ + B_{13}\rho_{v_0} \left(1 + \dfrac{A_{32} + k_{32}}{A_{21} + k_{21}}\right) \Big/ \left(1 + \dfrac{k_{12}}{A_{21} + k_{21}}\right)\Big]} \tag{49a}$$

and again, using Eqs. (47) and (30),

$$\frac{n_3}{n_T} = \frac{(g_3/g_1)\rho_{v_0} Y_{31} \Big/ \left(1 + \dfrac{k_{12}}{k_{21} + A_{21}}\right)}{D + \rho_{v_0} Y_{31}[1 + (g_3/g_1)M]} \tag{50}$$

where

$$M \equiv [1 + (A_{32} + k_{32})/(A_{21} + k_{21})]/[1 + k_{12}/(k_{21} + A_{21})] \tag{50a}$$

In the limit of extremely high radiation density, one can see that the ratio $n_3/n_T$ will now reach the limiting value $g_3/(g_1 + g_3)$, that is, 0.5 if $g_1 = g_3$, *only* if both ratios contained in Eq. (50a) are negligible with respect to unity.

The error made in considering a three-level system as a two-level system can be also estimated as follows. Let $n_1 + n_2 + n_3 = n_T$, so that $n_1 = n_T - n_3(1 + n_2/n_3)$. Thus, it can be seen that neglecting level 2 leads to an error which depends upon the ratio $n_2/n_3$. This ratio can be expressed by means of Eqs. (45) and (44) as

$$\frac{n_2}{n_3} = \frac{k_{12}(k_{31} + A_{31} + k_{32} + A_{32}) + k_{12}B_{31}\rho_{v_0} + B_{13}\rho_{v_0}(k_{32} + A_{32})}{B_{13}\rho_{v_0}(k_{21} + A_{21})} \tag{i}$$

In the limit of $\rho_{v_0}$ approaching infinity, Eq. (i) becomes

$$\frac{n_2}{n_3} = \frac{(g_1/g_3)k_{12} + (k_{32} + A_{32})}{k_{21} + A_{21}} = (g_1/g_3)\frac{k_{12}}{k_{21} + A_{21}} + \frac{k_{32} + A_{32}}{k_{21} + A_{21}} \tag{ii}$$

where use has been made of the relation $B_{31} = (g_1/g_3)B_{13}$. It is easy to see that

$$\frac{k_{12}}{k_{21} + A_{21}} = (1 - Y_{21})(g_2/g_1)\exp(-E_2/kT) \tag{iii}$$

and therefore the error made by neglecting level 2 is given by

$$\frac{n_2}{n_3} = (1 - Y_{21})(g_2/g_3)\exp(-E_2/kT) + \{(k_{32} + A_{32})/(k_{21} + A_{21})\} \tag{iv}$$

## 4.5 SHAPE OF CURVES OF GROWTH

The analytical practice of atomic fluorescence spectroscopy, and in general of all spectroscopic techniques, relies entirely upon the construction of the experimental curve of growth, which represents the plot of the analytical signal, or the log of the signal, versus the element concentration, or the log of the concentration. Extensive literature exists on the shape of such curves in atomic fluorescence (4,13,24). Generally, the lack of information on some experimental parameters does not allow the prediction of the exact shape of the curve. However, its general behavior is clearly understood; and therefore to perform the measurements in optimal conditions, the analyst should be aware of the shape that the curve of growth is expected to take.

Ideally, at low atomic concentrations (low optical densities) the integrated radiance of the fluorescence signal is linearly related to the atomic concentration, and this holds for continuum as well as for line sources of excitation. At high atomic concentrations (high optical densities), the relationship between the integrated fluorescence radiance and the atomic concentration becomes complex, the essential feature of the curve being a zero slope in the case of a continuum source and a negative slope in the case of a line source. An additional complication, which is typical of the fluorescence technique, derives from the geometry of illumination[4] and observation. A prefilter effect may be present because of the weakening of the excitation beam in a region that is not observed by the detector. Likewise, a postfilter effect occurs if there is an unexposed region between the illuminated volume and the detector, this effect being also called *self-reversal* (see Fig. 4.8). The worst case is of course given by the presence of both effects, which therefore alter the shapes of the curves of growth.

To describe the influence of the laser radiation upon the form of the analytical fluorescence curve on a qualitative basis, it is not necessary to complicate the treatment by taking into account pre- and postfilter effects. Therefore, only the phenomenon of self-absorption will be considered, with the idealized illumination geometry depicted in Fig. 4.9, where the absorption path $L$ is restricted to a vanishingly small interval while the fluorescence emission path $l$ toward the detector can be arbitrarily long. The assumption of homogeneity of atomic concentration

---

[4] In the present case, only right-angle illumination–observation is considered. However, front-surface illumination–observation (13) prevents such effects as prefilter and postfilter losses. Because the treatment given in reference 13 is generally satisfactory for laser excitation and because nothing is added to the present discussion to consider other geometries, only the conventional right-angle case is considered here.

**Post-Filter Effect**

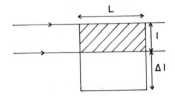

**Pre-Filter Effect**

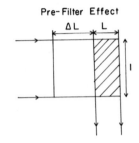

**Combined Post/Pre-Filter Effects**

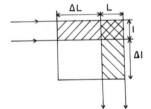

**Fig. 4.8.** Prefilter and postfilter effects in atomic fluorescence.

and source radiation throughout the observed region are retained, however.

Referring to Fig. 4.9, the contribution of section $dx$ to the fluorescence spectral radiance is given by

$$dB_F(\nu,x) = (A_{21}n_2h\nu/4\pi)g(\nu - \nu_0)\exp[-k(\nu)x]\,dx \qquad (51)$$

Integration of this equation over $x$ yields for the spectral radiance of the outgoing fluorescence radiation at frequency $\nu$

$$B_F(\nu) = (A_{21}n_2h\nu/4\pi)g(\nu - \nu_0)\{1 - \exp[-k(\nu)l]\}/k(\nu) \qquad (52)$$

Substitution for $k(\nu)$ in the denominator of the right-hand side of Eq. (52) the expression given by Eq. (24) gives

$$B_F(\nu) = (C/4\pi)(A_{21}/B)[n_2/(n_1 - n_2)]\{1 - \exp[-k(\nu)l]\} \qquad (53)$$

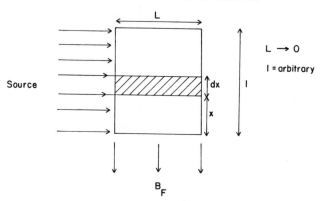

**Fig. 4.9.** Idealized geometry of illumination and observation of fluorescence radiance.

Integrating over frequency $\nu$, the fluorescence radiance of the whole atom line is given by

$$B_F = \int B_F(\nu) \, d\nu = (C/4\pi)(A_{21}/B)[n_2/(n_1 - n_2)]A_t \qquad (54)$$

where $A_t$ is called the total absorption factor and is defined as

$$A_t \equiv \int_{\text{line}} \{1 - \exp[-k(\nu)l]\} \, d\nu \qquad (55)$$

Equation (54) describes the influence of self-absorption on the radiance of the outgoing fluorescence radiation. When the laser radiation density is very high, the populations of the two levels coupled by the laser equalize, that is, $n_1 = n_2$. In this case, the absorption coefficient $k(\nu)$ will approach zero; compare Eq. (24). A series expansion of the factor $A_t$ then gives, at the limit of $k(\nu)l \to 0$,

$$A_t = l \int k(\nu) \, d\nu \qquad (56)$$

and Eq. (54) becomes

$$B_F = (C/4\pi)(A_{21}/B)[n_2/(n_1 - n_2)]l \int k(\nu) \, d\nu \qquad (57)$$

or, with the help of Eq. (10),

$$B_F = (l/4\pi)n_2 h\nu A_{21} = (l/4\pi)h\nu A_{21}(n_T/2) \qquad (58)$$

where the second expression holds for a two-level system with equal statistical weights.

)nclusion, if the system holds at saturation *for any value of the total concentration* $n_T$, the vapor becomes transparent to the resonance _____ diation, and therefore the *same expressions given in the case of negligible self-absorption result.*

These theoretical predictions have been qualitatively borne out experimentally using tunable dye lasers (25, 26). Figures 4.10 and 4.11 show the effect of decreasing the laser irradiance on the atomic fluorescence curves for thallium, indium, and strontium aspirated in an air–acetylene flame. Note from the insert in Fig. 4.10 that the laser beam was focused at the edge of the flame so as to minimize self-reversal effects, which if present

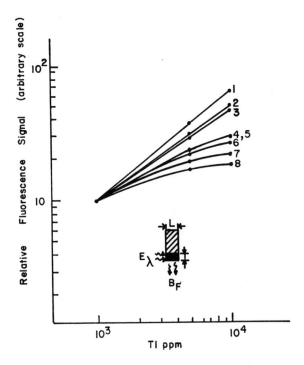

**Fig. 4.10.** Effect of source irradiance on the shape of resonance fluorescence curves for thallium (illumination geometry shown by insert). The fluorescence signals for each curve are relative to the signal from 1000 ppm of Tl for that specific irradiance: (1) focusing lens, no filters; (2) focusing lens, 50% transmittance filter between source and sample; (3) focusing lens, 10% transmittance filter between source and sample; (4) focusing lens, 0.1% transmittance filter between source and sample; (5) no lens, no filters; (6) no lens, 50% transmittance filter between source and sample; (7) no lens, 10% transmittance filter between source and sample; (8) no lens, 0.1% transmittance filter between source and sample. From N. Omenetto, L. P. Hart, P. Benetti, and J. D. Winefordner, *Spectrochim. Acta,* **28B,** 30 (1973). Reproduced by permission.

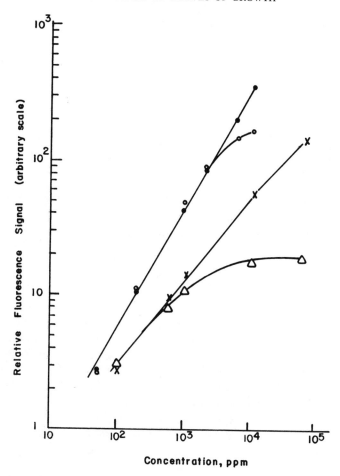

**Fig. 4.11.** Resonance fluorescence calibration curves for indium and strontium. The fluorescence signals for the different curves are not reported on the same scale: (●) indium, focusing lens; (○) indium, no lens; (×) strontium, focusing lens; (△) strontium, no lens. From N. Omenetto, L. P. Hart, P. Benetti, and J. D. Winefordner, *Spectrochim. Acta,* **28B,** 301 (1973). Reproduced by permission.

would have hindered the observation of the phenomenon. The same effect is shown in Fig. 4.12, which demonstrates the occurrence of saturation for magnesium atoms in an air–acetylene flame. As one can see, because of self-absorption the fluorescence due to 10 ppm is *higher* than that given by 100 ppm if the irradiance level is kept below $0.1 \, kW \, cm^{-2}$. At higher irradiances the opposite holds, demonstrating that the influence of self-absorption decreases.

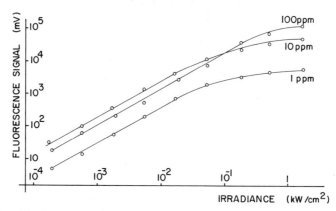

**Fig. 4.12.** Effect of laser irradiance on shape of saturation curve of magnesium at three different concentration levels. From J. Kuhl and H. Spitschan, *Opt. Commun.*, **7**, 256 (1973). Reproduced by permission.

## 4.6  LOCALIZED DIAGNOSTICS

Physical characterization of a plasma requires the knowledge of several fundamental parameters such as the plasma temperature, electron, ion and radical concentration, and so on. Most of the diagnostic techniques available use stainless steel or silica probes placed somewhere in the combustion mixture and coupled with conventional systems of analysis such as gas chromatography, infrared spectroscopy, and mass spectrometry. The insertion of a probe into the plasma allows for detailed spatial resolution of its state since sampling of a small volume can be realized. This is mandatory whenever strong inhomogeneities in composition and temperature exist due to the very nature of the combustion process.

Emission and absorption methods can be used to provide average "line of sight" measurements from which the measured quantity can be derived by somewhat elaborate procedures, such as the Abel inversion method (27). In principle, the fluorescence technique has to be considered a local sampling technique because only the region at the intersection of the excitation beam and the optical path of the detector is sampled. If the source and the observation region are narrow, the sampled volume can be very small, say, less than 1 mm$^3$. Early in 1962, fluorescence excited by a powerful thin electron beam (up to 30 keV) was used as a tool for local density measurements in a rarefied gas flow. The selected fluorescence volume was approximately equal to 1 mm$^3$ (28).

The advent of lasers tunable over most of the near-ultraviolet and visible range has opened new possibilities in the field of plasma diagnostics

(29–31). Raman spectroscopy has been successfully used as a means of investigating gaseous flame combustion products (32). Fluorescence from the (0,0) band of the $A^2\Delta$–$X^2\Pi$ electronic transition of CH in an oxyacetylene flame at atmospheric pressure was reported (33). Low concentrations of other radicals such as OH, CN, and SH have been also detected by resonance fluorescence excited with tunable dye lasers (34–36), and it is easy to foresee attractive analytical applications of molecular fluorescence in flames at atmospheric pressure (39).

The application of laser-excited fluorescence as a powerful tool for flame diagnostics is somewhat beyond the scope of the present chapter. However, it is interesting to speculate about the possibility of measuring the flame temperature and the quantum efficiency of atoms in flames by simply referring to the theory outlined in the previous paragraphs. Also, the measurement of spatial densities will be briefly discussed.

### 4.6.1  Flame Temperature

It has been demonstrated (37, 38) that the ratio of the anti-Stokes to the Stokes direct-line fluorescence signals obtained when a thallium solution is aspirated into a flame irradiated by a continuum source of excitation allows the calculation of the flame temperature. This method requires the calibration of the electro-optical setup as well as knowledge of the source spectral irradiances at the wavelengths of interest. If we use a laser whose power at both excitation wavelengths is sufficient to saturate the transitions, then the populations of both levels coupled by the laser will remain locked to the ratio of their respective degeneracies, and no knowledge of the source power is needed. Furthermore, if the laser is pulsed and we can measure the *peak of the time-resolved fluorescence waveform*, that is, at the point where $dB_F/dt = 0$, then the ratio of the signals emitted as resonance and anti-Stokes fluorescence yields directly the flame temperature.

Referring to Fig. 4.13, let the (homogeneous) laser beam, whose bandwidth is much larger than the absorption linewidth, be tuned to the transition $1 \rightarrow 3$ (377.6 nm), the fluorescence being measured at the same wavelength (resonance fluorescence). Under saturation conditions,

$$n_3 g_1 / n_1 g_3 = 1 \tag{59a}$$

and

$$n_1 + n_3 = (n_1)_{th} \tag{59b}$$

where $(n_1)_{th}$ is the population of the ground level *prior* to the laser irradiation. Equation (59a) then becomes

$$n_3 = [g_3/(g_1 + g_3)](n_1)_{th} \tag{60}$$

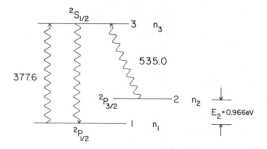

**Fig. 4.13.** Oversimplified sketch indicating the thallium transitions used for flame temperature measurements.

Now let the laser be tuned to the transition $2 \to 3$ (535.0 nm) while the fluorescence is still measured at 377.6 nm (anti-Stokes direct-line fluorescence). Again, for saturation conditions, the laser will equalize the weighted populations of levels 2 and 3, therefore giving

$$n_3' g_2 / n_2 g_3 = 1 \tag{61a}$$

and

$$n_3' = [g_3/(g_2 + g_3)](n_2)_{th} \tag{61b}$$

where $n_3'$ is now the population of level 3 induced by excitation at $\nu_{23}$ and $(n_2)_{th}$ is again the population of level 2 *prior* to the laser irradiation. If the system is in thermodynamic equilibrium before the irradiation, the ratio of Eqs. (60) and (61b) gives

$$n_3/n_3' = (g_2 + g_3)/(g_1 + g_3)(g_1/g_2) \exp(E_2/kT) \tag{62}$$

The left-hand side of Eq. (62) is simply the ratio of the resonance fluorescence signals to the anti-Stokes direct-line fluorescence signals. Because the same frequency is involved, no calibration of the electro-optical detection system is required. A similar scheme, involving the fluorescence of barium ion, was proposed by Measures as a means of evaluating the electron temperature of a low-temperature plasma ($\leq 5000$ K) early in 1968 (29).

It is worth stressing again here that this method can be applied only *if saturation occurs and if the fluorescence waveform can be resolved by the detection system. When the entire fluorescence waveform is integrated, as for example in a measurement with a boxcar whose gate aperture is similar to or larger than the fluorescence pulse, the usual complications arise because of the collisional terms which have to be taken into account in deriving a final relationship.* Preliminary attempts to obtain *relative* temperature profiles in nonshielded air–acetylene and nitrous oxide–acetylene flames at atmospheric pressure have been reported, demonstrating the expected existence of significant thermal gradients toward the outer, cooler zones of the flame (39).

If the three-level system of thallium is treated as in Sect. 4.4.4 and the collisional (de-)activation of level 2 with level 1 is considered, we come to the following approximate relationship:

$$n_3/n_3 = (k_{21}/k_{12})\left[\left(1+\frac{k_{31}+A_{31}}{k_{21}}\right)\middle/\left(2+\frac{k_{32}+A_{32}}{k_{21}}\right)\right]$$ (i)

where, for the sake of simplicity, the statistical weights have been taken as equal, that is, $g_1 = g_2 = g_3$. Equation (i) still shows the temperature dependence of the fluorescence ratio $n_3/n_3'$ via the ratio

$$k_{21}/k_{12} = \exp(E_2/kT)$$ (ii)

However, in this case it would not be possible to obtain an *absolute* value for $T$ because the collisional rate constants $k_{21}$, $k_{31}$, and $k_{32}$ are not known.

### 4.6.2 Quantum Efficiency

It has been shown in Sect. 4.4.3, see Eq. (35), that the saturation parameter $z$ contains the product of the laser spectral irradiance times the quantum efficiency of the transition. From Eq. (35), if $z$ is much greater than the first term in the numerator, it is easy to show that

$$n_T/n_2 = (g_1/g_2)(1/z) + [(g_1+g_2)/g_2]$$ (63)

It follows that the plot of $(B_F)^{-1}$, proportional to $(n_2)^{-1}$, as a function of $\Phi^{-1}$, proportional to $z^{-1}$, should yield a straight line whose slope varies with $Y^{-1}$. Moreover, the ordinate axis of this plot can be simply scaled in *absolute* values of $n_T/n_2$ because, for $z$ approaching infinity, $n_T/n_2$ approaches the ratio $(g_1+g_2)/g_2$.

Again, for a three-level system, expressing $n_1$ as a function of $n_T = n_1 + n_2 + n_3$ and proceeding as in Sect. 4.4.4, we have to modify Eq. (63) as follows:

$$(n_T/n_3) = (g_1/g_3)(1/z) + \left[\frac{(g_1+g_3)}{g_3} + (1-Y_{21})\frac{g_2}{g_3}\exp(-E_2/kT) + \frac{(k_{32}+A_{32})}{k_{21}}\right]$$

As a consequence, it is not possible to scale the ordinate values of this plot in absolute quantities because of our lack of knowledge of the collisional rate constants involving level 2 with the other two levels.

A similar possibility derives from Eq. (39), which shows that the quantum efficiency is inversely proportional to the saturation spectral irradiance of the source. If we make a plot of the fluorescence radiance versus laser power up to saturation, it is only necessary to interpolate to the power needed to reduce the maximal fluorescence radiance to one half. The quantum efficiency can therefore be calculated from Eq. (39), provided the spectral bandwidth of the laser and its cross section in the atomic system are known.

### 4.6.3 Concentrations

As shown before, see Eq. (23), it is possible to measure concentrations (species $cm^{-3}$) within the flame gases provided that saturation has been reached and the maximal fluorescence radiance is measured in absolute units. The same can be done with transient fluorescent signals, as speculated by Daily (40), who called the technique *pulsed resonance spectroscopy* (PRS). This technique has been utilized by Haas (41) for low-pressure laboratory plasma diagnostics. When the fluorescence waveform is·observed, the measurement of the decay time will give the quenching rate constant (assuming a two-level system and a simple exponential decay) while the measurement of the time-integrated signal will give the atomic concentration.

Daily (40) also derived an expression for the concentration, $n_T$, in terms of $S/N$ ratio, assuming that this ratio depends exclusively on the uncertainty in time in the estimation of the decay constant. Under these conditions, the detection limit was given by the following relationship:

$$(n_T)_{lim} \cong \left(\frac{2\pi}{\epsilon\eta\Omega V}\right)\left(\frac{g_1 + g_2}{g_2}\right)\left(\frac{k_{21} + A_{21}}{A_{21}}\right)\left(\frac{S}{N}\right)_{lim}^3 \qquad (64)$$

where $(S/N)_{lim}$ is the limiting detectable $S/N$ ratio, $\epsilon$ is the quantum efficiency of the photocathode of the detector, $\eta$ is the overall optical transfer efficiency, $\Omega$ is the solid angle of the collection optics, and $V$ is the effective focal volume of the collection optics. Typical values of these parameters, for $S/N = 10$, result in the possibility of detecting $\geq 10^9$ species $cm^{-3}$ via electronic (UV-visible fluorescence) resonance spectroscopy and $\geq 10^{13}$ species $cm^{-3}$ via vibrational (IR fluorescence) resonance spectroscopy. Therefore, PRS appears quite useful for concentration measurements of atoms and molecules in flames.

### 4.7  INSTRUMENTATION

#### 4.7.1  Dye Laser Characteristics

Analytical studies using laser excitation have been carried out with both pulsed and cw lasers. We emphasize here the use of pulsed tunable dye lasers, since cw results are reported elsewhere in this book (Chapter 8). Tunable dye lasers may be pumped by a flashlamp or by another laser. Typical values obtainable in terms of power, duty cycle, and spectral bandwidth are shown in Table 4.2, where the approximate ranges of tunability achieved with the most important classes of dyes are also reported. From this table one can see that the pulse characteristics are usually determined by the pumping source. With nitrogen, ruby, and neodymium lasers, the pulse length is short, few tens of nanoseconds and even less than 10 ns for nitrogen, while flash lamp excitation gives more energy per pulse due to the increased length of the pulse itself. However, with nitrogen laser excitation, the repetition rate can be much higher.

The laser used by Winefordner et al. (10, 11, 20) consisted of a commercial dye laser unit (AVCO Everett Research Laboratory, Massachusetts) with a grating in the cavity replacing one laser mirror, pumped by a nitrogen laser. The parameters characterizing the laser output were as follows: Nitrogen laser (337.1 nm): 120 kW peak power, pulse width 8–10 ns, repetition rate 1–100 Hz, beam size 3.2×51 mm, time jitter 1–2 ns. Dye laser: tunability range 357–650 nm, with ten interchangeable

**TABLE 4.2**
**Typical Values of Dye Laser Performances[a]**

| Pumping mode | Power, W Peak | Mean | Bandwidth, $cm^{-1}$[b] | Pulse length, ns | Repetition rate, $s^{-1}$ | Duty cycle |
|---|---|---|---|---|---|---|
| Nitrogen | $10^4$–$10^5$ | <0.05 | $10^{-2}$–$10^{-1}$ | 3–10 | 100–500 | <2×$10^{-6}$ |
| Ruby-Sh | $10^4$–$10^6$ | <$10^{-3}$ | $10^{-2}$–$10^{-1}$ | 10–30 | <0.1 | ≈$10^{-9}$ |
| Nd-Sh | $10^5$–$10^7$ | <$10^{-3}$ | $10^{-2}$–$10^{-1}$ | 10–30 | <0.1 | ≈$10^{-9}$ |
| Ruby | $10^5$–$10^8$ | <$10^{-3}$ | $10^{-2}$–$10^{-1}$ | 10–30 | <0.1 | ≈$10^{-9}$ |
| Nd | $10^4$–$10^7$ | <$10^{-3}$ | $10^{-2}$–$10^{-1}$ | 10–30 | <0.1 | ≈$10^{-9}$ |
| Flash lamp (50 ns) | $10^3$–$10^5$ | <$10^{-3}$ | $10^{-2}$–$10^{-1}$ | 100–200 | <0.1 | ≈$10^{-9}$ |
| Flash lamp (200 ns) | $10^4$–$10^5$ | 0.05–0.5 | $10^{-3}$–$10^{-1}$ | 300–1000 | 5–20 | ~$10^{-5}$ |
| Argon (cw) | — | 0.02–1 | $10^{-5}$–$10^{-3}$ | — | — | — |

0.3  0.4  0.5  0.6   0.7  0.8  0.9 1.
     $\lambda$, $\mu$m
—— scintillator dyes
——— coumarines
———— rhodamines
————— cresyl violet
—————— cyanines

[a] Values given in the Table do not represent ultimate performances achievable. Flash lamp pumping results are given both for 50 ns and 200 ns lamp rise times. Output characteristics of a cw argon ion laser are included for comparison purposes.

[b] $\Delta\bar{\nu}(cm^{-1}) \times \lambda^2(cm^2) \times 10^7(nm\ cm^{-1}) = \Delta\lambda(nm)$
$\Delta\bar{\nu}(cm^{-1}) \times 3 \times 10^{10}(cm\ s^{-1}) = \Delta\nu(s^{-1})$

**TABLE 4.3**
**Laser Spectral Output with Rhodamine 6G, $2.5 \times 10^{-4}M$ Solution in Methanol, Pump Energy 8.5 J, Reflectivity of the Laser Mirrors 99.8 and 50%**

| Laser cavity filter | Spectral half-width (FWHM), Å | Peak power, kW | Spectral peak power density, kW Å$^{-1}$ |
|---|---|---|---|
| No frequency-selective elements in the cavity | 61 | 17.5 | 0.29 |
| Interference filter (FWHM = 60 Å, peak transmission 96%) | 7 | 11.4 | 1.63 |
| Interference filter + Fabry-Perot etalon | $<10^{-2}$ | 7.6 | >760 |
| Interference filter (FWHM = 6 Å, peak transmission 84%) | 1 | 6.1 | 6.1 |
| Interference filter + Fabry-Perot etalon | $<10^{-2}$ | 5.9 | >590 |

[a] Data taken from J. Kuhl, G. Marowsky, P. Kunstmann, and W. Schmidt, Z. *Naturforsch.*, **27A,** 601 (1972). Reproduced by permission.

dyes, peak power greater than 10 kW, pulse width 2–8 ns, spectral half-width ranging from 0.1 to 1 nm, therefore justifying the assumption made (20) that the source could be assimilated to a spectral continuum. In most experiments, 10 Hz was used as the repetition rate.

The laser used by Kuhl et al. (40–42) was a flash lamp-pumped dye laser (Carl Zeiss, Oberkochen, West Germany) providing approximately 10 kW peak power with 200 ns pulse width for a $5 \times 10^{-5}M$ solution of rhodamine 6G in methanol. Care was taken by these authors to reduce the spectral bandwidth of the laser output by inserting frequency-selective elements in the laser cavity. Table 4.3 summarizes their results of power and linewidth measurements. As one can see, at the expense of a factor of 2–3 only for the output power, the spectral power density increased considerably. The emitted intensity per unit frequency interval was raised by a factor of 21 by the insertion of the interference filter alone and by at least 2000 when the filter and the Fabry-Perot interferometer were both present in the cavity. Compared to a grating, the interference filter in the cavity has the advantage of providing a high value of angular dispersion and negligible adjustment demands (42). With this system, laser bandwidths as small as $5 \times 10^{-4}$ nm were obtained.

It is known that, because of strong absorption of the fluorescence emission by molecules in the excited singlet state, it is impossible to find a highly efficient laser dye for wavelengths shorter than 300 nm. Because of this, all experiments requiring an excitation below 300 nm take advantage of the properties of nonlinear crystals such as ADP (ammonium dihydrogen phosphate) and KDP (potassium dihydrogen phosphate) to generate the second harmonic when visible laser radiation impinges on these crystals. Because this is a nonlinear process, the second harmonic output power goes with the square of the power in the fundamental frequency. When the beam spread of the laser is reduced by mode selection, peak power conversion efficiencies between 10 and 18% can be achieved even for laser output powers as low as 15–25 kW (43). With 10 kW peak power at 566.6 nm, the use of a KDP crystal 38 mm long, cut and polished for $\lambda = 560$ nm, resulted in the generation of the second harmonic at 283.3 nm with a peak power of 400 W (44). (The alternative to doubling the dye laser output is to utilize multiphoton excitation, that is, doubling within the sample cell itself.)

### 4.7.2 Atomizers

Atomizers used in analytical atomic fluorescence excited by conventional low-intensity sources include flames and electrothermal devices such as graphite cuvettes, carbon rods, and so on. One of the practical disadvantages of the fluorescence technique often stressed in the pertinent literature (4,13) is that high quantum efficiency flames have to be used to maximize the fluorescence signals. Therefore, combustion mixtures containing argon as a diluent, for example, oxygen–argon–hydrogen flames, have been preferred because of the small quenching cross section of argon. However, compared to air–acetylene and nitrous oxide–acetylene mixtures, the argon-diluted flames containing hydrogen as fuel have a relatively low atomization capability that can cause severe scattering[5] problems when analysis of practical samples is attempted by means of resonance fluorescence. On the other hand, even in the absence of

---

[5] It should be stressed that scattering occurs not only from particles in the flame gases as well as dust particles but also occurs from Rayleigh scattering of radiation from molecules and atoms, and therefore the latter phenomenon represents a fundamental limit of resonance fluorescence measurements. The Rayleigh scattering cross sections increase as $\lambda_i^{-4}$, where $\lambda_i$ is the incident laser wavelength. The Rayleigh scattering (due to atoms and molecules) contribution in flames corresponds to ~200–2000 counts s$^{-1}$ for a typical fluorescence setup in which a 150-W EIMAC cw conventional xenon arc lamp is used as the source of excitation. Therefore, assuming a 10-s integration, and assuming shot noise due to the scatter is limiting, the noise is ~100 counts, which is typically the limiting noise level in conventional AF with conventional light sources, particularly in the UV region. Of course, in the visible region ($\gtrsim 300$ nm), flame background noise may even exceed scatter noise in some spectral regions. In any event, the magnitude of scatter noise inherent in any fluorescent technique is sufficient to justify extensive studies of nonresonance fluorescence for analysis.

scattering, conventional lock-in detection does not allow taking full advantage of the good atomization efficiency of a high-temperature flame, such as the nitrous oxide–acetylene flame (~2900°K) for refractory elements (Al, V, Mo, etc.) because the prohibitive flame background in certain spectral regions results in highly noisy signals.

The use of a pulsed laser and gated detection (see below) should permit full exploitation of such atomization capabilities, and this despite the low value of the quantum efficiency of the fluorescence.

Flames used in laser experiments include air–hydrogen, air–acetylene, and nitrous oxide–acetylene mixtures, supported by conventional circularly or rectangularly shaped burners. Results have also been reported for carbon rod atomization (44).

### 4.7.3  Detection Systems and Signal-to-Noise Considerations

Detection systems for laser-excited fluorescence have consisted of photomultiplier tubes coupled with either a sampling oscilloscope or a boxcar integrator. It has been stressed (11,45,46) that photomultipliers suitable for pulsed operation must possess certain specific requirements such as short transit time of the electron cloud, small transit time spread, short rise time, and minimal parasitic capacitance. In addition, the tube should also be capable of sustaining high-peak anodic currents. For pulsed work, the dynode chain typical of normal operation is modified by placing capacitors between the dynodes. The laser output is usually monitored by a photodiode or a photocell. If the detector is calibrated, power measurements can be performed with the use of neutral density filters. Signals are fed into an oscilloscope and into a boxcar integrator.

The boxcar (45–47) is the analogue of a sampling oscilloscope with all the gadgets and flexibility that one might add externally for signal processing. Both "scan mode" of operation and "single point" measurements are possible with commercially available boxcars. Basically, the instrument performs a sample and hold operation, the sampling time being determined by an appropriate reference pulse that bears a definite relationship to the signal of interest. When pulse measurements are performed, the timing and width of the sampling window are adjusted so as to coincide with the occurrence of the fluorescence pulse. More details of boxcar operation and measurements are discussed in Chapter 7.

This gated operation results in a significant improvement in signal-to-noise ratio if the system under study is background noise[6] limited, since the detector is "on" only during the laser "on" time, (or a short delay

---

[6] The background must not be source induced such as molecular fluorescence.

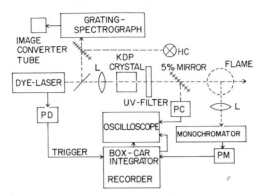

**Fig. 4.14.** Schematic diagram of a typical experimental setup for laser-excited fluorescence studies. J. Kuhl and H. Spitschan, *Opt. Commun.*, **7**, 256 (1973). Reproduced by permission.

after), so that background noise (shot and fluctuation), which is also measured only during this short time interval, will be greatly diminished. It has been shown (48) that if shot noise (source and background) dominates over fluctuation noise (source and background), the gain obtained when the $S/N$ ratio given by pulsing the source and gating the detector (po/go) over that obtained with cw sources and cw operation (cw/cw) is given by the relations

$$\psi \frac{\mathrm{po/go}}{\mathrm{cw/cw}} = \{B_{\mathrm{ave}}^{s,p}/B_{\mathrm{ave}}^{s}\}\{1/(ft_p)\}^{1/2} = (B_{\mathrm{peak}}^{s}/B_{\mathrm{ave}}^{s})(ft_p)^{1/2} \qquad (65a)$$

and

$$\psi \frac{\mathrm{po/go}}{\mathrm{cw/cw}} = (B_{\mathrm{ave}}^{s,p}/B_{\mathrm{ave}}^{s})^{1/2} \qquad (65b)$$

where $B_{\mathrm{ave}}^{s,p}$ represents the average radiance of the pulsed source, $B_{\mathrm{ave}}^{s}$ is the average radiance of the cw source, $f$ is the repetition rate, $t_p$ is the duration of the light pulse, and $B_{\mathrm{peak}}^{s}$ is the peak radiance of the pulsed source.

Equation (65a) holds if the system is background noise limited, while Eq. (65b) results when the limiting noise is source carried, that is, fluorescence or scattering. Clearly, no advantage will be obtained if the fluorescence measurements are plagued by scattering. Moreover, the equations derived assume that the fluorescence signal does follow linearly the source radiance, that is, no limiting value to the fluorescence due to the occurrence of saturation was considered.

Figure 4.14 shows a complete experimental setup for laser-induced flame fluorescence measurements. The flash lamp-pumped dye laser output is frequency doubled by a KDP crystal. A photodiode (PD) serves for triggering purposes. Beam splitters are used to either monitor the laser output power with a photocell (PC) or for spectral tuning of the generated ultraviolet radiation by observing the emission spectrum of a hollow-cathode lamp (HC) and that of the laser in the focal plane of a 1.5-m Czerny-Turner grating spectrograph by means of an UV image converter tube. The fluorescence signals are isolated by a monochromator, measured with a fast-response photomultiplier, and fed to a boxcar integrator.

## 4.8   ANALYTICAL RESULTS

It is helpful at this point to stress the analytical implications of the theory outlined in the previous paragraphs. Characteristic features of laser-excited fluorescence should include the following:

1.   *Extended tunability* over most of the UV-visible range with a spectral bandwidth variable from 1 nm down to less than $10^{-4}$ nm. This implies that the strongest absorption lines of most elements can be reached with approximately the same peak power (no filters are needed) and that scattering can be eliminated whenever nonresonance transitions are intense enough to be analytically useful. An example is shown in Fig. 4.15 for the atomic fluorescence of niobium excited in a nitrous oxide–acetylene flame, demonstrating the potential utility of fluorescence lines *other than* the resonance line where strong scattering would indeed significantly deteriorate the analytical results or even make the analysis impossible.

2.   *Independence of fluorescence signal upon source stability* if saturation is achieved. This outcome has been qualitatively seen in flames (20) where the peak-to-peak noise in the fluorescence signal obtained enlarging the beam into the flame (and therefore decreasing the source irradiance) compared with that given when the laser spot was focused (saturation more closely approached) increased more than the value which one would have predicted by taking into account the relative magnitude of the signals. However, this was not the case for electrothermal atomization (44) of lead where the relative standard deviation of the fluorescence signal, as high as 37%, did not show a significant decrease at high irradiances. This latter behavior was partially attributed to the pronounced inhomogeneity of the atomic cloud distribution during the carbon rod atomization cycle, with the result that clusters of analyte atoms were vaporized at different times causing rapid fluctuations in the fluorescence signal, which were resolved because of the short time constant ($<0.1$ $\mu$s) of the detection system. Moreover, considering the laser

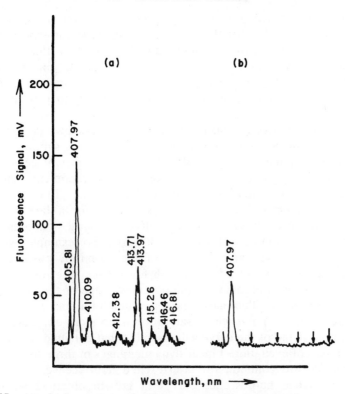

**Fig. 4.15.** Atomic fluorescence of niobium-wavelength scanning; $200\,\mu\text{m}$ slit (0.4 nm spectral band width). (*a*) Laser excitation set at 407.97 nm, 1000 ppm Nb; the intensity of the 407.97 nm line has been decreased 2.5 times as compared to the intensity of the other lines; (*b*) laser excitation set at 407.97 nm, 1000 ppm W; the intensity scattered at 407.97 nm has been also decreased 2.5 times and is therefore directly comparable to the Nb signal observed in (*a*) at the same wavelength. From N. Omenetto, N. N. Hatch, L. M. Fraser, and J. D. Winefordner, *Spectrochim. Acta*, **28B**, 65 (1973). Reproduced by permission.

bandwidth (<0.002 nm) and the length of the laser cavity, an estimation of the lead absorption linewidth of 0.0006 nm corresponded to roughly six mode spacings of a total of about 20 longitudinal modes. Since the oscillatory strength of these modes varied considerably from shot to shot and some may even have vanished, this caused the irradiance to vary considerably over the absorption linewidth.

3. *Less stringent dependence on the quantum efficiency of fluorescence.* This has been shown theoretically to hold for a two-level system and for a three-level system if a given definition of quantum efficiency is considered, compare Eqs. (47)–(48). The importance of such a consequence is obvious since it means that one can use atomizers with high atomization

efficiencies even if strong quenching is present. With reference to flames, it means, for example, that the fluorescence signal is not expected to be greater in an oxygen–hydrogen flame diluted with argon than the signal obtained in hydrocarbon flames containing nitrogen as the major constituent. This point, which clearly assumes identical atomization efficiencies for the two combustion mixtures, has not yet been proved experimentally.

     4. *Improved linearity of the calibration curves.* As shown in Sect. 4.5, if the source irradiance is such that the atomic system is saturated at any value of the population density, the absorption coefficient goes to zero and no self-absorption occurs, thus making the calibration curve obtained under idealized illumination conditions linear over all concentration ranges. This holds for a continuum as well as for a line source of excitation. However, unavoidable tradeoffs in practical situations do not allow the observation of such extended linearity. For example, when the laser beam is focused at the center of the atomizer, postfilter (self-reversal) effects in the observation path toward the detector have to be certainly included to account for the observed shape of the curve. If the beam is defocused to illuminate a larger volume of atoms, the irradiance can decrease to the point where saturation is no more approached, and therefore the shape of the curve will resemble that obtained with conventional sources of excitation. Linear dynamic ranges of three to five orders of magnitude are common for laser-excited fluorescence. Figure 4.16 compares relative fluorescence curves of growth obtained when lead vapor from a carbon rod atomizer is excited with a hollow-cathode lamp, an electrodeless discharge lamp, and a dye laser. The linear range, defined as the ratio of upper to lower concentration limits, found here was $1.25 \times 10^5$, $3.33 \times 10^3$, and $1 \times 10^3$ for dye laser, electrodeless discharge lamp, and hollow-cathode excitation, respectively. The lower concentration limit corresponds to the detection limit while the upper limit was defined as the lead concentration resulting in a signal whose deviation from linearity did not exceed 5%.

     Table 4.4 summarizes the limits of detection obtained with laser excitation in flames and compares them with those obtained by other conventional sources operated in the pulsed mode and with the ones given by cw sources. Limits of detection are usually defined on the basis of a $S/N$ (rms) ratio of 2.

     The column giving the laser results in Table 4.4 has more entries than the other columns; this is due to the fact that several elements such as the rare earths and some transition elements were never studied before by conventional fluorescence because they required both the nitrous oxide–acetylene flame and high-intensity sources, most of which were not

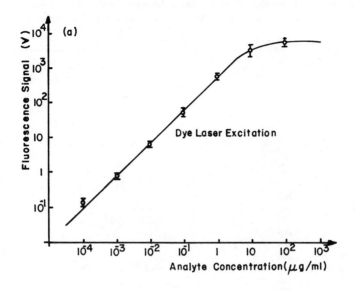

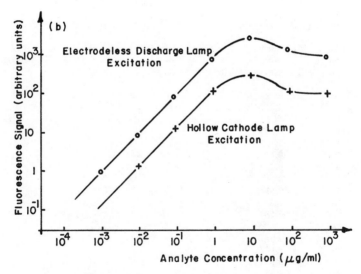

**Fig. 4.16.** Analytical curves for flameless atomic fluorescence of lead. Exciting light sources: (a) dye laser; (b) EDL and HCL. From S. Neumann and M. Kriese, *Spectrochim. Acta*, **29B,** 127 (1974). Reproduced by permission.

## TABLE 4.4
### Comparison Between Detection Limits Obtained with Dye Lasers and Conventional Sources in Atomic Fluorescence Flame Spectrometry[a]

| | Limits of detection, $\mu g\,ml^{-1}$ | | | | |
| | Pulsed sources | | | cw Sources | |
| Element | Laser | Line | Continuum | Line | Continuum (Eimac 150 W) |
|---|---|---|---|---|---|
| Ag | — | 0.004 | 0.02 | 0.0001 | 0.006 |
| Al | 0.005 | 0.07 | — | 0.1 | 0.2 |
| Au | — | — | — | 0.003 | — |
| Be | — | — | 0.2 | 0.01 | 0.07 |
| Ca | 0.005 | 0.0003 | — | 0.02 | — |
| Cd | — | 0.004 | 0.03 | 0.000001 | 0.01 |
| Ce | 0.5(I) | — | — | — | — |
| Co | 0.2 | 0.007 | 0.1 | 0.005 | 0.02 |
| Cr | 0.02 | 0.004 | — | 0.05 | 0.01 |
| Cu | — | 0.002 | — | 0.0005 | — |
| Dy | 0.3(I) | — | — | — | — |
| Er | 0.5 | — | — | — | — |
| Eu | 0.02 | — | — | — | — |
| Fe | 0.3 | 0.008 | — | 0.008 | — |
| Ga | 0.02 | — | — | 0.01 | — |
| Gd | 0.8(I) | — | — | — | — |
| Hf | 100. | — | — | — | — |
| Ho | 0.1 | — | — | — | — |
| In | 0.002 | — | — | 0.1 | — |
| Lu | 3. | — | — | — | — |
| Mg | 0.0003 | 0.001 | 0.004 | 0.001 | 0.0003 |
| Mn | 0.01 | 0.002 | 0.03 | 0.006 | 0.004 |
| Mo | 0.3 | 0.06 | — | 0.5 | 0.2 |
| Nb | 1. | — | — | — | — |
| Nd | 2. | — | — | — | — |
| Ni | 0.05 | 0.02 | — | 0.003 | — |
| Os | 150. | — | — | — | — |
| Pb | 0.03 | 0.07 | — | 0.01 | — |
| Pr | 1(I) | — | — | — | — |
| Rh | 0.1 | — | — | 3. | — |
| Ru | 0.5 | — | — | — | — |
| Sb | — | 0.05 | — | 0.05 | — |
| Sc | 0.01 | — | — | — | — |
| Se | — | 1. | — | 0.04 | 3. |
| Sm | 0.1(I) | — | — | — | — |
| Sn | — | 0.2 | — | 0.05 | — |
| Sr | 0.01 | 0.01 | — | 0.03 | — |
| Tb | 0.5(I) | — | — | — | — |
| Ti | 0.1 | — | — | 4. | 0.9 |
| Tl | 0.02 | — | — | 0.008 | — |
| Tm | 0.1 | — | — | — | — |
| V | 0.5 | 0.2 | — | 0.07 | 0.1 |
| Yb | 0.01 | — | — | — | — |
| Zn | — | 0.003 | 0.1 | 0.00001 | 0.006 |

[a] Table reprinted with permission from N. Omenetto, *Anal. Chem.* **48,** 75A (1976). Copyright by the American Chemical Society. Values are for analyte in aqueous solutions; (I) means ionic fluorescence.

212

available. On the other hand, it can be seen that the laser detection limits are not so spectacular as one would have predicted. The erratic quality of these results needs careful interpretation on the basis of parameters such as laser output variations (sometimes up to 25%) if saturation is not approached and limiting noise in the system. Unlike the fluorescence, the scattering signal keeps increasing with the laser irradiance, and therefore a compromise must be reached between scattering and signal stability. It is important to stress that most of the laser limits reported in Table 4.4 were obtained on nonresonance transitions, and this is a definite advantage from the scattering point of view. Another point is that in several cases (e.g., Mn, Co, Fe, Ni, Os, Ru, Rh, etc.), the most sensitive *absorption* lines could not be reached because frequency doubling was not available.

Certainly, more experimental work is needed in this respect, and far better results would be achieved if the range of tunability could be extended down into the 200–300 nm region. When resonance fluorescence is the only possibility, the use of a narrow bandwidth laser output is to be preferred. In this way, scattering can be compensated by recording the signal at two wavelengths, one tuned at the center and the other just outside the absorption profile.

## 4.9 CONCLUDING REMARKS

As we have repeatedly pointed out in the appropriate paragraphs, the theory given in this chapter *accounts only for the basic features* of the results obtained experimentally. Indeed, no quantitative agreement between theory and experimental results could be claimed in most cases. It is essential to stress that this is so because many complicating effects have been purposely neglected in the derivation given. Although a number of concomitant effects induced by extremely high radiation densities ($>10^7$ W cm$^{-2}$), such as self-focusing of the laser beam, multiple photon absorption, pulse shape distortion, and so on, would probably not alter the experimental results for flames, the parameters characterizing the laser output power do certainly require a more realistic evaluation. For example, it was always assumed that the laser beam was homogeneous and of constant irradiance while traversing the atomic vapor. In practice, however, the laser has both a *spatial* and a *temporal* profile with extremely large gradients. This causes the saturation effects to occur at different times and at different radii of the excited atom volume. Saturation of the transition will initially take place along the laser axis, where the power density is greatest, and then expand radially as the power increases. As a consequence, although the temporal variation of the fluorescence signal depends only on the local laser power and the local conditions of the atomic system, the signal detected by the photomultiplier will be due to the average effect over the whole observed excited volume (49). Therefore, *different* results are to be expected when time-integrated measurements are compared with signals measured at the peak of the time-resolved fluorescence waveform. In addition, the theory was considered for idealized geometries of excitation and observation of the fluorescence signals. It is almost impossible to exclude completely prefilter as well as postfilter effects in practical measurements when the concentration of the analyte becomes significant.

Doubtlessly, more experiments are needed to gain full insight into the physics of the process and to unambiguously compare theoretical predictions with experimental results. Up to now, analytical papers on laser-excited atomic fluorescence as a means of chemical analysis are still scanty, both because of the high cost required to set up on a laboratory scale a dye laser tunable over the whole spectral range of interest (strong commercial competition might improve this situation now) and because the results obtained for flames at atmospheric pressure are not significantly superior to those already achieved by conventional atomic absorption and emission spectroscopy.

Nevertheless, quite spectacular results have been achieved for cw laser excitation (see Chapter 8), and research on efficient frequency doubling down to 200 nm is under way.[7] Therefore, this field will receive increasing attention, and new results are expected in the future.

## References

1. A. C. G. Mitchell and M. W. Zemansky, *Resonance Radiation and Excited Atoms*, Cambridge University Press, New York, (1971).

2. C. Th. J. Alkemade, Proc. 10th CSI, Spartan Books, Washington, D.C. (1963), p. 143.

3. J. D. Winefordner and T. J. Vickers, *Anal. Chem.*, **36,** 161 (1964).

4. V. Sychra, V. Svoboda, and I. Rubeška, *Atomic Fluorescence Spectroscopy*, Van Nostrand-Reinhold, New York (1975).

5. R. F. Browner, B. M. Patel, T. H. Glenn, M. E. Rietta, and J. D. Winefordner, *Spectrosc. Lett.*, **5,** 311 (1972).

6. J. P. Webb, *Anal. Chem.*, **44,** 30A (1972).

7. J. R. Allkins, *Anal. Chem.*, **47,** 752A (1975).

8. J. I. Steinfeld, MIT Report, 1975; *CRC Crit. Rev. Anal. Chem.*, **5,** 225 (1975).

9. N. Omenetto and J. D. Winefordner, *Appl. Spectrosc.*, **26,** 555 (1972).

---

[7] The use of a KPB (potassium pentaborate) crystal allows tuning below 250 nm.

10. N. Omenetto, N. N. Hatch, L. M. Fraser, and J. D. Winefordner, *Spectrochim. Acta*, **28B**, 65 (1973).

11. L. M. Fraser and J. D. Winefordner, *Anal. Chem.*, **44**, 1444 (1972).

12. C. Th. J. Alkemade and P. J. Th. Zeegers, in *Spectrochemical Methods of Analysis*, J. D. Winefordner, Ed., Wiley, New York (1971).

13. J. D. Winefordner, S. G. Schulman, and T. C. O'Haver, "Luminescence Spectrometry in Analytical Chemistry," in *Chemical Analysis*, Vol. 38, P. J. Elving, J. D. Winefordner, Eds., Wiley, New York (1972).

14. P. L. Lijnse and R. J. Elsenaar, *J. Quant. Spectrosc. Radiat. Transfer*, **12**, 1115 (1972).

15. C. Th. J. Alkemade, *Pure Appl. Chem.*, **23**, 73 (1970).

16. M. Hercher, *Appl. Opt.*, **6**, 947 (1967).

17. L. Huff and L. G. DeShazer, *J. Opt. Soc. Amer.*, **60**, 157 (1970).

18. E. H. Piepmeier, *Spectrochim. Acta*, **27B**, 431 (1972).

19. E. H. Piepmeier, *Spectrochim. Acta*, **27B**, 445 (1972).

20. N. Omenetto, P. Benetti, L. P. Hart, J. D. Winefordner, and C. Th. J. Alkemade, *Spectrochim. Acta*, **28B**, 289 (1973).

21. J. Kuhl, S. Neuman, and M. Kriese, *Z. Naturforsch.*, **28A**, 273 (1973).

22. L. de Galan and H. C. Wagenaar, *Meth. Phys. Anal.*, 10 (Sept. 1971).

23. W. Lange, J. Luther, and A. Stendel, *Adv. Atom. Mol. Phys.*, **10**, 173 (1974).

24. H. P. Hooymayers, *Spectrochim. Acta*, **23B**, 567 (1968).

25. N. Omenetto, L. P. Hart, P. Benetti, and J. D. Winefordner, *Spectrochim. Acta*, **28B**, 301 (1973).

26. J. Kuhl and H. Spitschan, *Opt. Commun.*, **7**, 256 (1973).

27. H. R. Griem, *Plasma Spectroscopy*, McGraw-Hill, New York (1964).

28. E. O. Gadamer, UTIAS Report No. 83 *Institute of Aerophysics, University of Toronto* (March 1962).

29. R. M. Measures, *J. Appl. Phys.*, **39**, 5232 (1968).

30. D. E. Jensen and B. E. L. Travers, paper given at IUPAC International Symposium on Plasma Chemistry, Kiel, Germany (September 1973).

31. S. S. Penner and T. Jerskey, *Ann. Rev. Fluid Mech.*, **5**, 9 (1973).

32. W. M. Arden, T. B. Hirschfeld, S. M. Klainer, and W. A. Mueller, *Appl. Spectrosc.*, **28**, 554 (1974).

33. R. H. Barnes, C. E. Moeller, J. F. Kircher, and C. M. Verber, *Appl. Opt.*, **12**, 2531 (1973).

34. K. H. Becker, D. Haaks, and T. Tatarczyk, *Z. Naturforsch.*, **27A**, 1520 (1972).

35. K. H. Becker, D. Haaks, and T. Tatarczyk, *Z. Naturforsch.*, **28A**, 829 (1974).

36. C. C. Wang and L. J. Davis, Jr., *Appl. Phys. Lett.*, **25**, 34 (1974).

37. N. Omenetto, P. Benetti, and G. Rossi, *Spectrochim. Acta,* **27B,** 453 (1972).

38. N. Omenetto, R. F. Browner, J. D. Winefordner, G. Rossi, and P. Benetti, *Anal. Chem.,* **44,** 1683 (1972).

39. N. Omenetto, Plenary Lecture Given at 5th International Conference on Atomic Spectroscopy, Melbourne (August 1975).

40. J. W. Daily, *Appl. Opt.,* **15,** 955 (1976).

41. Y. Haas, Lawrence Berkeley Laboratory Report No. 14035, University of California, Berkeley (1975).

42. J. Kuhl, G. Marowsky, P. Kunstmann, and W. Schmidt, *Z. Naturforsch.,* **27A,** 601 (1972).

43. J. Kuhl and H. Spitschan, *Opt. Commun.* **5,** 382 (1972).

44. S. Neumann and M. Kriese, *Spectrochim. Acta,* **29B,** 127 (1974).

45. F. E. Lytle, *Anal. Chem.,* **46,** 545A (1974).

46. F. E. Lytle, *Anal. Chem.,* **46,** 817A (1974).

47. J. D. W. Abernethy, *Wireless World* (December 1970).

48. N. Omenetto, L. M. Fraser, and J. D. Winefordner, in *Applied Spectroscopy Reviews,* Vol. 7, E. G. Brame, Ed., Marcel Dekker, New York (1973), p. 147.

49. A. B. Rodrigo and R. M. Measures, *IEEE QE,* **9,** 972 (1973).

50. N. Omenetto, *Anal. Chem.,* **48,** 75A (1976).

## Additional References

Several interesting papers have appeared in the literature since the writing of this chapter and are reported below. Theoretical writings on laser-induced fluorescence deal with the spectral distribution of the light emitted under intense field excitation (51–56), with the derivation of a fluorescence expression under selfabsorption conditions (57), with the applicability of the rate equations approach to describing the optical interaction (58), with the saturation of the sodium fluorescence (59–61), with the feasibility of the technique in the field of plasma diagnostics (61–63), with its analytical sensitivity (64–65) and with the time and wavelength dependence of the fluorescence signal (66). A comparison between pulsed and continuous wave fluorescence in terms of analytical results has been given (67), and the fluorescence behavior of iron and lead has been discussed (68, 69).

51. R. E. Grove, F. Y. Wu, and S. Ezekiel, *Phys. Rev.* **A, 15,** 227 (1977).

52. E. Courtens and A. Szöke, *Phys. Rev.* **A, 15,** 1588 (1977).

53. P. Avan and C. Cohen-Tannoudji, *J. Phys. B: Atom. Mol. Phys.,* **10,** 155 (1977).

54. B. Renaud, R. M. Whitley, and C. R. Stroud Jr., *J. Phys. B: Atom. Mol. Phys.,* **10,** 19 (1977).

55. J. L. Carlsten, A. Szöke, and M. G. Raymer, *Phys. Rev.* **A, 15,** 1029 (1977).

56. C. Cohen-Tannoudji and S. Reynaud, *J. Phys. B: Atom. Mol. Phys.,* **10,** 345 (1977).

57. N. Omenetto, J. D. Winefordner, and C. Th. J. Alkemade, *Spectrochim. Acta,* **30B,** 335 (1975).

58. J. W. Daily, *Appl. Opt.,* **16,** 2322 (1977).

59. D. D. Burgess and M. J. Eckart, *J. Phys. B: Atom. Molec. Physics,* **9,** L519 (1976).

60. B. L. Sharp and A. Goldwasser, *Spectroch. Acta,* **31B,** 431 (1976).

61. B. Smith, J. D. Winefordner, and N. Omenetto, *J. Appl. Phys.,* **48,** 2676 (1977).

62. J. W. Daily, *Appl. Opt.* **16,** 568 (1977).

63. A. P. Baronavski and J. R. McDonald, *Appl. Opt.,* **16,** 1897 (1977).

64. J. P. Hohimer and P. J. Hargis Jr., *Appl. Phys. Lett.,* **30,** 344 (1977).

65. J. A. Gelbwachs, C. F. Klein, and J. E. Wessel, *Appl. Phys. Lett.,* **30,** 344 (1977).

66. H. L. Brod and E. S. Yeung, *Anal. Chem.,* **48,** 344 (1976).

67. N. Omenetto, G. D. Boutilier, S. J. Weeks, B. W. Smith, and J. D. Winefordner, *Anal. Chem.,* **49,** 1076 (1977).

68. M. A. Bolshov, A. V. Zybin, L. A. Zybina, V. G. Koloshnikov, and I. A. Majorov, *Spectrochim. Acta,* **31B,** 493 (1976).

69. M. A. Bolshov, A. V. Zybin, V. G. Koloshnikov, and K. N. Koshelev, *Spectrochim. Acta,* **32B,** 279 (1977).

# 5

# MOLECULAR ABSORPTION AND FLUORESCENCE SPECTROSCOPY WITH LASERS

## W. DEMTRÖDER

*Fachbereich Physik*
*Universität Kaiserslautern, Germany*

## 5.1  INTRODUCTION

Most of our present knowledge about molecular structure has been provided by molecular spectroscopy. The analysis of molecular spectra not only yields the different energy levels of molecules from measured wavenumbers of spectral lines but also allows to determine the configuration of the molecular frame such as bond angles and bond lengths (1). From spectral line intensities, transition probabilities and symmetries of molecular states can also be deduced. Measurements of Zeeman or Stark splittings enable determination of magnetic and electric dipole moments and give information about molecular polarizabilities.

Since the intensities of spectral lines are proportional to the number density of absorbing or emitting molecules, spectroscopy has long proved to be of considerable value as a tool in analytical chemistry. Crucial for any analytical method are attainable sensitivity, capability of unambiguous identification of the sample constituents, and last but not least facility of practical performance. We must examine how close spectroscopic methods meet these demands.

The accuracy and sensitivity of all spectroscopic investigations are substantially limited by the equipment used, in particular by the intensity and spectral purity of the light source. The introduction of lasers to spectroscopy has remarkably improved both sensitivity and spectral resolution, resulting in an unambiguous and highly sensitive detection of spurious components in the presence of large amounts of other constituents. This improvement is mainly due to the attainable spectral brightness of lasers exceeding that of incoherent light sources often by several orders of magnitude. In addition, the development of several new detection methods, used in connection with lasers, has furthermore increased the sensitivity and in favorable cases allows the safe detection of as few as $10^6$ molecules $cm^{-3}$. In Sect. 5.2 we therefore discuss at first these high-sensitivity methods. Some examples will illustrate the main application domains of the different techniques in order to bring about some feeling of the capabilities of these methods.

*For many analytical problems the spectral selectivity of the lasers is of at least equal significance as its intensity.* One example is the spectroscopic selection of different molecular isotopes. Using fixed-frequency lasers, however, one has to search for fortuitous coincidences between laser lines and molecular absorption lines. For most applications only tunable monochromatic lasers that can be tuned to the center of any desired molecular line take full advantage of the capabilities of laser spectroscopy. The development of tunable lasers has made rapid progress during the last years, and several reliable commercial devices are already available covering spectral regions from the ultraviolet to the infrared. The different tuning methods depending on the spectral range of interest are discussed in Sect. 5.2.2.

Clearly, before lasers can be successfully applied to analytical problems, the absorption spectra of the molecules under study have to be known. Section 5.2.3 brings some examples of laser applications in Doppler-limited absorption spectroscopy. The spectral resolution in these experiments is only limited by the Doppler width of the molecular absorption lines. The next section explains several techniques aimed at overcoming this resolution limit. Regarding spectral resolution these methods represent the most impressive progress achieved in molecular spectroscopy in the last years, and many detailed features of molecular spectra, hitherto masked within the Doppler width, have been revealed for the first time.

The last part of Sect. 5.2 illustrates some specific applications of laser absorption spectroscopy in analytical chemistry. Section 5.3 deals with optical pumping of molecules by lasers. This technique allows a great variety of different investigations of *excited* states with high resolution. Examples are the laser-induced fluorescence technique, measurements of excited-state lifetimes, level crossing spectroscopy, and optical double resonance. These experiments give some insight into details of excited molecular states and their internal dynamics. The spectroscopic investigations of collision processes have recently gained increased interest. A very important question in reaction kinetics, namely, the internal-state distribution of reaction products, has been successfully answered in some cases, using laser-induced fluorescence measurements. This is outlined in more detail in Sect. 5.3.6.

Besides molecular absorption and fluorescence spectroscopy where lasers were merely used as outstanding light sources, the spectroscopy of laser lines itself and the investigations of spectral and time behavior of pulsed molecular laser lines allowed much insight to be gained into excitation mechanisms and collision processes in molecular discharges, leading to the identification of complex spectra of radicals and vibrationally excited molecules. This subject is covered in Sect. 5.4.

Raman spectroscopy, one of the outstanding spectroscopic techniques for the determination of vibrational and rotational structure and symmetry of molecules, can hardly be imagined nowadays without lasers. Since laser Raman spectroscopy has been extensively treated in several textbooks (2–5) and recent reviews (6–8), Sect. 5.5 only briefly summarizes the progress achieved during the last years in this field. We discuss some new experimental techniques which have considerably increased detection sensitivity, allowing Raman spectra to be recorded with simple He–Ne lasers within several seconds.

Despite being a physicist, the author believes that one of the most exciting developments in laser spectroscopy—at least if viewed with a chemist's eyes—will be the rapidly expanding field of laser-induced chemical reactions. Starting from some more or less unspecified and not well-understood experiments in laser chemistry (labeled as laser alchemy by some people) where the laser simply replaced a conventional Bunsen burner to heat the sample, much progress has been made since then. In several detailed experiments selective excitation of molecular levels has resulted in controllable reactions with the future goal to enhance wanted reactions and to suppress unwanted ones. This subject is treated in Sect. 5.6.

Closely related to this topic is the possibility of isotope separation with lasers. This field has excited many researchers since last year the discovery of multiphoton dissociation of molecules has opened new possibilities of inexpensive separation methods. More about this field is given in Sect. 5.7.

Of course, because of the vastly increasing number of publications in the field of laser spectroscopy this review cannot be complete but tries to give an informative survey of some basic principles and recent developments. The examples are selected to illustrate the advantages and limitations of the different methods. We shall restrict the discussion mainly to spectroscopy of free molecules in the gas phase and shall not include the liquid or solid state, as, for instance, matrix isolation spectroscopy (9). Several reviews have been recently published (10–14) that cover some of the aspects discussed here or treat some subjects, only briefly mentioned here, in more detail. The reader is referred to the literature cited in the corresponding sections.

## 5.2  ABSORPTION SPECTROSCOPY WITH LASERS

The high spectral brightness and narrow frequency bandwidth of lasers, especially the possibility of tuning the monochromatic laser output continuously within a certain wavelength range, have brought such definite advantages to molecular absorption spectroscopy that, without exaggeration, the introduction of lasers to spectroscopy can be called the beginning of a new era in this field. Because of the increased resolution and sensitivity in laser spectroscopy, many experiments that could not be performed with incoherent light sources have proved to be feasible with lasers.

The main limitations of laser applications in chemistry for all spectral regions up to now are the lack of conveniently tunable commercial lasers in some spectral ranges. We shall see, however, that in principle the whole spectrum between 100 nm to 1000 $\mu$m can be spanned by nonlinear frequency mixing techniques. Some of these devices are still in the laboratory stage and have to be operated by people familiar with the special techniques. From extrapolation of the rapid development during recent years there is well-founded hope that commercial devices will soon fill these spectral gaps.

Instead of tuning the laser frequency, one may also tune the absorption lines of molecules into resonance with fixed laser frequencies by external electric or magnetic fields. This is especially useful in spectral regions where intense fixed-frequency lasers already operate but tunable devices with sufficient output power have not been developed.

Figure 5.1 illustrates the differences between measuring absorption spectra in a conventional way and using continuously tunable laser sources. In a conventional arrangement (Fig. 5.1a), the radiation from a source emitting a spectral continuum (high-pressure arc, flash lamp, etc.) passes the absorption cell. The transmitted intensity is recorded as a function of wavelength behind a spectrograph or an interferometer. The *spectral resolution* is limited by the resolving power of these instruments, and great efforts have been made to improve the resolution by using large grating spectrographs (up to 9 m in focal length) or high-dipersion inter-ferometers, such as Fourier spectrometers, which may approach the Doppler-limited resolution of molecular absorption lines. The *maximum sensitivity* is determined by the lowest still measurable light attenuation in the absorption cell. The lower limit is set by intensity fluctuations of the light source and by the detector noise. Because of light beam divergence problems (for instance, light scattering from the cell walls) the length of

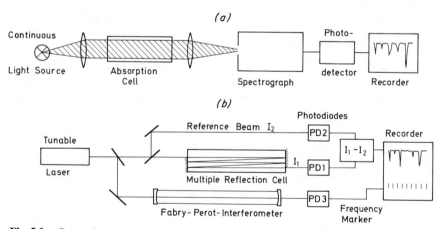

**Fig. 5.1.** Comparison between conventional spectroscopy (a) and tunable laser spectros-copy (b). The Fabry-Perot interferometer in (b) serves as frequency marker, which transmits intensity peaks at equal frequency intervals corresponding to the free spectral range of the FPI.

the absorption cell is limited, and faint absorption lines can only be detected by increasing the gas pressure. This results, however, in pressure broadening of the lines and therefore decreases the spectral resolution.

Using a monochromatic laser, on the other hand, which can be continuously tuned within the wanted spectral range, no extra dispersive instrument is necessary (Fig. 5.1b). The laser itself acts as monochromator and the spectral resolution is only limited by the linewidth of the molecular absorption lines, provided the laser linewidth is narrow enough. The attainable spatial collimation of laser beams offers the further advantage of multiple path absorption cells increasing the effective path length and consequently the sensitivity. This allows to perform absorption measurements at low pressure, avoiding pressure broadening of absorption lines. Intensity stabilization of the laser and use of a reference beam with difference detection techniques further increases sensitivity and accuracy. Because of the high laser intensity, detector noise plays a minor role in most experiments and can often be neglected.

Besides direct measurements of the attenuation of laser beams passing through an absorption cell outside the laser resonator, some other methods have been developed that may increase the sensitivity by several orders of magnitude. The first method employs intracavity absorption; the second uses laser-induced total fluorescence to monitor the absorption spectrum; and the third method, called optoacoustic spectroscopy, detects absorption by measuring the increase of pressure in the absorption cell. We now discuss these high-sensitivity methods in more detail.

### 5.2.1  High-Sensitivity Techniques in Absorption Spectroscopy

#### 5.2.1.1  Intracavity Absorption

The sensitivity of measuring small absorption coefficients may be greatly enhanced by placing the sample inside the laser resonator (15). This enhancement has several reasons: The first is simply due to the fact that the laser photons inside the cavity are being reflected back and forth by the cavity mirrors and therefore pass the absorption cell many times. If one of the mirrors is totally reflecting and the other has a reflectivity $R < 1$, the average number $m$ of transits for each laser photon is $m = 1/(1 - R)$. The effective absorption path length for a cell with length $L$ is then $mL$. With a typical value of $R = 0.99$, the enhancement factor is $m = 100$. This increased effective absorption path length is independent on the special characteristics of the laser, whether it is operated close to or high above threshold. Another way to explain the same fact is to consider the increased intensity inside the cavity which is $m$ times as high as the output intensity. Since the absorbed power is proportional to this intensity (as long as no saturation occurs), the enhancement factor is again $m$.

The second reason for increased sensitivity becomes important when the laser is operated close above threshold. The laser can only start to oscillate when the total gain overcomes the total losses (threshold condition). The gain decreases with increasing laser intensity because the induced emission reduces the population inversion of the active medium (gain saturation) (16). In steady-state operation the laser intensity always builds up to a limiting value where the gain $\alpha(I)$ at this intensity just equals the total losses. Close above threshold the gain saturation is very small, which means that large relative intensity changes cause only small changes in gain $\alpha(I)$. This in turn implies that already diminuitive changes in absorption losses may result in drastic changes in the laser output. For a detailed theoretical description see references 17 and 18. The subject is also treated in Chapter 8 of this book.

A further increase in sensitivity of intracavity absorption can be achieved with multimode dye lasers (19). This increase is due to the coupling of the laser modes by spatial saturation of the active medium with resulting mode competition. Each oscillating mode decreases the gain of the active medium. If a mode is suppressed through selective absorption by molecules inside the laser resonator, the gain for the other modes is increased. These modes will therefore oscillate with increased intensity. It can be shown that the detection sensitivity using a broadband dye laser is proportional to the number of oscillating modes (20). An increase in absorption sensitivity by a factor of $10^5$ (compared to a single-pass measurement) could be obtained with an iodine vapor cell inside the cavity of a cw dye laser (20). Extinction coefficients of $10^{-7}$ corresponding to an iodine vapor pressure of $1.5 \times 10^{-6}$ torr could be detected.

The method can be used for quantitative measurements of small absorption coefficients (21) and has been applied so far to the molecules $I_2$ (20), $H_2O$ (21), and HCN (22), and to the detection of free radicals such as $NH_2$, HCO (23), and OH (24).

### 5.2.1.2  Absorption Spectra Monitored by Laser-Excited Fluorescence

The most sensitive method of detecting weak absorption lines in the visible and ultraviolet region takes advantage of the fluorescence excited by the absorbed laser radiation. This method is especially advantageous at low partial pressure of the absorbing molecules, either in gas cells or in molecular beams. Because collision-induced radiationless transitions do not play a significant role under these conditions, *the quantum efficiency*

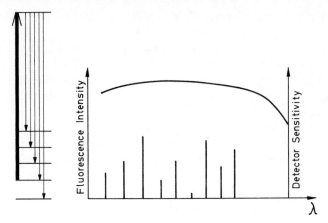

**Fig. 5.2.**    Energy level diagram of laser-excited fluorescence used for monitoring absorption spectra.

*generally is 100%. This means that each absorbed laser photon generates one fluorescence photon.* Since many different transitions from the excited state $E_i$ to several lower levels $E_m$ may be optically allowed, the fluorescence spectrum generally is spread out over a wide wavelength range (see Fig. 5.2). If the fluorescence detector has a wavelength-independent quantum detection efficiency in this range (this is, for instance, true for some modern photomultipliers), the detected signal will be strictly proportional to the absorbed laser intensity. When monitoring the fluorescence intensity as a function of laser wavelength while the laser is tuned over the absorption lines, one obtains, under the above assumptions, a true image of the absorption spectrum which is often called "excitation spectrum."

The sensitivity of excitation spectroscopy may be briefly estimated as follows: The absorption rate $n_{ik}$ (number of absorbed laser photons per second and per centimeter absorption path length) is given by

$$n_{ik} = N_i \sigma_{ik} P_L / (h \nu_{ik}) \tag{1}$$

where $N_i$ is the number density of molecules ($cm^{-3}$) in the absorbing level $E_i$, $\sigma_{ik}$ is the absorption cross section ($cm^2$) per molecule, $P_L$ is the average laser power per second at the absorption frequency $\nu_{ik}$ ($s^{-1}$), and $h$ is the Planck constant.

With a molecular quantum efficiency of 100%, the total fluorescence rate equals the absorption rate $n_{ik}$. Since the fluorescence is emitted into the solid angle $4\pi$ steradian, the detection optics will only collect a fraction $\beta$ of the total fluorescence. The counting rate $R$ of fluorescence

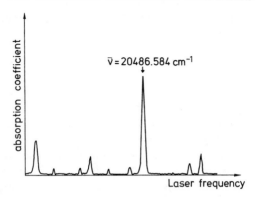

**Fig. 5.3.** Excitation spectrum of Na$_2$ molecules in a molecular beam. Beam divergence was about 1/20.

photons is then given by

$$R = n_{ik}\beta\eta \tag{2}$$

where $\eta$ is the quantum detection efficiency of the fluorescence detector. Modern photomultipliers reach $\eta = 0.2$ in the visible and UV region, and a well-designed light collecting optics will allow $\beta = 0.1$. Using photon counting techniques, counting rates of $R = 100$ counts s$^{-1}$ are already well above the thermal background noise, which may be pushed below 10 counts s$^{-1}$ with cooled photomultipliers. This means that a fluorescence rate of $n_{ik} = 5 \times 10^3$ s$^{-1}$ already yields a reasonable signal-to-noise ratio. For a laser with 1 W output power, which corresponds to $5 \times 10^{18}$ photons s$^{-1}$ at $\lambda = 500$ nm, an absorption rate of $5 \times 10^3$ photons s$^{-1}$ means a relative attenuation $\Delta P_L/P_L$ of $10^{-15}$!

To demonstrate the sensitivity of this method, Fig. 5.3 shows the excitation spectrum of Na$_2$ molecules in a molecular beam at a molecular pressure of about $10^{-6}$ torr, excited by a 0.5 W single-mode argon laser, which was tunable within a frequency interval of 10 GHz around $\lambda = 488$ nm. Taking into account that the Na$_2$ molcules are thermally distributed among roughly 300 different rotational-vibrational levels in the electronic ground state, the partial pressure of those molecules $N_i(v_i, J_i)$ that are in the right level $(v_i, J_i)$ for absorbing the laser line is below $10^{-8}$ torr, and the absorption path length is about 2 mm. The good signal-to-noise ratio in Fig. 5.3 shows that this concentration is not the detection limit by far.

The wavelength distribution of the fluorescence from the excited level $E_k$ to all lower levels $E_m$ depends on the spontaneous transition probabilities $A_{km}$. If the photodetector has a wavelength-dependent response, these $A_{km}$ values have to be known in order to deduce the relative absorption coefficients

$$\alpha_{ik} = N_i \sigma_{ik} \tag{3}$$

from the excitation spectrum.

Because of the high quantum efficiency of photodetectors in the visible and UV region and the 100% fluorescence yield of molecular gases at low pressure in this spectral range, *excitation spectroscopy is certainly the most sensitive technique for measuring absorption spectra.* One may even further increase the sensitivity by combining this detection technique with intracavity absorption by taking advantage of the increased laser power inside the cavity. Another way is to place the sample inside an external optical resonator that is matched to the external laser beam wavefront and tuned to the proper frequency. The laser field inside this resonator may be two orders of magnitude more intense than outside. The method of measuring absorption spectra via excitation spectra has already been widely used, and we discuss several examples in the following section.

In the infrared region, the sensitivity of this method is drastically degraded for two reasons. Firstly, the quantum detection efficiency and gain amplification of infrared detectors is considerably smaller than that of photomultipliers in the visible. The second restriction is imposed by the long radiative lifetimes of vibrationally or rotationally excited molecules in their electronic ground state. Because of these long lifetimes, the excited molecules may collide with other molecules or with the walls of the cell before they radiate, and they may transfer their excitation energy into translational energy. This reduces the molecular quantum efficiency. If the pressure is kept low enough to avoid gas phase collision, the molecules may diffuse out of the observation region before they radiate.

Fortunately, another high-sensitivity technique has been developed that takes advantage of the low molecular quantum efficiency and relies on the transfer of excitation energy to translational energy. It is called optoacoustic spectroscopy and has proved to be a very powerful tool in infrared spectroscopy. It is discussed in the next section.

### 5.2.1.3    *Optoacoustic Spectroscopy*

Suppose molecules in an absorption cell are optically pumped into excited levels by absorption of laser radiation. If the gas pressure in the cell is sufficiently high, collisions may cause the excited molecules to transfer their excitation energy into translational, rotational, or vibrational energy of the collision partners before they can release it by spontaneous emission. After thermal equilibrium has been reached, the excitation energy is equally distributed among all possible degrees of freedom. This results in an increase in temperature and pressure. *The absorbed laser energy heats the whole gas in the absorption cell.*

In case of vibrationally excited levels, the relaxation times for this collisional redistribution of excitation energy are of the order of $10^{-6}$ s at atmospheric pressure. If the laser intensity is modulated with a modulation period that is long compared to this relaxation time, the gas pressure in the absorption cell shows a corresponding modulation that can be monitored with a sensitive microphone placed at the wall of the sample chamber. When the laser frequency is tuned across a molecular absorption line, the microphone signal is proportional to the absorbed energy and therefore provides a measure for the absorption coefficient.

This technique of detecting absorption spectra through pressure changes is called optoacoustic spectroscopy, and the apparatus is known as a *spectraphone*. It has been used for many years with conventional light sources (25,26), but only the application of lasers to this method has made it a very sensitive and useful tool in infrared spectroscopy. The high power of infrared lasers increases the sensitivity, and the good beam collimation allows concentration of the excitation energy into a small sample volume. The high sensitivity of modern laser spectraphones allows the detection of spurious molecular constituents in gases down to concentrations far below the ppm range. The technique has therefore met with increasing interest during recent years, and several reviews have been published (27–29) which discuss the basic principles of the method, its advantages and limitations, together with several applications.

As long as saturation effects can be neglected (30), the measured acoustic signal $s$ is given by

$$s = A \cdot \sigma \cdot N \cdot \bar{P} \cdot B \qquad (4)$$

Equation (4) shows that $s$ is proportional to the absorption cross section $\sigma$, the density $N$ of absorbing molecules, the average laser power $\bar{P}$, and the sensitivity $B$ of the microphone. The proportionality factor $A$ depends on the cell geometry and the total pressure and also contains the efficiency of the transfer of excitation energy into translational energy. Modern electret microphones, which use a thin membrane as one of the two plates of a charged capacitor, in combination with low-noise FET preamplifiers followed by small-bandwidth lock-in amplifiers reach sensitivities of better than 1 V torr$^{-1}$. Typical noise figures are below 30 nV at integration times of 1 s (28). The frequency response of these electret microphones extends to 20 kHz.

The sensitivity is generally not limited by the electronic noise level, but mainly by a background signal caused by laser light reflected from the cell windows or scattered by aerosols in the cell. This light is partially absorbed by the cell walls resulting in a temperature rise and a corresponding pressure increase, which is, of course, also modulated at the signal frequency. Antireflection coatings reducing the reflection losses to about 0.2% are therefore desirable.

The sensitivity of the spectraphone may be considerably enhanced by chopping the laser beam at a frequency that corresponds to a natural acoustic "eigenresonance" of the sample cell. Dewey et al. (31) achieved acoustic amplification factors of 100 and more by this means. Moreover, this technique eliminates the difficulties of the nonresonant cell arising from absorption of laser radiation by the walls. Since the signal detection circuit can be tuned sharply to the resonance frequency of the desired acoustic mode (for instance, with nodes at the walls), absorption by the walls and cell windows which contribute primarily to acoustic waves with different spatial and phase characteristics does not contribute significantly to the detected signal. Another way of increasing the sensitivity is to use intracavity absorption. When the sample chamber is placed inside the laser resonator, the acoustic signal is enhanced because of the increased intracavity intensity. The enhancement factor may be as large as 100 (see Sect. 2.1.1).

The acoustic signal decreases with increasing molecular quantum efficiency because the spontaneous emission from the excited level takes away part of the absorbed energy without heating the gas. *The optoacoustic method is therefore best suited for monitoring vibrational-rotational absorption spectra* since the large spontaneous lifetimes of vibrational levels (typical of the order of $10^{-2}$–$10^{-4}$ s) allow nearly complete collision-induced vibrational energy transfer already at pressures of several torr (32).

In case of pure rotational excitation the energy of each absorbed laser quantum is very small. Burenin and Krupnow (33) have shown that it should still be possible to observe rotational spectra of polar molecules with a spectraphone, using a laser that emits at least some watts in the submillimeter region.

Even electronic molecular absorption spectra in the visible region have been recorded by optoacoustic spectroscopy (34). Using a jet stream cw dye laser with 250 mW output and a spectral bandwidth of 0.05 nm, Angus et al. (35) have proved that $NO_2$ concentrations around 10 ppb could be detected in air at a total pressure of 10 torr. This sensitivity nearly reaches that of excitation spectroscopy monitored by laser-induced fluorescence as discussed in the previous section. The technique is especially useful for the detection of pollutant gases in air at atmospheric pressure. The sensitivity is, however, higher in the infrared region because of the larger transfer efficiency of absorbed laser energy into translational energy.

This has been demonstrated by Kreutzer et al. (36) who used the different lines from CO and $CO_2$ lasers for sensitive detection of 10 different pollutant gases including NO, $NO_2$, $NH_3$, and ethylene. They were able to detect $NH_3$ concentrations down to 0.4 ppb, ethylene even down to 0.2 ppb, at total air pressure of 500 torr in the absorption cell. The extension to pollution measurements of automobile exhausts is obvious and should allow, for instance, the detection of NO concentration as low as 10 ppb.

If different absorbing molecular constituents with overlapping absorption lines are present, the unambiguous determination of the concentration of each component demands absorption measurements at several wavelengths. This allows the unraveling of the contributions of the different constituents to the total absorption, provided the absorption spectra of the molecules have been measured separately under comparable conditions. The minimum detectable amount of low-absorbing constituents in the presence of strongly absorbing molecules is, however, restricted (37).

With tunable lasers, such as the spin flip Raman laser (see next section), the application range as well as the sensitivity may be largely increased. Using a frequency modulation technique, where the laser frequency is swept periodically across a molecular absorption line, as little as $10^{-9}$ W of absorbed laser power could still be detected with good signal-to-noise ration (38). This technique also avoids the background problem caused by wall absorption.

Another interesting potential application of optoacoustic spectroscopy lies in the study of photoinduced chemical reaction and molecular dissociation (39). Tuning the laser frequency above the dissociation limit results in a drastic drop of the acoustic signal because most of the absorbed laser energy is now used for dissociation without heating the gas (35).

### 5.2.1.4 Comparison of the Different Methods

In the ultraviolet and visible spectral regions, where spontaneous transition probabilities are large, and where fluorescence detectors with a high quantum efficiency and low noise figures are available, the detection of molecular absorption via laser-excited fluorescence certainly provides the most sensitive method. In the infrared region this method loses sensitivity for two reasons: (1) The quantum efficiency and sensitivity of infrared detectors are smaller than those of photomultipliers. (2) Because of the long radiative lifetimes of excited vibrational levels, the excitation energy may be quenched by collisions before it is emitted by spontaneous radiation.

Therefore, optoacoustic spectroscopy, which utilizes this collisional energy transfer, provides the most sensitive technique in this spectral range. In contrast to the first method, it is limited to a minimum pressure of about 1 torr. This causes pressure broadening of the absorption lines, which may, however, be still smaller than the Doppler width and therefore cause no real disadvantage in Doppler-limited spectroscopy. Both methods may use intracavity absorption techniques or absorption cells inside passive resonators to increase the laser intensity incident on the sample and, with it, the sensitivity.

*The main advantage of these techniques is that the measured signal is directly proportional to the absorbed laser power,* wheras in conventional single- or multiple-path absorption measurements the difference between incident and transmitted intensity is recorded, which means a small difference of large quantities. Some theoretical considerations on the sensitivity limits of laser spectrometers may be found in a paper by Shimoda (40).

### 5.2.2  Tuning Methods in Absorption Spectroscopy

Tunable monochromatic coherent light sources can be obtained in different ways. One way, for instance, is to use active laser media that exhibit sufficiently high gain over a broad spectral range. The insertion of dispersive elements inside the laser resonator narrows the spectral output of the laser and tunes its wavelength to within the spectral gain-profile of the laser medium. Examples of such tunable devices are dye lasers and excimer lasers. Another way to achieve wavelength tuning uses the shift of energy levels of the active medium by external fields or other interactions. Examples are spin flip Raman lasers and diode lasers. A third possibility utilizes optical frequency mixing techniques in nonlinear media to obtain continuously tunable light sources. Parametric oscillators and difference- or sum-frequency mixing of the output from two lasers illustrate this group of tunable coherent sources.

The experimental realization and the efforts to achieve continuous tuning strongly depend on the spectral region and on the desired bandwidth. As long as this bandwidth is large compared to the mode separation of the laser–resonator (multimode operation), continuous tuning may be achieved by changing only one parameter that determines the laser wavelength, for instance, a prism or grating in dye lasers or the magnetic field in spin flip lasers). If continuously tunable *single-mode* operation is wanted, generally more than one frequency-selective element is necessary and one has to take care that synchronous tuning of all these elements, including the resonator length, is achieved.

In order to take advantage of the intrinsic monochromaticity of such single-mode tunable devices, any frequency jitter introduced by fluctuations of cavity lengths or other frequency-determining parameters has to be avoided. This demands careful design and in most cases electronic frequency control. Up to now only a few tunable single-mode devices have been commercially available.

We shall now discuss the most important tunable light sources and the different tuning methods in more detail. For recent reviews of tunable lasers, the reader is referred to references 12, 41–43, and 43a.

### 5.2.2.1 Infrared and Microwave Region

Since nearly all known polar molecules have their vibration–rotation spectra in the range between 2 and 20 $\mu$m, this spectral region, often called the fingerprint region of molecular spectroscopy, is most important for high-resolution spectroscopy. We therefore discuss first some tunable lasers operating in this range and then some submillimeter-wave devices that allow pure rotational spectroscopy.

*Semiconductor Diode Lasers*

Up to now the most widely used tunable laser for infrared spectroscopy has been the semiconductor diode laser (SDL). Since Hinkley (44) in 1970 first used a $Pb_{1-x}Sn_xTe$ laser to record a high-resolution absorption spectrum of $SF_6$ around 10 $\mu$m, a large number of laboratories have reported on spectroscopic results with SDLs (42).

The basic principle of semiconductor lasers has been explained in Chapter 1 and shall be summarized only as follows: If an electric current is sent in the forward direction through the semiconductor diode, spontaneous emission is produced in the $p$–$n$ junction due to radiative recombination of electrons and holes. The wavelength of this spontaneous emission is determined by the energy gap of the semiconductor and has a bandwidth of a few wavenumbers.

Laser emission within the spontaneous gain profile is achieved by multiple reflections between the parallel end faces of the semiconductor crystal. The laser wavelength $\lambda_L$ is determined by the optical length $nL$ of the cavity through the relation

$$m\lambda_L = 2nL \tag{5}$$

Since $L$ is normally about 1 mm, the free spectral range of the resonator is also a few wavenumbers, and the laser will oscillate only in one or at most a few cavity modes. The wavelength range of the gain profile may be changed through any parameter that influences the energy gap of the semiconductor. First of all, the composition of the semiconductor components and their percentages are chosen so as to determine the desired wavelength range, which can be anywhere between 300 and 4000 cm$^{-1}$.

With a SDL of given composition, wavelength tuning can be accomplished by any methods that alters the energy gap, such as varying the diode temperature (45) or applying a magnetic field (46) or external pressure (47) to the semiconductor. However, the laser frequency generally does not follow continuously the tuning of the spectral gain profile over the entire tuning range, but jumps after a few cm$^{-1}$ of continuous tuning by some cm$^{-1}$. The reason for these jumps is that the resonator "eigenfrequency" does not follow synchronously the maximum of the spontaneous gain profile. With temperature tuning, for instance, which can be performed by altering the electric current through the diode, the heating of the $p$–$n$ junction affects both the refractive index $n$ and the energy gap. The resulting frequency shift $\Delta\nu/\nu = \Delta n/n$ is, however, only about 0.1–0.2 of the shift of the gain profile. After a continuous tuning of some wavenumbers, the resonator frequency limps after the gain maximum and is therefore shifted to the slope of the gain profile. This favors mode hopping to another resonator mode that is closer to the gain maximum.

Larger tuning ranges of $Pb_{1-x}Sn_xSe$ diode lasers from 800 to 850 cm$^{-1}$ have been achieved by applying magnetic fields up to 7 kG to the diode (47$a$). Recent improvements using an advanced fabrication technique of double heterostructures (48) allow cw operation above 77 K (49) and for some semiconductor compounds even at room temperature (50). These advanced SDLs can be operated continuously on electrical closed-cycle refrigerators and can be quasi-continuously tuned (with mode hopping) over as much as 280 cm$^{-1}$ (8.2–10.6 $\mu$m) by varying the refrigerator temperature (51).

The mode hopping problem can be avoided by using an external cavity with a length that can be separately controlled. The endfaces of the semiconductor must have antireflection coatings. Since the cavity length is larger, extra mode-selective elements have to be inserted to provide single-mode operation.

Regarding the total costs, a complete SDL system is still much less expensive than any other infrared instrument of comparable resolution.

*Spin Flip Raman Laser*

Within the last years the spin flip Raman (SFR) laser has been developed to a widely tunable IR source with output powers of more than

1 W, exceeding that of SDLs considerably. Since several reviews of SFR lasers have appeared (52), we mention only the basic features that influence the tuning characteristics. The SFR laser is based on stimulated Raman scattering from conduction electrons in certain cryogenic cooled semiconductors that are placed in magnetic fields and are irradiated with focused light from a strong laser. As with the Zeeman effect in free atoms, the energy levels of these electrons are split by the magnetic field $B$ into Landau levels:

$$E \pm = E_0 \pm g\mu_B B \qquad (6)$$

owing to the interaction of the magnetic field with the magnetic moment of the electron spin, which can be parallel or antiparallel to the field. In Eq. (6), $g$ is the Landé factor and $\mu_B$ is the Bohr magneton. When the incident radiation is inelastically scattered by electrons in the lower energy level, the electrons are excited into the higher Landau level and the scattered radiation has just lost this excitation energy (Stokes radiation). Scattering by electrons in the upper level produces anti-Stokes radiation with a corresponding increase in frequency. The output frequency of the SFR laser can therefore be tuned by the magnetic field $B$ according to the relation

$$\nu = \nu_p \pm g\mu_B (B/h) \qquad (7)$$

The tuning rate depends on the Landé factor and is, for example, about $2.3 \, \text{cm}^{-1}$ per kG for InSb. To obtain wide tuning ranges, high magnetic fields are necessary, and superconducting magnets are often used that allow fields up to 200 kG. Its potential of wide-range tunability, high output power, and good mode quality makes the SRL laser very attractive for infrared spectroscopy. One disadvantage is its comparatively high cost and its cumbersome assembly due to the need of a pump laser and a cryogenic cooling system for the superconducting magnet. Pump sources are pulsed or cw CO lasers (53), $CO_2$ lasers (54), or chemical lasers such as the HF laser (55). In some favorable cases threshold pump powers of less than 5 mW have been reported with a CO pump laser (43).

The tuning region depends on the semiconductor material (56a); the linewidth is limited mainly by frequency jitter due to instabilities of the pump intensity, which causes thermal drifts, and by acoustic vibrations of the resonator system. One has furthermore to take into consideration that at tuning rates of $2 \, \text{cm}^{-1} \, \text{kG}^{-1}$ a frequency stability of 1 MHz already requires, at 20 kG, a stability of $\Delta I/I < 10^{-6}$ for the magnet current $I$.

In order to achieve continuous tuning without mode hopping, one has to use external cavity mirrors instead of the end faces of the semiconductor crystal (56). Brueck and Mooradian (57) attained an output spectral width of $0.027 \text{ cm}^{-1}$ using an external cavity with a germanium etalon as additional mode selector. Several cavity modes were simultaneously excited. The tuning range of a single mode was $0.13 \text{ cm}^{-1}$.

## Optical Parametric Oscillators

If laser light of sufficient intensity is focused onto a nonlinear birefringent crystal, a parametric interaction between the pump light and the molecules of the crystal may take place, where the pump photon $\nu_p$ is split into two photons $\nu_i$ and $\nu_s$ (58). Energy conservation demands that $\nu_i + \nu_s = \nu_p$, and momentum conservation yields $\mathbf{K}_p = \mathbf{K}_i + \mathbf{K}_s$ for the corresponding wave vectors of pump wave, signal wave, and idler wave. This parametric process may occur at every molecule of the crystal that is irradiated by the pump light. If the refractive index of the crystal has the proper value for the three frequencies $\nu_p$, $\nu_i$, and $\nu_s$ in the directions of $\mathbf{K}_p$, $\mathbf{K}_s$, and $\mathbf{K}_i$, coherent superposition of the different microscopic contributions may add up in phase (phase matching) resulting in high-intensity monochromatic coherent waves with frequencies $\nu_i$ and $\nu_s$. By rotating the nonlinear crystal or by varying its refractive index through temperature changes, a tunable output at $\nu_s$ or $\nu_p$ can be achieved. For instance, pumping a $LiNbO_3$ crystal with a frequency-doubled YAG laser at $\lambda_p = 0.56 \,\mu\text{m}$ allows tuning ranges between 0.6 and 0.85 $\mu$m for the signal wave and between 1.6 and 3.6 $\mu$m for the idler wave by varying the temperature of the crystal from 200 to 430°C (59). With pulsed pump lasers, parametric oscillator output peak powers of 300 W have been obtained around 2.5 $\mu$m. Commercial devices with these characteristics are already available (59).

Using proustite ($Ag_3AsS_3$) as nonlinear crystal, Hanna et al. (60) even achieved a tuning range between 1.2 and 8.5 $\mu$m by rotating the crystal. Byer et al. (61) recently demonstrated that through improved crystal growth techniques, longer $LiNbO_3$ crystals with high optical quality could be produced allowing the construction of a high-gain efficient parametric oscillator that can be used as a widely tunable narrow-band infrared spectrometer (62). One advantage is the possibility of rapid tuning, since only a ±4° external crystal rotation angle is required for tuning over the entire available spectral range from 1.4 to 4.4 $\mu$m (63). Energy conversion up to 40% from pump wave to signal wave has been achieved. With a tilted etalon within the resonator, a linewidth of about $0.001 \text{ cm}^{-1}$ could be obtained (63a).

## Frequency Mixing Techniques

If the outputs from two lasers with frequencies $\nu_1$ and $\nu_2$ are mixed in a nonlinear optical crystal, the difference frequency $\nu_3 = \nu_1 - \nu_2$ may be generated with high efficiency if the phase-matching condition is fulfilled. This means that the wave vectors of the three waves in the crystal must match the condition $K_1 = K_2 + K_3$ (64). Choosing the second laser to be tunable results in a tunable difference frequency $\nu_3$. This has been demonstrated by Dewey and Hocker (65) who produced up to 6 kW infrared radiation by mixing a pulsed dye laser with a ruby laser in $LiNbO_3$. Infrared tuning was achieved by tuning the dye laser and simultaneously tilting the $LiNbO_3$ crystal to maintain proper phase matching. The tuning range was restricted to 2–4 $\mu$m by the phase-matching range of the crystal.

Pine (66) constructed a cw difference frequency spectrometer that operates in the 2.2–4.2 $\mu$m region by mixing an argon laser with a cw dye laser. With single-mode operation of two visible lasers, Pine achieved a tunable infrared output with 1 $\mu$W power, a spectral resolution of $5 \times 10^{-4}$ cm$^{-1}$ (15 MHz), and a continuous electronic scan in excess of 1 cm$^{-1}$.

Byer (61) pointed out that mixing in $AgGaSe_2$ should allow continuous tuning from 3 to 12 $\mu$m, in CdSe from 10 to 25 $\mu$m, in GaP from 20 to 200 $\mu$m, and in $LiNbO_3$ from 170 $\mu$m to 1 cm. With these crystals, the whole infrared and millimeter range from 1 $\mu$m to 1 cm can be opened to tunable coherent narrow-band spectrometers.

One disadvantage of mixing the two frequencies in nonlinear crystals is the severe restriction of tuning ranges imposed by the phase-matching condition. This may be overcome by a novel method of resonantly enhanced four-wave parametric mixing in alkali metal vapors (67). The output beams of two independently tunable dye lasers, simultaneously pumped by a nitrogen laser, are combined in a collinear beam that is focused into a heat pipe oven (68) containing alkali metal atoms. One of the lasers ($\nu_L$) is tuned to the vicinity of an alkali resonance line and acts as a pump for stimulated electronic Raman scattering at $\nu_s$. The frequency $\nu_p$ of the second laser mixes with $\nu_L$ and $\nu_s$ in the alkali vapor to generate the difference frequency $\nu_{IR} = \nu_L - \nu_s - \nu_p$. This technique allows a broad, smooth tuning range which presently spans the 2–25 $\mu$m range but may soon be expanded to longer wavelengths.

Mixing the output from a tunable spin flip Raman laser with that of the $CO_2$ pump laser in indium antimonide, efficient generation of far-infrared radiation tunable from 90 to 110 cm$^{-1}$ has been reported (69).

### High-Pressure Gas Lasers

Most high-power gas lasers, such as the $CO_2$ and $N_2O$ lasers, oscillate on many closely spaced vibrational–rotational lines. If the linewidths of these individual lines are pressure broadened to such an extent that the widths become larger than the line spacings, the gain profile of the molecular laser merges into a spectral continuum, and continuously tunable laser oscillation may be obtained (70). For the $CO_2$ laser, about 10 atmospheres are necessary to achieve this continuum. When mixtures of different isotopes are used, the line density becomes larger and the continuum starts already at lower pressures.

Most of these high-pressure lasers constructed up to now are pumped by high-energy electron beams (71) or by transverse high-voltage gas discharges using preionization (71a). Recently, optical pumping of $CO_2$ with a HBr laser has been reported (72), where the $CO_2$ excitation energy was transferred to $N_2O$ at 33 atmospheres and a continuous tuning of the $N_2O$ laser output between 9 and 11 $\mu$m was obtainable. Because of their high output power, the tunable lasers can be efficiently used for difference frequency generation in the far-IR by mixing their output in nonlinear crystals with that of a fixed-frequency $CO_2$ laser.

Harris et al. (72a) constructed a compact electron beam-controlled $CO_2$ laser at 15 atmospheres, with a tuning range of $70 \text{ cm}^{-1}$. Two intracavity etalons narrowed the laser output to a linewidth of $0.03 \text{ cm}^{-1}$. Output pulses had 100 mJ energy at pulse durations of 100 ns.

*Laser Magnetic Resonance and Stark Spectroscopy*

Instead of tuning the laser frequency across the molecular absorption lines, the molecular levels may often be shifted by external magnetic or electric fields to generate coincidences between fixed laser lines and rotational or rotational–vibrational transitions (73,74). The rotational levels of paramagnetic molecules placed inside a laser resonator can be, for instance, Zeeman shifted by a magnetic field into coincidences between far-infrared laser lines to produce resonance absorption of the laser radiation. The resonance is monitored by the resultant decrease in laser power. This laser magnetic resonance spectroscopy (LMR) provides one of the most sensitive techniques for the detection of molecules and free radicals. It was first demonstrated with $O_2$ (75) and has subsequently been successfully applied to other stable molecules such as NO, $NO_2$, and $H_2O$ and to transient free radicals such as OH, CH, and HCO (76).

The range over which molecules can be tuned depends on the transition involved and may be, in favorable cases, as large as $2 \text{ cm}^{-1}$ at magnetic fields up to 2 teslas. In most cases the Zeeman pattern can be calculated from rotational and fine structure parameters obtained from microwave or electronic spectroscopy, and therefore assignments of the LMR transitions can be made.

The sensitivity of laser magnetic resonance spectroscopy is about two orders of magnitude higher than that of electron paramagnetic resonance or microwave spectroscopy, where the detection limits are radical densities of about $5 \times 10^{10}$ cm$^{-3}$. This extremely high sensitivity has been demonstrated in LMR studies of gas phase reactions of OH with CO, NO, and $NO_2$ (77) where OH radical densities of $2 \times 10^8$ cm$^{-3}$ could be detected.

These examples have shown that the entire spectral range in the IR can already be covered by coherent monochromatic, tunable sources. Not all of them are convenient and easy to operate, but this situation will certainly improve if we continue with the rapid progress made during the last years.

### 5.2.2.2  Visible and Ultraviolet Region

In this spectral range the dye laser with its different modifications does certainly provide the most important tunable spectroscopic light source. By using different dyes or dye mixtures, the spectral range between 340 nm and 1 $\mu$m can be completely covered. The three most commonly used dye lasers are (78) (i) $N_2$ laser-pumped dye lasers with pulse lengths of 1–10 ns, peak powers up to 1 MW, and repetition rates up to 1 kHz (78a,78b); (2) flashlamp-pumped dye lasers with pulse durations of 0.1–100 $\mu$s, peak powers in the kilowatt range, and repetition rates up to 50 pulses s$^{-1}$ (79); (3) argon or krypton laser-pumped cw dye lasers with continuous outputs between a few milliwatts to several watts, depending on pump power and dye laser wavelength (80). The laser-pumped dye lasers allow the widest tuning range because they exhibit the largest gain.

In all of the three versions, the bandwidth can be narrowed by inserting etalons inside the laser cavity (81). With cw lasers, single-mode operation has been achieved with output powers of 100 mW and more (82). By using frequency stabilization techniques the frequency jitter may be kept below 1 MHz, which allows high-resolution Doppler-free spectroscopy (83). Continuous tuning of single-mode dye lasers without mode hopping demands synchronous drive of all frequency-determining resonator elements such as the cavity length and two tilted etalons (84). Some commercial devices already allow continuous tuning, free of mode hopping, over more than 1 cm$^{-1}$ (85).

In the UV range, wavelength tuning can be obtained by optical frequency doubling of tunable dye lasers (85a). Since the efficiency of second harmonic generation increases with the intensity of the fundamental wave, focusing of pulsed dye laser outputs into nonlinear crystals such as KDP or RDP (rubidium dihydrogen phosphate) has resulted in high peak-power UV pulses with conversion efficiencies up to 52% (86) at 7 MW input power. Placing the nonlinear crystal inside the laser cavity of a cw dye laser, continuously tunable UV radiation in the milliwatt range has been produced (87). With RDP crystals and a cw dye laser power of 100 W inside a ring laser cavity, even more than 100 mW cw UV radiation has been achieved (88).

Below 200 nm all crystals known today for optical frequency doubling begin to absorb strongly and cannot be used in this range. A new technique of third harmonic generation in phase-matched mixtures of metal vapors and inert gases has been developed by Harris and co-workers (89), that allows efficient generation of coherent light in the UV down to 88 nm.

By mixing two visible dye lasers in atomic and ionic vapors, resonance-enhanced tunable vacuum ultraviolet sum frequency generation has been obtained if one dye laser was tuned to a two-photon resonance with an excited state of vapor atoms (90).

The discovery of excimer lasers has opened a large, new class of active media for tunable lasers from the visible to the vacuum ultraviolet (91). Excimers are quasi-molecules that have bound excited electronic states but a dissociative ground state which is characterized by a repulsive potential curve with a shallow van der Waals minimum. The depth of this minimum is small compared to thermal energy $kT$ at room temperature (see Fig. 5.4). Examples of atoms that form homonuclear excimers are the noble gases He to Xe. Heteronuclear excimers are, for instance, alkali–noble gas molecules (92) such as NaXe or LiAr, and the noble gas halides such as KrF (93).

Excimers are ideal candidates for efficient high-power tunable lasers (94) because inversion is easily maintained for laser transitions from an upper bound state to the lower repulsive state which is automatically depopulated by rapid dissociation. The lowest excited bound states of the different excimers range from 1 to 15 eV above the repulsive ground

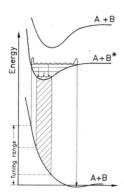

**Fig. 5.4.** Schematic diagram of potential curves and continuous emission from excited bound states of excimers into repulsive ground states.

state. Possible laser wavelengths may therefore extend from the near-infrared to the vacuum ultraviolet region.

Much effort has been made to optimize efficient excitation of the upper excimer state (95). High-energy electron excitation or transverse discharges have been used so far as pump mechanisms (96). Tuning of the laser wavelengths has been achieved by fused quartz or CaF prisms inside the laser resonator (97). The bandwidth could be further decreased by an etalon. Tuning ranges of $2500 \, cm^{-1}$ are comparable with those of dye lasers (94).

### 5.2.3 Doppler-Limited Absorption Spectroscopy

In this section several applications of the different tuning methods discussed above to molecular absorption spectroscopy are illustrated by examples. In most experiments presented here, *the spectral resolution is mainly limited by the Doppler widths or the pressure broadened-line profiles of absorption lines.* This implies that the laser linewidth has to be smaller than the molecular linewidth. Some tunable lasers, such as the diode laser or the spin flip laser, often simultaneously oscillate on several modes. However, since the mode spacing is of the order of $1 \, cm^{-1}$, a single mode can be selected out of the laser output by a monochromator with medium resolving power.

One of the major problems in tunable laser absorption spectroscopy is the determination of the absolute wavelengths for the different absorption lines. Often, part of the laser output is sent through an external Fabry-Perot interferometer, which gives frequency marks every time the laser frequency is tuned across a transmission maximum of the interferometer (see Fig. 5.1b). By this means the *frequency separation* of different lines can be measured with high precision, provided that the free spectral range of the Fabry-Perot has been accurately determined. In order to get the *absolute* wavelength, either a calibration spectrum from molecules with known absorption lines is simultaneously recorded by sending part of the laser output through a reference cell, or the laser output is mixed in a nonlinear crystal with a fixed-frequency laser of known wavelength and the difference frequency is measured (heterodyne technique).

Among the numerous applications of diode lasers, the following three shall serve as examples: With a single magnetic and current-tuned PbSe diode laser, Allario et al. (98) measured the completely resolved spectra of $SO_2$ between 1176 and $1265 \, cm^{-1}$. The combination of the two separate tuning mechanisms allows this large tuning range of nearly $90 \, cm^{-1}$. The tuning range of individual laser modes was about $2 \, cm^{-1}$, which is larger than the mode spacing. Therefore, a complete coverage of the whole spectral range could be achieved. The $SO_2$ sample was maintained at a pressure of 1 torr in a 45-cm-long cell.

As many as 169 individual transitions in the band of ethylene around $942.2\,cm^{-1}$ could be resolved using a PbSnTe diode laser (99). The accuracy of measuring line distances was mainly limited by random fluctuations of the laser output and by lack of suitable frequency marks. However, the frequency splittings between different lines could still be measured reproducibly to $\pm 0.008\,cm^{-1}$, allowing accurate determination of deviation from the symmetric top configuration. Owing to the attainable narrow linewidth of the diode laser, true lineshapes of molecular absorption lines can be determined. This has been demonstrated by Aronson et al. (100) who used a current-tuned PbSnSe diode laser to measure lineshapes and intensities of several rotational lines in the Q-branch of the $CO_2$ $\nu_2$ band as a function of temperature and pressure. Since water absorption plays a crucial role in atmospheric transmission, knowledge of pressure-broadened absorption profiles is necessary for quantitative analysis of atmospheric water vapor concentration. The experimental results about pressure broadening of water lines at 5 $\mu$m by oxygen, nitrogen, and noble gases obtained by Eng et al. (101) with a PbSSe diode laser and by Guerra et al. (102) with a spin flip laser confirmed that the wings of the absorption profiles show non-Lorentzian behavior, which increases the absorption at the wings.

Walther and his group (103) performed high-resolution spectroscopy of ethylene by means of a spin flip Raman laser. With a 20-cm absorption cell and a gas pressure of 30 torr, the resolution was limited by the combined effects of 1200 MHz pressure broadening, 130 MHz Doppler width, and 600 MHz laser linewidth. The $K$ doubling due to the slightly asymmetric top configuration of the molecule could be clearly resolved, and the asymmetry parameter as well as the rotational constants could be determined.

Pine (104) used a cw difference frequency spectrometer that operated in the 2.2–4.2 $\mu$m range by mixing a tunable cw dye laser with a fixed-frequency argon laser in a $LiNbO_3$ crystal (66). With single-mode operation of the two visible lasers, Pine achieved tunable infrared output with 1 $\mu$W power, a spectral resolution of $5 \times 10^{-4}\,cm^{-1}$ (15 MHz), and a continuous electronic scan in excess of $1\,cm^{-1}$. With this device, Doppler-limited spectra in the $\nu_3$ vibrational band of $^{12}CH_4$ and $^{13}CH_4$ have been measured. The tetrahedral splittings in the $P$ and $R$ branches could be examined with high accuracy. The high resolution obtained is essential to distinguish between two different models proposed for the higher-order vibration–rotation interaction that causes the rotational level splittings. Pressure broadening and shift of the molecular lines at low and high pressures (atmospheric air pressure) were also investigated (105).

With a tunable spin flip Raman laser, Butcher et al. (106) detected the optoacoustic spectrum of a $Q$-branch in the fundamental NO vibrational band. At low NO pressures, a Doppler-limited linewidth of 127 MHz could be achieved. The $\Lambda$ doubling (about 700 MHz) appeared as a gross splitting at this resolution, and even the hyperfine splittings of each $\Lambda$ component could be partially resolved. With an incident laser power of about 10 mW the entire spectrum (about 2.5 $cm^{-1}$) was swept in 10 minutes at a resolution better than $2 \times 10^5$ and signal-to-noise ratios up to 200:1. Using the same techniques the rotational constants of OCS have been derived (106). With a spectraphone, absorption coefficients down to $10^{-11} cm^{-1}$ have been measured for NO and $CH_4$ at their atmospheric concentration levels in the 3.8-$\mu$m region at 17 different DF laser wavelengths (107).

As an example of visible absorption spectroscopy with dye lasers, we mention the measurements of $NO_2$ absorption lines in the region of 5935 Å by Stevens and Zare (108). The $NO_2$ molecules were excited inside the resonator of a narrowband (0.035 $cm^{-1}$ bandwidth) tunable pulsed dye laser which was tuned between 5939 and 5941 Å. The absorption lines were identified by recording the fluorescence spectra from the corresponding upper levels. Compared to conventional absorption spectroscopy in case of dense and badly overlapping spectra, this "excitation spectroscopy," that is, the combination of absorption and fluorescence measurements, has the advantage of unambiguous indentification of the absorption lines through their induced fluorescence spectra. This work provided the first successful rotational analysis of a visible band in the extremely complicated $NO_2$ spectrum. From the fluorescence analysis it could be proved, for instance, that the upper state has $^2B_2$ symmetry and that the equilibrium configuration in this state is $r_0 = 1.31$ Å and $\theta_0 = 111°$.

Excitation spectroscopy of the $A^2\Pi \leftarrow X^2\Sigma$ transition of CaF has been performed by Field et al. (109), who used a cw dye laser with 10 mW output and a bandwidth of $\leq 0.001 cm^{-1}$ (30 MHz) which could be tuned in 0.003-$cm^{-1}$ steps. The molecular constants of CaF including $\Lambda$ doubling could be determined with high precision. The superiority of cw dye laser "spectrometers" has been tested by Green et al. (110) who compared the absorption spectra of molecular iodine around 5890 Å taken with a high-dispersion 7.3-m Ebert grating spectrograph used in 10th order with those obtained with the dye laser. Needless to say, better results were obtained with the laser.

A limited number of papers have been reported so far on ultraviolet absorption spectroscopy with tunable lasers. Either frequency-doubled dye lasers have been used or two-photon spectroscopy with visible lasers has been performed. Thompson et al. (111) presented measurements of

$SO_2$ absorption coefficients in the wavelength interval of 296–301 nm using a frequency-doubled pulsed dye laser with a 0.02-nm bandwidth as excitation source. The resolution was limited by the residual laser linewidth and could be improved by further narrowing the dye laser bandwidth through an additional etalon. Anyway, for atmospheric $SO_2$ detection, the molecular lines are pressure broadened resulting in quasi-continuous absorption spectra where the rotational lines can hardly be resolved.

Very interesting information about molecular structure can be obtained by the technique of two-photon spectroscopy. The transition probability $R_{if}$ of a two-photon transition connecting levels i and f is analogous to a Raman transition (111a):

$$R_{if} \propto \left| \sum_k \frac{M_{ik}M_{kf}}{\Delta\omega_k} \right|^2 \cdot \frac{\gamma}{\delta\omega^2 + (\gamma/2)^2} ; \qquad \delta\omega = 2\omega_L - \omega_{if} \qquad (8)$$

where the summation extends over all molecular states; $\omega_L$ is the laser frequency and $M_{ik}$ and $M_{kf}$ are the matrix elements of normal one-photon transitions between levels $i \rightarrow k$ and $k \rightarrow f$ (see Fig. 5.5 and reference 112). If the mismatch $\Delta\omega_k = \omega_L - \omega_{ik}$ is large, the transition probability will be very low, and intense pulsed lasers will have to be employed. From Eq. (8) one sees that the selection rules for two-photon transitions are the same as those in Raman spectroscopy. With two-photon transitions, therefore, excited molecular states can be reached which are not connected to the initial state by allowed dipole transition, meaning that they cannot be populated by linear spectroscopy. The excitation can be monitored by allowed fluorescence transitions from the excited state $f$ to lower molecular states.

This technique, therefore, allows the discovery of new excited states

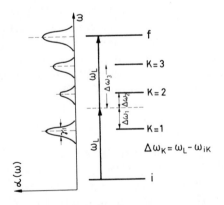

**Fig. 5.5.** Energy level diagram of two-photon excitation.

and the elucidation of coupling phenomena such as radiationless transitions and intersystem crossings. The first experimental results were obtained with nitrogen laser-pumped dye lasers for benzene (113–115) and NO (115,116). The symmetry of the state that can be reached by two-photon absorption depends on the polarization of the two laser waves used for excitation. This facilitates the identification of the upper state considerably (117).

### 5.2.4 Doppler-Free Absorption Spectroscopy

Many molecular absorption spectra in the visible and UV corresponding to electronic transitions exhibit such a high density of lines that the mean line spacing is much smaller than their Doppler widths. This entails that they always appear as quasi-continuous spectra when recorded by means of Doppler-limited spectroscopy. All information about molecular structure, which could be in principle derived from line spacings and positions, is in these cases masked by the Doppler width. In order to resolve the different lines and to investigate the structure of the excited electronic state, *Doppler-free methods have been developed which are only limited by the natural linewidth.* Since in the visible and UV region the natural linewidth is about two to three orders of magnitude smaller than the Doppler width, the progress in resolution achieved by these methods is quite impressive. We briefly discuss below the most important Doppler-free techniques in laser absorption spectroscopy.

### 5.2.4.1 *Spectroscopy in Collimated Molecular Beams*

One obvious method for reducing the Doppler width uses well-collimated molecular beams where the velocity components perpendicular to the molecular beam axis are reduced by a factor depending on the collimation ratio $d/L$ (see Fig. 5.6). With a collimation ratio of $1:200$, which is easily achieved (e.g., with two collimating slits 1 mm wide at a distance of 20 cm), the transverse velocity distribution is narrowed by a factor of 200. If a monochromatic tunable laser is crossed perpendicular with the molecular beam and tuned across the molecular absorption profiles, the Doppler width of the absorption lines is reduced by the same factor. For many molecular transitions this residual Doppler width is already below the natural linewidth. In order to take full advantage of this reduction of Doppler width, the laser linewidth has to be even smaller. This implies that single-mode frequency-controlled tunable lasers must be employed (118). Pine and Nill (119) achieved high-resolution spectra of NO and CO infrared spectra around 5 $\mu$m using a tunable

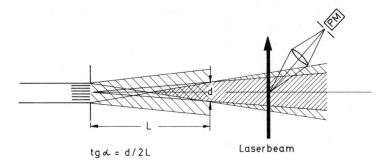

$$tg \, \alpha = d / 2L \qquad \text{Laserbeam}$$

**Fig. 5.6.** Reduction of Doppler width by absorption of monochromatic tunable lasers in collimated molecular beams.

PbSSe diode laser and a molecular beam from a multichannel capillary array nozzle. The beam divergence was still 10° resulting in a collimation ratio of about 1:20. Infrared absorption spectroscopy of $NH_3$ and $CH_3F$ in better collimated molecular beams performed by Chu and Oka (119$a$) yielded molecular linewidths of 1.3 MHz.

At higher collimation ratios the molecular density in the beam is generally very low; in the crossing volume of molecular beam and perpendicular laser beam, the partial density of molecules in the absorbing level ($v'',J''$) is often below $10^8$ cm$^{-3}$. Since the absorption path length is about 1 mm, the relative attenuation of the laser beam through absorption in the molecular beam may be less than $10^{-12}$. The absorption spectrum is therefore generally measured through the total fluorescence from the laser-excited molecular levels (excitation spectrum, see Sect. 5.2.1.2).

Some examples will illustrate the technique and its advantages over Doppler-limited spectroscopy. Figure 5.7 shows part of the $NO_2$ absorption spectrum monitored through the laser-excited total fluorescence when the output from a single-mode stabilized, tunable argon laser is frequency tuned around $\lambda = 5145 \, \text{Å}$ (120). The upper curve has been taken from $NO_2$ molecules in a glass cell at room temperature and a pressure of about 1 torr. The middle spectrum was obtained by crossing the laser beam perpendicularly with a $NO_2$ beam having a divergence of about 1:100. The many resolved lines are hyperfine structure components of visible absorption lines (121) corresponding to different transitions, (o,o,o,$N''$) → ($v_1,v_2,v_3,N'$), between rotational levels in the ground state and in an electronically excited state, respectively. These lines are completely hidden in the Doppler-limited spectrum. From these different, still partly overlapping, transitions (middle spectrum, Fig. 5.7), single

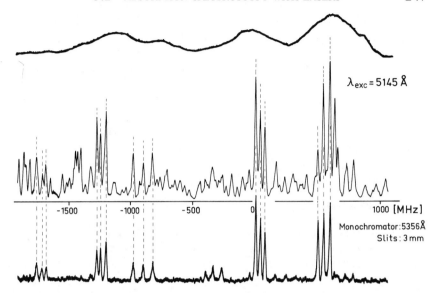

$\lambda_{exc} = 5145 \text{ Å}$

-1500    -1000    -500    0    1000 [MHz]

Monochromator: 5356 Å
Slits: 3 mm

**Fig. 5.7.** High-resolution excitation spectrum of $NO_2$ obtained with a single-mode argon laser tunable around $\lambda = 5145$ Å. Upper trace, $NO_2$ in a cell at $10^{-3}$ torr pressure; middle trace, $NO_2$ in a collimated beam, total fluorescence observed; lower trace, selected fluorescence terminating at the $(0,2,0)$ vibrational level of the electronic ground state. From Demtröder (14), reproduced by permission.

transitions such as, for instance, $(o,o,o,N_i) \to (v_{1k}, v_{2k}, v_{3k}, N_k)$ between specific lower and upper rotational levels can be selected by observing the fluorescence as a function of laser frequency with a monochromator set on the $v_2$ fluorescence band emitted from upper rotational levels in a specific vibrational state. The lower trace in Fig. 5.7 shows such a selected excitation spectrum indicating three *hfs* levels for each rotational level. The three hyperfine structure components correspond to $\Delta F = \Delta J = 0$ transitions between the three *hfs* levels of rotational levels due to the nuclear spin $1 \cdot \hbar$ of the $^{14}N$ nucleus. The combination of high-resolution absorption spectroscopy and laser-induced fluorescence spectroscopy allows to identify the many lines of the absorption spectrum and to perform a complete analysis of the spectrum (122). Due to the relatively long upper-state lifetime, the natural linewidth of the $NO_2$ transition should be below 100 kHz. The observed residual linewidth of about 10 MHz is mainly due to the divergence of the molecular beam.

With a highly stabilized argon laser tunable around 514.5 nm, Ezekiel and his group have obtained absorption linewidths of less than 3 MHz in collimated iodine beams (123). They could resolve the hyperfine structure of $I_2$ achieving a resolution of 5 parts in $10^9$. A comparable resolution was obtained with a single-mode jet stream cw dye laser which had the additional advantage of larger tuning ranges (124). With an angular divergence of $2 \times 10^{-3}$ rad of the $I_2$ molecular beam, the residual Doppler width is about 500 kHz, while the natural linewidth can be estimated from measured lifetimes to be 70 kHz.

An improved version of the experiment employed a 5145-Å argon laser which was long term stabilized (1 part in $10^{14}$) to one hyperfine structure component in a molecular $I_2$ beam (125). The stabilized laser frequency was precisely tuned by an external acousto-optical scheme to excite a neighboring hyperfine component in a second $I_2$-molecular beam. The measured linewidth was 150 kHz (FWHM).

### 5.2.4.2  Saturation Spectroscopy

This Doppler-free spectroscopic technique utilizes the selective saturation of molecular transitions caused by monochromatic lasers with sufficient intensity. Its basic principles may be briefly outlined as follows (125a):

If molecules in the ground state level 1 absorb light, the population density $n_1$ of level 1 will decrease by an amount which depends on the ratio of the absorption rate to the total repopulation rate, the latter being due to fluorescence or collisional relaxation (Fig. 5.8a). At high intensities, as attainable with lasers, the absorption rate may become large enough to cause a considerable depletion of level 1. This can be monitored, for instance, by observing the fluorescence intensity $I_F$ emitted from level 2 as a function of laser intensity $I_L$. Since $I_F$ is proportional to the product $n_1 I_L$, the fluorescence intensity $I_F$ will increase linearly with $I_L$ as long as saturation is negligible, which means that $n_1$ remains constant. As soon as saturation becomes notable, $n_1$ starts to decrease and the slope of $I_F(I_L)$ decreases. The absorption deviates from Beer's law and becomes nonlinearly dependent on $I_L$. From Fig. 5.8b, which demonstrates this saturation for excitation of $NO_2$ by the 5145-Å argon laser line at different pressures, the influence of the collisional relaxation rate on saturation can be clearly seen.

Assume a monochromatic laser wave with frequency $\nu_L$ in the laboratory frame to interact with molecules in the gas phase having velocity components $v_z$ in the direction of light propagation. The laser frequency in the rest frame of the molecule is Doppler shifted to a frequency $\nu$ given by

$$\nu = \nu_L(1 - v_z/c) \tag{9}$$

The molecule can absorb only if $\nu$ is within the natural width $\Delta \nu_n$ comprising the corresponding molecular transition frequency $\nu_{12}$. At a

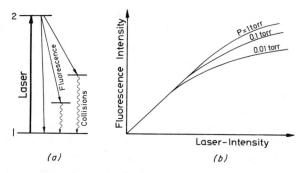

**Fig. 5.8.** Saturation of molecular transitions: (a) level diagram; (b) fluorescence intensity as a function of laser intensity at different pressures. Curves are obtained from $NO_2$ excitation with $\lambda = 5145$ Å argon laser.

fixed laser frequency $\nu_L$, therefore, only those molecules can absorb the laser light that are Doppler shifted into resonance with $\nu_L$. These are all the molecules with velocity components given by

$$v_z \pm \Delta v_z = (c/\nu_L)(\nu_{12} - \nu_L \pm \Delta \nu_n) \tag{10}$$

where $\Delta \nu_n$ is the natural linewidth. If the laser wave has sufficiently high intensity, it will saturate the molecular transitions for all molecules in the velocity interval $v_z \pm \Delta v_z$. In the velocity distribution $n_1(v_z)$ of the ground state molecules, this group of molecules is partly missing. In other words, the laser has burnt a hole around $v_z$ into the velocity distribution (Fig. 5.9a) with a width $\Delta v_z = (\Delta \nu_n / \nu_L)c$ which depends on the natural

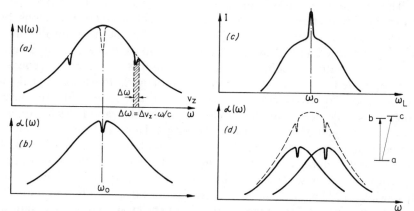

**Fig. 5.9.** Hole burning in Doppler-broadened absorption profiles: (a) by a monochromatic wave propagating into one direction; (b) absorption coefficient including the Lamb dip, (c) by a standing wave as a superposition of two waves traveling in opposite directions; (d) resolution of closely spaced molecular lines by Lamb dip spectroscopy.

linewidth of the molecular transition. For a laser wave traveling in the opposite direction, the Doppler shift $\nu = \nu_L(1 + v_z/c)$ is just opposite, and the hole appears on the other side of the Doppler distribution around $v_z$. The total absorption of both waves is proportional to $n(E_i, + v_z) + n(E_i, - v_z)$. If the laser frequency $\nu_L$ is tuned to the line center of the Doppler-broadened absorption profile, both holes coincide because both waves are now absorbed by the same group of molecules around $v_z = 0$. The total absorption decreases in this case because this group of molecules is absorbing twice the intensity and therefore suffers larger saturation, which means that the hole becomes deeper.

If both waves travel simultaneously in opposite directions through a gaseous absorbing probe, the absorption coefficient $\alpha(\nu)$ will therefore exhibit a small dip (Fig. 5.9b), called a Lamb dip (126), at the center of the Doppler-broadened line profile. This dip can be monitored in different ways: If the probe is placed inside the laser resonator, the laser intensity will show a sharp increase when the laser frequency is tuned across the absorption line center, because the losses have a minimum (see Fig. 5.9c). Since the total fluorescence is proportional to the ground state density, the Lamb dip can be monitored through the corresponding dip in the fluorescence intensity when the laser frequency is tuned across the line center. The importance of this Lamb-dip spectroscopy for Doppler-free resolution is illustrated in Fig. 5.9d. Here, two transitions between a common ground state and two closely spaced upper levels overlap within their Doppler width. The Lamb dips, however, with a half-width some orders of magnitude smaller, are clearly separated.

The advantage of saturation spectroscopy over molecular beam spectroscopy lies in the possibility to measure collision cross sections by observing the broadening of the Doppler-free linewidths when the pressure is increased (127,128). A further benefit is the relatively simple vacuum apparatus needed, less expensive than that necessary for generating collimated molecular beams. Its disavantage is the lower signal-to-noise ratio. To avoid power broadening of the Lamb dip, the laser intensity should not be too high. Therefore, the saturation reaches only a few percent and the noise from the large, Doppler-broadened linear signal is superimposed on the Doppler-free nonlinear saturation signal. A technique for improving signal-to-noise ratio and to extend saturation spectroscopy to low pressures has been demonstrated by Sorem and Schawlow (127) who chopped the two laser beams traveling in opposite directions at two different modulation frequencies $\omega_1$ and $\omega_2$. The saturation dip, which is caused by the combined absorption of both waves by the same group of molecules, is proportional to $I_1 I_2$. The corresponding decrease in absorption is therefore modulated at the sum frequency and can be monitored by observing that part of the total fluorescence from the upper level that is modulated at this frequency.

Figure 5.10 shows as an example of this intermodulated fluorescence technique—a saturation spectrum of $I_2$ taken in our laboratory (129) with a single-mode argon laser tunable around $\lambda = 5145$ Å. The upper curve is obtained when the total fluorescence is observed as a function of laser frequency. The Lamb dips corresponding to different *hfs* transitions within the rotational transition can be clearly seen. The lower curve has been obtained with the intermodulated fluorescence technique, which suppresses the linear Doppler-broadened background.

Most of the work in the visible spectrum demonstrating the capability of saturation spectroscopy in molecular physics has been performed on different iodine transitions (130–134). The nuclear quadrupole coupling of both upper and lower electronic states of $I_2$ could be determined from hyperfine structure splittings (134). Even with a free-running single-mode dye laser without frequency stabilization, saturation spectra of several iodine lines have been measured over the tuning range of the rhodamine 6G laser (5800–6200 Å) (135). Using the intermodulated fluorescence technique, Murihead et al. (136) measured the hyperfine splitting of one of the transitions in $BO_2$ coincident with the 5145 Å argon line. From *hfs* splittings in $NH_2$, obtained with the same technique using a cw dye laser, Hills et al. (137) determined the Fermi contact term and the dipole-dipole coupling constants for both the ground state and the $\Pi(0,10,0)$ vibrational level of the excited $A^2A_1$ state.

In the infrared region, the first experiments utilized coincidences between strong laser lines and molecular absorption lines. The laser lines could be tuned only in a narrow range within their gain profile. In some experiments the laser frequency remained fixed and the molecular absorption lines were tuned by electric fields. Goldberg and Yusek (138) could resolve parts of the very dense absorption spectrum of $SF_6$ in the 10.4-$\mu$m region. They observed 43 $SF_6$ absorption lines coincident with the P(18) line of the $CO_2$ laser and 24 lines coincident with the P(16) laser line. Line spacings of the order of 1 MHz could be resolved and were measured by a heterodyne technique (see Sect. 5.4.2).

Linewidths of 100 kHz have been observed in saturation spectroscopy of $CH_3F$ with a He–Ne laser at 3.39 $\mu$m (139). The molecular lines were Stark-tuned, and the electric dipole moment of $CH_3F$ in the ground state and in excited vibrational states could be determined (140). Using Zeeman splittings of the methane line coincident with the He–Ne laser line at 3.3 $\mu$m, the rotational magnetic moment of $CH_4$ could be measured (141). Saturation spectroscopy of the $H_2O$ line at 1889.58 cm$^{-1}$ has been performed with a tunable spin flip laser by Patel (142). The

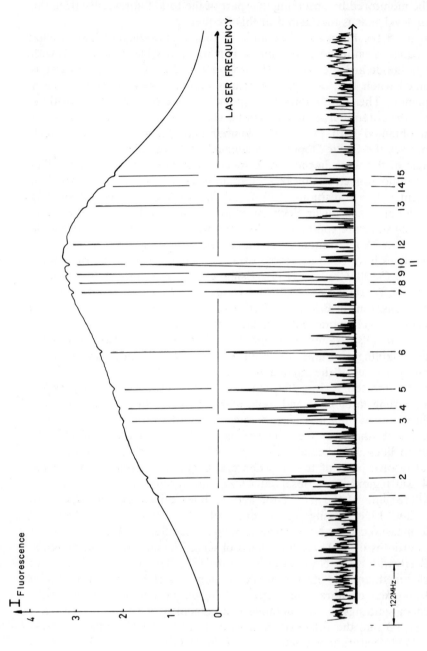

**Fig. 5.10.** Saturation spectrum of the $I_2$ transition $B^3\Pi_{0_u^+} \leftarrow X^1\Sigma_g^+$ around $\lambda = 5145$ Å: (*a*) Lamb dips in the Doppler-broadened absorption profiles; (*b*) isolation of Lamb dips from the Doppler profile by using the intermodulated fluorescence technique. From Demtröder (14), reproduced by permission.

observed linewidths of 200 kHz demonstrates the usefulness of the SFR laser for ultrahigh resolution spectroscopy.

These examples may have shown the capability of saturation spectroscopy to gain information about molecules with dense spectra not resolvable with Doppler-limited techniques (142a).

### 5.2.4.3 Optical Double-Resonance Techniques

These Doppler-free techniques utilize the simultaneous interaction of molecules with two electromagnetic fields. They have been successfully used in both the visible and the infrared region (143). Suppose a monochromatic laser wave 1 is tuned to the center frequency $\nu_{12}$ of a molecular transition (see Fig. 5.11). Only molecules with velocity components $v_z = 0 \pm \Delta v_z$ are excited into the upper level 2 (see Fig. 5.9a and Sect. 5.2.4.2). If a second electromagnetic field (which can be either another laser, a microwave, or a rf field) is applied to the molecules and is tuned to the frequency $\nu_{23}$, the molecules in level 2 can be further excited to level 3. This double resonance can be detected, for instance, through the fluorescence from level 3 or through the change in absorption of level 2. Since the absorption of the second field is due to molecules in level 2 with a narrow interval of velocity components $\Delta v_z$, the double-resonance signal will be essentially Doppler free if both waves travel parallel or antiparallel through the sample. Instead of using the velocity-selective *population* of the upper level 2 by the monochromatic laser wave 1, the selective *depopulation* (hole burning) of the lower level 1 can be also employed for double resonance (see Fig. 5.11b).

This double-resonance technique, where two transitions sharing a common level are coupled through the population of the common level, has

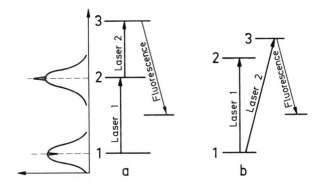

**Fig. 5.11.** Optical double-resonance spectroscopy with two lasers.

proved to be eminently useful for identification of complex molecular spectra (144). Assume laser 1 is intensity modulated at a frequency $\omega$. Then all transitions to other levels k starting from level 1 or 2 are also modulated where the modulation phase of transitions $1 \rightarrow k$ is opposite to that of transition $2 \rightarrow k$. Tuning the unmodulated laser 2 across the molecular absorption spectrum will yield a signal at the modulation frequency $\omega$ each time the laser coincides with a transition $1 \rightarrow k$ or $2 \rightarrow k$. The upper or lower level of the unknown transition can therefore be labeled, which considerably facilitates the identification of the other level.

Infrared microwave double-resonance experiments have been meanwhile performed on quite a few molecules. With an infrared laser a specific rotational–vibrational level was selectively populated and the absorption or amplification of a microwave traveling through the laser-pumped sample was monitored. Inversion lines in $NH_3$ (145), K-type doubling transitions in $H_2CO$ (146), and rotational transitions of vibrational excited molecules (147) have been investigated.

Using an argon laser and a tunable dye laser for excitation, absorption lines of the $X^1\Sigma \rightarrow A^1\Sigma$ transition of the $Na_2$ molecule (144) and in the BaO spectrum (148) have been identified with the double-resonance technique.

### 5.2.4.4  Doppler-Free Two-Photon Spectroscopy

Two-photon spectroscopy can be performed without Doppler broadening if the two photons that are simultaneously absorbed to excite the molecular transition are supplied by two different laser waves propagating in opposite $\pm z$ directions. In the reference frame of a molecule moving with a velocity component $v_z$, the frequencies of the two laser waves are Doppler shifted to

$$\nu_+ = \nu_L(1 + v_z/c) \quad \text{and} \quad \nu_- = \nu_L(1 - v_z/c) \tag{11}$$

If a two-photon transition is induced with one photon contributed by each wave, the sum of both energies, as seen in the molecular frame, is

$$\Delta E = h(\nu_+ + \nu_-) = 2h\nu_L \tag{12}$$

This energy is independent of the velocity components $v_z$! This means that if the laser frequency $\nu_L$ is tuned to $\nu_L = (E_i - E_k)/2h$, all molecules in the lower level $E_k$, notwithstanding their velocity, can absorb the two photons at the same laser frequency $\nu_L$. The width of the Doppler-free two-photon signal is equal to the sum of the natural linewidths of levels $E_i$ and $E_k$. *In contrast to saturation spectroscopy, where only a small* fraction $\eta$ of all molecules in the absorbing state within the narrow velocity interval $\Delta v_z$ contribute to the Doppler-free transition, *in Doppler-free two-photon transitions all molecules in the absorbing level have equal probability of absorbing the photons.* This increase in $\eta$ from about 0.01 to 1 may outweight the smaller transition probability $R_{ik}$ for two-photon transitions, see Eq. (8).

There is, of course, also a chance that both photons come from the same laser beam. Since there is no cancellation of the Doppler shift in this case, a Doppler-broadened background is present. It can be shown that the area under the Doppler-free curve is just twice as large as that under the Doppler-broadened profile, which means that the latter can often be neglected (148a) (see Fig. 5.12).

This novel technique was first applied to molecules by Bischel et al. (149) who used two opposite beams from a fixed-frequency $CO_2$ laser and tuned the energy levels of $CH_3F$ by an external electric field into two-photon resonances with the laser frequency. The Doppler-free resolution allowed measurement of the pressure broadening and pressure shifts of the $0 \rightarrow 2\nu_3$ transition in $CH_3F$.

Doppler-free two-photon electronic state absorption spectra of NO and benzene have been recorded by Gelbwachs et al. (150). The linewidth of the two-photon signals monitored through the fluorescence from the upper level was still limited by the width of the nitrogen laser-pumped dye laser used for excitation. Recently, the Doppler-free two-photon transitions in the $Na_2$ molecule have been observed (151). The method has been widely used in atomic spectroscopy (152) and will soon gain

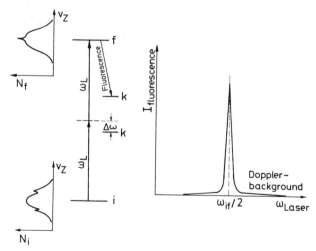

**Fig. 5.12.** Schematic illustration of Doppler-free two-photon spectroscopy.

much more attention in molecular physics, since it allows ultrahigh-resolution UV spectroscopy with visible lasers and the discovery of new molecular states with symmetries that do not allow their excitation by linear spectroscopy.

### 5.2.5  Analytical Applications of Laser Absorption Spectroscopy

In this section we briefly cover the benefits to analytical chemistry by using the different techniques of laser spectroscopy discussed in the previous sections. Alobaidi and Hills (153) used a 2-mW He-Ne laser as a source at 3.39 $\mu$m to measure alcohol vapor concentrations. With a double-beam arrangement with absorption cell and reference cell and phase-sensitive detection, the sensitivity reached a few ppm. Applications of this instrument to analysis of alcohol vapor concentrations in the breath are obvious. Intracavity absorption (Sect. 5.2.1.1) with broad-band pulsed dye lasers has been developed to a high-sensitivity technique which greatly improves the conventional broad-band absorption method used, for example, in flash photolysis (154). Latz et al. (155) could photograph absorption spectra of $NO_2$ with a single broadband laser pulse at low pressures in a 15-cm absorption cell inside the cavity of a flash lamp-pumped dye laser. Instead of using photographic plates, one can also employ polychromatic photoelectric recording. Horlick and Codding (156) mounted a photodiode array of 256 diodes, with a spatial separation of about 25 $\mu$m, at the exit of a spectrograph. With a dispersion of 25 Å mm$^{-1}$ the whole array covers a spectral range of 125 Å at a resolution of about 0.5 Å, which can be simultaneously recorded. This opens the way for qualitative detection of spurious transient chemical species such as radicals produced photolytically or for tracing the formation and decay of short-lived products in chemical reactions. Atkinson et al. (23) used this technique to measure the time-dependent concentrations of $NH_2$ and HCO radicals following flash photolysis of $NH_3$ and acetaldehyde, respectively.

Excitation spectroscopy (Sect. 5.2.1.2) has been used to detect low concentrations of spurious components in low-pressure gases. Especially the technique of laser-induced fluorescence (Sect. 5.3) has proved to be very helpful for monitoring concentrations of reactive products in flames or from chemical reactions and to determine their internal state distribution (157) (Sect. 5.3.6). In cases where the absorption coefficient is large enough but may change on a time scale of less than $10^{-6}$ s, direct absorption spectroscopy outside the cavity of a rapidly tunable cw dye laser may be useful, as has been proved by Telle and Tang (158). These authors employed electronic tuning of a cw dye laser to scan repeatedly every 10 $\mu$s the absorption spectrum of a sample, which could be monitored on a scope. The system is capable of measuring spectral changes in times as short as $10^{-7}$ s and is therefore of considerable interest for transient spectroscopy of short-lived excited molecular states of radicals.

The technique of laser magnetic resonance (Sect. 5.2.2.1), recently developed to high perfection (73), allows detection of molecular densities as small as $2 \times 10^8$ cm$^{-3}$. This method combines both high sensitivity and high spectral resolution and can be applied to all molecules or radicals that can be Zeeman tuned into resonance with existing laser lines. Many radicals, especially those of astrophysical interest, have been measured by LMR. Examples are CH (159), OH (160), HCO (161), and NH$_2$ (162).

A wide field of analytical laser applications is the detection of air pollution. Several techniques have been employed so far. Either vibrational–rotational absorption spectra are used for identification of molecular constituents (163) or laser-induced fluorescence in the visible or ultraviolet region is monitored (164). In some cases Raman spectroscopy was employed for detection (165) although the sensitivity was lower because of the low signal intensity in Raman scattering.

Optoacoustic spectroscopy (Sect. 2.1.3) seems to be especially suited for sensitive detection of low concentrations of the pollutant gases at atmospheric air pressure because of the high transfer efficiency of the excitation energy into translational energy. This has been demonstrated by Patel (166) who measured NO and H$_2$O concentrations in the stratosphere with an airborne spin flip Raman laser.

An interesting method for remote sensing at atmospheric pollution has been reported by Walther and his group (167) who used a combination of direct absorption by the pollutant molecules and Mie scattering of dye laser radiation from aerosols in the atmosphere for detection of NO$_2$ concentration. Short pulses from a tunable dye laser were sent through a beam expanding telescope, and the collected intensity of Mie-scattered light was measured as a function of time delay while the laser was alternatingly tuned to a NO$_2$ absorption line and to an absorption-free spectral region. This differential technique allowed to map NO$_2$ concentration in the ppm range up to distances of several kilometers. For a more detailed discussion of remote sensing with lasers the reader is referred to Chapter 6.

## 5.3 LASER-INDUCED FLUORESCENCE

### 5.3.1 Optical Pumping of Molecules

Optical pumping means the preparation of atoms or molecules in well-defined quantum states by selective absorption of narrow-band radiation. While optical pumping techniques have been successfully used for many years in atomic spectroscopy, their application to molecules has been more reluctant. The reasons lie in the more complex molecular spectra and the lack of suitable excitation sources that could be used for sufficiently strong but selective excitation. The situation changed drastically after the introduction of lasers to molecular spectroscopy, and a rapidly increasing number of publications on optical pumping of molecules can be found in the literature (168).

There are different aspects of this preparation of selected molecular levels. The first is the selective population of specified excited levels $E_k$, which can be achieved if the laser is tuned to the molecular transition $E_i \rightarrow E_k$. Due to the attainable high laser intensity, a rather large population density of level $E_k$ can be generated, comparable to that of the lower absorbing state $E_i$. A number of experiments have been performed which rely on this selective level population, such as measurements of fluorescence spectra emitted from $E_k$, lifetime measurements, or spectroscopic investigations of collision processes.

The second aspect refers to the possibility of producing alignment or polarization of the molecules in the optically pumped level. Since the absorption probability $P$ of molecules in the lower level $E_k$ is proportional to the scalar product

$$P \propto \boldsymbol{\mu}_{ik} \mathbf{E} \tag{13}$$

of the transition dipole moment $\boldsymbol{\mu}_{ik}$ and the electrical vector $\mathbf{E}$ of the exciting laser wave, those molecules with $\boldsymbol{\mu}_{ik} \| \mathbf{E}$ will preferentially absorb. This results in a partial polarization of molecules in the upper state if the laser is polarized. Depending on the laser polarization, definite magnetic sublevels can therefore be selectively populated. This can be used to measure depolarization cross sections or to perform level crossing experiments.

At sufficiently high laser intensities, the lower level of the absorbing transition may be completely depopulated. This selective disturbance of thermal equilibrium can be empoyed to study inelastic collision processes in the electronic ground state. The following example illustrates this complete depletion of definite molecular levels by selective optical pumping and its application for measuring the velocity distribution of molecules in supersonic beams.

The experimental arrangement (169) is shown in Fig. 5.13a. The laser beam is split into a strong pump beam (no. 1) and a weak detection beam (no. 2). Both beams are crossed perpendicularly by a molecular beam. If the laser is tuned to a molecular transition $E_i \rightarrow E_k$, all molecules passing beam 1 will be optically pumped and the level $E_i$ will be completely depleted. At the crossing point $P_2$ with laser beam 2, there are no

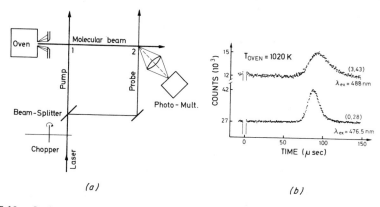

*(a)* *(b)*

**Fig. 5.13.** Optical pumping of molecules in a molecular beam: (a) experimental arrangement; (b) time resolved fluorescence signal at crossing point 2 from which the velocity distribution for molecules in quantum states $(v'' = 0, J'' = 28)$ and $(v'' = 3, J'' = 43)$ respectively, has been obtained. From Bergmann et al. (239a), reproduced by permission.

molecules in $E_i$. This means that beam 2 shows no absorption and therefore no laser-excited fluorescence. If laser beam 1 is interrupted for a short time interval $\Delta t$, all molecules passing $P_1$ during $\Delta t$ will not be pumped and will therefore cause a fluorescence signal at $P_2$. Because of their different velocities, these molecules will arrive at different times at $P_2$. Monitoring the fluorescence signal in $P_2$ as a function of time delay with respect to the interruption pulse yields the velocity distribution of molecules in defined quantum states. Figure 5.13b shows the result where both beams have been interrupted at $t = 0$ in order to mark the time $t = 0$ at $P_2$. Thus, optical pumping acts as a velocity selector that has definite advantages compared with mechanical selectors (170). The method has been applied to $Na_2$ molecules in supersonic beams and yielded the result that molecules in different vibrational–rotational levels had distinctively different velocity distributions. This gives information about the formation of $Na_2$ molecules during supersonic expansion and about their collisional deactivation (171).

We now discuss some experimental techniques using laser-induced fluorescence that rely on optical pumping of molecules.

### 5.3.2 Molecular Spectroscopy with Laser-Induced Fluorescence

The selective excitation of a single upper level results in a fluorescence spectrum that is relatively simple compared to an emission spectrum generated by broad-band excitation. This facilitates the identification of fluorescence lines considerably. If the separation of molecular absorption

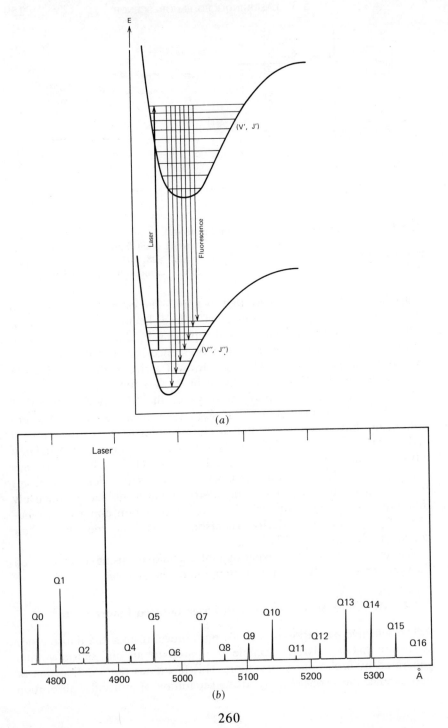

*(a)*

*(b)*

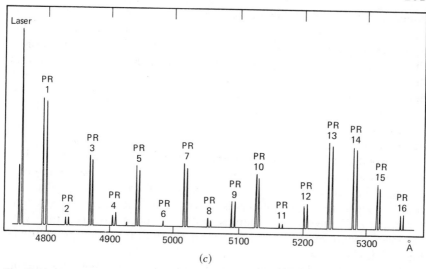

**Fig. 5.14.** Energy level diagram (*a*) and laser induced fluorescence spectrum of Na$_2$-molecules excited by the argon laser lines $\lambda = 4880$ Å (*b*) and 4765 Å (*c*).

lines is larger than the Doppler width, excitation of the sample may be performed in a cell or oven containing the molecular vapor; and for some small molecules even multimode lasers can provide a sufficiently high resolution. In case of more complex absorption spectra, Doppler-free excitation with narrow-band lasers is necessary to avoid overlapping of different absorption lines which would result in simultaneous population of different upper levels. Figure 5.14 shows for illustration purposes a fluorescence spectrum of Na$_2$ excited by the argon laser lines at $\lambda = 4765$ Å (*c*) and 4888 Å (*b*). From the rotational and vibrational line spacings, the molecular constants can be derived. In most cases it is necessary to analyze at least two different fluorescence progressions from different upper levels to achieve an unambiguous identification of the levels involved and a correct determination of the molecular constants. The advantage of laser-induced fluorescence spectroscopy can be summarized as follows:

1. The relative simplicity of the fluorescence spectra enables fast identification of lines and therefore reduces considerably the time spent for spectrum analysis. In addition, the reduction of the manifold of fluorescence lines allows to completely resolve the fluorescence spectrum even with medium-sized spectrometers.

2. Depending on the Franck-Condon factors, fluorescence transitions to very high vibrational levels ($v_k''$) may be observed that are not thermally

populated at feasible temperatures and therefore not accessible to absorption spectroscopy. With high laser intensity, a large population in the upper level $(v'_i, J'_i)$ is obtainable which is comparable to the population in the absorbing level $(v''_k, J''_k)$, see Fig. 5.14$a$. This allows detection of fluorescence lines even with Franck-Condon factors down to $10^{-6}$. The determination of energy levels close to the dissociation limit enables the accurate construction of the potential curve over a wide range of internuclear distances and allows a more reliable extrapolation to obtain the dissociation energy (172).

3.   Relative intensity measurements of the different fluorescence lines in a fluorescence progression from a single upper level allow the determination of Franck-Condon factors. Comparisons with the values calculated from the measured potential curves provide a very sensitive check for the consistency of the former analysis from line positions. Together with lifetime measurements they further yield absolute transition probabilities for individual transitions (173).

4.   The increased sensitivity achieved because of the high laser intensity enables one to investigate molecular fluorescence from radicals and unstable molecules (174,175). Measurements of the internal state distribution in chemical reaction products and analysis of scarce compounds in analytical chemistry have become possible (176,177).

A combination of a spectrafluorometer with a tunable dye laser has been constructed by Harrington and Malmstadt (177$a$) and proved to be very useful in the practice of analytical chemistry. The design simultaneously employs one monochromator in the dual role of tuning element for the dye laser and as the dispersive system for the fluorescence spectra. With standard gratings for the monochromator, a 1-nm bandwidth of the dye laser was obtained while a fluorescence spectral range of 100 nm could be simultaneously photographed at a dispersion of 2 nm mm.$^{-1}$

Most of the molecules studied so far by the technique of laser-induced fluorescence have been compiled in the reviews (10,14). Some recent publications report spectroscopic investigations of LiH and LiD (172), KH (178), CsH (175), $Bi_2$ (179), $Te_2$ (180), $BO_2$ (181), $NO_2$ (182), $SO_2$ (183), $NH_2$ (184), and $CSCl_2$ (185).

### 5.3.3   Lifetime Measurements of Excited Molecular Levels

While measurements of spectral line positions yield energy levels, lifetime measurements of excited molecular states provide a valuable tool for obtaining information about the oscillator strength of individual molecular transitions and Franck-Condon factors. In some cases they also give a sensitive check on the molecular wave functions and their dependence on internuclear distance $r$ in diatomic molecules through measurements of the transition moment $R(r)$ as a function of $r$.

In polyatomic molecules nonexponential decays have been observed due to nonradiative transitions caused by couplings between Born-Oppenheimer states. Lifetime measurements may elucidate in these cases the coupling dynamics in nonadiabatic intramolecular transitions.

Essentially, three different experimental techniques are used:

1. *The Pulse Excitation Method.* The molecules are excited by a short optical or electron pulse, and the time behavior of the subsequent fluorescent decay is observed. This technique has the advantage that the molecules are not disturbed by the excitation while their fluorescence is measured. Pulsed lasers or mode-locked lasers (186) are ideally suited for this technique. With high-power pulsed lasers, which generally have a low repetition rate, many fluorescence photons are detected per pulse. The fluorescence decay can be either directly viewed on a scope (187) or may be stored in a transient recorder (188) or displayed by a signal averager.

With mode-locked cw lasers, having medium peak powers (some watts) but a very high repetition rate, the single-photon counting technique proves to be favorable. With this method the detection probability of a fluorescence photon per excitation pulse is kept small compared to 1. The probability distribution for the time delay between a fluorescence photon and the excitation pulse, averaged over many excitation pulses, yields the fluorescent decay. The accuracy of this method can be better than 1% for lifetimes in the nanosecond range (189).

2. *The Phase Shift Method.* If the exciting light $I = I_0(1 + \cos \omega t)$ is intensity modulated at a frequency $\omega$, the fluorescence also shows a modulation at the same frequency which is phase shifted against the phase of the primary modulation. In case of pure exponential decay, the time-dependent fluorescence intensity is given by

$$I_{Fl} = b\left[1 + \frac{\cos(\omega t - \phi)}{(1 + \omega^2 \tau^2)^{1/2}}\right] \tag{14}$$

and the phase shift $\phi$ is connected to the mean lifetime $\tau$ by tg $\phi = \omega \tau$. The constant $b$ depends on laser intensity and on the detection efficiency of the system.

If the molecular decay is not purely exponential (for instance, in case of cascade effects or if the fluorescence from several levels with different lifetimes overlaps), measurements at different modulation frequencies are necessary to separate the different contributions to the total fluorescence. If the intensity modulation of the exciting light is not strictly sinusoidal, a

Fourier analysis of the fluorescence has to be performed either by a computer fit or by electronic filtering of the first harmonic in the detection system. The disadvantage of the phase-shifted method, especially in connection with high-intensity lasers as excitation sources, is the perturbation of the molecule by the laser field which may influence the phase shift by stimulated emission and falsify the spontaneous lifetime results (190).

3. The third group of experiments is based on lifetime determination from *measurements of the natural width of the excited level.* This method can be utilized by Doppler-free spectroscopic techniques such as level crossing, double resonance, and saturation spectroscopy.

The natural linewidth $\Delta \nu_n$ of a transition between two levels $E_i$ and $E_k$ is connected to the lifetimes $\tau_i$ and $\tau_k$ of these levels by

$$2\pi \, \Delta \nu_n = (1/\tau_i) + (1/\tau_k) \tag{15}$$

This reduces to $\Delta \nu_n = 1/2\pi\tau_i$ if the lower level is a ground state level with $\tau_k = \infty$.

Linewidth measurements in collimated molecular beams (Sect. 5.2.4.1) have to take into account the molecular time of flight, $\tau_f$, through the laser beam, which may affect the line width if $\tau_f$ is comparable to $\tau_{rad}$.

If the lifetimes of individual rotational levels in excited molecular states are to be measured, one has to take care that only the fluorescence from this level is monitored. This may be achieved either by selective excitation or, in case of broad-band excitation, by selecting a special fluorescence line emitted from this level (e.g., by a monochromator in the detection path) (189).

Lifetime measurements of individual rotation–vibration levels in excited molecular states allow determination of the variation in the electronic transition moment with the internuclear distance $r$ in diatomic molecules. This variation is generally small (several percent) in homonuclear molecules without a permanent dipole moment but may be quite considerable in heteropolar molecules. Since the transition moment $R$ is very sensitive to changes in the upper state wave function, accurate measurements of $R(r)$ provide excellent tests for the quality of excited state molecular orbitals, which can be calculated only with much less accuracy than ground state functions (191). Such measurements and computations of $R(r)$ have been performed for OH (192,193) and $Na_2$ (189). In the $A^2\Sigma^+$ state of OH, the lifetime increases by 8% with increasing rotational quantum number from $N' = 1$ to $N' = 8$ for the same vibrational level $v' = 0$ (194).

In case of perturbations of molecular levels by nearby levels of other electronic states, the lifetime may be influenced considerably. This has been found, for example, for the CN molecule where different rotational levels of the $B^2\Sigma^+$ state have been excited by a nitrogen laser-pumped tunable dye laser (195). The radiative lifetime of the unperturbed levels is $65.6 \pm 1.0$ ns. A longer lifetime is observed, however, for the $K' = 4$ level which is perturbed by a doublet of the $A^2\Pi_{3/2}$ state.

An even more drastic change in the lifetimes of different energetically close levels has been measured in excited states of $NO_2$ (196). In this molecule, which has an extremely complicated absorption spectrum in the visible region, several electronic states overlap in the energy range of visible excitation. This overlap causes strong perturbations of vibronic levels, and lifetimes between 3 and 100 $\mu$s have been found for different vibronic levels in a narrow energy range (197). Even when exciting a single hyperfine structure level of a vibronic state in the free $NO_2$ molecule, nonexponential decay is observed, indicating the occurrence of radiationless transitions in the excited molecule. In these nonradiative processes the total excitation energy, which is of course preserved in a free molecule, is redistributed among electronic, vibrational, and rotational parts. Similar observations have been made in the $SO_2$ molecule (198). These experiments are very useful for studying molecular states that cannot be adequately described by the Born-Oppenheimer approximation.

A comprehensive study of radiative lifetimes of the alkaline earth monohalide molecules has been performed by Zare and his group (199) in order to gain more insight into the nature of the bonding in these molecules. The authors used a pulsed tunable dye laser to pump the MX molecules (M = Ca, Sr, or Ba; X = F, Cl, Br, or I) into selected vibrational levels in the electronic $A_1$, $A_2$, or B states, respectively. It was found that a consistent explanation of the transition moments (deduced from the lifetime results) can be obtained by assuming that the transition occurs between two nonbonding orbitals centered on the metal atom.

In case of complex spectra, where the excitation into several electronic states may overlap, the different lifetimes of these states can be used to separate their contributions to the total fluorescence. Such lifetime-separated spectroscopy has been performed by Pruett and Zare (200) for the rotational analysis of the BaO $A'^1\Pi$ state. The $A'^1\Pi$ and $A^1\Sigma^+$ states are simultaneously excited by a pulsed tunable dye laser with a 0.2 Å bandwidth. The weak, long lifetime A'-X fluorescence ($9 \pm 1$ $\mu$s) is separated from the strong, short lifetime A-X fluorescence ($<1$ $\mu$s) by delaying the observation of the A'-X emission until the A-X emission dies away. In this way the spectrum could be unambiguously identified and the vibrational and rotational constants of the $A'^1\Pi$ state determined.

Finally, some examples shall be mentioned where lifetimes have been deduced from the measured width of level crossing signals. German et al. (201) obtained a lifetime of $0.58 \pm 0.05$ $\mu$s for the $v' = 0$, $N' = 2$ level in the $A^2\Sigma^+$ state of OD. Broyer et al. (202) could separately determine lifetime and Landé factor of a rotational level in the $B^3\Pi_{ou}^+$ state of $I_2$ by a computer fit of the Hanle signal from the different *hfs* sublevels of this level. Hanle measurements of different rotational levels in the $A^1\Pi$ state of CS molecules (203) yielded lifetimes of 0.2 $\mu$s, while the *g*-values could be calculated. Combining the information from Hanle signals and double resonance in the $N = 1$ and $N = 2$ levels of different vibrational levels in the $1s3p^3\Pi_u$ state of the parahydrogen molecule excited by electron impact, Marechal et al. (204) could determine *g* and $\tau$ separately, and found $\tau = 3.1 \times 10^{-8}$ s, independent on the vibrational number.

These few examples should give an impression about the progress achieved during recent years in experimental techniques for measuring lifetimes of individual levels in excited molecular states. For a compilation of molecules studied in this way see Walther (10, page 18).

### 5.3.4   Doppler-Free Techniques Using Optical Pumping

The high population density in excited molecular states obtainable through selective excitation with lasers greatly facilitates the application of high-resolution Doppler-free methods to investigate these states in more detail. Since excited states play an important role in many chemical reactions, a thorough knowledge of their coupling schemes and internal dynamics may be useful in reaction kinetics. In this section we briefly mention the basic principles of the following methods: *quantum beat spectroscopy, molecular level crossing, and double resonance.*

If two molecular levels with an energy difference $\Delta E = E_i - E_k$ are simultaneously excited by a short laser pulse with a spectral width $\Delta \nu \geq E/h$, the time-dependent wave functions of both levels,

$$\psi_i = \psi_i^{(0)} e^{i\omega_i t} e^{-(\gamma/2)t} \quad \text{and} \quad \psi_k = \psi_k^{(0)} e^{i\omega_k t} e^{-(\gamma/2)t} \tag{16}$$

are initially in phase because of the coherent excitation. The fluorescence intensity $I_F$ of the subsequent radiative decay is obtained by the squared coherent superposition of both amplitudes:

$$\begin{aligned} I_F \propto |\psi_i(t) + \psi_k(t)|^2 &= [|\psi_i^0|^2 + |\psi_k^0|^2 \\ &\quad + \psi_k^{(0)} \psi_i^{(0)} (e^{i(\omega_i - \omega_k)t} + e^{-i(\omega_i - \omega_k)t})] e^{-\gamma t} \\ &= e^{-\gamma t} [|\psi_i^{(0)}|^2 + |\psi_k^{(0)}|^2 + 2\psi_i^{(0)} \psi_k^{(0)} \cos(\omega_i - \omega_k)t] \end{aligned} \tag{17}$$

The exponential decay of the fluorescence is modulated at the difference frequency $\omega_i - \omega_k$ (quantum beats). From the modulation period $T = 2\pi/\Delta\omega$, the energy difference $\Delta E = \Delta\omega\hbar$ of both levels can be measured, even if $\Delta E$ is smaller than the Doppler width.

The time resolution of the detection electronics must be, of course, high enough to resolve the modulation. This limits the applications to level separations below 1000 MHz. Larger level separations can be already resolved with Doppler-limited techniques. The two levels may be Zeeman sublevels or hyperfine structure levels. The first study of molecular quantum beats in the fluorescence from individual rotational levels of the $B^3\Pi_{ou}^+$ state of the iodine molecule $I_2$ has been performed by Wallenstein et al. (205).

Level crossing spectroscopy (206) is another Doppler-free technique which also relies on the coherent excitation of closely spaced molecular levels. Unlike quantum beat spectroscopy, where the excitation is performed with short pulses and the interference effects are observed on a time scale, here cw excitation is preferable and the interference between the fluorescence amplitudes from the different coherently excited levels influences the spatial intensity distribution and polarization of the total fluorescence. The phase relations between the fluorescence amplitudes that cause the interference effects depend on the level separation. If this separation becomes large compared to the natural widths of the level, the phase differences change so rapidly during the level lifetime that the interference effects vanish.

The technique can be applied either to degenerate Zeeman or Stark sublevels which are split by external magnetic or electric fields [zero field level crossing or Hanle effect (207)] or to closely spaced hyperfine structure levels which are differently shifted by the external field (see Fig. 5.15) and which therefore may cross at certain values of the field (high field level crossing) (208). Observing the polarization of fluorescence intensity in a selected direction as a function of external field results in a "level crossing signal" around the field where the levels cross. The

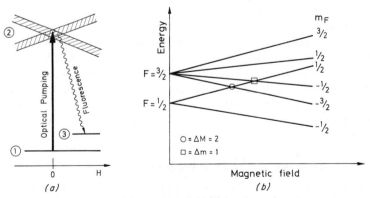

**Fig. 5.15.** Zero-field (a) and High-field (b) level crossing spectroscopy.

half-width of this signal reflects the sum of the natural level widths. Since this Doppler-free linewidth is achieved with broad-band excitation, multimode lasers can be used as pumping sources.

Many molecules have been meanwhile investigated by level crossing spectroscopy. A compilation of publications up to 1973 can be found in the reviews of Zare (209), Levy (210), and Novikov (211).

In addition to the *spontaneous* level crossing experiments, where the angular redistribution of the spontaneous emission is observed, *stimulated* level crossing spectroscopy has been performed (212), which is based on saturation of molecular transitions by intense monochromatic laser fields. If the molecular sample is placed inside the laser resonator and the laser frequency is tuned to a molecular transition $1 \rightarrow 2$ between a ground state 1 and a degenerate excited state 2, a sharp decrease in attenuation (or gain) of the laser occurs if several sublevels of 2 approach each other within their natural linewidth. This method is especially useful for level crossing experiments in excited vibrational levels of the electronic ground state, which show a weak fluorescence because of their long spontaneous lifetimes. It has been applied, for instance, to $CH_4$ (212,213), $I_2$ (214), and $CD_3I$ (215). The merits of these level crossing experiments lie in the information they yield about the Landé factors, that is, the coupling schemes of angular momenta in excited molecular states. This often allows classification of the state (216) and helps elucidate the molecular structure.

In addition to the optical–optical double-resonance techniques, which are based on selective absorption of two laser waves by two molecular transitions sharing a common level, discussed in Sect. 2.4.3, optical rf or optical microwave double-resonance can be also performed with broadband excitation to achieve high-resolution spectroscopy in excited molecular states. The application of this technique to $NO_2$ molecules illustrates the method as follows: With different lines from an argon laser (217) or with tunable cw dye lasers (218), rovibronic levels in the $^2B_2$ state of $NO_2$ could be selectively populated. A simultaneously applied microwave field was frequency tuned and induced transitions to other rotational levels (see Fig. 5.16). These transitions could be monitored by the fluorescence from these levels. Besides facilitating the analysis of the complex $NO_2$ spectrum considerably, this double-resonance technique allows to transfer the accuracy of microwave spectroscopy, hitherto proved for the molecular ground states, also to excited states.

### 5.3.5   Spectroscopy of Collision Processes

Spectroscopy can contribute in an outstanding way to investigations of various kinds of collision processes where ground state or excited state

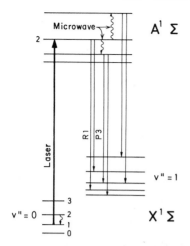

**Fig. 5.16.** Optical microwave double-resonance for studies of rotational structure in electronically excited molecular states.

molecules are involved. Integral information about *elastic* collisions can be obtained from measurements of spectral line profiles and their pressure-dependent broadenings or shifts (219). Doppler-free laser spectroscopy is especially sensitive to small effects at low pressures where the collision broadening is still smaller than the Doppler width (220). It allows one to study in more detail the influence of far distance collisions and to probe the outer part of the interaction potential (221).

Chemists are generally more interested in *inelastic* and reactive collisions. Some examples will illustrate the amount of information about collision rates, cross sections, and initial and final states of the collision partners that can be obtained from laser spectroscopy of collision processes. One may divide the different methods into those where cw lasers are used and steady-state conditions are studied and those that measure the time behavior of collision kinetics using pulsed lasers.

Molecules in the gas phase at thermal equilibrium that are optically pumped into a selected excited level (v',J') can release their excitation energy by spontaneous emission generating a fluorescence spectrum such as that shown in Fig. 5.14. If the excited molecules, however, suffer inelastic collisions with other atoms or molecules, they may lose or gain internal energy from the collision process, which means that collision-induced transitions to neighboring rotational or vibrational levels or even to other electronic states occur (Fig. 5.17). These collisionally populated excited levels can also emit fluorescence, resulting in the appearance of new lines in the fluorescence spectrum (Fig. 5.17b). The atomic counterpart of these processes, generally referred to as nonresonance fluorescence, is discussed in Chapter 4 of this book. From the intensity of these collision-induced "satellite lines" the cross sections of the corresponding

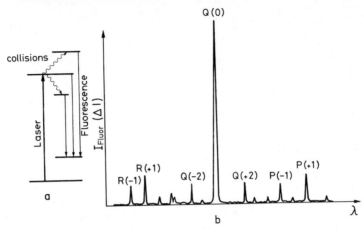

**Fig. 5.17.** Spectroscopic investigations of inelastic collisions: (*a*) energy level diagram; (*b*) part of a Na$_2$ fluorescence spectrum showing collision-induced satellite lines.

inelastic collisions can be obtained. This may be seen as follows: With the steady-state population density $N_i$ of the optically pumped level $E_i$, the corresponding population $N_k$ of the collisionally populated level $E_k$ is

$$N_k = \sigma_{ik} n N_i \bar{v} \tau_{sp} \qquad (18)$$

where $n = p/kT$ is the total density at the absolute temperature $T$ and the pressure $p$. The mean relative velocity $\bar{v}$ of the colliding particles at thermal equilibrium can be expressed as

$$\bar{v} = (8kT/\mu\pi)^{1/2} \qquad (19)$$

where $\mu$ is the reduced mass of the collision partners and $k$ is the Boltzmann's constant.

The intensity ratio of satellite line $E_k \rightarrow E_n$ to the "parent line" $E_i \rightarrow E_m$ is determined by the corresponding radiative transition probabilities $A_{kn}$ and $A_{im}$, respectively, and can be written as

$$\frac{I_{k \rightarrow n}}{I_{i \rightarrow m}} = \frac{A_{kn} N_k}{A_{im} N_i} = \frac{A_{kn}}{A_{im}} \sigma_{ik} n \bar{v} \tau_{sp} \qquad (20)$$

Measurements of this intensity ratio, of pressure $p$, and of the spontaneous lifetime $\tau_{sp}$ allow determination of the absolute value of the cross section $\sigma_{ik}$. If the experiments are performed in a gas cell at thermal equilibrium, in fact a mean value $\langle \sigma_{ik}(v) \rangle$ is determined as the average over all velocities $v$. Measurements in crossed beams with velocity selection allow, however, to determine the velocity dependence of differential cross sections. Such experiments are underway in several laboratories.

This technique of collision-induced satellite lines in laser-excited fluorescence spectra has been applied to measurements of rotational or vibrational energy transfer in many molecules. Examples are investigations of $I_2$ (222), $Na_2$ (223), and $Li_2$ (224).

Chemical kinetics have benefited greatly from time-resolved laser spectroscopy of collision processes (225). Here, short laser pulses are used to selectively excite molecular levels. The collisional deactivation of these levels and the energy transfer to other collision partners can be monitored through the time-resolved fluorescence from the primarily excited level or from other levels populated by collisions (226). Vibrational and rotational energy transfer (227), reactive collisions (228), and photochemical decomposition of molecules (229) have been studied with this method. For more information about this technique and for intensive literature compilation see the reviews by Steinfeld (229a), Lee and Roun (229b), and Moore and Zittel (230) in this field.

The relaxation process at high pressure or in molecular liquids may be very fast and can be studied only on a picosecond time scale. Selected single pulses from mode-locked lasers allow to investigate ultrafast vibrational relaxation and energy transfer in liquids (231), electron transfer processes (232), and absorption recovery times after saturation of molecular transitions in liquids (233).

Another sensitive technique for studying relaxation phenomena in molecular gases such as rotational energy transfer relies on time-resolved double-resonance spectroscopy (234) as has been proved for $H_2CO$ (235), $NH_3$ (236), and other molecules.

### 5.3.6 Internal-State Distribution of Reaction Products

Laser absorption spectroscopy monitored through the total fluorescence from the excited level (see Sect. 5.2.1.2) is a very sensitive tool for measuring the internal state distribution of molecules that are formed under nonequilibrium conditions. If the laser excites molecules from the initial state into the upper level $(v'_k, J'_k)$ the total fluorescence intensity $I_k$ from the upper level is proportional to

$$I_k \propto I_L N_i B_{ik} A_k \eta_k \qquad (21)$$

where $B_{ik}$ stands for the absorption probability of the absorbing transition $i \rightarrow k$, $A_k$ is the total spontaneous transition probability of the total fluorescence, and $\eta_k$ is the quantum efficiency, that is, the ratio of spontaneous and collisional deactivation of level k.

Tuning the laser with intensity $I_L$ to two different transitions $i \rightarrow k$ and $j \rightarrow m$ allows to determine the relative populations of the two absorbing ground states i and j from

$$N_i/N_j = (I_k/I_m)(B_{im}/B_{ik})(A_m/A_k) \tag{22}$$

provided the quantum efficiencies are equal for both upper levels. If the relative transition probabilities are known from a spectroscopic analysis of the molecule, the relative populations can be determined from the measured fluorescence intensities $I_k$ and $I_m$. This laser-induced fluorescence technique has been used by Becker et al. (237$a$) to monitor the $C_2$ ($a^3\Pi_u$) radical concentration at different locations in an acetylene–oxygen flame. The good laser beam collimation has the advantage that the geometric resolution is very high. Barnes et al. (157) could localize the laser-excited fluorescence from CH radicals in flames within 0.5 mm geometric resolution, and Wang and Davies (237$b$) determined the internal state distribution of OH radicals with a tunable UV laser.

Another example is the measurement of total energy distributions among translational rotational and vibrational degrees of freedom for molecules in supersonic beams (170). Using the laser-induced fluorescence technique, Schultz et al. (238) could prove for a supersonic $Na_2$ beam that substantial internal cooling happens for the molecules that convert part of their rotational–vibrational energy into directed translational energy (239). This internal cooling has been used in $NO_2$ supersonic beams to simplify the complicated absorption spectrum (240).

A very important information necessary for a detailed understanding of reactive collisions is the internal energy distribution of the reaction products. The interest in a thorough knowledge of this distribution in exothermic chemical reaction has greatly increased since chemical lasers have been developed that are based on inverted population distributions generated by some favorable chemical reactions (241). Up to now inversion could only be achieved between vibrational or rotational levels in the electronic ground state of the reaction products (242). There may, however, exist reactions that form preferably electronically excited products, which open up the possibility of visible chemical lasers (243).

Zare and his group (244–246) have performed extensive studies of the internal state distribution of molecular reaction products formed in the reactions

$$Ba + O_2 \rightarrow BaO^* + O$$
$$Al + O_2 \rightarrow AlO^* + O$$
$$Ba + HX \rightarrow BaX^* + H$$

where X is one of the halogen atoms F, Cl, Br, or I. For the crossed beam reaction

$$Ba + LiCl \rightarrow BaCl^* + Li$$

even a coarse angular distribution of the BaCl products could be measured (247).

The interesting question how the internal energy of the reactants affects the internal-state distribution of the products has been studied by Pruett and Zare (248) for the reaction

$$Ba + HF^* \rightarrow BaF^* + H$$

The HF molecules could be vibrationally excited by a HF laser, and the difference in the internal-state distribution of BaF* formed from HF* (v = 1) and HF (v = 0) was measured.

For further aspects and more details, the reader is referred to Zare (249) and Zare and Dagdigian (250).

## 5.4 EMISSION SPECTROSCOPY OF LASER LINES

### 5.4.1 Spectroscopy of Molecular Lasers

The spectroscopic investigation of molecular gas lasers in the infrared and submillimeter region has added many details to our knowledge of molecular spectra of excited vibrational–rotational levels and also of collision processes and chemical kinetics such as the formation and recombination of transient products in chemical reactions. In particular, detailed studies of the time behavior of induced emission in chemical lasers allowed much insight to be gained into the reaction paths and internal energy distribution of the different reaction products (251).

With regard to high-resolution spectroscopy, the wavelength measurements of molecular laser lines have several definite advantages. Because of the comparatively high intensity of most laser lines, excellent signal-to-noise ratios are obtainable which allow an accurate wavelength determination also in spectral regions where detector sensitivities are low. The laser resonator itself may act as a long distance Fabry Perot. The high intensity has the additional advantage that frequency mixing techniques can be applied to measure absolute laser frequencies. The laser output at a frequency $\nu_L$ is mixed in a nonlinear crystal or a metal point contact diode with a second calibrated laser line $\nu_c$. This calibration laser line is selected in such a way that the difference frequency $\nu_D = \nu_L - m\nu_c$ between $\nu_L$ and the $m$th harmonics of $\nu_c$ lies in the microwave region and therefore can be digitally counted. In this way a frequency calibration chain has been established that spans the whole range from microwave to optical frequencies (252) and enables absolute frequency measurements and therefore also determinations of molecular energy level spacings with accuracies that are several orders of magnitude higher than before.

Whereas the first far-infrared molecular lasers, namely, $H_2O$ and HCN (255) and $SO_2$ and $H_2S$ (256), had been pumped in electrical discharges, optical pumping with $CO_2$, $N_2O$, CO, or HF lasers has succeeded in achieving laser oscillation on numerous rotational transitions in many molecules such as $CH_3F$, $CH_3OH$, $CH_2$, CN, $NH_3$, etc. (257, page 242). The measurement of optically pumped laser lines may be regarded as an extension of double-resonance techniques to induced emission, which has facilitated the assignment of molecular transitions and has increased the accuracy of wavelength determination. This technique has proved to be especially useful in cases of strong perturbations where attempts of assignment had failed with conventional absorption spectroscopy (254).

Energy transfer lasers, in which the inversion of the lasing medium is achieved by selective collisional energy transfer from optically pumped molecules added to the lasing medium, have experienced increasing interest (258). These lasers not only allow efficient conversion of pump energy into induced emission, but they furthermore enable thorough studies of resonant and near-resonant transfer of electronic or vibrational energy (259).

Several lasers make use, as active medium, of excited radicals that have been formed either by photolysis or as intermediate products in chemical reactions (260). Measurements of laser wavelengths and time-resolved studies of induced emission permit accurate spectroscopy of formation and decay of the radical states involved.

### 5.4.2  Heterodyne Spectroscopy

This modern branch of laser spectroscopy *probably represents the most accurate method for determining level spacings in molecules.* The basic principle is simple: Two lasers are independently stabilized on two molecular transitions $\nu_1$ and $\nu_2$ sharing a common level (see Fig. 5.18). The frequency stabilization of the lasers may, for instance, use the narrow Lamb dips at the center of each Doppler-broadened absorption line (see Sect. 5.2.4.2). The outputs of both lasers are superimposed and mixed in a nonlinear detector which generates the difference frequency $\nu_1 - \nu_2$. This difference frequency can be directly counted if it is below $10^9$ Hz. At higher difference frequencies electronic mixing techniques can be employed for absolute frequency measurements. Since both lasers are stabilized on the center of each molecular line, the energy separation $\Delta E = (\nu_1 - \nu_2)/h$ can be directly obtained. The selection of the nonlinear detector depends on the spectral range in which the two lasers oscillate and on the frequency range of the difference frequency. In the visible and

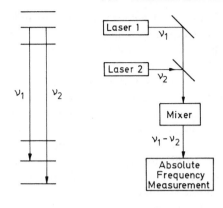

**Fig. 5.18.** Heterodyne spectroscopy with two lasers that are stabilized on two molecular transitions sharing a common level.

UV region, multiplier cathodes may be used as nonlinear mixers; in the infrared, either fast diodes or nonlinear crystals are used.

The method was first employed by Bridges and Chang (261) who stabilized two $CO_2$ lasers on the line centers of different vibrational–rotational transitions of $CO_2$. The frequency mixing was achieved in a bulk GaAs crystal, and the beat frequencies ranging between 50 and 80 GHz were measured for 37 pairs of transitions to better than 1 MHz. The deduced rotational constants for the relevant vibrational levels are 25–200 times more accurate than the best previous results from conventional spectroscopy. The experimental technique was improved by Petersen et al. (262) who focused the two $CO_2$ laser beams onto a Josephson junction and mixed it with the $n$th harmonic of an applied X-band frequency. Thus, the junction served both as the microwave harmonic generator and as the infrared mixer, producing a beat frequency around 60 MHz. This technique allowed the measurement of laser difference frequencies ranging from 32 to 63 GHz with a single microwave oscillator. Similar measurements of the $N_2O$ (00°1–10°0) band have been performed by the group of Javan (263). Heterodyne spectroscopy has been applied to the visible spectrum by Ezekiel and his group (264) who stabilized two single-mode argon lasers onto different hyperfine structure components of the line P(13) in the $(43 \leftarrow 0)$ $B^3\Pi_u \leftarrow X^1\Sigma$ band of $I_2$. Doppler-free spectroscopy could be performed by directing the laser beam perpendicularly to a well-collimated molecular beam. The difference frequency could be measured accurately to $10^{-11}$ ($\hat{=} 5$ kHz)! This extremely high accuracy allowed to determine even the octupole interaction in $I_2$.

Instead of using two lasers one may also perform heterodyne spectroscopy with a single laser stabilized on a molecular transition $\nu_1$. Intensity modulation of the laser output at a frequency $f$ generates two sidebands at $\nu_1 \pm f$. One may tune the modulation frequency $f$ until $\nu_1 \pm f$ coincides with another molecular transition, which can be monitored either by the resulting change in absorption or by fluorescence.

Different molecular transitions have been measured by Hall et al. (265) using a special difference frequency technique called "frequency offset locking" technique. One laser is stabilized on the Lamb dip of a molecular transition. A second laser is frequency locked to this laser by an electronic circuit that allows introduction of a controllable frequency offset between both lasers. This second, electronically tunable laser is tuned across molecular absorption lines. The absolute frequency differences of these lines measured against the frequency standard provided by the first laser yield molecular level spacings with extremely high accuracy.

## 5.5  LASER RAMAN SPECTROSCOPY

Because of the low scattering intensities in Raman spectroscopy, *most of the investigations performed with conventional light sources have been restricted to molecules in the liquid or solid state.* In Raman spectroscopy of gases, the molecular linewidths are much narrower and the spectral resolution is limited mainly by the spectral width of the light source. Lasers as excitation sources in Raman spectroscopy have the advantage of high power and excellent beam collimation. Their particular merits for Raman spectroscopy of gases are their narrow linewidth and their polarization characteristics which permit accurate measurements of low depolarization ratios of almost completely polarized Raman vibrational lines (8).

Since the field of laser Raman spectroscopy has been extensively treated in different text books and reviews (2–8), we discuss here only some recent progress of interest for applications in analytical chemistry (266). The examples given below illustrate the different techniques of laser-excited spontaneous Raman scattering with fixed-frequency and tunable lasers, coherent anti-Stokes Raman scattering, and inverse Raman scattering inside the laser resonator.

Spontaneous Raman scattering has been widely used to detect pollutant constituents in atmospheric gases. Accurate measurements of mole fractions of the major species, namely $N_2$, $O_2$, $H_2O$, $CO$, $CO_2$, etc., in the exhaust flow of combustion emission have been performed with a nitrogen laser at $\lambda = 337$ nm (267). Special transmission optics focused the laser beam to the desired volume of the exhaust gases to be analyzed, and a receiver optics provided efficient collection of the Raman-scattered photons. In order to localize the analyzed volume the laser beam and the field of view of the collector should have a sharply peaked overlap to discriminate against intense background radiation from other parts of the exhaust. From the intensity distribution of $N_2$ Raman lines, the radial profiles of the exhaust temperature could be determined (268).

A serious problem in Raman spectroscopy of gas mixtures is caused by the laser-excited fluorescence emitted from the molecular constituents which have absorption bands at the laser wavelength and which may exceed the Raman intensity by several orders of magnitude. Since the fluorescence has a decay time corresponding to the lifetime of the upper level while the Raman intensity follows promptly the excitation pulse, gating of the detector will reduce the fluorescence intensity but not the Raman intensity.

Remote Raman spectroscopy of air pollutants at ppm levels has become feasible up to distances in the km range with pulsed frequency-doubled ruby laser (269) or pulsed UV lasers. Time-controlled gated detectors allow a range resolution of some meters. Photon counting techniques may further increase the sensitivity (270).

Raman spectroscopy of very sensitive detection employs image intensifier phototubes and a high-sensitivity television camera (271). This combination is able to record simultaneously all the spectral lines and allows to obtain a complete Raman spectrum with one shot of a pulsed laser. The incident light energy necessary to record a complete Raman spectrum is, for example, only 160 $\mu J$ for $GeCl_4$ and 8 mJ for $C_2H_4$, $CO_2$, or $O_2$ at atmospheric pressures.

An elegant technique for recording Raman spectra of liquid samples makes use of the efficient light collection in optical waveguides (272). If the refractive index of the sample is larger than that of the glass envelope of the waveguide, the laser light introduced at one end as well as the Raman light will be guided by internal total reflection to the other end of the waveguide where it can be imaged onto the entrance slit of a monochromator. Because of the large possible length of the waveguide the sensitivity is very high. Raman spectra of different liquids have been recorded with a 1-mW He–Ne laser within some seconds (273), whereas many hours of exposure time had been necessary with conventional techniques.

Instead of using fixed-frequency lasers, it is often advantageous to excite the Raman spectrum with tunable lasers. The laser wavelength can be tuned in such a way that the Raman lines always appear at the same wavelength (274). This has the advantage that the monochromator can be replaced by a narrowband interference filter whose higher transmission increases the sensitivity. A very narrow band filter may be, for example, a cell filled with an atomic alkali vapor. If the Raman line coincides with the atomic resonance line, it will be absorbed and the resulting fluorescence can be used as monitor.

If the Raman sample is placed inside the cavity of a broadband dye laser but is simultaneously pumped by a monochromatic laser, the population change of the ground state molecules induced by the Raman transition can be detected with high sensitivity through the intensity change of the dye laser (see Sect. 5.2.1.1). With this technique of "inverse Raman scattering" nearly complete Raman spectra of samples at concentration down to $10^{-3}$ mole liter$^{-1}$ could be obtained within 30 ns. (275).

A recently developed very sensitive method is the coherent anti-Stokes Raman scattering (CARS) (276) where the gaseous sample is illuminated by two collinear laser beams at frequencies $\nu_1$ and $\nu_2$ such that the difference $\nu_1 - \nu_2 = \nu_v$ equals a frequency $\nu_v$ of a Raman-active vibrational transition. Since dispersion is negligible in gases, satisfactory phase matching can be achieved for collinear beams, and side bands at the combination frequencies $2\nu_1 - \nu_2$ and $2\nu_2 - \nu_1$ are generated in the same direction as the incoming beams. The anti-Stokes side band at frequency $\nu_a = 2\nu_1 - \nu_2$ is detected.

CARS is a four-wave mixing process. Its advantages over the incoherent spontaneous Raman scattering stem from the parametric nature of this process and are the following: The scattered light is well collimated within a cone angle of typically $10^{-3}$. Modest pump powers between 1 kW and 1 MW are sufficient, which can be easily obtained with moderate pulsed lasers. The intensity is between five to 10 orders of magnitude higher than in spontaneous Raman scattering. The main disadvantage lies in the increased cost because a tunable laser is necessary in addition to the fixed-frequency laser.

The main application areas of CARS lie in the concentration measurements of gaseous constituents and temperature evaluation from the population differences of different rotational levels (277). $H_2$, 100 ppm in air, at a total pressure of $10^{-5}$ atmosphere could be detected with one shot, that is, in 20 ns (278).

Another field that has developed rapidly is laser-excited resonance Raman spectroscopy (279). Here, the laser line coincides with a molecular transition to an excited electronic state. In the gas phase this excitation usually produces fluorescence. In the liquid or solid phase, however, this fluorescence often is quenched and the resonance Raman effect can be observed. Local heating effects caused by the absorption of laser radiation can be prevented by rotating sample techniques (280). Since the scattering cross section for the resonance case is by orders of magnitude larger than in off-resonance Raman scattering, the sensitivity is high and short-lived radicals can be detected (281).

## 5.6 LASER-INDUCED MOLECULAR REACTIONS AND ISOTOPE SEPARATION

The probability of molecular dissociation at a given temperature and the distribution of fragmentation products can be strongly influenced by the excitation of certain vibrational modes that lead to breaking of selected chemical bonds. Due to the attainable high intensity and monochromaticity of lasers, large population densities in selectively excited molecular levels can be achieved. Intense infrared laser lines that coincide with molecular vibrational bands may therefore be used to enhance molecular dissociation and to obtain controlled formation of the fragmentation products. The dissociation may proceed without collisions or may be aided by collisional energy transfer.

Chemical reactions of molecules that take place because of collisions with other molecules or atoms may also strongly depend on the internal energy of the reactant molecule. If the reaction rate $W$ depends on the temperature $T$ and on the activation energy $E_a$ according to the Arrhenius equation

$$W \propto A \, e^{-E_a/(RT)} \tag{23}$$

$W$ is increased by a factor $B = e^{h\nu/kT}$ through selective excitation with a photon of energy $h\nu$. In case of $h\nu \gg kT$, this enhancement factor $B$ may be very large. The crucial point, however, is that the maximum value of $B$ can only be realized if the reaction of the selectively excited molecule proceeds faster than other relaxation processes which tend to redistribute the excitation energy among all possible degrees of freedom. These relaxation processes would therefore lead to a temperature increase of the whole sample and to a useless consumption of laser energy which is converted into heat without considerably enhancing the reaction rate (282). In other words, the excitation energy of a laser-excited molecule must come close to the activation energy $E_a$ in order to make the reaction rate fast compared to the relaxation rate.

If $E_a \gg h\nu$, either many photons have to be simultaneously absorbed by the molecule (multiphoton excitation) or a stepwise excitation along the vibrational level ladder by successive photon absorption can help to meet this requirement. Another important excitation mechanism is collisional V–V transfer between laser-excited molecules, where inelastic collisions

$$M^*(v=1) + M^*(v=1) \to M^{**}(v=2) + M(v=0)$$

may increase the vibrational excitation of one collision partner.

Some examples will illustrate these different mechanisms. The population of high energy levels by infrared radiation has been proved, for instance, by the observation of visible fluorescence of $BCl_3$ molecules excited by a $CO_2$ laser (282). It was shown that there are two time regions for the luminescence: the instantaneous region, which arises simultaneously with the laser pulse, and a delayed luminescence, which is pressure dependent.

With short pulses ($\tau < 100$ ns) of a $CO_2$ laser, excitation of high vibrational levels of $SF_6$ was achieved that led to the dissociation of the molecule (284). It could be proved that the process proceeds already at pressures of a few torr and that collisions cannot play a significant role for the excitation, which therefore must have been caused by multiphoton processes. In a similar way chemical reactions of $BCl_3$ with oxygen have been induced by a $CO_2$ laser. By tuning of the laser frequency to absorption lines of either the $^{10}BCl_3$ or the $^{11}BCl_3$ isotope, selective excitation could be achieved which resulted in enhanced formation of the corresponding $^{10}BO$ and $^{11}BO$ isotope, respectively. Application of this technique to isotope selective dissociation of $^{32}SF_6$ yielded an enrichment factor of 2800 (285).

Detailed investigations of laser-specific reactions of boranes induced by monochromatic low-power radiation from a cw $CO_2$ laser were performed by Bachmann et al. (286). The experiments showed that a chain reaction was initiated in diborane, $B_2H_6$, resulting in the formation of $B_{20}H_{16}$ as the main product. Contrary to the first example, in this experiment collisions seem to have been essential for initiating the chain reaction; at a $B_2H_6$ pressure of 200 torr a laser power of 1.5 W was sufficient to initiate the reaction, whereas below 50 torr no reaction could be observed even at laser powers of 8 W.

Within the last year multiphoton excitation followed by dissociation has been successfully used in several laboratories to separate the isotopes of S, B, C, Si (287), Os (287a), and H (287b). The polyatomic molecules $SF_6$, $BCl_3$, $CF_2Cl_2$, and $SiF_4$ were irradiated in the presence of $H_2$ or $N_2$ using intense focused $CO_2$ laser pulses. The dissociation products reacted with $H_2$ to form stable molecules. $N_2$ was only employed as a buffer gas to aid rotational thermalization.

Observation of the visible fluorescence of the SiF radical in $SiF_4$ gas irradiated by $CO_2$ laser pulses (288) revealed that the molecules were dissociated into electronically excited fragments. A clear distinction was observed between the fluorescence produced by the direct action of the intense laser field and that associated with thermalization of the incident energy. This indicates that besides the direct photodissociation by multiphoton processes, undissociated molecules are left after the laser pulse which are in highly excited vibrational levels and may be dissociated by collisional energy transfer. The dissociation into electronically excited fragments may be only a minor path in comparison to channels leading directly to fragments in the electronic ground state. Since these fragments cannot emit fluorescence, their detection is more difficult.

Welge and his group (289) used the technique of laser-induced fluorescence (see Sect. 5.3.6) to detect ground state reaction products $NH_2$ produced from the multiphoton dissociation of $NH_3$ by a pulsed $CO_2$ laser. Time-resolved laser fluorescence spectroscopy together with observation of luminescence from the excited dissociation products allowed to estimate the relation between the amount of $NH_2$ ($^2B_1$) and excited $NH_2$ ($^2A_1$) produced under virtually collision-free conditions by multiphoton dissociation. The results indicated that the yield of ground state fragments is about $5 \times 10^5$ times larger than that of electronically excited fragments.

The laser-enhanced reaction of NO with $O_3$ is another example of how selective excitation can influence chemical reactions in a controllable way. Excitation of the $O_3$ (001) vibrational level by a $CO_2$ laser enhanced the reaction

$$O_3 + NO \rightarrow NO_2 \, (^2A_1) + O_2$$

by a factor of 17 compared to the reaction rate at room temperature in the absence of laser excitation (290). However, the excitation of the first vibrational level of NO with a CO laser yielded a much smaller enhancement factor. These and additional observations are consistent with the assumption that the (001) vibrational level is mainly responsible for the rate constant enhancement (291).

A last example of chemical reaction induced by visible laser radiation concludes this section. In a mixture of hydrogen gas with alkali vapors illuminated with lines from an argon laser, a dense cloud of small particles forms in the laser beam that fall out of the laser beam and cover the bottom of the cell ["laser snow" (293)]. Spectroscopic studies of the fluorescence from the particles' region showed that alkali hydride diatomic molecules have been formed at concentrations far in excess of thermal equilibrium values, which therefore condensate to microscopic droplets. The formation probably proceeds by reactions of laser-excited alkali dimers with hydrogen molecules. This new phenomenon of laser-induced particle formation shows considerable promise for possible inexpensive and rapid separation of alkali isotopes.

Laser-induced reactions of methyl halides $CH_3X$ (X = F, Cl, Br) have been investigated by Earl and Ronn (292). There are two possibilities for the reaction path:

$$CH_3X + Cl_2 \nearrow^{CH_3Cl + XCl} \qquad (a)$$

$$\searrow_{CH_2XCl + HCl} \qquad (b)$$

In no case did the reactions proceed unaided at room temperature. In case of $CH_3F$, the $CO_2$ laser excites the $\nu_3$ C–F stretching vibration, in case of $CH_3Br$ the $\nu_2$ deformation and in $CH_3Cl$ the $\nu_6$ methyl rocking vibration. Thus, by monitoring the product branching ratios as a function of the excited species concentration, some information concerning mode specificity of the reaction can be obtained. The experimental results indicate clearly that the reaction proceeds primarily via reaction ($b$) but that it probably may not be bond specific.

## 5.7  CONCLUSIONS

The examples presented in the last sections may have illustrated that many laser applications in chemistry require a thorough knowledge of molecular spectra. The goal of this chapter was to show that lasers are ideally suited to provide this information. The development of high-sensitivity methods and ultrahigh-resolution techniques enable molecular spectroscopists to elucidate spectra even of complex molecules. Since this rapidly expanding field of laser spectroscopy is just at its beginning, its further development will certainly be very fruitful for analytical chemistry and may considerably influence analytical methods. Optical pumping methods, for instance, will open new fields for excited-state chemistry with all its possible capabilities of controlled chemical reactions and isotope separation.

In order to provide the necessary knowledge about molecular states and their internal dynamics in excited polyatomic molecules, much more spectroscopic work has to be done. In spite of the impressive progress achieved within the last years in laser spectroscopy, there is still a long way to go until we may understand chemical processes from a physical standpoint looking at the detailed molecular processes involved.

## References

1.  G. Herzberg, *Molecular Spectra and Molecular Structure*, Vols. I–III, Van Nostrand, Princeton (1967).

2.  M. C. Tobin, *Laser Raman Spectroscopy*, Wiley-Interscience, New York (1971).

3. T. R. Gilson and P. J. Hendra, *Laser Raman Spectroscopy*, Wiley, New York (1970).

4. S. K. Freeman, *Applications of Laser Raman Spectroscopy*, Wiley, New York (1974).

5. M. Lapp and C. M. Penney, Eds., *Laser Raman Gas Diagnostics*, Plenum Press, New York (1974).

6. J. A. Koningstein; *Ann. Rev. Phys. Chem.*, **24**, 121 (1973).

7. Y. S. Bobovich, *Sov. Phys. USPEKHi*, **15**, 671 (1973).

8. J. M. Cherlow and S. P. S. Porto, in *Laser Spectroscopy of Atoms and Molecules*, H. Walther, Ed., Springer-Verlag, Berlin (1976), pp. 253ff.

9. H. E. Hallam, *Vibrational Spectroscopy of Trapped Species*, Wiley, London (1973).

10. H. Walther, Ed., *Laser Spectroscopy of Atoms and Molecules*, Springer-Verlag, Berlin, (1976).

11. R. A. Smith, Ed., *Very High Resolution Spectroscopy*, Academic Press, New York (1976).

12. S. Jacobs, M. Sargent III, J. F. Scott, and M. O. Scully, Eds., "Laser Applications to Optics and Spectroscopy" in *Physics of Quantum Electronics*, Vol. 2, Addison-Wesley, London (1975).

13. C. B. Moore, *Chemical and Biochemical Applications of Lasers*, Academic Press, New York (1974).

    C. B. Moore, *Advan. Chem. Phys.*, **23**, 41 (1973).

13a. J. I. Steinfeld, "Tunable Lasers and their application in analytical chemistry," in *CRC Crit. Rev. Anal. Chem.*, **5**, 225 (1975).

14. W. Demtröder, "Investigations of Small Molecules by Modern Spectroscopic Techniques," in *Case Studies in Atomic Physics*, M. R. C. McDowell and E. W. McDaniel, Eds., Vol. 6, North-Holland, Amsterdam (1976).

15. N. C. Peterson, M. J. Kurylo, W. Braun, A. M. Buss, and R. Keller, *J. Opt. Soc. Amer.*, **61**, 746 (1971).

16. A. Yariv, *Quantum Electronics*, Wiley-Interscience, New York (1975).

17. H. K. Holt, *Phys. Rev.*, **A11**, 625 (1975).

18. K. Tohma, *Opt. Commun.*, **15**, 17 (1975).

19. R. A. Keller, J. D. Simmons, and B. A. Jennings, *J. Opt. Soc. Amer.*, **63**, 1552 (1973).

20. Y. W. Hänsch, A. L. Schawlow, and P. E. Toschek, *IEEE J. Quant. Electr.* **8**, 802 (1972).

21. E. N. Antonov, V. G. Koloshinikov, and V. R. Mironenko, *Opt. Commun.*, **15**, 99 (1975).

22. T. P. Belikova, E. A. Sviridenko, and A. F. Suchkov, *Opt. Spectrosc.*, **37**, 372 (1974).

23. G. H. Atkinson, A. H. Laufer, and M. J. Kurylo, *J. Chem. Phys.*, **59**, 350 (1973).

24.  Ch. C. Wang and L. I. Davis, *Appl. Phys. Lett.*, **25,** 34 (1974).

25.  A. W. Read, *Advan. Mol. Relaxation Proc.*, **1,** 257 (1967).

26.  H. J. Bauer, *J. Chem. Phys.*, **57,** 3130 (1972).

27.  L. B. Kreutzer, *Anal. Chem.*, **46,** 239A (1974).

28.  C. Forbes Dewey, Jr., "Impact of Lasers in Spectroscopy," *Proc. Soc. Photo-Opt. Instr. Eng.* 13 (1974).

29.  A. Rosenwaig, *Anal. Chem.*, **47,** 592A (1975).

30   A. B. Antipov and Y. N. Ponomarev, *Sov. J. Quant. Electron.*, **4,** 740 (1974).

31.  C. F. Dewey, Jr., R. D. Kamin, and C. E. Hackett, *Appl. Phys. Lett.*, **23,** 633 (1973).

32.  C. K. N. Patel, *Science*, **184,** 1173 (1974).

33.  A. V. Burenin and A. F. Krupnov, *Sov. Phys. JETP*, **40,** 252 (1975).

34.  G. Stella, J. Gelfand, and H. W. Smith, *Chem. Phys. Lett.*, **39,** 146 (1976).

35.  A. M. Angus, E. E. Marinero, and M. J. Colles, *Opt. Commun.*, **14,** 223 (1975).

36.  L. B. Kreutzer, N. D. Kenyon, and C. K. N. Patel, *Science*, **177,** 347 (1972).

37.  J. Gelbwachs, *Appl. Opt.*, **13,** 1005 (1974).

38.  C. K. N. Patel, in *Laser Spectroscopy*, R. G. Brewer and A. Mooradian, Eds., Plenum Press, New York (1974), pp. 471ff.

39.  W. R. Harschberger and M. H. Robin, *Act. Chem. Res.* **6,** 329 (1973).

40.  K. Shimoda, *Appl. Phys.*, **1,** 77 (1973).

41.  J. Kuhl and W. Schmidt, *Appl. Phys.*, **3,** 251 (1974).

42.  E. D. Hinkley, K. W. Nill, and F. A. Blum, "Infrared Spectroscopy with Tunable Lasers," in *Laser Spectroscopy of Atoms and Molecules*, H. Walther, Ed., Springer-Verlag, Berlin (1976) pp. 125ff.

43.  A. Mooradian, "High Resolution Tunable Infrared Lasers," in *Laser Spectroscopy*, R. G. Brewer and A. Mooradian, Eds., Plenum Press, New York (1974), p. 223.

43a. M. J. Colles and C. R. Pidgeon, "Tunable Lasers," *Rept. Progr. Phys.*, **38,** 329 (1975).

44.  E. D. Hinkley, *Appl. Phys. Lett.*, **16,** 351 (1970).

45.  T. C. Harman et al., *Appl. Phys. Lett.*, **14,** 333 (1969).

46.  K. W. Nill et al., *Appl. Phys. Lett.*, **21,** 132 (1972).

47.  V. J. Corcoran et al., *Appl. Phys. Lett.*, **16,** 316 (1970).

47a. H. Flicker and N. Nereson, *IEEE J. Quant. Electron.*, **12,** 326 (1976).

48.  H. Preier et al., *Appl. Phys. Lett.*, **28,** 669 (1976).

49.  S. H. Groves, K. W. Nill, and A. J. Strauss, *Appl. Phys. Lett.*, **25,** 331 (1974).

50.  J. J. Hsieh, J. A. Rossi, and J. P. Ponnelly, *Appl. Phys. Lett.*, **28,** 709 (1976); H. Kressel and F. Z. Hawrylo, *Appl. Phys. Lett.*, **28,** 598 (1976).

51.  K. W. Nill, "Tunable Infrared Lasers: Prospects for Instrument Applications," *Proc. Soc. Photo-Opt. Instrum. Eng.*, **49,** 56 (1974).

52.  H. G. Häfele, *Appl. Phys.*, **5,** 97 (1974); J. F. Scott, in *Laser Applications to Optics and Spectroscopy*, S. Jacobs et al., Eds., Addison-Wesley, London (1975), pp. 123ff.

53.  A. Mooradian, S. R. J. Brueck, and F. A. Blum, *Appl. Phys. Lett.*, **17,** 481 (1970).

54.  C. K. Patel and E. D. Shaw, *Phys. Rev.*, **B3,** 1279 (1971).

55.  R. S. Eng and D. L. Spears, *Appl. Phys. Lett.*, **27,** 650 (1975).

56.  H. A. McKenzis, S. D. Smith, and R. B. Dennis, *Opt. Commun.*, **15,** 151 (1975); J. T. Ganley, F. B. Harrison, and W. T. Leland, *J. Appl. Phys.*, **45,** 4980 (1974); S. D. Smith and R. B. Dennis, in *Laser Spectroscopy*, S. Haroche et. al, Eds., Springer-Verlag, Berlin (1975), p. 79.

56a. C. K. N. Patel, T. Y. Chang, and V. T. Nguyen, *Appl. Phys. Lett.*, **28,** 603 (1976).

57.  S. R. J. Brueck and A. Mooradian, *IEEE J. Quant. Electron.*, **QE12,** 201 (1976).

58.  S. E. Harris, *Proc. IEEE*, **57,** 2096 (1969).

59.  Chromatix Information Sheet on Parametric Oscillators, 1145 Terra Bella Avenue, Mountain View, California.

60.  D. C. Hanna, B. Luther-Davies, R. C. Smith, and R. Wyatt, *Appl. Phys. Lett.*, **22,** 440 (1973).

61.  R. L. Byer, R. L. Herbst, and R. N. Fleming, in *Laser Spectroscopy*, S. Haroche et al., Eds., Springer-Verlag, Berlin (1975), pp. 207ff.

62.  R. L. Herbst, R. N. Fleming, and R. L. Byer, *Appl. Phys. Lett.*, **25,** 520 (1974).

63.  R. L. Byer, in *Laser Spectroscopy*, Eds. R. G. Brewer and A. Mooradian, Eds., Plenum Press New York (1974) pp. 77ff.

63a. J. Pinard and J. F. Young, *Opt. Commun.*, **4,** 425 (1972).

64.  A. Yariv and J. E. Pearson, "Parametric Processes," *Progr. Quant. Electron.*, **1,** 1 (1969).

65.  C. F. Dewey and L. O. Hocker, *Appl. Phys. Lett.*, **18,** 58 (1971).

66.  A. S. Pine, *J. Opt. Soc. Amer.*, **64,** 1683 (1974).

67.  J. J. Wynne, P. Sorokin, and J. R. Lankard, in *Laser Spectroscopy*, R. G. Brewer and A. Mooradian, Eds., Plenum Press, New York (1974), p. 103.

68.  C. R. Vidal and J. Cooper, *J. Appl. Phys.*, **40,** 3370 (1969).

69.  T. Nguyen and T. J. Bridges, in *Laser Spectroscopy*, R. G. Brewer and A. Mooradian, Eds., Plenum Press, New York (1969), p. 513.

70.  V. N. Bagratashvil, I. N. Knyazev, and V. V. Lobko, *Sov. Phys. J. Quant. Electron.*, **5,** 857 (1975).

71.  F. O. Neill and W. T. Whitney, *Appl. Phys. Lett.*, **28**, 539 (1976).

71a. N. G. Basov, E. M. Blenov, V. A. Danilychev, and A. F. Suckkov, *Sov. Phys. USPEKHI*, **17**, 705 (1975).

72.  T. Y. Chang and O. R. Wood, *Appl. Phys. Lett.*, **23**, 370 (1973).

72a. N. W. Harris, F. O. Neill, and W. T. Whitney, *Opt. Commun.*, **16**, 57 (1976).

73.  P. B. Davies and K. M. Evenson, in *Laser Spectroscopy*, S. Haroche et al., Eds., Springer-Verlag, Berlin (1975), p. 132.

74.  K. Ueda and K. Shimoda, *ibid.*, p. 186.

75.  K. M. Evenson et al., *Phys. Rev. Lett.*, **21**, 1038 (1968); *ibid.*, **29**, 831 (1972).

76.  K. M. Evenson and C. J. Howard, in *Laser Spectroscopy*, R. G. Brewer and A. Mooradian, Eds., Plenum Press, New York (1974), p. 535.

77.  C. J. Howard and K. M. Evenson, *J. Chem. Phys.*, **61**, 1943 (1974).

78.  F. P. Schäfer, Ed., *Dye Lasers*, Springer-Verlag, Berlin, Heidelberg, New York (1973); A. Dienes, "Dye Lasers," in *Laser Applications to Optics and Spectroscopy*, S. F. Jacobs et al., Eds., Addison-Wesley, London (1975).

78a. D. Basting, F. P. Schäfer, and B. Steyer, *Appl. Phys.*, **3**, 81 (1974).

78b. F. B. Dunning and R. F. Stebbings, *Opt. Commun.*, **11**, 112 (1974).

79.  G. Marowsky, *Rev. Sci. Instrum.*, **44**, 890 (1973).

80.  H. Gerhardt and F. K. Tittel, *Opt. Commun.*, **16**, 307 (1976); R. E. Grove et al., *Appl. Phys. Lett.*, **23**, 442 (1973).

81.  G. K. Klauminzer, Opt. Eng., **13**, 528 (1974); *Proc. Soc. Photo-Opt., Instr. Eng.*, **49**, 71 (1974).

82.  B. Wellegehausen and H. Welling, *Appl. Phys.*, **3**, 387 (1974).

83.  M. Steiner, H. Walther, and C. Zygan, *Opt. Commun.*, **18**, 2 (1976).

84.  J. P. Goldshorough, *Opt. Eng.*, **13**, 523 (1974).

85.  Information sheets on cw dye laser from Coherent Radiation and Spectra Physics, Palo Alto, Calif.

85a. J. Kuhl and H. Spitschan, *Opt. Commun.*, **13**, 6 (1975).

86.  K. Kato, *J. Appl. Phys.*, **46**, 2721 (1975).

87.  C. Gabel and M. Hercher; *IEEE J. Quant. Electron.*, **QE8**, 850 (1972).

88.  D. Fröhlich, L. Hein, W. Schröder, and H. Welling, *Appl. Phys.*, **11**, 97 (1976).

89.  S. E. Harris et al., "Generation of Ultraviolet and Vacuum Ultraviolet Generation," in *Laser Applications to Optics and Spectroscopy*, S. F. Jacobs et al., Eds., Addison-Wesley, London (1975).

90.  P. P. Sorokin et al., in *Laser Spectroscopy*, S. Haroche et al., Eds., Springer-Verlag, Berlin (1975), p. 46.

91.  D. C. Lorents and D. L. Huestis, in *Laser Spectroscopy*, S. Haroche et al., Eds., Springer-Verlag, Berlin (1975), p. 100.

92. A. C. Tam et al., *Opt. Commun.*, **16**, 376 (1976).

93. J. M. Hoffmann, A. K. Hays, and G. C. Tigune, *Appl. Phys. Lett.*, **28**, 538 (1976).

94. D. J. Bradley, in *Laser Spectroscopy*, S. Haroche et al., Eds., Springer-Verlag, Berlin (1975), pp. 55ff.

95. C. K. Rhodes, *IEEE J. Quant. Electron.*, **QE10**, 153 (1974).

96. C. P. Wang et al., *Appl. Phys. Lett.*, **28**, 326 (1976).

97. D. J. Bradley et al., *Opt. Commun.*, **14**, 1 (1975).

98. F. Allario, C. H. Bair, and J. F. Butler, *IEEE J. Quant. Electron.* **QE11**, 205 (1975).

99. G. P. Montgomery and J. C. Hill, *J. Opt. Soc. Amer.*, **65**, 579 (1975).

100. J. R. Aronson, P. C. von Thuna, and J. F. Butler, *Appl. Opt.*, **14**, 1120 (1975).

101. R. S. Eng et al., *Mol. Phys.* **28**, 653 (1974).

102. M. A. Guerra et al., *J. Chem. Phys.*, **63**, 1317 (1975).

103. J. Jäger, W. Hinz, and H. Walther, *Appl. Phys.*, **2**, 35 (1976).

104. A. S. Pine, *J. Mol. Spectrosc.*, **54**, 132 (1975).

105. A. S. Pine, *J. Opt. Soc. Amer.*, **66**, 97 (1976).

106. R. J. Butcher, R. B. Dennis, and S. D. Smith, *Proc. Roy. Soc. London*, **344**, 541 (1975).

107. T. F. Deaton, D. A. Depatie, and T. W. Walker, *Appl. Phys. Lett.*, **26**, 300 (1975).

108. C. G. Stevens and R. N. Zare, *J. Mol. Spectrosc.*, **56**, 167 (1975).

109. R. W. Field, D. O. Harris, and T. Tanaka, *J. Mol. Spectrosc.*, **57**, 107 (1975).

110. J. M. Green, J. P. Hohimer, and F. K. Tittel, *Opt. Commun.*, **9**, 407 (1973).

111. R. T. Thompson Jr., J. M. Hoell, and W. R. Wade, *J. Appl. Phys.*, **46**, 3040 (1975).

111a. B. Cagnac, G. Grynberg, and F. Biraben, *J. Physique*, **34**, 845 (1973).

112. R. M. Hochstrasser and J. E. Wessel, *Chem. Phys. Lett.*, **24**, 1 (1974).

113. L. Wunsch, H. J. Neusser, and E. W. Schlag, *Chem. Phys. Lett.*, **31**, 433 (1975); *ibid.*, **32**, 210 (1975).

114. R. Wallenstein and T. W. Hänsch, *Opt. Commun.*, **14**, 353 (1975).

115. R. G. Bray, R. M. Hochstrasser, and H. N. Sung, *Chem. Phys. Lett.*, **33**, 1 (1975).

116. R. G. Bray, R. M. Hochstrasser, and J. E. Wessel, *Chem. Phys. Lett.*, **27**, 167 (1974).

117. L. Wunsch, H. J. Neusser, and E. W. Schlag, *Chem. Phys. Lett.*, **38**, 216 (1976).

118. K. M. Baird and G. R. Hanes, *Rep. Progr. Phys.* **37**, 127 (1974).

119. A. S. Pine and K. W. Nill, IXth Int. Conf. Quant. Electron., Amsterdam, 1976, *Opt. Commun.*, **18,** 57 (1976).

119a. F. Y. Chu and T. Oka, *J. Appl. Phys.*, **46,** 1204 (1975).

120. R. Schmiedl, Thesis, Department of Physics, University of Kaiserslautern (1976).

121. W. Demtröder, F. Paech, and R. Schmiedl, *Chem. Phys. Lett.*, **26,** 381 (1974).

122. R. Schmiedl, R. Bomilla, F. Paech, W. Demtröder, *J. Mol. Spectrosc.*, **68,** 236 (1977).

123. D. G. Youmans, L. A. Haeckel, and S. Ezekiel, *J. Appl. Phys.*, **44,** 2319 (1973).

124. R. E. Growe et al., *Appl. Phys. Lett.*, **23,** 442 (1973).

125. L. A. Haeckel, D. G. Youmans, and S. Ezekiel, *J. Opt. Soc. Amer.*, **64,** 1387 (1974).

125a. V. S. Letokhov and V. P. Chebotayev: *Nonlinear Laser Spectroscopy.* Springer Series in Optical Sciences, Vol. 4, Springer-Verlag Berlin, (1977).

126. W. Lamb, *Phys. Rev.*, **134A,** 1429 (1964).

127. M. S. Sorem and A. L. Schawlow, *Opt. Commun.*, **5,** 148 (1972).

128. A. Javan, in *Laser Spectroscopy*, S. Haroche et al., Eds., Springer-Verlag, Berlin (1975), p. 439.

129. J. Foth, Diplom. Thesis, Department of Physics, University of Kaiserslautern (1976).

130. M. Kroll, *Phys. Rev. Lett.*, **23,** 631 (1969).

131. G. R. Hanes, J. Lapierre, P. R. Bunker, and K. C. Shotton, *J. Mol. Spectrosc.*, **39,** 506 (1971).

132. G. R. Hanes and G. E. Dahlström, *Appl. Phys. Lett.*, **14,** 362 (1969).

133. M. S. Sorem, M. D. Levenson, and A. L. Schawlow, *Phys. Lett.*, **37A,** 33 (1971).

134. M. S. Sorem, T. W. Hänsch, and A. L. Schawlow, *Chem. Phys. Lett.*, **17,** 300 (1972).

135. B. Couillaud and A. Ducasse, *Opt. Commun.*, **13,** 398 (1975).

136. A. Murihead, K. V. Sastry, R. F. Corl, J. Cook, and F. K. Tittel, *Chem. Phys. Lett.*, **24,** 208 (1974).

137. G. W. Hills and D. L. Philen, *Chem. Phys.*, **12,** 107 (1976).

138. M. W. Goldberg and R. Yusek, *Appl. Phys. Lett.*, **17,** 349 (1970).

139. A. C. Luntz, J. D. Swalen, and R. G. Brewer, *Chem. Phys. Lett.*, **14,** 512 (1972).

140. R. G. Brewer, *Phys. Rev. Lett.*, **25,** 1639 (1970).

141. E. E. Uzgiris, J. L. Hall, and R. L. Barger, *Phys. Rev. Lett.*, **26,** 289 (1971).

142. C. K. N. Patel, *Appl. Phys. Lett.*, **25,** 112 (1974).

142a. Ch. Bordé and J. L. Hall, in *Laser Spectroscopy*, R. G. Brewer and A. Mooradian, Eds., Plenum Press, New York (1974), pp. 125ff.

143. K. Shimoda, in *Laser Spectroscopy of Atoms and Molecules*, H. Walther, Ed., Springer-Verlag, Berlin (1976), pp. 197ff.

144. M. E. Kaminsky, R. T. Hawkins, F. V. Kowalski, and A. L. Schawlow, *Phys. Rev. Lett.*, **36**, 671 (1976).

145. M. Fourrier, M. Redon, A. van Lerherghe, and C. Bordé, *C. R.*, **270**, 537 (1970).

146. M. Takami and K. Shimoda, *Jap. J. Appl. Phys.*, **10**, 658 (1971).

147. M. Takami, K. Shimoda, *Jap. J. Appl. Phys.*, **12**, 603 (1973).

148. R. W. Fields, G. H. Capelle, and M. A. Revelli, *J. Chem. Phys.*, **63**, 3228 (1975).

148a. B. Cagnac, G. Grynberg and F. Biraben, *J. Physique*, **34**, 845, (1973).

149. W. Bischel, P. Kelley, and Ch. K. Rhodes, *Phys. Rev. Lett.*, **34**, 300 (1975).

150. J. A. Gelbwachs, P. F. Jones, and J. E. Wessel, *Appl. Phys. Lett.*, **27**, 551 (1975).

151. J. P. Woerdman, Post Deadline Paper U2, Int. Conf. Quant. Electron. Amsterdam, 1976, *Opt. Commun.*, **18**, (1976).

152. B. Cagnac, in *Laser Spectroscopy*, S. Haroche et al., Eds., Springer-Verlag, Berlin (1975), p. 165.

153. T. A. Alobaidi and D. W. Hills; *J. Phys.* **E8**, 30 (1975).

154. R. G. W. Norrish and G. Porter, *Nature*, **164**, 658 (1949).

155. H. W. Latz, H. W. Wyles, and R. B. Green, *Anal. Chem.*, **45**, 2405 (1973).

156. G. Horlick and E. G. Codding, *Anal. Chem.*, **46**, 133 (1974).

157. R. H. Barnes, C. E. Moeller, J. F. Kircher, and C. M. Verber, *Appl. Opt.*, **12**, 2531 (1973).

158. J. M. Telle and C. L. Tang, *Opt. Commun.*, **11**, 251 (1974).

159. K. M. Evenson, H. E. Radford, and M. Moran, *Appl. Phys. Lett.*, **18**, 426 (1971).

160. K. M. Evenson, J. S. Wells, and H. E. Radford, *Phys. Rev. Lett.*, **25**, 199 (1970).

161. J. M. Cooke et al., *J. Chem. Phys.*, **64**, 1381 (1976).

162. P. B. Davies et al., *J. Chem. Phys.*, **62**, 3739 (1975).

163. L. B. Kreutzer and C. K. N. Patel, *Science*, **173**, 45 (1971).

164. A. W. Tucker, A. Peterson, and M. Birnbaum, *Appl. Opt.*, **12**, 2036 (1973).
A. Tucker, M. Birnbaum, and C. L. Flincher, *Proc. Soc. Photo-Opt. Instrum. Eng.*, **49**, 103 (1974).

165. D. A. Leonard, *ibid.*, p. 110.

166. C. K. N. Patel, in *Laser Spectroscopy*; S. Haroche et al., Eds., Springer-Verlag, Berlin (1975), pp. 71ff.

167. K. W. Rothe, U. Brinkmann, and H. Walther, *Appl. Phys.* **3**, 115 (1974); *ibid.*, **4**, 181 (1974).

168.  R. N. Zare, in Proc. Int. Conf. on Doppler-free Spectroscopy, Aussois, France, 1973, CNRS, Paris, France (1973).

169.  K. Bergmann, U. Hefter, and P. Hering, J. Chem. Phys., 65, 488 (1976).

170.  K. Bergmann, W. Demtröder, and P. Hering, Appl. Phys., 8, 65 (1975).

171.  K. Bergmann and P. Hering, to be published.

172.  G. Ennen and C. Ottinger, Chem. Phys. Lett., 36, 16 (1975).

173.  M. M. Hessel, E. W. Smith, and R. E. Drullinger, Phys. Rev. Lett., 33, 1251 (1974).

174.  W. N. Jackson, J. Chem. Phys., 59, 160 (1973).

175.  A. C. Tam and W. Happer, J. Chem. Phys., 64, 2456 (1976).

176.  M. P. Sinha, A. Schultz and R. N. Zare, J. Chem. Phys., 58, 549 (1973).

177.  R. N. Zare; Laser Fluorimetry, in Laser Spectroscopy, S. Haroche et al., Eds., Springer-Verlag, Berlin (1975), pp. 112ff.

177a. D. C. Harrington and H. V. Malmstadt, Anal. Chem., 47, 271 (1975).

178.  J. H. Cruse and R. N. Zare, J. Chem. Phys., 60, 1182 (1974).

179.  G. Gerber, K. Sakurai, and H. P. Broida, J. Chem. Phys., 64, 3410 and 3423 (1976).

180.  E. O. Degenkolb and J. I. Steinfeld, Chem. Phys. Lett., 8, 288 (1971).

181.  D. K. Russel, M. Kroll, D. A. Pows, and R. A. Beandet, Chem. Phys. Lett., 20, 153 (1973).

182.  S. E. Schwartz and G. I. Senuin, Chem. Phys. Lett., 32, 569 (1975); H. D. Bist and J. C. D. Brand, J. Mol. Spectrosc., 62, 60 (1976); R. Schmiedl and W. Demtröder, J. Mol. Spectrosc., 68, 236 (1977); F. Paech, R. Schmiedl, and W. Demtröder, J. Chem. Phys., 63, 4369 (1975).

183.  J. C. D. Brand et al., Can. J. Phys., 51, 530 (1973).

184.  G. Hancock, W. Lange, M. Lenzi, and K. H. Welge, Chem. Phys. Lett., 33, 168 (1975).

185.  J. C. D. Brand, J. L. Hardwick, and K. E. Teo, J. Mol. Spectrosc., 57, 215 (1975).

186.  P. W. Smith, M. A. Duguay, and E. P. Ippen, "Mode-locking of lasers," in Progress in Quantum Electronics, Vol. 3, Part 2, J. H. Sanders and S. S. Stenholm, Eds., Pergamon Press, London (1974).

187.  S. E. Johnson, J. Chem. Phys., 56, 149 (1972).

188.  J. A. Paisner and R. A. Wallenstein, J. Chem. Phys., 61, 4317 (1974).

189.  W. Demtröder, W. Stetzenbach, M. Stock, and J. Witt, J. Mol. Spectrosc., 61, 382 (1976).

190.  L. Armstrong, Jr., and S. Feneville, J. Phys. B, Atom. Mol. Phys., 8, 546 (1975).

191.  R. Hefferlin, J. Quant. Spectrosc. Radiat. Transfer, 15, 925 (1975).

192.  D. R. Crosley and R. K. Lengel, J. Quant. Spectrosc. Radiat. Transfer, 15, 579 (1975).

193. W. H. Hennecker and H. E. Popkie, *J. Chem. Phys.*, **54**, 1763 (1971).

194. K. R. German, *J. Chem. Phys.*, **62**, 2584 (1975).

195. W. M. Jackson, *J. Chem. Phys.*, **61**, 4177 (1974).

196. C. G. Stevens, M. W. Swagel, R. Wallace, and R. N. Zare, *Chem. Phys. Lett.*, **18**, 465 (1973).

197. F. Paech, R. Schmiedl, and W. Demtröder, *J. Chem. Phys.*, **63**, 4369 (1975).

198. J. G. Calvert, *Chem. Phys. Lett.*, **20**, 484 (1973).

199. P. J. Dagdigian, H. W. Cruse, and R. N. Zare, *J. Chem. Phys.*, **60**, 2320 (1974).

200. J. G. Pruett and R. N. Zare, *J. Chem. Phys.*, **62**, 2050 (1975).

201. K. R. German, T. H. Bergeman, E. M. Weinstock, and R. N. Zare, *J. Chem. Phys.*, **58**, 4304 (1973).

202. M. Broyer and J. C. Lehmann, *Phys. Lett.*, **40A**, 43 (1972).

203. S. J. Silvers and C. L. Chiu, *J. Chem. Phys.*, **56**, 5663 (1972).

204. M. A. Marechal, R. Rost, and M. Lombardi, *Phys. Rev.*, **A5**, 732 (1972).

205. R. Wallenstein, J. A. Paisner, and A. L. Schawlow, *Phys. Rev. Lett.*, **32**, 1333 (1974).

206. P. Franken, *Phys. Rev.*, **121**, 508 (1961).

207. R. N. Zare, *J. Chem. Phys.*, **45**, 4510 (197

208. D. H. Levy, *J. Chem. Phys.*, **56**, 5493 (1972).

209. R. N. Zare, in Proc. Int. Conf. Doppler-free Spectroscopy, Aussois, France, 1973, CNRS, Paris, France (1973) p. 29.

210. D. H. Levy, *Advan. Magnet. Resonance*, **6**, (1973).

211. L. Novikov et al., *Sov. Phys. USPEKHI*, **17**, 542 (1975).

212. A. C. Luntz, R. G. Brewer, K. L. Foster, and J. P. Swalen, *Phys. Rev. Lett.*, **23**, 951 (1969).

213. E. E. Uzgiris, J. L. Hall, and R. L. Barger, *Phys. Rev. Lett.*, **26**, 289 (1971).

214. R. Solarz and D. H. Levy, *Chem. Phys. Lett.*, **17**, 35 (1972).

215. J. Sakai and M. Katayama, *Appl. Phys. Lett.*, **28**, 119 (1976).

216. D. W. Robinson, *J. Mol. Spectrosc.*, **35**, 1 (1970).

217. R. Solarz and D. Levy, *J. Chem. Phys.*, **58**, 4026 (1973); *ibid.*, **60**, 1158 (1974).

218. T. Tanaka et al., *J. Chem. Phys.*, **59**, 5217 (1973); *J. Mol. Spectrosc.*, **49**, 310 (1974); *ibid.*, **56**, 188 (1975).

219. F. Schuler and W. Behmenburg, *Phys. Rep.*, **124**, 275 (1974).

220. T. W. Hänsch, M. P. Levenson, and A. L. Schawlow, *Phys. Rev. Lett.*, **26**, 946 (1971).

221. J. L. Hall, in *Atomic Physics*, Vol. 3, S. J. Smith and G. K. Walters, Eds., Plenum Press, New York (1973), p. 615.

222. R. Kurzel, J. Steinfeld, D. Hatzenbuhler, and G. Leroi, *J. Chem. Phys.*, **55**, 4822 (1971).

223. K. Bergmann and W. Demtröder, *J. Phys. B, Atom. Mol. Phys.*, **5**, 1386 and 2098 (1972); *ibid.*, **7**, 2036:4).

224. G. Ennen and C. Ottinger, *Chem. Phys.*, **3**, 404 (1974).

225. P. L. Huston, in *Impacts of Lasers in Spectroscopy*, Proc. Soc. Photo-Opt. Instr. Eng., Palos Verdes Estates, California (1975), p. 35.

226. C. B. Moore, *Advan. Chem. Phys.*, **23**, 41 (1973).

227. P. F. Zittel and C. B. Moore *J. Chem. Phys.*, **58**, 2922 (1973).

228. T. J. Odiorne, P. R. Brooks, and J. V. Kasper, *J. Chem. Phys.*, **55**, 1980 (1971).

229. S. J. Riley, R. K. Sander, and K. R. Wilson, in *Laser Spectroscopy*, R. Brewer and A. Mooradian, Eds., Plenum Press, New York (1974), p. 597.

229a. J. I. Steinfeld, in *Int. Rev. Science, Phys. Chemistry Series* 1, Vol. 9, Butterworth, London (1972) pp. 247ff.

229b. S. M. Lee and A. M. Roun, *Spectrosc. Lett.*, **8**, 915 (1975).

230. C. B. Moore and P. F. Zittel, *Science*, **182**, 541 (1973).

231. W. Kaiser, A. Lauberau, in *Laser Spectroscopy*, S. Haroche et al., Eds., Springer-Verlag, Berlin (1975), p. 380.

232. K. B. Eisenthal, *ibid.*, p. 390.

233. C. V. Shank and E. P. Ippen, *ibid.*, p. 408.

234. K. Shimoda, in *Laser Spectroscopy of Atoms and Molecules*, H. Walther, Ed., Springer-Verlag, Berlin (1976), p. 197.

235. M. Takami and K. Shimoda, *Jap. J. Appl. Phys.*, **12**, 934 (1973).

236. J. M. Levy et al., *Chem. Phys. Lett.*, **21**, 598 (1973).

237a. K. H. Becker, P. Haaks, and T. Tantavczyk, *Z. Naturforsch.*, **29A**, 829 (1974).

237b. Ch. C. Wang and L. I. Davis, Jr., *Appl. Phys. Lett.*, **25**, 34 (1974).

238. M. P. Sinha, A. Schultz, and R. N. Zare, *J. Chem. Phys.*, **58**, 549 (1973).

239. K. Bergmann et al., *J. Chem. Phys.*, **65**, 488 (1976).

239a. K. Bergmann, U. Hefter, P. Hering, University of Kaiserlautern, internal report.

240. R. E. Smalley et al., *Opt. Commun.*, **18**, 59 (1976); *J. Chem. Phys.*, **63**, 4977 (1975).

241. K. L. Kompa, *Chemical Lasers* (Topics in Current Chemistry, Vol. 37), Springer-Verlag, Berlin (1973).

242. G. A. West and M. J. Benny, *Opt. Commun.*, **18**, 128 (1976).

243. C. R. Jones and H. P. Broida, *Laser Focus*, **10**, 37 (1974).

244. P. J. Dagdigian et al., *J. Chem. Phys.*, **61**, 4450 (1974).

245. P. J. Dagdigian et al., *J. Chem. Phys.*, **62**, 1824 (1975).

246. H. W. Cruse, P. J. Dagdigian, and R. N. Zare; *Faraday Disc. Chem. Soc.*, **55**, 277 (1973).

247. P. J. Dagdigian and R. N. Zare, *J. Chem. Phys.*, **61**, 1964 (1974).

248. J. G. Pruett and R. N. Zare, *J. Chem. Phys.*, **64**, 1774 (1976).

249. R. N. Zare, "Laser Fluorimetry," in *Laser Spectroscopy* S. Haroche et al., Eds., Springer-Verlag, Berlin (1975), p. 112.

250. R. N. Zare and P. Dagdigian, *Science*, **185**, 739 (1974).

251. A. B. Petersen, C. Wittig, and S. R. Leune, *Opt. Commun.*, **18**, 125 (1976).

252. K. M. Evenson and F. R. Petersen, in *Laser Spectroscopy of Atoms and Molecules*, H. Walther, Ed., Springer-Verlag, Berlin (1976), pp. 349ff.

253. T. Y. Chans and T. J. Bridges, *Opt. Commun.*, **1**, 423 (1969); *ibid.*, **2**, 77 (1970); *Appl. Phys. Lett.*, **17**, 249 (1970).

254. G. Hübner, J. C. Hassler, and P. D. Coleman; *Appl. Phys. Lett.*, **18**, 511 (1971).

255. L. E. S. Mathias and J. T. Parker, *Appl. Phys. Lett.*, **3**, 16 (1963).

256. P. D. Coleman; *IEEE J. Quant. Electron.*, **QE9**, 130 (1973).

257. V. J. Cororan, in *Applied Spectroscopy Review*, E. G. Brame, Jr., Ed., Marcel Dekker, New York (1974), pp. 215ff.

258. T. A. Cool, *IEEE J. Quant. Electron.*, **QE9**, 72 (1973).

259. A. P. Petersen, C. Wittig, and S. R. Leone, *Appl. Phys. Lett.*, **27**, 305 (1975).

260. See collection of papers on chemical lasers in *Appl. Opt.*, **12**, (Febr. 1973).

261. T. J. Bridges and T. Y. Chang, *Phys. Rev. Lett.*, **22**, 811 (1969).

262. F. R. Petersen et al., in *Laser Spectroscopy*, R. G. Brewer and A. Mooradian, Eds., Plenum Press, New York (1974), p. 555ff.

263. D. R. Sokoloff and A. Javan, *J. Chem. Phys.*, **56**, 4028 (1972).

264. L. A. Hackel, K. H. Casleton, S. G. Kukolich, and S. Ezekiel, *Phys. Rev. Lett.*, **35**, 568 (1975).

265. R. L. Barger and J. C. Hall; *Phys. Rev. Lett.*, **22**, 4 (1969).

266. W. E. L. Grossman, "Raman Spectrometry," *Anal. Chem.*, **47**, 345R (1975).

267. D. A. Leonard, *Proc. Soc. Photo-Opt. Instrum. Eng.*, **49**, 110 (1974).

268. J. J. Barret and H. B. Harvey, *J. Opt. Soc. Amer.*, **65**, 392 (1975).

269. T. Hirschfeld, E. R. Schildkraut, H. Tannenbaum, and D. Tannenbaum, *Appl. Phys. Lett.*, **22**, 38 (1973).

270. J. Levatter, R. Sandstrom, and S. Chitin, *J. Appl. Phys.*, **44**, 3273 (1973).

271. M. Bridoux et al., in *Laser Raman Gas Diagnostics*, M. Lapp and C. M. Penney, Eds., Plenum Press, New York, (1974), p. 249; *Advan. Raman Spectrosc.*, **1**, 65 (1973).

272. J. Stone, *Appl. Phys. Lett.*, **6a**, 163 (1975).

273. W. J. Schmid, private communication; also described in R. J. H. Clark, and R. E. Hester, Eds., *Advances in Infrared and Raman Spectroscopy*, Vol. 3, Heyden, London (1976).

274. G. T. McNice, *Appl. Opt.*, **11**, 699 (1972).

294 MOLECULAR ABSORPTION AND FLUORESCENCE SPECTROSCOPY

275. W. Werneke, J. Klein, A. Lau, K. Lenz, and G. Hunsalz, *Opt. Commun.*, **11,** 159 (1974).

276. P. Regnier and J. P. E. Taran, in *Laser Raman Gas Diagnostics*, M. Lapp and C. M. Penney, Eds., Plenum Press, New York (1974), pp. 87ff.

277. R. A. Hill, A. J. Mulay, and P. R. Smith, *Appl. Spectrosc.*, **30,** 183 (1976).

278. F. Moya, S. A. J. Druet, and J. P. E. Taran, in *Laser Spectroscopy*, S. Haroche et al., Eds., Springer-Verlag, Berlin (1975), pp. 66ff; *Opt. Commun.*, **13,** 169 (1975).

279. W. Kiefer, *Appl. Spectrosc.*, **28,** 115 (1974).

280. W. Kiefer and H. J. Bernstein, *Appl. Spectrosc.*, **25,** 500 (1971).

281. R. Wilbrandt, P. Pagsberg, K. B. Hansen, and K. V. Weisberg, Proc. IXth Int. Conf. Quant. Electronics, Amsterdam, 1976, Paper 05 (1976), p. 135.

282. N. G. Basov, A. N. Oraevsky, and A. V. Pankratov, in *Chemical and Biochemical Applications of Lasers*, C. B. Moore; Ed., Academic Press, New York (1974), pp. 203ff.

283. R. V. Ambartzumian et al., *Chem. Phys. Lett.*, **25,** 515 (1974).

284. A. N. Oraevsky and A. N. Pankratov, in *Laser Spectroscopy*, S. Haroche et al., Eds., Springer-Verlag, Berlin (1975), pp. 304ff.

285. R. V. Ambartzumyan et al., *Sov. Phys. JETP Lett.*, **21,** 171 (1975).

286. H. R. Bachmann, H. Nöth, R. Rinck, and K. L. Kompa, *Chem. Phys. Lett.*, **29,** 627 (1974).

287. J. L. Lyman and S. D. Rockwood, *Opt. Commun.*, **18,** 29 (1976); *J. Appl. Phys.*, **47,** 595 (1976).

287a. R. V. Ambartzumyan et al., *Sov. Phys. JETP Lett.*, **22,** 43 (1975).

287b. A. Yogev and R. N. J. Benmair, *J. Amer. Chem. Soc.*, **97,** 4430 (1975).

288. N. R. Isenor, V. Merchant, R. S. Hallsworth, and M. C. Richardson, *Can. J. Phys.*, **51,** 1281 (1973).

289. J. Campbell, H. Hancock, J. Halpern, and K. H. Welge, IXth Int. Conf. Quant. Electronics, Amsterdam, 1976, Conference Digest (1976), p. 34.

290. R. Gordon and M. C. Lin, *Chem. Phys. Lett.*, **22,** 262 (1973).

291. J. C. Stephenson and S. M. Freund, *Opt. Commun.*, **18,** 93 (1976).

292. B. L. Earl and A. M. Roun, *Opt. Commun.*, **18,** 96 (1976).

293. W. Happer, *Opt. Commun.*, **18,** 93 (1976).

# ANALYTICAL USE OF LASERS
# IN REMOTE SENSING

## R. M. MEASURES

*Institute for Aerospace Studies*
*University of Toronto, Canada*

## 6.1  INTRODUCTION

The concurrence of technology with man's expanding need to study his environment is timely. Within the last decade and a half we have witnessed the invention of the laser and its remarkable development while at the same time we have grown acutely aware of the finite nature of the earth and the fragile balance of its ecosystems. The current issue regarding the influence of fluorocarbons and nitric oxide on the earth's protective shield of ozone (1) is a good case in point.

Although Townes and Schawlow (2) and Maiman (3) deserve recognition for their respective roles in creating the first laser, it was the giant pulse technique invented by McClung and Hellwarth (4) that made remote optical probing really attractive. The first laser studies of the atmosphere were undertaken in 1963 by Fiocco and Smullein (5) who recorded laser echoes from the upper regions of the atmosphere and by Ligda (6) who probed the troposphere.

In the decade that followed this pioneering work, great strides were taken both in the development of laser radar (or lidar) systems and in the sophisticated way in which they were employed (lidar is an acronym for Light Detection And Ranging). It was immediately appreciated that, as with radar, lidar systems could provide spatially resolved measurements in "real time." Recognition that the radiation detected at wavelengths different from that of the laser's output contained highly specific information that could be used to determine the composition of the target region represented one of the most important advances. The specificity of this approach was further enhanced by the broad selection of laser wavelengths that became available and, in the case of resonance excitation, by the precise tuning made possible with certain classes of laser. This ability of lidar systems to perform an effective spectral analysis of a distant target has added a new dimension to remote sensing and has made possible an extraordinary variety of applications. These range from mapping of trace constituents in the tenuous outer reaches of the atmosphere

(7–11) to airborne fingerprinting of an oil slick at sea in order to prosecute those responsible (12–15).

The atmosphere was one of the first arenas in which the laser's special properties of high power, monochromaticity, short duration, and beam collimation were put to the test. The resulting triumph of the laser has been well documented in several review articles (16–23). Direct laser measurement of the atmospheric parameters important in evaluating the composition (24–27), structure (16,28–31), properties (32–34), and dynamic behavior (35–36) of the atmosphere have been undertaken. The results have led to speculation that, in the future, lidar systems could play a key role in providing the base information pertinent to man's understanding of the atmosphere and his ability to both predict and modify the weather. Laser probing of the atmosphere has also revealed that lidar systems can detect and quantify trace constituents that arise both naturally (7–11,37–39) and as unfortunate by-products of our technological society (19,20,22,40–46).

The range of processes available include Rayleigh scattering, Mie scattering, Raman scattering, resonance scattering, fluorescence, absorption, and differential absorption and scattering (DAS). A brief description of each is provided in Table 6.1, and the range of cross sections observed for each process is schematically presented in Fig. 6.1. It is evident from this figure that the cross section for Mie scattering can be so large that only a few appropriately sized scatters could give rise to a scattered signal that completely swamps any Rayleigh-scattered component. This implies that quite low concentrations (or changes in concentration) of dust or aerosols can be detected. Although resonance scattering, sometimes referred to as atomic or resonance fluorescence, also has an inherently large cross section, collision quenching with the more abundant atmospheric species generally ensures that the detected signal is small; consequently this technique is used to best effect in studies of the trace constituents in the upper atmosphere (7–11). The influence of collision quenching on molecular fluorescence can be equally detrimental, particularly where long-lived states are involved (46). The broadband nature of molecular fluorescence invariably leads to a low value for the signal-to-noise ratio where background radiation forms the major component of noise (17,20,46). In the event that the fluorescence is long-lived, spatial resolution can be degraded.

Raman scattering represents an inelastic scattering process whereby the laser radiation may be thought to raise the molecule to a virtual level from which it immediately decays ($<10^{-14}$ s), with the subsequent emission of radiation having a different wavelength. The difference in energy between the incident and emitted photons is a characteristic of the

## TABLE 6.1
### Optical Interactions of Relevance to Laser Environmental Sensing

| Technique | Physical description |
|---|---|
| Rayleigh scattering | laser radiation elastically sca*tered from atoms or molecules is observed with no change of frequency |

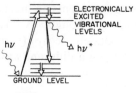

| | |
|---|---|
| Mie scattering | laser radiation elastically scattered from small particulates or aerosols (of size comparable to wavelength of radiation) is observed with no change in frequency |

| | |
|---|---|
| Raman scattering | laser radiation inelastically scattered from molecules is observed with a frequency shift characteristic of the molecule ($h\nu - h\nu^* = E$) |

| | |
|---|---|
| Resonance scattering | laser radiation matched in frequency to that of a specific atomic transition is scattered by a large cross section and observed with no change in frequency |

| | |
|---|---|
| Fluorescence | laser radiation matched to a specific electronic transition of atom or molecule suffers absorption and subsequent emission at lower frequency; collision quenching can reduce effective cross section of this process; broadband emission is observed with molecules |

298

**TABLE 6.1 (*Continued*)**

| Technique | Physical description |
|---|---|
| Absorption | observe attenuation of laser beam when frequency matched to the absorption band of given molecule |

| Differential absorption and scattering (DAS) | the differential attenuation of two laser beams is evaluated from their backscattered signals when the frequency of one beam is closely matched to a given molecular transition while the other's frequency is somewhat detuned from the transition |

irradiated molecule and usually corresponds to a change of one vibrational quanta. The frequency shifts of the $Q$-branch of vibrational–rotational Raman spectra are summarized for a large group of molecules in Fig. 6.2. It is evident that most of the molecules likely to be of interest in any air pollution study are included. Unfortunately, the Raman cross sections are so small that the range and sensitivity of this technique are rather limited. Raman scattering is consequently most likely to be employed for remote monitoring of effluent plumes where the concentrations can be quite high, tens to hundreds of ppm, as opposed to the few ppm or parts of ppm more typical of dispersed contaminants. As seen from Fig. 6.1, some degree of enhancement (at least a factor of $10^3$) in cross section can be achieved if the excitation frequency is made to closely coincide with an allowed transition. A particularly attractive feature of Raman scattering relates to the ease with which the concentration of any species relative to some reference species, such as nitrogen, can be evaluated from the ratio of the respective Raman signals provided the relevant cross section ratio is known (47–50).

The cross section for absorption of radiation is, in general, much greater than either the effective (quenched) fluorescence cross section or the cross sections for Raman scattering. Consequently, the attenuation of a beam of suitably tuned laser radiation is viewed as a sensitive method of

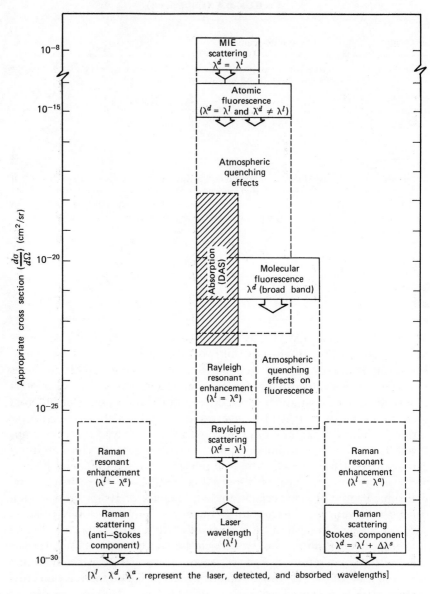

**Fig. 6.1.** Optical interaction processes of relevance to laser environmental sensing.

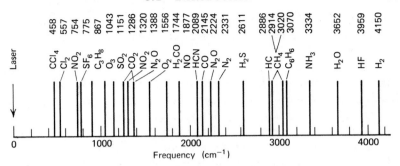

**Fig. 6.2.** Frequency shifts of $Q$-branch of vibrational–rotational Raman spectra of typical molecular species present in polluted as well as ordinary atmosphere relative to the exciting laser frequency. From H. Inaba and T. Kobayasi, *Opto-Electron.*, **4**, 101 (1972). Reproduced by permission.

evaluating the mean density of a given constituent. In order to separate absorption by the molecule of interest from other causes of attenuation, a differential approach is often adopted. In this instance two frequencies are employed, one centered on a line within the absorption band, the other detuned into the wing of the line. With a few notable exceptions, most of the absorption bands of interest lie in the infrared and correspond to vibrational–rotational transitions (51). Although in principle this approach involves a bistatic arrangement,[1] clever use of retroreflectors or topographical reflectors allows the more convenient monostatic configuration to be employed (19, 42, 52–56). Lack of spatial resolution and poor infrared detector sensitivity represent the major drawbacks of this approach.

High sensitivity with good spatial resolution can be achieved by the combination of differential absorption and scattering (DAS). In this approach a comparison is made between the atmospheric backscattered laser radiation monitored, when the frequency of the laser is tuned to closely match that of an absorption line (within the molecule of interest) and then when it is detuned to lie in the wing of the line. In this way, the large Mie scattering cross section is employed to provide spatial resolution and to ensure a strong return signal at both frequencies, while the ratio of the signals yields the required degree of specificity due to differential absorption. These advantages appear to bestow upon the DAS technique the greatest sensitivity for long-range monitoring of specific molecular constituents (20, 46, 57–62). At present, detector sensitivity makes both fluorescence and the DAS techniques more amenable to those molecules which possess an absorption band in the visible or

---

[1] A configuration involving a considerable separation between the laser transmitter and the receiver optics, see Sect. 6.2.1.

near-ultraviolet part of the spectrum. Improvements in infrared detector sensitivity could give them more universal appeal, as indicated recently by Murray et al. (63).

Toward the end of the laser's first decade some consideration was given to the development of earth-oriented laser sensors that could be used from mobile platforms such as aircraft and helicopters. In the initial applications, these downward-pointing laser systems were operated in a mode somewhat analogous to radar, where surface scattering and reflection represented the dominant form of interaction. Surface wave studies and bathymetric measurements in coastal waters were the first topics to be given serious consideration (64–66). The possibility of undertaking studies of water turbidity grew naturally from the latter series of experiments (67).

An important advance was made with the realization that use of a short-wavelength laser could broaden the spectrum of applications as a result of laser-induced fluorescence and led to the development of a new form of remote sensor termed a "laser fluorosensor" (12, 68, 69). The fluorescence induced in thin oil films was of particular interest as it provided the motivation for developing this new instrument in the hope that it would be capable of detecting and classifying an oil slick (12, 13, 68–72). Detailed spectroscopic studies (12–15) of both crude oils and petroleum products have indicated that an airborne laser fluorosensor with high spectral resolution may indeed be capable of classifying an oil slick with sufficient precision for this information to be entered as evidence in a court of law. The recent discovery of fluorescent decay spectra (14, 15) should allow an improvement in the design of laser fluorosensors by providing adequate identification capability with somewhat lower spectral resolution.

Early in these studies laser-induced fluorescence from natural bodies of water was considered to constitute a source of background emission that could interfere with the oil fluorescence signal (12, 68–70). Further studies have diminished this worry but have discovered that this water fluorescence signal might serve to indicate the presence of high organic contamination and enable the dispersion of an effluent plume to be remotely mapped (73). The fluorescent nature of chlorophyll has long been known, and consequently the possibility of employing a laser fluorosensor to remotely map the chlorophyll concentration of natural bodies of water has also been studied (74–76).

It is clear that *the scope of lasers in environmental sensing is extensive.* They can be used to undertake (i) concentration measurements of both major and minor constituents, well suited for pollution monitoring; (ii) evaluation of thermal, structural, and dynamic properties; (iii) threshold detection of specific constituents, well suited for alarm purposes; (iv) mapping of effluent plume dispersal, and (v) spectral fingerprinting of a specific target such as an oil slick.

Furthermore, these observations can be made remotely with both spatial and temporal resolution. Laser sensors can be operated from the ground or from boats, helicopters, aircraft, or satellites. Satellite observations, however, are likely to be limited to studies of the outer atmosphere, except where reflection from the earth is employed. In this instance it is proposed that the atmospheric burden of the constituent, integrated between the satellite and the earth, would be determined by differential absorption using the earth as a diffuse reflector for the two-probe wavelengths (77–79).

In this section we have attempted to present a broad survey of the field of laser environmental sensing. In the next three sections we endeavor to lay much of the groundwork of the subject, while in the subsequent sections we consider in more detail both atmospheric and hydrographic applications. In keeping with the flavor of the remainder of this book we emphasize the spectroscopic aspects of the subject. A broad review of the nonspectroscopic aspects of atmospheric probing is presented by Hall (16), while Byer (20) has prepared a useful review of laser atmospheric pollution measurements. Recently, E. D. Hinkley has edited a book entitled, "Laser Monitoring of the Atmosphere" (Springer-Verlag, N.Y., 1977).

## 6.2  LASER SYSTEMS AS REMOTE SENSORS

### 6.2.1  Basic Laser Environmental Sensor System

Several names have been coined for laser-based environmental sensors; these include among others, laser radar, optical radar, lidar, laser fluorosensor, and laser bathymeter. In order to discuss the more general features of this new class of remote sensor, I shall introduce a new term, *lasor* (short for <u>la</u>ser <u>sensor</u>), to cover all such instruments. Almost all remote environmental "lasors" comprise the same functional elements and operate in a similar manner. This is schematically illustrated in Fig. 6.3. An intense pulse of optical energy emitted by a laser is directed through some appropriate output optics toward the target of interest. The function of the output optics can be threefold—improve the beam collimation, provide spatial filtering, and block the transmission of any broadband emission. Often a small fraction of this pulse is sampled to provide both a zero-time marker and a reference signal with which the return signal can be normalized in the event that the laser's output reproducibility is inadequate.

The radiation gathered by the receiver optics is passed through some

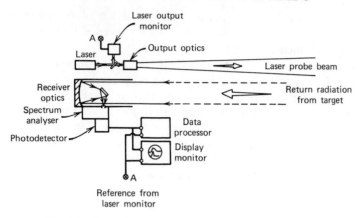

**Fig. 6.3.** Essential elements of laser environmental sensor.

form of spectrum analyzer on its way to the photodetection system. The spectrum analyzer serves to select the observation wavelength interval and thereby discriminate against background radiation at other wavelengths. It can take the form of a monochromator, a polychromator, or a set of narrowband spectral filters together with a laser wavelength-blocking filter (unless elastic scattered light is of interest). The choice of photodetector is often dictated by the spectral region of interest, which in turn is determined by the kind of application and the type of laser employed.

In principle, there are two basic lasor configurations. The *bistatic* arrangement involves a considerable separation of the transmitter and receiver to achieve spatial resolution in optical probing studies. Today, nanosecond lasers are capable of providing spatial resolution of a few feet, and so in most instances a *monostatic* configuration is employed. Under these circumstances the transmitter and receiver are at the same location so that, in effect, one has a single-ended system. A monostatic lasor can either be colinear or offset. In a colinear system the axis of the laser beam is coincident with the axis of the receiver optics, while in the offset arrangement the laser beam only enters the field of view of the receiver optics beyond some predetermined range. This configuration avoids the problem of near-field backscattered radiation saturating the photodetection but is optically not quite as efficient as the colinear approach. The near-field backscattering problem in the colinear case can be overcome by either gating of the photodetector or use of a fast shutter (80).

Newtonian and Cassegrainian reflecting telescopes form the mainstay of the receiver optics to date and are illustrated in Fig. 6.4. An offset

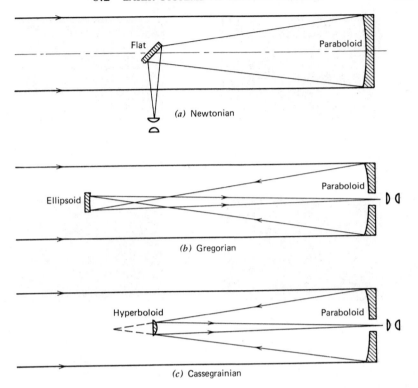

**Fig. 6.4.** Telescope configurations: (*a*) Newtonian; (*b*) Gregorian; (*c*) Cassegrainian. From M. Ross, *Laser Receivers* (*Devices, Techniques, and Systems*), Wiley, New York (1966). Reproduced by permission.

Newtonian system is portrayed in Fig. 6.3. The combined virtues of compact design and long focal length provide the Cassegrainian system with growing popularity. Telescopes based on large plastic Fresnel lenses may offer some advantages with regard to cost, weight, and size (81) and thereby be of particular interest in the development of operational airborne lasors. The size of the receiver's aperture depends to a large extent on the technique and range involved. Observations based on Raman scattering appear to be the most demanding, with 30- or 40-in. collecting telescopes not being uncommon. Hirschfeld et al. (82) used a 36-in. telescope to detect $CO_2$, $H_2O$, and $SO_2$ at a range of 200 m.

The signal from the photodetector may be processed via analog or digital techniques. The early work invariably involved the former, A-scope, approach where the backscattered signal intensity was displayed as a function of elapsed time (proportional to range) on a wide bandwidth

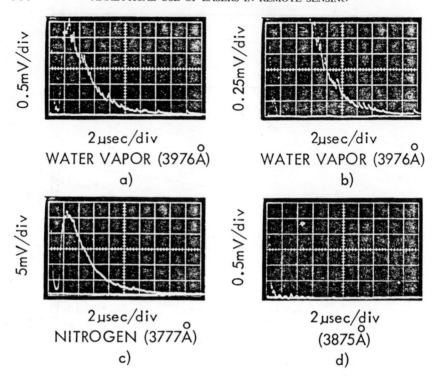

**Fig. 6.5.** (a) and (b) Typical Raman backscatter returns from the atmosphere vs. time after laser emission; oscillograms taken with monochromator set at 3976 Å (3654.5 cm$^{-1}$ band of water molecule). (c) Oscillogram taken with monochromator set at 3777 Å (2330.7 cm$^{-1}$ band of nitrogen molecule). (d) Oscillogram taken with monochromator set at 3875 Å (no Raman line is predicted from the atmosphere at this wavelength). From S. H. Melfi, J. D. Lawrence, Jr., and M. P. McCormick, *Appl. Phys. Lett.*, **15**, 295 (1969). Reproduced by permission.

oscilloscope and photographed. The set of photographs presented in Fig. 6.5 is taken from Melfi et al. (83) and represents the Raman returns from water vapor and nitrogen, respectively. From such photographs the data could be processed manually or the curves digitized and a computer employed. The development of very fast waveform digitizers (84) has made real-time data processing possible. Uthe and Allen (85) have provided a brief review of the data handling techniques employed in atmospheric probing and have detailed a real-time data recording system currently in operation.

The very brief spike at the far left of the photographs shown in Fig. 6.5 represents the zero-time marker alluded to earlier. The initial rise of the return signal results from the offset configuration, and the fall of the signal is principally a reflection of the inverse square diminution of the signal with range. This $1/R^2$ dependence leads in many applications to a signal amplitude dynamic range that extends over several decades. Wideband logarithmic amplifiers and gain switching techniques can be used to compress this range so that the signals are compatible with recording and display electronics (85–86). Alternatively, Hirschfeld (87) has suggested that a variable focus system could be used to significantly reduce the range dependence of the signal. When the return signal is very weak, as obtained from Raman experiments or studies of the upper atmosphere, integration of many pulses may be necessary, and under these circumstances photon counting (88) is usually employed. In this mode of operation the current pulses arising from the detection of single photons are recorded and counted. Such an approach lends itself naturally to digital data processing.

### 6.2.2 Detection Systems and Spectroscopic Instrumentation

The basic characteristics instrumental in determining the choice of a photodetector include spectral response, quantum efficiency, frequency response, current gain, and dark current. Sometimes, other considerations such as physical size, ruggedness, and cost may also be important. In most instances, the wavelength of the signal to be detected constitutes the primary factor in selecting the class of photodector to be employed in any application. For wavelengths that lie between 200 nm and 1 $\mu$m (ultraviolet to near infrared), photomultipliers are generally preferred because of their high gain and low noise specifications. Indeed, the single-photon detection capability of these devices has led to low light level detection schemes based on counting of individual photons (88).

In general, the performance of a photomultiplier is determined by (i) the spectral response of its photocathode, (ii) the dark current characteristics of its photocathode, (iii) the gain of the dynode chain, (iv) time dispersal effects of the electrons moving through the dynode chain, and (v) the transit time of the electrons between the last dynode and the anode. Table 6.2 provides the kind of data required for a selection of photomultipliers currently available. Since a photomultiplier represents a current source, the observed signal is limited at high frequencies to the voltage that can be generated across the cable impedance (i.e., $Z_0 <$ 1000 $\Omega$ usually). (Although cables with impedance $Z_0 > 100$ $\Omega$ are available, in general their high-frequency loss characteristics limit their usefulness.) Consequently, current gain constitutes an important consideration when high-frequency response is required. Photomultipliers with high

**TABLE 6.2**
**Representative Characteristics for a Selection of Photomultiplier Tubes[a]**

| Anode pulse rise time at maximum supply voltage, ns | Spectral response | Radiant sensitivity Anode, A W⁻¹ | Cathode, mA W⁻¹ | Amplifi- cation (approx.) ×10⁶ | Anode Dark Current, n A (approx.) | RCA Type no. |
|---|---|---|---|---|---|---|
| ≤1.5 | 110(S-20) | 6,900 | 64 | 0.11 | 3 | 8644 |
| | 101(S-1) | 235 | 1.9 | 0.125 | 300 | C31004A |
| | 140 | 1,700 | 43 | 0.04 | 1 | C31025K |
| | 142 | 1,700 | 30 | 0.06 | 1 | C31025N |
| ≤2 | 101(S-1) | 310 | 2.8 | 0.11 | 800 | C70102B |
| | 110(S-20) | 6,900 | 64 | 0.11 | 3 | 8645 |
| | 125 | 3,000 at 253.7 nm | 15 at 253.7 nm | 0.2 | 0.5 at 3000 A W⁻¹ | C70128 |
| ≤2.5 | 101(S-1) | 940 | 2.8 | 0.33 | 400 | C70007A |
| | 110(S-20) | 37,500 | 64 | 0.59 | 3 | 7326 |
| | 133 | 710,000 | 97 | 7.3 | 0.6 | C31000M |
| | 140 | 14,000 | 57 | 0.25 | 10 | C31034B |
| | 142 | 17,000 | 42 | 0.4 | 10 | C31034D |

[a] From RCA catalogue.

gain reach almost ideal quantum noise-limited sensitivity for the detection of weak light signals. The large range of photocathode materials currently available offer a wide choice of spectral response characteristics, as illustrated by the selection of curves presented in Fig. 6.6. Melchior (89) and Ross (90) have prepared comprehensive reviews on demodulation and photodetection techniques, while Poultney (88) has provided a detailed discussion of fast-response photomultipliers of interest in photon counting.

Channel multipliers (91) are tiny, hollow glass tubes with a secondary emitter coating deposited on the inner walls. When a voltage is applied to such a tube and an electron is injected (from a photocathode), current gains of $10^7$ can be achieved. The small size, fast response, high gain, and low noise characteristics of these devices make them suitable for single-photon counting and offer exciting new possibilities with regard to photo-detection arrays (92). Wafers comprising tens of thousands of such microchannels are commercially available.

Infrared detectors can, broadly speaking, be divided into two classes—photodetectors and thermal detectors. The most sensitive infrared detectors are semiconductors in which the incident radiation creates charge carriers via a quantum interaction. These photodetectors may be further divided into photovoltaic and photoconductive devices. Of these, photodiodes (photovoltaic devices) are the more popular for environmental sensing. Although some photodiodes can be used in the visible, they come into their own at longer wavelengths, where their high quantum efficiency (30–80%) becomes important. Unfortunately, the output of a photodiode must be externally amplified so that its sensitivity is often limited by thermal noise. Some photodiodes, when operated at a high reverse bias, develop internal gain through a process of carrier multiplication. These avalanche photodiodes are similar to photomultipliers in the sense that their sensitivity is no longer determined by thermal noise of the detector and output circuit (93). One of the most widely used and most sensitive infrared detectors for the 1- to 5.5-$\mu$m range appears to be the liquid nitrogen-cooled InSb photodiode (17,56). A recent overview of infrared detectors has been prepared by Emmons, Hawkins, and Cuff (94).

The spectrum analyzer is used to select the wavelength interval of observation and to provide an adequate rejection of all off-frequency radiation, whether this be laser-scattered radiation, solar background radiation, or any other form of radiation having a wavelength different from that of the signal. In general, this is accomplished with the aid of one or more spectral components. The components fall into one of three basic categories: absorption filters, interferometric elements, and dispersive systems. Absorption filters can take the form of colored glass, gelatine, or a liquid solution (95) and are employed to attenuate the incident intensity, separate interfering spectral orders, or block all wavelengths that are shorter, or longer, than those of interest. Long-wavelength pass filters (sometimes called short-wavelength blocking filters) are available from 250 nm to 1 $\mu$m (95,96). However, short-wavelength pass filters are less abundant.

An important component of many environmental lasors is a vapor-deposited dielectric interference filter. These filters, made of alternating layers of high and low refractive index, are useful over most of the UV to intermediate IR region of the spectrum. Their transmission profiles are similar to those of a low-order Fabry-Perot interferometer, being dispersive in nature and having a bandwidth that can be close to 1 nm. In almost all instances interference filters are designed for use with collimated radiation incident normal to the surface of the filter. If the filter is tilted by 45°, the center passband can shift to shorter wavelengths by as much as 2–3%. These filters can also be temperature sensitive. The side band transparency of such a filter is close to $10^{-3}$ times the center band transparency, giving an off-frequency stray light rejection ratio of $10^{-3}$. In many instances a stack of two or more such filters is used to improve this factor and narrow the spectral width of the passband. However, this is always achieved at the expense of the passband transmission coefficient and an enhanced sensitivity to tipping. The best, but most expensive, narrowband filters are made of birefringent materials and are called Lyot filters (97).

For those applications that require high spectral resolution, the choice often lies between a Fabry-Perot interferometer and a grating mono-chromator (98). Of these, the Fabry-Perot etalon is usually the cheaper, can provide the higher resolving power, and has the greater light through-put. Indeed, it would be ideal for many applications if it were not for its major drawback—many overlapping orders. This difficulty can be over-come by prefiltering with an interference filter, a second, wider-passband Fabry-Perot interferometer, or a dispersive element. Each approach has its own limitation. A detailed comparison of prism, grating, and Fabry-Perot etalon spectrometers has been given by Jacquinot (99). The pass-band of a Fabry-Perot etalon can be scanned by varying the pressure of gas (100) between the interferometer plates or by displacing one plate relative to the other (101).

In those situations where measurement of a spectral profile is important or where many wavelengths are of interest, a grating monochromator offers some advantages. Of the wide array of monochromator systems available, the Czerny-Turner arrangement appears to be one of the most popular, being capable of providing a stray light rejection ratio of typically $10^{-6}$. For those applications (such as those involving Raman backscattered signals) where this feature is of critical importance, double monochromators are often employed. Stray light rejection ratios of $10^{-6}$ (single) and $10^{-12}$ (double) can be achieved for displacements of about $60 \, cm^{-1}$ from the exciting line (102). The recent development of holo-graphic gratings has led to the manufacture of compact, simplified, and cheaper instruments possessing even higher stray light rejection capability (103). For applications where a fixed number of well-defined spectral features are to be monitored, the monochromator may be replaced with a polychromator that provides a series of preselected wavelength intervals. The poor optical coupling between the entrance slit of any dispersive system and the receiver optics constitutes one of the major weaknesses of this class of spectral analyzer. A significant improvement in the through-put has been achieved by employing shaped fiber-optic bundles (104) or an image slicer (105). Some degree of challenge to monochromators and spectrometers has come from wedge filters for which the passband wavelength linearly varies with position along the length, or around the circumference of a disc. The compact nature, mechanical stability, in line optics and high throughput of such devices are particularly relevant features for certain classes of remote environmental lasors.

### 6.2.3 Types and Attributes of Lasers Relevant to Remote Sensing

Certain types of lasers are capable of emitting pulses of optical energy that possess very high peak power, narrow bandwidth, and short duration and propagate with a low degree of divergence. Lasers of this nature are close to ideal for probing the environment, but must also be capable of operating at a high repetition rate for most airborne missions and for those atmospheric applications in which the return signal is very weak.

The range of lasers available for environmental sensing can be gauged from the selection of commercial pulsed lasers presented in Table 6.3. A much wider spectrum of possibilities exists if use is made of second harmonic generation and parametric conversion. In the past, fixed-frequency solid-state lasers offered the highest values of peak power and could be thought of as the workhorses of atmospheric lasors. Such lasers were quite adequate for experiments involving Rayleigh, Mie, and Raman scattering (16,106). However, the development of new lasers operating over a broader range of frequencies encouraged exploration of a variety of additional approaches, including resonance excitation, differential absorption, and fluorescence (see Table 6.1).

Of particular relevance to the airborne aspect of environmental sensing was the development of the high-power, high repetition-rate nitrogen laser operating at 337.1 nm (107,108). This device is really nothing more than a gas channel pumped by an ultrafast transverse discharge to become superradiant. As such, it employs virtually no collimating optics and consequently suffers considerable output divergence, as indicated in Table 6.3. The output character of this device can be improved, at the expense of some power, by introducing a degree of mode control (109). The short wavelength of the nitrogen laser makes it suitable for exciting fluorescence in a variety of materials, the high repetition rate makes airborne surveillance with good spatial resolution possible, the divergence and wavelength are favorable from an eye safety standpoint, and the short duration (110) attained with the Blumlein system (111) make it ideal for measurement of fluorescent decay times (13,15).

The development of tunable organic dye lasers (112) has provided the means to excite specific atomic and molecular electronic transitions and thereby exploit resonance scattering and differential absorption techniques for the purpose of remote sensing. As indicated in Table 6.3, tunable organic dyes are available commercially, covering the wavelength range from the near ultraviolet to the near infrared. A population

**TABLE 6.3**

**Selection of Pulsed Lasers Commercially Available and Suitable for Laser Environmental Sensing [Source-Laser Focus 1975 Buyers' Guide]**

| Wavelength, μm | Output, J TEM$_{00}$ | Multimode | Pulse-width, μs | Pulse rate, pps | Beam size, mm×mrad | Manufacturer |
|---|---|---|---|---|---|---|
| *Gas Lasers Pulsed* | | | | | | |
| Argon/krypton | | | | | | |
| 0.351 to 0.647 | $10^{-6}$ | | 0.015 to 1 | 1 to $5\times10^6$ | 1.5×0.5 | Spectra-Physics |
| Carbon dioxide | | | | | | |
| 9.1 to 11 | 3 | 15 | 0.1 | 1 to 5 | 30×0.6 | Lumonics Research |
| 9.4 to 10.8 | 0.25 | 1 | 0.1 | 1 | 25×4 | Israel Electro-Optical |
| Carbon monoxide | | | | | | |
| 5 to 6 | 0.008 | 0.04 | 1 to 4 | 1 to 5 | 30×0.6 | Lumonics Research |
| 5.3 | 0.0007 | 0.003 | 3 | 50 | 10×25×2×6 | Molectron |
| Copper vapor | | | | | | |
| 0.5106, 0.5782 | $4\times10^{-4}$ | 0.001 | 0.02 | 1,000 | 35×1 | Laser Consultants |
| Deuterium fluoride | | | | | | |
| 3.5 to 4 | 0.2 | 0.7 | 0.5 | 1 to 2 | 30×0.5 | Lumonics Research |
| Hydrogen Fluoride | | | | | | |
| 2.8 to 5 | 0.005 | 1 | 3 | 25×4 | | Israel Electro-Optical |
| 2.8 to 3 | 0.3 | 1 | 0.5 | 1 to 2 | 30×0.3 | Lumonics Research |
| Neon | | | | | | |
| 0.5401 | | $3\times10^{-5}$ | 0.003 | 2 to 100 | 3.2×50 | Avco Everett |
| Nitrogen | | | | | | |
| 0.3371 | | 0.01 | 0.01 | 50 | 6×25×1×7 | Molectron |
| 0.3371 | | 0.003 | 0.003 | 100 | 7×15×4×10 | Lambda Physik |
| 0.3371 | | $10^{-4}$ to $10^{-3}$ | 0.01 | 6 to 100 | 3×1 | Laser Energy |
| 0.3371 | 0.001 | 0.01 | | 10 to 500 | 3.2×50 | Avco Everett |
| Nitrous oxide | | | | | | |
| 10.5 to 11 | 0.3 | 1 | 0.1 | 1 to 5 | 30×0.6 | Lumonics Research |

## Solid-State Lasers Pulsed

| Wave length, μm | Peak power, W | Diode or array | Rep. rate, Hz | Pulse width, ms | Line width, μm | Beam divergence, mrad | Manufacturer |
|---|---|---|---|---|---|---|---|
| **Neodymium glass** | | | | | | | |
| 0.53 | 2 | | | 1/30 | 0.03 | 30E3 | Raytheon, Microwave |
| 1.06 | 1 | | | 4/60 | 0.015 | 9E3 to 5 | Holobeam Laser |
| 1.06 | 4 | | | 1/30 | 0.02 | 15E3 | Apollo Lasers |
| 1.06 | 100 | | | 1/240 | 0.005 | 15E2 | Apollo Lasers |
| 1.06, 0.53 | 0.3 | | | 0.03 | $10^{-5}$ | 10E2 | Block Engineering |
| **Neodymium YAG** | | | | | | | |
| 0.265 | 0.02 | | | 20 | 0.015 | 6E3 to 5 | Holobeam Laser |
| 1.06 | 1.2 to 1.5 | | | up to 30 | 0.015 | 45E0.8 | Intl Laser Systems |
| **Ruby** | | | | | | | |
| 0.6943 | 2 | | | 1/15 | 0.02 | 10E3 to 5 | Apollo Lasers |
| 0.6943 | 2 | | | 6/60 | 0.015 | 9E3 to 5 | Holobeam Laser |
| 0.6943 | 4 | | | 1/60 | 0.003 to 0.005 | 18E1.2 | Korad |
| 0.6943 | 16 | | | 1/30 | 0.02 | 20E2 | Raytheon, Microwave |
| 0.6943 | 3 to 30 | | | 1 | 0.02 | 15E2 to 5 | Apollo Lasers |
| | 0.05 to 1 | | | | | | |

## Semiconductor Lasers

| Wave length, μm | Peak power, W | Diode or array | Rep. rate, Hz | Pulse width, ms | Line width, μm | Beam divergence, mrad | Manufacturer |
|---|---|---|---|---|---|---|---|
| 0.85 | 3 | diode | 1,000 | $10^{-4}$ | | | RCA Elec Components |
| 0.85 | 5 to 75 | array | 2,000 | 0.0001 | 0.0035 | 125 | Laser Diode Labs |
| 0.85 | 250 | array | 1,000 | $10^{-4}$ | | | RCA Elec Components |
| 0.85 | 1,000 | array | 1,000 | 0.0001 | 0.0035 | 125 | Laser Diode Labs |
| 0.9 | 1 | diode | 1,000,000 | $10^{-5}$ | 0.002 | 125 | Power Technology |
| 0.904 | 60 to 360 | array | 1,000 | 0.0002 | 0.0035 | 125 | Laser Diode Labs |
| 0.905 | 25 | diode | 12,000 | $5 \times 10^{-5}$ | 0.002 | 125 | Power Technology |
| 0.905 | 150 | | $5 \times 10^5$ | $10^{-4}$ | | | Meret |

**TABLE 6.3(Continued)**

*Tunable Lasers*

| Tunable spectral range, Å | Power, W | Rep. rate pps | Pumping method | Pulse ns | Beam size, mm☰mrad | Manufacturer |
|---|---|---|---|---|---|---|
| 2,300 to 9,600 | $0.5 \times 10^6$ | 10 | Coax flash lamp | 250 | 5☰3 | Phase-R |
| 2,500 to 7,800 | 150,000 | 100 | $N_2$ laser | 1 to 3 | 1☰4 | Lambda Physik |
| 2,600 to 7,500 | $2 \times 10^5$ | 1,000 | pulsed $N_2$ laser | 5 to 7 | 0.5☰2 | Molectron |
| 3,600 to 6,700 | 10 to 1,000 | 10 to 1,000 | $N_2$ laser | 2 to 8 | 1☰4 | Avco Everett |
| 3,700 to 7,000 | 10,000 | 6 to 100 | pulsed $N_2$ laser | 6 to 8 | 1☰1 | Laser Energy |
| 4,200 to 8,500 | $6 \times 10^6$ | 5 | Coax flash lamp | 250 | 15☰3 | Phase-R |

314

inversion is created within the dye by optical pumping with either a flash lamp or another laser. For pulsed operation, a nitrogen laser is most often employed, while continuous wave (cw) operation is achieved by pumping with a tightly focused argon laser. The flash lamp-pumped dye lasers have, in general, provided the greater energy per pulse, but their duration is rather long (hundreds of nanoseconds) for measurements with reasonable spatial resolution. Nevertheless, such a system lends itself to the oscillator–amplifier mode of operation and is ideal for probing the outer regions of the atmosphere (7). Pumping with a nitrogen laser, on the other hand, produces a pulse of less energy but with a duration of only a few nanoseconds and a repetition rate of up to several hundred pulses per second. Spectral narrowing and tuning across the broad emission band of a dye is achieved by means of a dispersive element such as a prism or a grating. Hänsch (113) demonstrated that good spectral condensation, better than 10 pm or $10^{-1}$ Å, could be achieved with a diffraction grating and an intracavity beam-expanding telescope. Further spectral narrowing can be achieved with the inclusion of a Fabry-Perot etalon into the dye laser cavity.

A small degree of tunability can also be achieved with many high-pressure gas lasers and with some semiconductor lasers. A comprehensive review of high-pressure pulsed molecular lasers is provided by Wood (114), while Nill (115) and Hinkley (23) present useful reviews of tunable infrared lasers. From an environmental sensing standpoint, tunable infrared lasers have the advantage that most materials possess vibrational–rotational transitions that can be selectively excited by infrared radiation (63). Unfortunately, poor sensitivity of the corresponding detectors presently limits the remote sensing possibilities of this part of the spectrum. At short wavelengths (below 300 nm) there is again a rich selection of transitions for a variety of materials, however, the lack of a convenient laser limits the scope of remote sensing applications. Tunable laser radiation down to 230 nm is available (Table 6.3) with a frequency-doubled dye laser, but such a system is somewhat elaborate and limited to fairly low output power.

New developments in the area of rare gas halide lasers (116,117) could provide a new range of lasers that are capable of supplying high power below 337 nm. Such excimer lasers are inherently tunable, albeit over a small spectral interval. For example, 100-MW pulses have been generated at 248.4 nm from a krypton fluoride laser, and it is believed that the output might be tunable across 4 nm (116). Although such developments offer exciting new possibilities with regard to remote sensing, due to the absence of solar background at these wavelengths and the possibility of achieving resonance Raman scattering (118), the extreme sensitivity of living material to this radiation (119) could prevent its realization, except in rather limited situations.

## 6.3  INTERACTION PROCESSES AND RELEVANT CROSS SECTIONS

### 6.3.1  Atmospheric Transmission Characteristics

As a laser beam propagates through the atmosphere it spreads and suffers attenuation as a result of elastic scattering, inelastic scattering, and absorption. For a well-defined beam of optical radiation[2] propagating in the $z$ direction we may write

$$I(z) = I(0) \exp\left\{ - \int_0^z \epsilon \, dz \right\} \tag{1}$$

where $I(z)$ represents the intensity of the beam after propagating a distance $z$ through a medium possessing a volume attenuation coefficient $\epsilon$, and $I(0)$ is the incident intensity. In general, we may write

$$\epsilon = \epsilon^R + \epsilon^M + \epsilon^{RAM} + \epsilon^A \tag{2}$$

where $\epsilon^R$ and $\epsilon^M$ represent, respectively, the Rayleigh and Mie elastic scattering contribution to the net volume attenuation coefficient, $\epsilon^{RAM}$ represents the inelastic Raman scattering contribution (which in most instances is negligible), and $\epsilon^A$ represents the effect of absorption. "Elastic scattering" represents a process in which the incident radiation is simply redirected, whereas "inelastic scattering" represents an interaction that involves both a redirection and a change in the wavelength of the radiation.

The latter term, $\epsilon^A$, can dominate in Eq. (2) if the wavelength of the radiation coincides with a strong absorption line or band of even a minor constituent of the atmosphere (46). For wavelengths less than about 180 nm, the atmosphere is totally opaque as a result of the Schumann-Runge bands of molecular oxygen $O_2$. The absorption due to $O_2$ decreases with increasing wavelength such that beyond 250 nm it is unimportant and likely to be exceeded by the presence of small quantities of ozone (120). In the infrared part of the spectrum many atmospheric constituents such as $H_2O$ and $CO_2$ contribute to the absorption (51,121), leaving only a few spectral windows through which optical probing is possible (Fig. 6.7). Between 300 nm and 1 $\mu$m there are few absorption

---

[2] If the intensity of the beam is sufficiently high, nonlinear effects can significantly modify this relation.

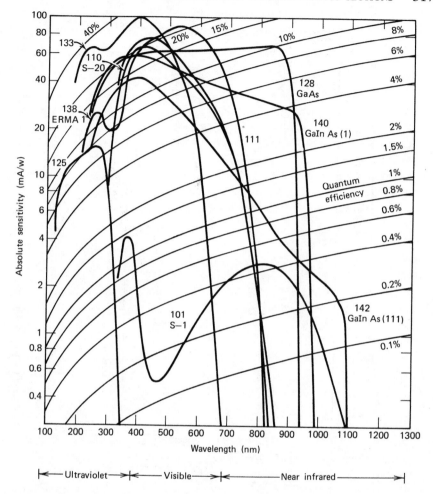

**Fig. 6.6.** Typical photocathode spectral response characteristics. From RCA photomultiplier brochure. Reproduced by permission.

bands, and under clear sky conditions it is the elastic scattering terms $\epsilon^R$ and $\epsilon^M$ that determine the attenuation characteristics of the atmosphere.

In the early days of environmental laser probing, the use of fixed-frequency lasers, operating in the red or infrared, restricted the range of activity to studying Rayleigh and Mie backscattering signals from the atmosphere. The subsequent explosive development in laser technology considerably expanded the repertoire of laser interactions that can be exploited for environmental sensing. At this point it will be helpful to review these various processes.

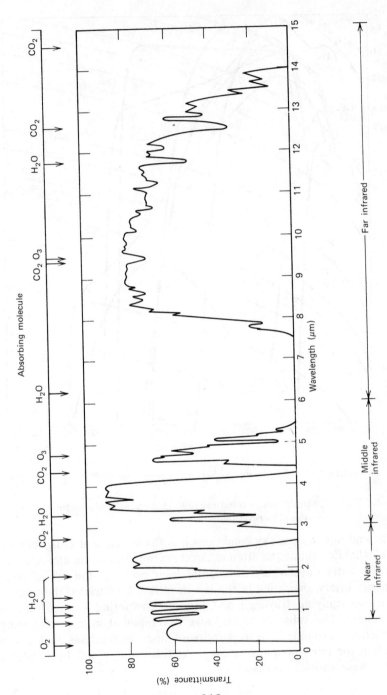

**Fig. 6.7.** Transmittance through the earth's atmosphere (horizontal path at sea level, length 1828 m). From R. D. Hudson, Jr., and J. W. Hudson. *Proc. IEEE.* **63.** 104 (1975). Reproduced by permission.

318

## 6.3.2  Rayleigh and Mie Scattering

In general, the scattering of light from molecular or particulate matter is not isotropic, so that reference is made to the differential cross section $\sigma(\theta,\varphi) = d\sigma/d\Omega$, where $\theta$ is the scattering angle and $\varphi$ is the polarization angle as indicated in Fig. 6.8. For isotropic scattering, $d\sigma/d\Omega = \sigma/4\pi$, where $\sigma$ represents the total cross section of the relevant scattering process. With "molecular scattering," the appropriate differential cross section for plane-polarized laser radiation is

$$\frac{d\sigma^R}{d\Omega} = \frac{\pi^2(\mu^2-1)^2}{N^2\lambda^4}\{\cos^2\theta\cos^2\varphi + \sin^2\varphi\} \tag{3}$$

where $\mu$ represents the molecule's refractive index relative to the background medium, $N$ is the molecular number density, and $\lambda$ is the wavelength of the incident radiation. If the laser beam is "unpolarized," the angular factor averaged over $\varphi$ reduces to $(\cos^2\theta + 1)/2$. The total Rayleigh scattering cross section, integrated over $4\pi$ steradians, is

$$\sigma^R(\lambda) = \frac{8\pi}{3}\left[\frac{\pi^2(\mu^2-1)^2}{N^2\lambda^4}\right] \tag{4}$$

and the backscattered cross section is

$$\sigma^R(\pi,\lambda) = \frac{\pi^2(\mu^2-1)^2}{N^2\lambda^4} = \frac{3}{8\pi}\sigma^R(\lambda) \tag{5}$$

A typical value for $d\sigma^R/d\Omega$ in the lower atmosphere, at around 700 nm, is $2\times10^{-26}$ cm$^2$ sr$^{-1}$. This is shown schematically in Fig. 6.1.

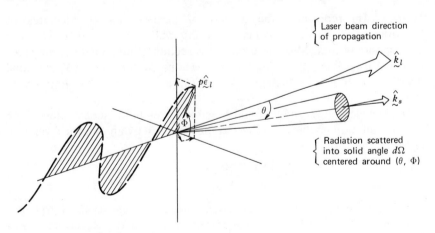

**Fig. 6.8.**  Angular distribution of laser-scattered radiation.

In general, the scattering from aerosols and particulates is characterized by a scale parameter $a \equiv 2\pi r/\lambda$, where $r$ is the appropriate radius. In the limit of very small dielectric spheres, such as Aitken nuclei, where $a < 0.5$, Rayleigh scattering adequately describes the interaction (122). Under these circumstances,

$$\frac{d\sigma^R}{d\Omega} = r^2 \left(\frac{2\pi r}{\lambda}\right)^4 \left(\frac{\mu^2 - 1}{\mu^2 + 2}\right)^2 \{\cos^2 \theta \cos^2 \varphi + \sin^2 \varphi\} \tag{6}$$

where $\mu$ is the effective refractive index of the material. When $a \gtrsim 1$, the cross section calculation becomes quite complicated, involving an infinite series as first shown by Mie (123), and the scattering pattern loses its symmetry. Mie scattering has been treated extensively in the literature (122,98), the results indicating a critical dependence on the shape and size of the particle. Even a sphere has a complex scattering pattern, with the strongest scattering occurring in the forward and backward directions. As seen from Fig. 6.1, the range of cross section attributed to Mie scattering spans several orders of magnitude depending on the value of $a$.

As can be seen from Eq. (4), Rayleigh scattering is inversely proportional to the fourth power of the wavelength, whereas Mie scattering is much less dependent on wavelength. Consequently, Rayleigh scattering tends to be more important at the short wavelength end of the spectrum, while Mie scattering dominates in the visible and near infrared. Thus, although the atmospheric elastic scattering attenuation coefficient $\epsilon^s$ comprises the sum of $\epsilon^R$ and $\epsilon^M$, for wavelengths beyond the ultraviolet we may approximate $\epsilon^s$ with $\epsilon^M$ where

$$\epsilon^M(\lambda) = \int_0^r N_M(r) \, \sigma^M(r,\lambda) \, dr \tag{7}$$

Here $N_M(r)$ represents the density of aerosols and particulates or "aeroticulates" having a radius in the interval $(r,dr)$, and $\sigma^M(r,\lambda)$ is the corresponding total Mie scattering cross section. Unfortunately, the complex nature of $\sigma^M$ and the local variations in the aeroticulate radial distribution function, $N_M(r)$, makes $\epsilon^M(\lambda)$ a highly variable entity with range.

Nevertheless, it has been found possible to relate the mean value of the Mie attenuation coefficient $\bar{\epsilon}^M(\lambda)$ to the visibility by an empirical relation (124) of the form

$$\bar{\epsilon}^M(\lambda) = \frac{3.91}{V} \left(\frac{550}{\lambda}\right)^{0.585 V^{1/3}} \quad (km^{-1}) \tag{8}$$

where $\lambda$ is in nm and $V$ is the visual range (defined as the horizontal range at which the transmission at 550 nm is 2%) in km. Recent measurements (125) have questioned the reliability of this expression in the case of laser radiation, and Schuster (126) has recently suggeste separation of the scattering from the aerosol and molecular com might be achieved by a method of two-angle multiwavelength inte tion.

### 6.3.3 Raman Scattering

In Raman (inelastic) scattering the cross section is about three orders of magnitude smaller than the corresponding Rayleigh cross section, and the scattered signal consists of radiation that has suffered a frequency shift that is characteristic of the stationary energy states of the irradiated molecule. Raman spectroscopy represents a particularly powerful tool with regard to remote sensing because it enables a trace constituent to be *both identified* and *quantified* relative to the major constituents of a mixture.

In the event that the molecule gains energy from the radiation field, the resulting lower-frequency scattered radiation is termed the *Stokes* component. If the molecule loses energy to the radiation field, the scattered radiation is referred to as the *anti-Stokes* component. These scattering processes are schematically illustrated in Fig. 6.9.

The structure of the Raman spectrum is often quite complex, even for diatomic molecules. In the special case of diatomic molecules possessing zero electron angular momentum around the internuclear axis ($\Lambda = 0$), the selection rules (127) allow vibrational–rotational transitions for which the change in the molecular rotational quantum number $J$ can be only 0 or $\pm 2$, and the change in the vibrational quantum number $v$ can be only 0 or $\pm 1$:

$$\Delta v = 0, \pm 1$$

$$\Delta J = 0, \pm 2 \qquad (9)$$

Under these circumstances the Raman spectrum consists of three branches: $S$ ($\Delta J = +2$), $Q$ ($\Delta J = 0$), and $O$ ($\Delta J = -2$) plus a pure rotational

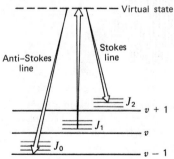

**Fig. 6.9.** Schematic representation of Raman vibrational ($\Delta v \pm 1$) Stokes and anti-Stokes scattering.

structure centered about the exciting wavelength. Inaba and Kobayasi (22) have computed the theoretical distribution of vibrational–rotational Raman lines corresponding to the $\Delta v = +1$ (Stokes shift) spectra of the $N_2$ molecule at 300 K. Their results are presented in Fig. 6.10. The ordinate gives the value of the differential scattering cross section for each of the Raman components corresponding to the vibrational transition $v = 0 \rightarrow 1$. As can be seen in Fig. 6.10, all lines in the $Q$-branch (for which $\Delta J = 0$) lie very close to each other and are not normally resolved. The $S$- and $O$-branches ($\Delta J = \pm 2$) are well separated and appear as side bands of the intense $\Delta J = 0$ line. It should be noted that although temperature variations do influence the intensity of the $S$- and $O$-branches, their effect on the $Q$-branch is often negligible.

Unfortunately, the $S$- and $O$-Raman side bands of a strongly scattering, high-concentration species may overlap and mask weaker $Q$-branch scattering from trace constituents in the atmosphere. This problem could

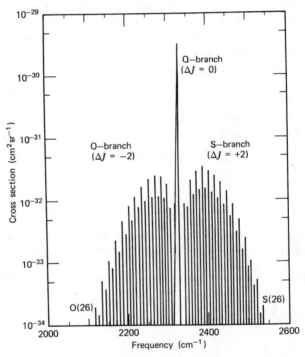

**Fig. 6.10.** Theoretical distribution of vibrational–rotational Raman spectrum ($v = 0 \rightarrow 1$ vibrational transition) at 300 K, showing the $O$-, $Q$- and $S$-branch structures, and differential Raman scattering cross section for $N_2$ molecules. From H. Inaba and T. Kobayasi, *Opto-Electron.*, **4**, 101 (1972). Reproduced by permission.

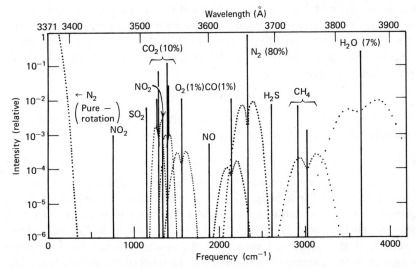

**Fig. 6.11.** Theoretical distribution of Raman volume backscattering coefficient due to a molecular mixture contained in a typical oil smoke as a function of Raman-shifted frequency. From H. Inaba and T. Kobayasi, *Opto-Electron.*, **4**, 101 (1972). Reproduced by permission.

arise in remote pollution monitoring. Inaba and Kobayasi (22) have illustrated this problem with the theoretical calculation of the spectral distribution of the Raman scattering to be expected from a typical oil smoke plume (Fig. 6.11). They assumed that the partial pressures of the plume constituents were $[N_2] = 0.80$ atm, $[CO_2] = 0.10$ atm, $[H_2O] = 0.07$ atm, $[O_2] = [CO] = 10^{-2}$ atm, $[NO] = [SO_2] = [H_2S] = [CH_4] = 10^{-3}$ atm, and $[NO_2] = 10^{-4}$ atm. The solid line indicates the $Q$-branch, and the dotted lines correspond to the $O$- and $S$-branches of the vibrational–rotational Raman band for each species excited by a $N_2$ laser operating at 337.1 nm.

An expression for the Raman differential cross section can be derived from a theoretical treatment of the polarizability of molecules exposed to a radiation field. According to the Placzek theory (128),

$$\frac{d\sigma_j}{d\Omega} = \frac{b_j^2(\omega_l - \omega_j)^4 g_j}{c^4[1 - \exp(-\hbar\omega_j/kT)]} \left\{ \overset{\circ}{\alpha}_j^2 + \frac{7}{45}\overset{\circ}{\gamma}_j^2 \right\} \tag{10}$$

when the Raman scattering is observed in a direction perpendicular to the polarization direction of linearly polarized incident light. Here, $\omega_l$ and $\omega_j$ represent the respective angular frequencies of the incident laser radiation and that of the $j$th vibrational mode of the molecule, $b_j [=(\hbar/2\omega_j)^{1/2}]$ represents the zero-point vibrational amplitude of this mode, $g_j$ its degree of degeneracy, and $3\mathring{\alpha}_j$ and $\mathring{\gamma}_j^2$ correspond to the trace and anisotropy of the "derived polarizability tensor" associated with the normal coordinate, $\hbar = h/2\pi$, where $h$ is the Planck constant, $k$ is the Boltzmann constant, $T$ is the vibrational temperature of the molecules, and $c$ is the velocity of light in vacuum. The factor

$$a_j = g_j(45\mathring{\alpha}_j^2 + 7\mathring{\gamma}_j^2) \qquad (11)$$

has been referred to as the "scattering activity" by Murphy, Holzer and Bernstein who relate it to the "depolarization ratio":

$$\rho_{nj} = 6\mathring{\gamma}_j^2 g_j/a_j \qquad (12)$$

Experimental measurements of $\rho_{nj}$ and $a_j$ have enabled both $\mathring{\alpha}_j^2$ and $\mathring{\gamma}_j^2$ to be determined for a large variety of molecules. An extensive list of "prelaser" vibrational Raman cross section data has been prepared by Murphy, Holzer, and Bernstein (129). Inaba and Kobayasi (22), have summarized the measured values of differential Raman backscattering cross section for a range of molecular constituents of the atmosphere. The recent interest in $O_3$ has prompted Schwiesow and Abshire (130) to determine its Raman cross section, while Stephenson (48) has evaluated the Raman cross section for a large number of hydrocarbon and freon gases. In all instances *these cross sections are experimentally evaluated relative to the* $2331 \text{ cm}^{-1}$ *vibrational Q-branch Raman cross section of nitrogen.* The absolute value of this cross section has been measured by several different methods (129,131,132) so that we can write, with some degree of confidence,

$$\frac{d\sigma}{d\Omega} \cong (4.3 \pm 0.2) \times 10^{-31} \qquad (\text{cm}^2 \text{ sr}^{-1}) \qquad (13)$$

for the $0 \rightarrow 1$ vibrational Q-branch Raman transition in $N_2$ gas when excited at 514.5 nm. This corresponds to the $2.8 \times 10^{-30} \text{ cm}^2 \text{ sr}^{-1}$ value quoted by Inaba and Kobayasi (22) for 337.1 nm excitation when allowance is made for the $\omega_s^4$ dependence of $d\sigma/d\Omega$.

In Table 6.4 we have prepared an updated list of Raman cross sections, corresponding wavenumber shifts, and wavelengths appropriate to excitation by the nitrogen laser operating at 337.1 nm. Where the sign $Q$ appears in the table, the value of the Q-branch vibrational Raman backscattering cross section is given instead of the total cross section.

Placzek (128) has also used a quantum treatment of the polarizability theory to derive the differential scattering cross section for the rotational Raman effect in diatomic molecules for the case of plane-polarized incident light. An example of a pure rotational Raman spectra is presented as Fig. 6.12. Several advantages could accrue from use of the pure

TABLE 6.4
**List of Raman Wavenumber Shift and Measured Differential Raman Backscattering Cross Section Appropriate for 337.1 nm Excitation**[a]

| Molecule | Raman shift, cm⁻¹ | Raman shifted wave-length, nm | Raman differential cross section $(cm^2\ sr^{-1}) \times 10^{30}$ | Cross section relative to Q-branch of $N_2$ | Reference no. |
|---|---|---|---|---|---|
| Freon 114 (1,2-dichloro-tetrafluoro ethane | 442 | 342.2 | 4.2(P) | 1.49(P) | 48 |
| $CCl_4$ | 459 | 342.4 | 26.0 | 9.3 | 129 |
| Freon C-318 (octafluoro-cyclobutane) | 699 | 345.2 | 7.8(P) | 2.77(P) | 48 |
| $NO_2\ (\nu_2)$ | 754 | 345.7 | 24.0 | 8.6 | 22 |
| $SF_6$ | 775 | 346.1 | 12.0 | 4.3 | 129 |
| Freon 116 (hexafluoro-ethane) | 807 | 346.5 | 7.3(P) | 2.6(P) | 48 |
| Freon 114 (tetrafluoro-methane) | 908 | 347.7 | 5.3(P) | 1.9(P) | 48 |
| $C_6H_6\ (\nu_2)$ | 991 | 348.7 | 44.0 | 15.7 | 22, 129 |
| $O_3$ | 1103.3 | 350.2 | 6.4 | 2.3 | 130 |
| $SO_2$ | 1151.5 | 350.8 | 17.0 | 6.1 | 22, 49 |
| $CO_2\ (2\nu_2)$ | 1285 | 352.5 | 3.1 | 1.1 | 22, 129 |
| $NO_2\ (\nu_1)$ | 1320 | 352.8 | 51.0 | 18.2 | 22 |
| $CO_2\ (\nu_1)$ | 1388 | 353.7 | 4.2 | 1.5 | 22, 129 |
| $O_2$ | 1556 | 355.9 | 4.6 | 1.6 | 22, 129 |
|  |  |  | 3.3(Q) | 1.2(Q) | 22, 129 |
| $C_2H_4\ (\nu_2)$ | 1623 | 356.6 | 5.4(Q) | 1.9(Q) | 129 |
| NO | 1877 | 360.0 | 1.5 | 0.54 | 22, 49 |
| CO | 2145 | 363.5 | 3.6 | 1.3 | 22, 129 |
| $N_2$ | 2330.7 | 365.9 | 3.5 | 1.3 | 129, 132 |
|  |  |  | 2.8(Q) | 1.0(Q) | 129, 132 |
| $H_2S$ | 2611 | 369.7 | 19.0 | 6.8 | 49, 129 |
| $CH_3OH\ (\nu_2)$ | 2846 | 372.8 | 14.0 | 5.0 | 22, 129 |
| $C_5H_{12}$ | 2885 | 373.4 | 124.0(C) | 44.3(C) | 48 |
| $C_3H_8$ | 2886 | 373.4 | 81.8(C) | 29.2(C) | 48 |
| $C_6H_{14}$ | 2886 | 373.4 | 134.0(C) | 48.0(C) | 48 |
| $C_4H_{10}$ | 2890 | 373.5 | 93.5(C) | 33.4(C) | 48 |
| $CH_4$ | 2914 | 373.8 | 32.2(C) | 11.5(C) | 48 |
| $-(\nu_1)$ |  |  | 21.0 | 7.5 | 22, 129, 132 |
| $C_5H_{10}$ | 2941 | 374.2 | 102.5(C) | 36.6(C) | 48 |
| $C_3H_6$ | 2942 | 374.2 | 63.6 | 22.7 | 48 |

## TABLE 6.4(*Continued*)

| Molecule | Raman shift, cm$^{-1}$ | Raman shifted wave-length, nm | Raman differential cross section, (cm$^2$ sr$^{-1}$)$\times 10^{30}$ | Cross section relative to Q-branch of $N_2$ | Reference no. |
|---|---|---|---|---|---|
| $C_2H_5OH$ | 2943 | 374.2 | 19.0 | 6.8 | 22, 129 |
| $CH_3OH$ ($2\nu_6$) | 2955 | 374.4 | 7.5 | 2.7 | 22, 129 |
| $C_4H_8$ | 3010 | 375.2 | 89.6(C) | 32.0 | 48 |
| $CH_4$ ($\nu_3$) | 3017 | 375.3 | 14.0 | 5.0 | 22, 129 |
| $C_2H_4$ | 3020 | 375.3 | 28.6 | 10.2 | 48 |
| $-(\nu_1)$ | | | 16(Q) | 5.7(Q) | 129 |
| $C_8H_{10}$ | 3064 | 375.9 | 87.9(C) | 31.4(C) | 48 |
| $C_6H_6$ ($\nu_1$) | 3070 | 376.0 | 30.0 | 10.7 | 22, 129 |
| $C_6H_6$ | 3072 | 376.0 | 65.2 | 23.3 | 48 |
| $NH_3$ | 3334 | 379.8 | 11.0 | 3.9 | 129 |
| $C_2H_2$ | 3372 | 380.3 | 3.36 | 1.2 | 48 |
| $H_2O$ | 3651.7 | 384.4 | 7.8(Q) | 2.8(Q) | 22, 132 |
| $H_2$ | 4160.2 | 392.2 | 8.7 | 3.1 | 22, 129 |

[a] $Q$ indicates the value of the $Q$-branch vibrational Raman backscattering cross section; $C$ indicates a broad multipeaked structure associated with the C–H stretch mode; $P$ indicates a cross section based on a ratio of peak intensites rather than spectrally integrated signals.

rotational Raman spectra (133,134). In the first place, the rotational scattering cross section of a molecule is usually larger (sometimes by two orders of magnitude) than its vibrational–rotational Raman cross section (47). Second, through the use of a Fabry-Perot interferometer as a comb filter, it has been possible to utilize all of the rotational Raman lines as a single signal (133). Third, the higher luminosity of a Fabry-Perot interferometer is greater than that of a grating spectrometer for the same resolution (99). The overall gain in signal could be three orders of magnitude, and such a technique has great potential for monitoring gas properties remotely.

The difficulty with using rotational Raman scattering for pollution detection lies in the overlapping of the rotational transitions of the major atmospheric constituents and the large Rayleigh-Mie backscattered signal. Both Barrett (133) and Smith (134) have indicated that these problems could be overcome with careful design of the Fabry-Perot interferometer.

### 6.3.4   Theoretical Foundations

For an idealized molecular system of nondegenerate levels, the Raman differential scattering cross section, corresponding to the Stokes transition

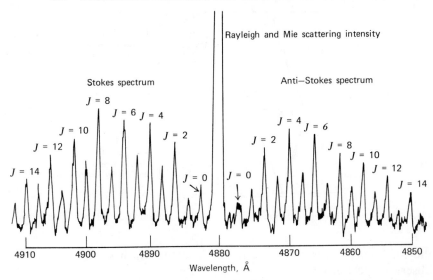

**Fig. 6.12.** Experimentally measured pure rotational Raman spectrum of nitrogen. Laser excitation wavelength 488.0 nm. From J. A. Salzman, "Low Temperature Measurements by Rotational Raman Scattering," in *Laser Raman Gas Diagnostics*, M. Lapp and C. M. Penney, Eds., Plenum Press, New York (1975), pp. 179–188. Reproduced by permission.

from an initial quantum state $|0\rangle$ to a final quantum state $|f\rangle$ via an intermediate state $|i\rangle$, can be expressed in the form

$$\left[\frac{d\sigma}{d\Omega}\right]_{l,s} = \frac{\omega_s^4}{\hbar^2 c^4} \sum_0 F_0 \sum_f \left| \sum_i \left( \frac{\langle f| \hat{e}_s \cdot d\, |i\rangle\langle i| \hat{e}_l \cdot d\, |0\rangle}{\omega_{i0} - \omega_l} + \frac{\langle f| \hat{e}_l \cdot d\, |i\rangle\langle i| \hat{e}_s \cdot d\, |0\rangle}{\omega_{if} + \omega_l} \right) \right|^2 \quad (14)$$

where $\omega_s (=\omega_l - \omega_{of})$ represents the Stokes angular frequency of the scattered radiation, $\omega_l$ represents the angular frequency of the incident laser radiation, $\omega_{of}$ represents the angular frequency corresponding to the 0 to f transition, $\omega_{0i}$ represents the angular frequency corresponding to the 0 to i virtual transition, $\omega_{if}$ represents the angular frequency corresponding to the i to f virtual transition, $F_0$ represents the fraction of molecules initially residing in the $|0\rangle$ state, $\hat{e}_l$ and $\hat{e}_s$ represent the electric field polarization unit vectors for the incident and scattered radiation, respectively, and $d$ represents the electric dipole moment operator. A schematic illustration of the interaction corresponding to one of each of the two kinds of terms is presented in Fig. 6.13. The sum over the intermediate states $|i\rangle$ is intended to include an integral over the continuum of positive energy states (dissociated and ionized). The sum over

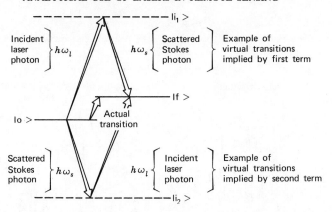

**Fig. 6.13.** Schematic representation of Raman (Stokes) scattering. Actual transition is from $|0\rangle$ to $|f\rangle$.

the final states $|f\rangle$ is restricted to those states which contribute to the observed spectral line.

It should be borne in mind that Eq. (14) is derived from quantum perturbation theory and that the summation arises as a mathematical consequence of expanding the perturbed wave function in terms of an infinite series of unperturbed wave functions. For this reason the transitions implied to arise *via* the intermediate states are termed "virtual" transitions (135). Although Eq. (14) has been used for $H_2$, its general application is made difficult by a lack of knowledge concerning the magnitude and sign of all the possible terms that could contribute and by a possible interference of terms that might have different phases (135,136).

### 6.3.5 Resonance Raman Scattering

When the exciting frequency approaches an isolated absorption line, one transition (from a particular initial level g to an intermediate level r) can dominate the sum in Eq. (14). Under these circumstances the gr transition corresponds to an allowed transition, so that the transition moments become large and allowance has to be made for the finite lifetime $\Gamma^{-1}$ of the r state. In which case we can write

$$\left[\frac{d\sigma}{d\Omega}\right]_{l,s} = \frac{\omega_s^4 F_g}{\hbar^2 c^4} \frac{|\langle f|\, \hat{e}_s \cdot d\, |r\rangle|^2\, |\langle r|\, \hat{e}_l \cdot d\, |g\rangle|^2}{[(\omega_0 - \omega_l)^2 + \Gamma^2]} \tag{15}$$

where $\omega_0$ is introduced as the center frequency for the resonance gr transition. We have assumed that the resonance is associated with one of the first kind of terms within the absolute square of Eq. (14), as a resonance for one of the second kind of terms could occur only for highly excited systems where the intermediate state lies below both the initial and final state (136).

The total (near resonance) scattering cross section integrated over all scattering angles and light polarizations can be expressed in the form

$$\sigma_{gf} = \frac{8\pi\omega_s^4 F_g}{9\hbar^2 c^4} \frac{|d_{fr}|^2 |d_{rg}|^2}{[(\omega_0 - \omega_l)^2 + \Gamma^2]} \tag{16}$$

where $d_{fr}$ and $d_{rg}$ represent the dipole matrix elements for the rf and gr transitions, respectively. The radiative decay rate $\Gamma_R$ of the rf transition can be related to the square of the appropriate transition moment:

$$\Gamma_R = \frac{4\omega_s^3}{3\hbar c^3} |d_{fr}|^2 \tag{17}$$

and the Milne absorption coefficient $B_{gr}$ of the gr transition can be related to the square of its transition moment:

$$B_{gr} = \frac{8\pi^2}{3\hbar^2 c} |d_{rg}|^2 \tag{18}$$

Consequently we can write

$$\sigma_{gf} = \frac{\hbar\omega_s B_{gr} F_g}{4\pi} \cdot \frac{\Gamma_R}{[(\omega_0 - \omega_l)^2 + \Gamma^2]} \tag{19}$$

or

$$\sigma_{gf} = \sigma_{gr}^A \phi_{rf}^F \left(\frac{\omega_s}{\omega_l}\right) \tag{20}$$

where the absorption coefficient corresponding to the gr transition is

$$\sigma_{gr}^A = \frac{\hbar\omega_l B_{gr}}{4\pi} \left\{ \frac{\Gamma}{(\omega_0 - \omega_l)^2 + \Gamma^2} \right\} F_g \tag{21}$$

and the quantum yield for fluorescence into the rf transition is

$$\phi_{rf}^F = \left(\frac{\Gamma - \Gamma_Q}{\Gamma}\right) \delta_{rf} \tag{22}$$

Here, $\Gamma_Q$ represents the collisional quenching rate coefficient for the r state and $\delta_{rf}$ is the radiative branching ratio for the rf transition. $\phi_{rf}^F$ essentially expresses the probability that an atom or molecule in the g state will undergo a radiative rf transition (viz., emit a photon of energy $\hbar\omega_s$) for each laser photon $\hbar\omega_l$ absorbed.

There is an interesting controversy associated with this analysis, for it appears that under resonance conditions there is no distinction between the scattering cross section $\sigma_{gf}$ and the fluorescence cross section $\sigma_{gf}^F$. The term fluorescence is normally used to refer to that process for which the decay time is in the nanosecond (or longer) range at low pressures and decreases due to collisional quenching at pressures greater than a few torr. On the other hand, Raman scattering is generally taken to refer to the two-photon process that is relatively instantaneous and suffers no quenching (i.e., its intensity per molecule is insensitive to gas composition or pressure, at least up to several atmospheres). Equation (16) appears to describe scattering at frequency separations that are large compared to the total linewidth, yet also can be seen to describe fluorescence close to an allowed transition. At high pressures the influence of homogeneous and inhomogeneous broadening and quenching confuse the issue. The pressure sensitivity and time-dependent nature of the cross sections have only been partially resolved by theory and experiment (128,137).

We see that, for atoms, Eq. (15) predicts large cross sections at near resonance for strong isolated atomic lines. In the case of aluminum atoms where the Raman shift is about 2 nm, an enhancement of $10^5$ has been predicted for a spectral misalignment of about 5 nm (138). In the case of vibrational Raman scattering within molecules, Eq. (16) can be rewritten in the form

$$\sigma_{gf} = \frac{8\pi\omega_s^4 F_g}{9\hbar^2 c^4} \cdot \frac{|d_{m,v_g''}^{n,v_r'}|^2 |d_{m,v_f''}^{n,v_r'}|^2}{[(\omega_0 - \omega_l)^2 + \Gamma^2]} \tag{23}$$

where $d_{m,v_g''}^{n,v_r'}$ and $d_{m,v_f''}^{n,v_r'}$ represent the dipole matrix elements for the m, $v_g'' \to$ n, $v_r'$ and n, $v_r' \to$ m, $v_f''$ transitions, n and m refer to the intermediate and lower electronic state, while $v_g''$, $v_r'$, and $v_f''$ represent the vibrational quantum numbers of the initial, intermediate, and final states, respectively. If we express these dipole matrix elements in terms of the mn electronic absorption oscillator strength $f_{mn}$ and the appropriate Franck-Condon factors $q(v_r', v_g'')$ and $q(v_r', v_f'')$, namely,

$$|d_{m,v_g''}^{n,v_r}|^2 = \tfrac{3}{2} r_0 \frac{\hbar c^2}{\omega_0} f_{mn} q(v_r, v_g'') \tag{24}$$

and

$$|d_{m,v_f''}^{n,v_r}|^2 = \tfrac{3}{2} r_0 \frac{\hbar c^2}{\omega_s} f_{mn} q(v_r, v_f'') \tag{25}$$

then we may write

$$\sigma_{gf} = 2\pi r_0^2 \left(\frac{\omega_s}{\omega_0}\right)^3 f_{mn}^2 q(v_r, v_g'') q(v_r, v_f'') \cdot \frac{\omega_0^2}{[(\omega_0 - \omega_l)^2 + \Gamma^2]} \tag{26}$$

where $r_0 (\equiv e^2/mc^2)$ represents the classical electron radius.

With molecules the variety of vibrational–rotational transitions possible ensure that the potential enhancement is much less. The reduction in the enhancement can be attributed to four factors (136):

1. The fraction of molecules capable of near resonance, $F_g$, may be much less than unity due to the rotational level spread of the vibrational ground state's population.

2. The average molecular transition moment can be several orders of magnitude smaller than the corresponding atomic transition moment due to the large number of nondegenerate transitions to the manifold of rotational–vibrational states associated with the excited electronic state.

3. There are a large number of alternative radiative decay modes associated with the intermediate level at resonance.

4. The closer the exciting frequency approaches the resonance condition, the greater is the possibility of collisions that can introduce quenching effects, as observed in fluorescence.

Nevertheless, Fouche and Chang (139) and St. Peters et al. (140,141) have reported strong enhancement when $I_2$ is excited within 0.01 nm of individual molecular lines. Although there is some disagreement (139,141) as to the degree of quenching observed in those experiments and the subsequent interpretation in terms of resonance Raman scattering or resonance fluorescence, in general, this emission appears to have a weaker quenching dependence upon background gas pressure than normally observed for fluorescence. This observation has been explained, however, by a near balance between the opposing effects of collisional broadening (which increases absorption in the wings of the line) and quenching (which reduces the probability of reemission following absorption) (137,140).

It is clear that the ambiguity inherent in quenching data tends to confuse the issue. Indeed, St. Peter and Silverstein (140) point out that quenching data can only be used to determine scattering time if elastic collisional broadening is unimportant. In order to avoid this difficulty, Williams, Rousseau, and Sworetsky (142) measured directly the scattering time associated with detuning a single-mode argon laser from an $I_2$ resonance line. They observed a marked shortening of this scattering time when the laser was detuned off-resonance by as much as 0.002 nm.

### 6.3.6 Fluorescence and Absorption

Although there may be some uncertainty as to the operative process (resonance Raman scattering or fluorescence) in cases of near resonance, there is little doubt that *fluorescence is the relevant process when the exciting wavelength approaches to within the linewidth of the absorption feature.* Under these circumstances the cross section per unit wavelength interval is given by an equation similar to Eq. (20) and can be expressed in the form

$$\sigma^F(\lambda,\lambda_l) = \sigma^F(\lambda_l)\mathscr{L}^F(\lambda) \tag{27}$$

where

$$\sigma^F(\lambda_l) = \lambda_l \sigma^A(\lambda_l)\phi^F/\lambda \tag{28}$$

and represents the spectrally integrated cross section for fluorescence when excited at wavelength $\lambda_l$. $\mathscr{L}^F(\lambda)\,\Delta\lambda$ represents the fraction of the total fluorescence emitted into the wavelength interval $(\lambda,\Delta\lambda)$ and $\sigma^A(\lambda_l)$ represents the absorption coefficient for the laser radiation. $\phi^F$ ($\equiv \Gamma_R/\Gamma$) is termed the quantum yield factor and represents the probability that a photon of wavelength $\lambda$ will be emitted for each laser photon absorbed. Since fluorescence is usually isotropic, the appropriate differential cross section per unit wavelength interval is given by

$$\frac{d\sigma^F\}(\lambda,\lambda_l)}{d\Omega} = \frac{\sigma^F(\lambda_l)\mathscr{L}^F(\lambda)}{4\pi}$$

In the case of resonance scattering, $\lambda = \lambda_l$ and

$$\frac{d\sigma(\lambda_l)}{d\Omega} = \frac{\sigma^A(\lambda_l)}{4\pi}\left(\frac{1}{1+\Gamma_Q/\Gamma_R}\right) \tag{29}$$

where $\Gamma_Q/\Gamma_R$ represents the ratio of the collision quenching rate to the radiative decay rate for the excited transition.

In general, $\phi^F \leq 1$ (indeed, for situations where the quenching is severe, $\phi^F \ll 1$) and $\lambda_l < \lambda$, so that invariably

$$\sigma^F(\lambda_l) < \sigma^A(\lambda_l)$$

It is evident that absorption represents by far the stronger form of interaction on the molecular level, and absolute measurements can be achieved with the aid of a reference cell containing a known concentration of the species of interest. Unfortunately, at long wavelengths (beyond 1 $\mu$m) the absorption band of a trace constituent can be overshadowed by the absorption band of a more abundant species, while at shorter wavelengths scattering can contribute to the attenuation of the laser beam. Although these difficulties can be overcome by operation at two or more wavelengths, such an approach lacks spatial resolution even though monostatic operation is made possible by use of retroreflectors or topological targets.

The sensitivity of selective absorption can be attained with spatial resolution if "aeroticulate" scattering is employed as a distributed reflector. In this instance it is the large Rayleigh-Mie backscattering cross section that determines the magnitude of the detected signal. Clearly this differential absorption and scattering (DAS) technique offers considerable promise, for it combines spectral selectivity with substantial sensitivity, spatial resolution being obtained by the usual time-of-flight technique. Fig. 6.1 provides an indication of the range of values to be expected for each of the cross sections discussed.

## 6.4 LASOR EQUATION AND SIGNAL-TO-NOISE CONSIDERATIONS

### 6.4.1 Lasor Equations Appropriate for Scattering and DAS

In the case of a pulsed system operating in a monostatic mode the signal power received by the photodetector and originating from the range interval $(R, \Delta R)$ can be expressed in the form

$$P(\lambda, R) = I(R)\xi(R)A_E(R)\, \Delta R \frac{A_r}{R^2}$$

$$\sum_i N_i(R)\left[\frac{d\sigma(\lambda, \lambda_l)}{d\Omega}\right]_i \int T_r(\lambda')T(\lambda', R)\mathscr{L}(\lambda')\, d\lambda'$$

where $I(R)$ ($\text{W cm}^{-2}$) represents the power density of the laser beam at range $R$ (cm); $N_i(R)$ ($\text{cm}^{-3}$) represents the i species number density at $R$; $[d\sigma(\lambda, \lambda_l)/d\Omega]_i$($\text{cm}^2\,\text{sr}^{-1}$) represents the spectrally integrated differential cross section of the i species for emission into a band centred about $\lambda$ on being irradiated at the laser wavelength $\lambda_l$; $\mathscr{L}(\lambda')\, d\lambda'$ represents the fraction of this emission that lies within the interval $(\lambda', d\lambda')$; $\xi(R)$ represents the geometric overlap factor that accounts for the fraction of the laser beam that lies within the field of view at $R$; $A_E(R)$ ($\text{cm}^2$) represents the area of irradiation at $R$; $\Delta R$ (cm) represents the effective range increment over which radiation returned from the target arrives simultaneously at the detector; $A_r$ ($\text{cm}^2$) represents the effective aperture of the receiver optics. $T(\lambda', R)$ represents the atmospheric transmission factor at wavelength $\lambda'$; and $T_r(\lambda')$ represents the receiver optics transmission efficiency at wavelength $\lambda'$. The basic geometric configuration for a lasor is illustrated in Fig. 6.14.

For those instances when the bandwidth of the receiver optics is much greater than that of the interaction process (almost always true for scattering or absorption events), the profile function $\mathscr{L}(\lambda')$ can be replaced by a delta function, namely, $\mathscr{L}(\lambda') \equiv \delta(\lambda - \lambda')$, in which case the

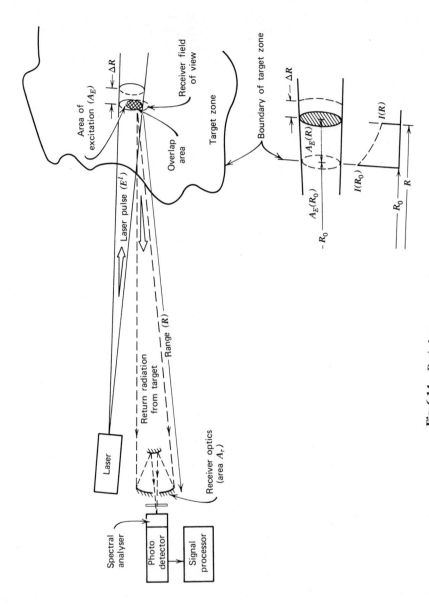

**Fig. 6.14.** Basic lasor geometric arrangement.

signal power received by the photodetector is

$$P(\lambda,R) = I(R)\xi(R)A_E(R)T_r(\lambda)T(\lambda,R) \Delta R \frac{A_r}{R^2} \sum_i N_i(R)$$
$$\times \left\{\frac{d\sigma(\lambda,\lambda_l)}{d\Omega}\right\}_i \quad \text{(W)} \quad (30)$$

If the laser's output can be represented by a pulse of energy $E_l$ and effective duration $\tau_l$, then we can write

$$I(R) = \frac{E_l T(\lambda_l,R)}{\tau_l A_E(R)} \quad \text{(W cm}^{-2}) \quad (31)$$

where, $T(\lambda_l,R)$ corresponds to the atmospheric transmission factor at the laser wavelength, namely,

$$T(\lambda_l,R) = \exp\left\{-\int_0^R \epsilon(\lambda_l,R) \, dR\right\}$$

and $(32)$

$$T(\lambda,R) = \exp\left\{-\int_0^R \epsilon(\lambda,R) \, dR\right\}$$

$\epsilon(\lambda_l,R)$ and $\epsilon(\lambda,R)$ represent the respective atmospheric attenuation coefficients per unit length at $\lambda_l$ and $\lambda$.

For instantaneous, narrowband phenomena, such as Rayleigh, Raman, and Mie scattering, the effective range element $\Delta R$ is $c\tau_l/2$ (where $c$ is the velocity of light), or half the laser pulse length, as can be seen by reference to Fig. 6.15. Making use of this fact and of Eq. (31), we may rewrite Eq. (30) in the form

$$P(\lambda,R) = \frac{c}{2} E_l T(\lambda_l,R) T(\lambda,R) T_r(\lambda)\xi(R) \frac{A_r}{R^2} \sum_i N_i(R) \left[\frac{d\sigma(\lambda,\lambda_l)}{d\Omega}\right]_i \quad (33)$$

However, since all detectors have a finite response time, it is more pertinent to evaluate the radiative energy $E(\lambda,R)$ received by the detector within the time interval $(t, \tau_d)$, where $\tau_d$ is the detector integration time and $t \ (\equiv 2R/c)$ is the time interval between firing the laser and observing the return signal from a range $R$, namely,

$$E(\lambda,R) = \int_t^{t+\tau_d} P(\lambda,R) \, dt \cong P(\lambda,R)\tau_d \quad \text{(J)}$$

or

$$E(\lambda,R) = E_l \frac{c\tau_d}{2} T(\lambda_l,R) T(\lambda,R) T_r(\lambda) \frac{A_r}{R^2} \xi(R) \sum_i N_i(R) \left\{\frac{d\sigma(\lambda,\lambda_l)}{d\Omega}\right\}_i$$

$$(34)$$

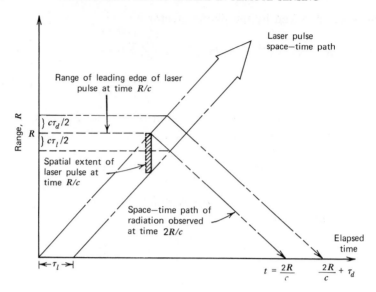

**Fig. 6.15.** Spatial resolution for scattering phenomenon as seen from space–time diagram of propagating rectangular shaped laser pulse.

The range resolution under these circumstances is given by $c(\tau_d + \tau_l)/2$, as can be seen by reference to Fig. 6.15.

In the event that we are interested in a semiinfinite target medium, extending from range $R_0$, possessing only one species i that interacts with the laser radiation, and having a total attenuation coefficient $\epsilon \equiv \epsilon(\lambda) + \epsilon(\lambda_l)$ that is independent of location, then, if the scattering process is isotropic,

$$E(\lambda,R) = E_l T_r(\lambda) T(R_0) \xi(R) \frac{c\tau_d}{2} \frac{A_r}{R^2} N_i \frac{\sigma_i(\lambda,\lambda_l)}{4\pi} \exp[-\epsilon(R - R_0)]$$

$$(35)$$

where $\sigma_i(\lambda,\lambda_l)$ represents the total scattering cross section,

$$T(\lambda,R)T(\lambda_l,R) = T(R_0) \exp[-\epsilon(R - R_0)]$$

and
$$(36)$$

$$T(R_0) = T(\lambda,R_0)T(\lambda_l,R_0)$$

In the case of Mie scattering, we replace $N_i\sigma_i(\lambda,\lambda_l)$ with the equivalent Mie volume backscattering coefficient $\beta^M(\lambda_l)$ so that we may write

$$E(\lambda_l,R) = E_l \frac{c\tau_d}{2} \frac{A_r}{R^2} T_r(\lambda_l)\xi(R) \frac{\beta^M(\lambda_l)}{4\pi} \exp\left\{-2\int_0^R \epsilon(\lambda_l)\, dR\right\} \quad (37)$$

since in this instance the return radiation is unchanged in wavelength. $\epsilon(\lambda_l)$ represents the total attenuation coefficient at $\lambda_l$ and, as we have seen earlier, comprises three main components:

$$\epsilon(\lambda_l) = \epsilon^R(\lambda_l) + \epsilon^M(\lambda_l) + \sum \epsilon_i^A(\lambda_l)$$

In the DAS approach, two laser wavelengths $\lambda_l$ and $\lambda_l + \delta\lambda$ are selected such that $\lambda_l$ corresponds to the center wavelength of some prominent absorption line, while $\lambda_l + \delta\lambda$ lies in the far wing of the line. Under these circumstances we may write

$$\epsilon(\lambda_l) = \bar{\epsilon}(\lambda_l) + \epsilon_i^A(\lambda_l)$$

$$\epsilon(\lambda_l + \delta\lambda) = \bar{\epsilon}(\lambda_l + \delta\lambda) + \epsilon_i^A(\lambda_l + \delta\lambda) \cong \bar{\epsilon}(\lambda_l) + \epsilon_i^A(\lambda_l + \delta\lambda) \qquad (38)$$

where $\bar{\epsilon}(\lambda_l)$ represents the total attenuation coefficient exclusive of the absorption contribution from the species of interest. We see that the value of $\lambda_l$ and the separation in wavelength between the two lasers, $\delta\lambda$, is chosen such that the only term that effectively depends upon $\delta\lambda$ is the absorption coefficient of the i species. We also assume that both the receiver transmission efficiency $T_r(\lambda)$ and the atmospheric volume back-scattering coefficient $\beta$ are virtually independent of $\lambda$ over the small interval $\delta\lambda$:

$$T_r(\lambda_l + \delta\lambda) \simeq T_r(\lambda_l)$$

and

$$\beta(\lambda_l + \delta\lambda) \simeq \beta(\lambda_l)$$

Under these circumstances we may write

$$\frac{E(\lambda_l + \delta\lambda, R)}{E(\lambda_l, R)} = \exp\left\{-2 \int_0^R [\epsilon_i^A(\lambda_l + \delta\lambda) - \epsilon_i^A(\lambda_l)] \, dR\right\} \qquad (39)$$

If we express this ratio of signals received at the two wavelengths as $J(R, \lambda_l, \delta\lambda)$, then we can arrive at an expression for the density of species i:

$$N_i(R) = \frac{1}{2\kappa_i(\lambda_l, \delta\lambda)} \frac{d[\ln J(R, \lambda_l, \delta\lambda)]}{dR} \qquad (40)$$

where the differential absorption cross section

$$\kappa_i(\lambda_l, \delta\lambda) = \sigma_i^A(\lambda_l)\left[1 - \frac{\mathscr{L}^A(\lambda_l + \delta\lambda)}{\mathscr{L}^A(\lambda_l)}\right] \qquad (41)$$

In this instance, $\sigma_i^A(\lambda_l)$ corresponds to the peak absorption cross section corresponding to the laser wavelength $\lambda_l$ and $\mathscr{L}^A(\lambda_l + \delta\lambda)/\mathscr{L}^A(\lambda_l)$ represents the ratio of the absorption cross section at the two wavelengths,

$\mathscr{L}^A(\lambda)$ being the absorption profile function. Schotland, who originally conceived this approach for studying the atmospheric water vapor profile (143), has made an error analysis (144) which indicates that the uncertainty in the density measurement arises as a result of the frequency instability of the laser at short range, while signal fluctuations associated with the measurement are responsible in the case of long-range observations.

A considerable improvement in sensitivity can be achieved if this differential absorption technique is used in conjunction with a topographical scatterer. However, this gain in sensitivity is achieved at the expense of range resolution so that this technique is only applicable in situations where the integrated concentration of the trace constituents along the path of the laser beam is worth evaluating. Under these circumstances the lasor equation takes the form

$$E(\lambda_l, R_0) = E_l T_r(\lambda_l) \xi_s \frac{\rho}{\pi} \frac{A_r \tau_d}{R_0^2 \tau_l} \exp\left\{-2 \int_0^{R_0} \epsilon(\lambda_l) \, dR\right\} \qquad (42)$$

where we have assumed $\tau_d \leq \tau_l$ and we introduce $\xi_s$ to represent the overlap factor and $\rho$ to represent the scattering efficiency of the topographical target; $R_0$ is the range to the topographical target. Values of $\rho$ can range from 0.1 in the visible to 1 in the infrared (145). In the event that $\tau_d > \tau_l$, the factor $\tau_d/\tau_l$ is replaced with unity. In order to have optimum temporal discrimination against any solar background illumination of the target, $\tau_d$ should be chosen to be as close to $\tau_l$ as possible. At certain locations of known pollution emission, a retroflector might be located so as to maximize the system sensitivity. Under these conditions the factor $\rho A_r / \pi R^2$ in Eq. (42) is replaced with $\xi_r$, the receiver collection efficiency. This can amount to an improvement of several orders of magnitude depending primarily upon the range.

As for the DAS technique, two closely spaced laser wavelengths must be employed if contributions other than from the specie of interest are to be eliminated. With the same reasoning as above and using Eq. (39), we can express the integrated concentration of the constituent along the path of the laser beam as

$$\int_0^{R_0} N_i(R) \, dR = \frac{1}{2\kappa_i(\lambda_l, \delta\lambda)} \ln\left[\frac{E(\lambda_l + \delta\lambda, R_0)}{E(\lambda_l, R_0)}\right] \qquad (43)$$

In many cases, the pollutant of concern is normally present in the atmosphere so that the increased loading is the entity to be measured.

Under these circumstances an additional measurement has to be undertaken either prior to release of pollution or at a different orientation so that the path of the laser beam misses the effluent plume. This second measurement provides the reference background level that has to be subtracted from the measurement across the plume. In such cases Byer and Garbuny (54) have indicated that the criterion for minimum transmitted energy does not correspond to use of a laser wavelength that coincides with the peak absorption cross section. Their results show that some degree of detuning may be necessary for optimization. On the other hand, Measures and Pilon (46) have drawn attention to the severe attenuation of the laser beam that can occur if the laser wavelength is chosen to maximize absorption.

### 6.4.2  Lasor Equation in the Case of a Fluorescent Target

If the laser is capable of exciting fluorescence within the target, then finite relaxation effects have to be taken into consideration. Kildal and Byer (146) were the first to recognize the significance of this feature and to illustrate its influence on the return signal. The following discussion stems from a more detailed analysis of this problem by Measures (147).

In the case of fluorescence the signal power received by the lasor photodetector is of a similar form to that indicated in Eq. (30), except that in this instance the radiance of the target element arises from the emission of excited molecules, in which case we may write

$$P(\lambda,R) = \xi(R)\, A_E(R)\, \Delta R \frac{A_r}{4\pi R^2} \frac{N^*(R)}{\tau_{RAD}} \frac{hc}{\lambda}$$

$$\int_{\Delta\lambda} \mathscr{L}^F(\lambda')\, T_r(\lambda')\, T(\lambda',R)\, d\lambda' \quad \text{(W)}$$

where $\Delta\lambda$ represents the spectral window of the receiver optics and includes the influence of the spectral filtering elements, $\mathscr{L}^F(\lambda')\, d\lambda'$ represents the fraction of fluorescence emitted into $(\lambda',d\lambda')$, $N^*(R)$ represents the number density of laser excited molecules at $R$, $R$ represents the range of the target element $\Delta R$, $t$ represents the elapsed time between the firing of the laser and the moment of observation, and $\tau_{RAD}$ represents the radiative lifetime for the excited molecules. All the other quantities have been previously defined. If we assume that the ground state population of the fluorescing species, $N_0$, is constant throughout the target and that the effective lifetime of the excited state is $\tau$, then the temporal variation of the excited state population at range $R'$ can be expressed in the form

$$N^*(R',t') = \frac{\lambda_l N_0 \sigma^A(\lambda_l)}{hc} \exp(-t'/\tau) \int_0^{t'} I(R',x) \exp(x/\tau)\, dx \qquad (44)$$

where $x$ is the dummy time variable and $t' \equiv 2(R - R')/c$, $I(R',x)$ represents the temporal variation of the laser power density at range $R'$, and $R (= ct/2)$ corresponds to the range of the laser pulse leading edge at half of the observed time interval $t$. Under these circumstances the signal power received by the laser photodetector takes the form

$$P(\lambda,R) = \int_{R_0}^{R} dR'\, \xi(R')\, A_E(R') \frac{A_r}{4\pi R'^2} \frac{N_0}{\tau} \exp(-t'/\tau) \int_0^{t'} I(R',x) \exp(x/\tau)\, dx$$

$$\times\, \sigma^F(\lambda_l) \int_{\Delta\lambda} \mathscr{L}^F(\lambda') T_r(\lambda') T(\lambda',R')\, d\lambda' \quad (45)$$

where Eq. (28) has been used to simplify Eq. (45) and $\sigma^F(\lambda_l)$ represents the spectrally integrated cross section for fluorescence when excited at wavelength $\lambda_l$.

The range integration in Eq. (45) is along the space–time path of the observed ray, path A to B of Fig. 6.16. In most cases of interest the variation in $T(\lambda',R')$ over the spectral bandwidth of the receiver optics, $\Delta\lambda$, is small enough so that the wavelength integral takes the form

$$\sigma^F(\lambda_l)\, T(\lambda,R) \int_{\Delta\lambda} \mathscr{L}^F(\lambda')\, T_r(\lambda')\, d\lambda$$

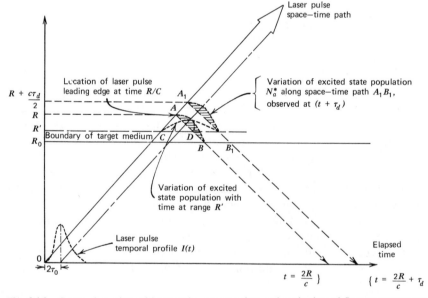

**Fig. 6.16.** Space–time view of laser pulse propagation and excitation of fluorescent target medium. From R. M. Measures, "Lidar Equation Analysis," *Appl. Opt.*, **16**, 1092 (1977). Reproduced by permission.

Most practical situations can be approximated by one of two limiting cases. If the bandwidth of the receiver optics is made adequate to accept the entire fluorescence profile, then we have a situation akin to that of scattering, where the wavelength integral can be replaced by the factor $T(\lambda, R') \sigma^F(\lambda_l) T_r(\lambda)$. Under these circumstances the wavelength $\lambda$ corresponds to the center value of the fluorescence profile.

If, as is often the situation, the spectral window of the photodetection system is small compared to the spectral width of the observed radiation, the wavelength integral can be approximated as

$$\sigma^F(\lambda_l) \int_{\Delta\lambda} \mathscr{L}^F(\lambda') \, T_r(\lambda') \, T(\lambda',R') \, d\lambda = T(\lambda,R') \, \sigma^F(\lambda_l) \, \mathscr{L}^F(\lambda) \, T_r(\lambda) \, \Delta\lambda_r \tag{46}$$

where the product

$$T_r(\lambda) \, \Delta\lambda_r \equiv \int_{\Delta\lambda} T_r(\lambda') \, d\lambda' \tag{47}$$

is often termed the "filter function" and the wavelength $\lambda$ indicated on the righthand side (RHS) of Eq. (46) corresponds to the value at the center of the receiver bandwidth; $\Delta\lambda_r$ represents the effective bandwidth which would transmit, at constant transmission efficiency $T_r(\lambda)$, the same fraction of the fluorescence as achieved by the real system.

We may also assume that the laser power density at range $R'$ and time $x$ is given by

$$I(R',x) = \frac{E_l T(\lambda_l, R')}{A_E(R')} \, j(x) \quad (\text{W cm}^{-2})$$

where

$$\int_0^\infty j(x) \, dx = 1$$

Furthermore, if the total attenuation coefficient $\epsilon \equiv \epsilon(\lambda) + \epsilon(\lambda_l)$ is again regarded as a constant within the target medium, the radiative energy received by the photodetector in the time interval $(t, \tau_d)$ is

$$E(\lambda,R) = E_l T(R_0) T_r(\lambda) \, \Delta\lambda_r \, \xi \frac{A_r N_0 \sigma^F(\lambda_l) \, \mathscr{L}^F(\lambda)}{4\pi R^2 \tau} \int_0^{t+\tau_d} H(R) \, dt \tag{48}$$

where

$$H(R) = \int_{R_0}^{R} dR' \exp[-\epsilon(R'-R_0)] \exp(-t'/\tau) \int_0^{t'} j(x) \exp(x/\tau) \, dx \tag{49}$$

To obtain this result, the fluorescence spectral width has been taken to be much greater than the receiver bandwidth, so that the approximation indicated by Eq. (46) has been employed. Furthermore, we have again used the simplifications represented by Eq. (36) and have assumed that the overlap factor $\xi(R')$ is only weakly dependent upon the range and so can be approximated by the value $\xi$. We have also assumed that the range of integration is sufficiently small compared to $R$ so that the $1/R^2$ factor can be taken out of the range integration. A somewhat better approximation can be obtained by replacing the $1/R^2$ factor by the weighted average $1/\{RR_0\}$.

We can determine the influence of the finite lifetime of the excited state upon the lasor equation by evaluating Eq. (48) for a representative laser pulse shape. A close approximation to many real pulses can be obtained by

$$
j(t) = \left[ \left( \frac{t}{\tau_0} \right)^n \exp(-t/\tau_0) \right] \Big/ [\tau_0 \Gamma(n+1)] \tag{50}
$$

where $\Gamma(n)$ represents the gamma function, $n$ being an integer. Under these circumstances

$$
E(\lambda, R) = E_0 \left\{ \frac{(1-A)c\tau_0/2}{\tau_0 \Gamma(n+1)A^{n+1}} \right\} \int_t^{t+\tau_d} dt
$$
$$
\times \int_0^Z \exp\{-\zeta Z' - (Z-Z')(1-A)\}\, dZ \int_0^{(Z-Z')A} y^n \exp(-y)\, dy \tag{51}
$$

where

$$
E_0 \equiv \frac{E_l T(R_0)\, T_r(\lambda)\, \xi A_r N_0\, \sigma^F(\lambda_l)\, \mathscr{L}^F(\lambda)\, \Delta\lambda_r}{4\pi R^2} \tag{52}
$$

$A \equiv (1 - (\tau_0/\tau))$; $\zeta \equiv \epsilon c\tau_0/2$ represents the nondimensional ratio of the laser characteristic pulse length $c\tau_0/2$ to the net target attenuation length $1/\epsilon$; $Z' \equiv (R' - R_0)/(c\tau_0/2)$ represents the penetration depth into the target medium in terms of the laser characteristic length $c\tau_0/2$; and $y$ represents the dummy time variable normalized by $\tau_0$.

In order to determine the influence of the finite lifetime of the excited state, Eq. (51) has to be evaluated. The resulting lasor equation can be expressed in the form

$$
E(\lambda, R) = E_l T(R_0)\, T_r(\lambda)\, \xi \frac{A_r}{R^2}\, N_0\, \frac{\sigma^F(\lambda_l)\mathscr{L}^F(\lambda)\, \Delta\lambda_r}{4\pi}\, \frac{c\tau_d}{2}\, \gamma(R)
$$
$$
\times \exp[-\epsilon(R - R_0)] \tag{53}
$$

where $\gamma(R)$ is a "correction factor" that includes the effects of finite lifetime for the excited molecules, the laser pulse duration, the detector integration period, and the laser pulse shape.

A comparison of Eqs. (53) and (35) reveals a striking similarity of form. The basic differences involve introduction of the correction factor $\gamma(R)$ and use of the factor $\mathscr{L}^F(\lambda) \Delta\lambda_r$ to account for the fraction of the emitted radiation accepted by the spectrally selective elements of the receiver optics. In the case of a representative laser pulse [$n = 2$ in Eq. (50)] Measures (147) has shown that the "correction factor" $\gamma(R)$ approaches unity for large penetration depths in an optically thin target medium, as would be typical for atmospheric applications. This is illustrated in Fig. 6.17a where the correction factor $\gamma(Z^*)$ is plotted against the target penetration depth normalized by the laser pulse length, namely, $Z^* \equiv (R - R_0)/L$, for several values of $T^* (\equiv \tau_l/\tau)$ (the laser pulse duration/fluorescence decay time). Here, $R$ corresponds to the range of the leading edge of the laser pulse at half the observed time interval and $R_0$ represents the range of the target boundary. The laser pulse length $L$ is identical to $c\tau_l/2$. In this instance $\tau_l$ ($\equiv 5\tau_0$) is approximately the period between the 20% points of the laser profile and thereby corresponds approximately to the pulse duration. For small values of $T^*$, it is clear that $\gamma(Z^*) \ll 1$ even for quite considerable penetration depths.

If $\tau_l \gg \tau$ (i.e., $T^*$ large) the value of $\gamma(Z^*)$ approaches unity for target penetration depths of about one laser pulse length. Nevertheless, the fact that the "correction factor" $\gamma(Z^*)$ can be much less than unity even for large values of $T^*$—corresponding to the "scattering limit"—indicates that in effect the scattering lasor Eq. (35) overestimates the expected signal in situations where the return arises close to a sharp boundary. The results presented in Fig. 6.17a were for $\tau_d/\tau_l \approx 0.2$. If, however, $\tau_d = \tau_l$, a somewhat similar behavior is observed except that $\gamma(Z^*)$ has a finite value at $Z^* = 0$. This can be understood in terms of the spatial resolution of the system.

In the "scattering limit" (i.e., $\tau \approx 0$), the spatial resolution is determined solely by the laser pulse duration provided $\tau_d \ll \tau_l$. If, on the other hand, $\tau_d \gtrsim \tau_l$, the range resolution can be approximated by $c(\tau_l + \tau_d)/2$. For those situations where the excited-state lifetime has to be taken into account, the range resolution is better approximated by $c(\tau_d + \tau_l + \tau)/2$. Results very similar to those indicated in Fig. 6.17a and b can be obtained for a rectangular-shaped laser pulse (147).

In the case of hydrographic work the optical depth of the target is usually large, that is, $\epsilon L \gtrsim 1$. In this case, combining the exponential factor of Eq. (53) with the lifetime correction factor $\gamma(Z^*)$, we derive a new correction factor termed the "optical thick correction factor":

$$\bar{\gamma}(Z^*) \equiv \gamma(Z^*) \exp(-\epsilon L Z^*) \qquad (54)$$

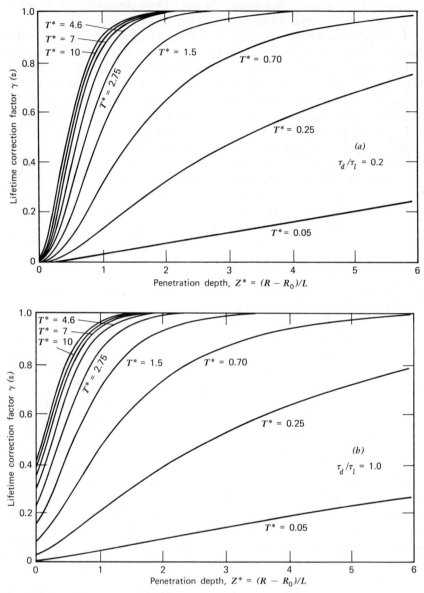

**Fig. 6.17.** Variation of "optical thin correction factor" $\gamma(Z^*)$ with normalized penetration $Z^*$ into a fluorescent target for several values of $T^*$ (ratio of laser pulse duration to fluorescence lifetime, $\tau_l/\tau$); $R$ represents the range of the laser pulse leading edge, $R_0$ is the range of the target boundary, and $L$ is the laser pulse length ($c\tau_l/2$). Laser pulse length to attenuation length $\epsilon L = 0.005$ for each case. From R. M. Measures, "Lidar Equation Analysis," *Appl. Opt.*, **16**, 1092 (1977). Reproduced by permission.

344

The variation of $\bar{\gamma}(Z^*)$ with penetration depth $Z^*$ has been evaluated for a similar range of $T^*$ values, but with $\tau_d = 0.2\tau_l$ and optical depth values $\epsilon L$ of 2.5 and 25, respectively. The results are presented in Figs. 6.18$a$ and $b$. It is quite apparent that whereas $\gamma(Z^*)$ tends to unity for large penetration depths in an optically thin medium, $\bar{\gamma}(Z^*)$ reaches a maximum and then decays to zero for large penetrations. It is also evident from a study of Figs. 6.17$a$ and $b$ that the influence of long relaxation times is merely to prevent $\gamma(Z^*)$ from reaching unity until considerable penetration of the target has been attained. By contrast, the maximum value of $\bar{\gamma}(Z^*)$ is reduced considerably by long fluorescent lifetimes in an optically thick media. For most atmospheric applications involving electronic transitions, $\gamma(Z^*)$ will be close to unity as typical values for $\tau$ are a few nanoseconds. This may not be true for infrared fluorescent studies, where the lifetime of vibrational–rotational transitions can be much longer.

With hydrographic work, the targets are nearly always optically thick, and Measures (147) has shown that, if $\epsilon L > 5$, $\tau_d > 0.2\tau_l$ and $\tau_l > \tau$ (short fluorescence lifetimes compared to laser pulse duration), the limiting value of the correction factor is given by

$$\bar{\gamma}(Z^*) \simeq \frac{2}{\epsilon c \tau_d} \tag{55}$$

Under these circumstances the lasor equation, represented by Eq. (53), takes the form

$$E(\lambda, R) = E_l T(R_0)\, T_r(\lambda)\, \xi \frac{A_r}{R^2}\, N_0\, \frac{\sigma^F(\lambda_l)}{4\pi}\, \frac{\mathscr{L}^F(\lambda)\, \Delta\lambda_r}{[\epsilon(\lambda_l) + \epsilon(\lambda)]} \tag{56}$$

If we assume that $\epsilon(\lambda_l) \gg \epsilon(\lambda)$, as often is the case, and further that $\epsilon(\lambda_l) \simeq N_0 \sigma^A(\lambda_l)$, then we arrive at an equation of the form

$$E(\lambda, R) = E_l T(R_0)\, T_r(\lambda)\, \xi \frac{A_r}{R^2}\, F(\lambda, \lambda_l)\, \Delta\lambda_r \tag{57}$$

where we have introduced the "target fluorescence efficiency"

$$F(\lambda, \lambda_l) \equiv \frac{\lambda_l \phi^F \mathscr{L}^F(\lambda)}{4\pi\lambda} \tag{58}$$

by using Eq. (28). Under such circumstances the lasor equation, as expressed by Eq. (37), becomes identical to the laser fluorosensor equation as suggested by Measures (73), and the return signal is no longer capable of providing any information regarding the concentration of the fluorescent species within the target zone. However, since the target

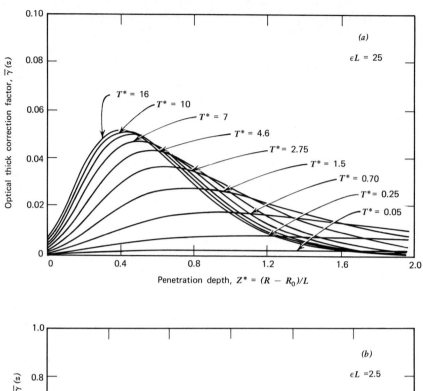

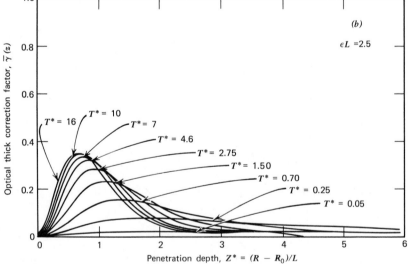

**Fig. 6.18.** Variation of "optically thick correction factor" $\bar{\gamma}(Z^*)$ with normalized penetration $Z^*$ into a fluorescent target, for several values of $T^*$; $\tau_d/\tau_l = 0.2$ for each case. From R. M. Measures, "Lidar Equation, Analysis," *Appl. Opt.*, **16,** 1092 (1977). Reproduced by permission.

fluorescence efficiency $F(\lambda, \lambda_l)$ is proportional to the emission profile, Eq. (58), identification of the target is still possible from a spectral scan of the fluorescence signal.

### 6.4.3   Sources of Noise

Of crucial importance to any discussion on remote sensing is the question of signal-to-noise ratio. Noise, in this context, may be thought of as false signals that can reduce the accuracy of a given measurement or even obscure the true signal completely. Noise can, in general, have either an optical or a thermal origin. In the context of laser environmental sensing, there are four important kinds of noise. These are listed in Table 6.5. The first three represent different forms of shot noise. Under daytime operation solar scattered radiation arising from either the sky or the ground can often dominate all other forms of noise. However, it is important to realize that for both Raman and fluorescence measurements the background radiation may include a laser-scattered component if adequate spectral rejection is not provided.

The increment of radiative energy (arising from natural sources) accepted by the receiving optics in the detection time $\tau_d$ can be expressed in the form (148)

$$E_b^N(\lambda) = \int_{\Delta\lambda} S_b(\lambda') \, T_r(\lambda') \, \Omega_r A_r \tau_d \, d\lambda' \qquad (59)$$

where $S_b(\lambda')$ represents the spectral radiance of the sky background (W cm$^{-2}$ nm$^{-1}$ sr$^{-1}$), $\Delta\lambda$ again refers to the spectral window available to the photodetection system, and $\Omega_r$ represents the acceptance solid angle of the receiver optics. For a good optical system the "etendue" matched condition implies $\Omega_r A_r = \Omega_s A_s$, where $\Omega_s$ represents the acceptance solid

**TABLE 6.5**
**Kinds of Noise Relevant to Laser Environmental Sensing**

| Kind of noise | Physical mechanism |
| --- | --- |
| Noise in signal (quantum noise) | statistical fluctuations of signal radiation |
| Background radiation noise | statistical fluctuations of background radiation |
| Dark current noise | thermal generation of current carriers in the absence of an optical signal |
| Thermal (Johnson, Nyquist) noise | thermal agitation of current carriers |

angle and $A_s$ the entrance aperture of the spectrometer, respectively. For example, if a $f$-7 spectrometer is used with an entrance slit of 0.1 cm² and $A_r = 10^3$ cm², then for a well-matched system $\Omega_r = 2 \times 10^{-6}$ sr. The spectral radiance of the clear daytime sky peaks in the visible due to scattered solar radiation and can attain a value of close to $10^{-5}$ W cm$^{-2}$ nm$^{-1}$ sr$^{-1}$ (90, 148). Toward 300 nm, this background decreases rapidly because of attenuation within the ozone shield above the earth. The more gradual decline in the solar spectral radiance at the infrared end of the spectrum is punctuated by many absorption bands (see Fig. 6.19).

In order to reduce the solar background, the receiver optics bandwidth $\Delta\lambda$ is always adjusted to be as small as compatible with the spectral width of the signal of interest. For all scattering interactions, $\Delta\lambda$ is sufficiently narrow that it can effectively be equated to $\Delta\lambda_r$ defined by Eq. (47), in

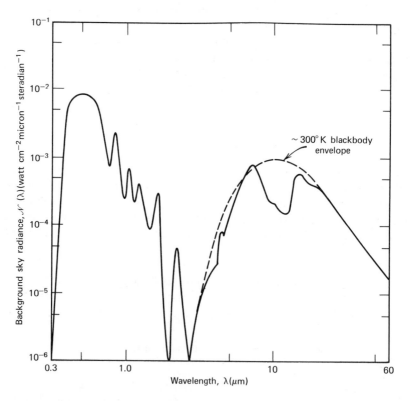

**Fig. 6.19.** Diffuse component of typical background radiance from sea level, zenith angle 45°, excellent visibility. From W. K. Pratt, *Lidar Communication Systems*, Wiley, New York (1969). Reproduced by permission.

which case we can write

$$E_b^N(\lambda) = S_b(\lambda)\Omega_r A_r \tau_d \, T_r(\lambda) \, \Delta\lambda \tag{60}$$

where $\lambda$ corresponds to the wavelength of the scattered radiation. In the case of fluorescence, $\Delta\lambda$ will be larger, though probably still small enough to neglect the variation of the solar spectral radiance over the range of integration. In this case the $T_r(\lambda) \, \Delta\lambda$ factor in Eq. (60) should be replaced by the "filter function" $T_r(\lambda) \, \Delta\lambda_r$.

The noise associated with a photomultiplier is determined by several factors: the type of photomultiplier, the characteristics of the photocathode, the operating gain, and the usage history of the particular photomultiplier selected. The predominant form of photomultiplier noise is associated with the release of single electrons from the photocathode in the absence of any incident light. This so-called "dark current" arises from thermal and field emission processes and can be anywhere from $10^{-13}$ to $10^{-17}$ amp at room temperature. Cooling of the device is a popular method of reducing this component of noise and has been reviewed by Foord, Jones, Oliver, and Pike (150). A good overall description of photomultiplier noise problems is provided by Poultney (88).

### 6.4.4 Signal-to-Noise Ratio

In the case of a photomultiplier the arrival of the signal pulse of energy $E_s(\lambda)$ corresponding to range $ct/2$ gives rise to a momentary photocathode current

$$i_s(t) = \frac{\lambda \, \eta(\lambda) \, e \, E_s(\lambda)}{hc\tau_d} \tag{61}$$

where $\eta(\lambda)$ represents the quantum efficiency of the photocathode at wavelength $\lambda$ and $e$ represents the electronic charge. It is worth noting that $\lambda E_s(\lambda)/hc\tau_d$ represents the mean rate of arrival of signal photons (energy $hc/\lambda$) at the photocathode. The equivalent noise photocathode current during the same time interval is given by

$$i_n = \frac{e\sqrt{\langle \delta N_e^2 \rangle}}{\tau_d} \tag{62}$$

where $\sqrt{\langle \delta N_e^2 \rangle}$ represents the root mean square in the fluctuation of the total number of photocathode electrons created in the time interval $\tau_d$.

If, as is most often the case, this burst of photocathode electrons can be described by Poisson statistics, then we can write (149)

$$[\langle \delta N_e^2 \rangle]^{1/2} = [\langle N_e \rangle]^{1/2} \tag{63}$$

where $\langle N_e \rangle$ represents the mean number of photocathode-generated electrons produced in the interval $\tau_d$:

$$\langle N_e \rangle = \frac{1}{hc} \{\lambda \eta(\lambda)[E_s(\lambda) + E_b^N(\lambda)] + \lambda_l \eta(\lambda_l) E_b(\lambda_l)\} + \langle N_e^d \rangle \tag{64}$$

where $\langle N_e^d \rangle$ represents the mean number of photocathode electrons created in the interval $\tau_d$ in the absence of any light; $E_b^N(\lambda)$ and $E_b(\lambda_l)$ respectively represent the natural background and laser-scattered background pulses of radiant energy incident upon the photocathode in the period $\tau_d$. If the wavelength of the photodetection system is set to $\lambda_l$, then of course the first and third term within the braces of Eq. (64) coalesce.

The photocathode signal-to-noise ratio can thus be expressed in the form

$$(SNR)_c = \frac{e\lambda \eta(\lambda) E_s(\lambda)}{hc\tau_d \left\{ \dfrac{e^2}{hc\tau_d^2}[\lambda\eta(\lambda)\{E_s(\lambda) + E_b^N(\lambda) + v_\eta E_b(\lambda_l)\}] + \dfrac{ei_d}{\tau_d} \right\}^{1/2}} \tag{65}$$

where $i_d$ represents the mean photocathode dark current $(e\langle N_e^d \rangle / \tau_d)$ and $v_\eta \equiv \lambda_l \eta(\lambda_l)/\lambda\eta(\lambda)$. An alternative form of this expression is used by some authors (22,88,151) who are involved in electronically gating the photomultiplier in order to account for the natural background and dark current contributions to the noise of the system. This is of particular relevance in photon counting (88). Under these circumstances one assumes that the photomultiplier is gated "on" for a duration $\tau_g^s$, during the period of signal return and again for an additional period $\tau_g^b$ in the absence of any signal. In this case the gated signal-to-noise ratio is

$$(SNR)_c^g = \frac{e\lambda \eta(\lambda) E_s(\lambda) (\tau_g^s/\tau_d)}{hc\tau_d \left\{ \dfrac{e^2}{hc\tau_d^3}[\lambda\eta(\lambda)\{E_s(\lambda)\tau_g^s + E_b^N(\lambda)(\tau_g^s + \tau_g^b) + \right.}$$
$$\left. + v_\eta E_b(\lambda_l)\tau_g^s\}] + \dfrac{ei_d}{\tau_d}\left(\dfrac{\tau_g^s + \tau_g^b}{\tau_d}\right) \right\}^{1/2} \tag{66}$$

For the sake of simplicity we have assumed that when the signal arises from fluorescence, the correction factor $\gamma$ is very close to unity and so independent of $\tau_g^s$.

As indicated earlier, in Eq. (61), the equivalent signal current can be related to the signal energy received by the photodetector, and by the same reasoning the equivalent background current can be expressed in the form

$$i_b \equiv e\{\lambda\eta(\lambda) E_b^N(\lambda) + \lambda_l \eta(\lambda_l) E_b(\lambda_l)\}/hc\tau_d \tag{67}$$

If we introduce the photodetection bandwidth

$$B \equiv \frac{1}{2\tau_d} \tag{68}$$

then we can rewrite Eq. (65) in the more conventional form

$$(SNR)_c = \frac{i_s}{[2eB(i_s + i_b + i_d)]^{1/2}} \tag{69}$$

where we see that the denominator represents the shot noise in the form first shown by Schottky (152):

$$i_n = [2eB(i_s + i_b + i_d)]^{1/2} \tag{70}$$

If we introduce $G$ as the "dynode chain gain," then the anode signal current is

$$I_s = G\xi_e i_s \tag{71}$$

where $\xi_e$ ($<1$) describes the collection efficiency of the electrostatic focusing, that is to say, $\xi_e$ defines what fraction of the photoelectrons, created at the cathode, arrive at the first dynode. The corresponding noise current at the anode is given by

$$I_n = eG\delta \frac{[\xi_e \langle N_e \rangle]^{1/2}}{\tau_d} \tag{72}$$

where $\delta$ ($\simeq 1$) accounts for the statistical fluctuations in the emission of secondary electrons from the dynodes (153) and can be thought of as a noise factor associated with the gain. Consequently, we may write the "signal-to-noise ratio" for the anode current as

$$(SNR)_a = \frac{I_s}{I_n} = \frac{G\xi_e i_s}{[2eBG^2 F_G \xi_e (i_s + i_b + i_d)]^{1/2}} \tag{73}$$

where we have introduced $F_G$ to represent $\delta^2$.

It is clear that in the case of a photomultiplier the current multiplication factor $G$ does not itself enter into the output $SNR$. However, a somewhat similar analysis in the case of an avalanche photodiode would produce an output $SNR$ of the form (89)

$$SNR = \frac{G\xi_e i_s}{\left[\dfrac{4kTB}{R_{eq}} + 2eBG^2 F_G \xi_e (i_s + i_b + i_d)\right]^{1/2}} \tag{74}$$

where the first term within the square brackets represents the Johnson or thermal noise current associated with the equivalent load resistance of the output circuit ($R_{eq}$), and $i_d$ in this case refers to the bulk leakage current of the avalanche photodiode; $\xi_e$ in this instance can be taken to be very close to unity. The magnitude of $G$ is so large ($10^5$–$10^7$) in the case of a photomultiplier that thermal noise never imposes a real limit. However, the value of $F_G$ can range from 1 to about 2.5 for both high-gain photomultipliers and photodiodes (89). In general, the signal-to-noise ratio improves by averaging over many pulses. This improvement is proportional to the square root of the number of pulses averaged.

The expression for the output current signal-to-noise ratio, Eq. (74), has found general application although several effects are omitted. These include the contribution to the dark current from electrons that originate on the dynodes and the possibility of leakage current at the output socket of the device (particularly relevant to cooled photomultipliers). The influences of cosmic radiation and natural radioactivity have also been neglected—with justification, one might add, in most situations. These additional noise contributions can be taken into account by appropriately modifying the dark current term (153).

### 6.4.5   Signal-to-Noise Ratio Limiting Cases and Improvements

In order to see where improvements to the $SNR$ can be made, we shall express the output signal-to-noise ratio (on the basis of a single pulse) in the general form

$$SNR = \frac{E_s(\lambda)[\xi_e\lambda\eta(\lambda)]^{1/2}}{\left[F_G hc\left\{E_s(\lambda) + E_b^N(\lambda) + \nu_\eta E_b(\lambda_l) + \frac{(i_d + i_J)}{2BS_d}\right\}\right]^{1/2}} \tag{75}$$

where

$$S_d \equiv \frac{e\lambda\eta(\lambda)}{hc} \tag{76}$$

and

$$i_J \equiv \frac{2kT}{eG^2\xi_e R_{eq}F_G} \tag{77}$$

represents the effective Johnson noise current. As we have seen, $i_J$ is only important where there is negligible internal gain.

It is immediately apparent that a careful choice of the detector and its mode of operation can optimize the $SNR$, all other things being equal. In the case of a photomultiplier, the photocathode should be selected on the basis of maximum quantum efficiency at the wavelength of interest. The noise factor associated with the gain and the collection efficiency of the first dynode can both be optimized by application of a suitable voltage between the photocathode and the first dynode (153). Thermionic and field emission release of electrons from the dynodes can represent an important component of the dark current noise in photomultipliers. Consequently, in very low light level situations where dark current noise is important, photon counting can significantly improve the SNR due to discrimination against single electron pulses that have not acquired the full gain (88,153). A further increase in the SNR under these conditions can be obtained by focusing of the incident light so that only a small area of the photocathode is illuminated while ensuring that all electrons that originate from the unilluminated regions of the photocathode are defocused magnetically. Cooling also helps since thermionic emission is strongly temperature dependent. Topp et al. (153) demonstrated that use of these techniques enabled signals with a power of less than $10^{-17}$ W (35 photons s$^{-1}$) to be detected with a time constant of 1 s at 650 nm.

The general output signal-to-noise ratio, Eq. (75), can be expressed in another form:

$$(SNR)^2 = \frac{E_s^2(\lambda)/E(\lambda)}{E_s(\lambda) + E_b^*(\lambda) + (i_d + i_j)\dfrac{E(\lambda)}{2eB^*}} \tag{78}$$

where

$$E_b^*(\lambda) \equiv E_b^N(\lambda) + \nu_n E_b(\lambda_l) \tag{79}$$

$$B^* \equiv BF_G/\xi_e \tag{80}$$

and

$$E(\lambda) \equiv \frac{hcF_G}{\lambda\eta(\lambda)\xi_e} \tag{81}$$

which corresponds to the signal energy that would just give a unit value to the signal-to-noise ratio in the signal shot noise limit. Equation (78) represents a simple quadratic expression in terms of $E_s(\lambda)$ and has the solution

$$E_s(\lambda) = \tfrac{1}{2}(SNR)^2 E(\lambda)\left[1 + \left\{1 + 4\left[\frac{E_b^*(\lambda)}{E(\lambda)} + \frac{(i_d + i_j)}{2eB^*}\right]\Big/(SNR)^2\right\}^{1/2}\right] \tag{82}$$

*In essence, there are four limiting situations that cover most laser sensing situations.* Three of them presuppose that the system's spectral discrimination is adequate to enable the laser scattered component of the background signal to be neglected, that is, $E_b^*(\lambda) \simeq E_b^N(\lambda)$. We shall consider these first. We shall also assume that $i_j$ can be neglected, in which case the highest sensitivity is achieved if the detectable signal is limited only by the quantum fluctuations of the signal itself, namely,

$$E_b^N(\lambda) + \frac{i_d E(\lambda)}{2eB^*} \ll E_s(\lambda)$$

Under these circumstances (*signal shot noise limit*) the minimum detectable energy (*MDE*) is

$$E_s^{min}(\lambda) \simeq [(SNR)^{min}]^2 E(\lambda) \tag{83}$$

If we assume that $F_G/\xi_e \simeq 1$ and that an acceptable value for $(SNR)^{min} \simeq 1.5$, we find that for $\eta(\lambda) \simeq 0.2$ the minimum detectable number of photons is close to 10 per pulse. Using the appropriate form of the lasor equation, this can be translated into either a minimum laser energy pulse for detection of a given density of a species at a given range or a threshold density of a species that can be detected at a given range for a given transmitted laser energy.

In the case of a scattering interaction and a rectangular shaped laser pulse, Eqs. (35), (81), and (83) may be employed to determine the minimum density of the i species that can be detected at range $R$ for a given pulse of laser energy $E_l$ and threshold value of $SNR$, as follows:

$$[N]^{min} \simeq \frac{R^2 [(SNR)^{min}]^2 \exp\{\epsilon(R - R_0)\}}{E_l U(\lambda) \, \xi(R) \, T(R_0)(\sigma_i/4\pi)} \tag{84}$$

Here, we have introduced the system parameter

$$U(\lambda) \equiv \frac{T_r(\lambda)\eta(\lambda)\xi_e A_r \lambda \tau_d}{2hF_G} \tag{85}$$

which clearly needs to be maximized in order to attain the highest sensitivity. In the case of fluorescence, this system parameter would take on a slightly different form due to use of Eq. (53) as opposed to Eq. (35).

It is evident from Table 6.4 that the differential Raman cross section for a wide range of atmospheric constituents has a value that lies between $10^{-30}$ and $10^{-29}$ cm$^2$ sr$^{-1}$. In order to illustrate the performance that might be expected from a system that is signal shot noise limited, let us consider the following set of system characteristics: $\lambda = 364$ nm, $(\sigma^{RAM}/4\pi) = 3.6 \times 10^{-30}$ cm$^2$ sr$^{-1}$ (appropriate to CO), $T_r(\lambda) = 0.5$, $\eta(\lambda) = 0.2$, $\xi_e = 1.0$, $F_G = 1.0$, $\tau_d = 10^{-7}$ s, and $A_r = 10^3$ cm$^2$. In this case, from Eq. (85), $U(364 \text{ nm}) = 2.74 \times 10^{23}$ cm$^3$ J$^{-1}$ so that, if we assume that $\xi(R) = T(R_0) = 1$ and $\epsilon = 0$, then for $(SNR)^{min} = 1.5$,

$$[N_{CO}]^{min} = \frac{2.29 \times 10^6 R^2 \text{ (cm}^2)}{E_l \text{ (J)}}$$

If we assume that a nitrogen laser with a 1 J output energy is available, then the minimum density of CO that could be detected with this system using Raman scattering at a range of 200 m is about $9.2 \times 10^{14}$ cm$^{-3}$, or 36 ppm.

In daytime operation, the level of background radiation can be so high that

$$E_b^N(\lambda) \gg \frac{i_d E(\lambda)}{2eB^*} \qquad (86)$$

and we speak of the *background noise limit*. Under these circumstances

$$E_s^{min}(\lambda) = \left\{ \frac{F_G hc T_r(\lambda) \Omega_r A_r \tau_d S_b(\lambda) \Delta\lambda}{\xi_e \lambda \eta(\lambda)} \right\}^{1/2} (SNR)^{min} \qquad (87)$$

and one could again employ Eq. (35) to derive an expression for the threshold concentration of the i species that would be detected at range $R$ with a given pulse of laser energy $E_l$:

$$[N_i]^{min} \simeq \frac{2R^2 \exp[\epsilon(R - R_0)]}{E_l \xi(R) T(R_0)(\sigma_i/4\pi)} \left\{ \frac{\Omega_r S_b(\lambda) \Delta\lambda}{2U(\lambda)c} \right\}^{1/2} (SNR)^{min} \qquad (88)$$

It is evident from Eqs. (85) and (88) that improvement of sensitivity in the background-limited situation necessitates the use of small values of $\Omega_r$ and $\Delta\lambda$ with large values of $T_r(\lambda)$, $\eta(\lambda)$, $A_r$, and $\tau_d$.

If we again consider the example of Raman scattering for CO, assuming the same system characteristics, but with $\Omega_r = 2 \times 10^{-6}$ sr and $\Delta\lambda = 1$ nm, then

$$[N_{CO}]^{min} \simeq \frac{1.8 \times 10^7 R^2 (\text{cm}^2)}{E_l(\text{J})}$$

where we have again assumed $(SNR)^{min} = 1.5$ and have taken a fairly conservative value of $4 \times 10^{-6}$ W cm$^{-2}$ nm$^{-1}$ sr$^{-1}$ for the sky spectral radiance at 364 nm. Unfortunately, the radiance of sunlit clouds can, for longer wavelengths, be close to an order of magnitude larger than the above figure (148). On the other hand, the radiance of a clear (moonless) night sky is typically $10^{-7}$ smaller.

In the *dark current-limited situation* we assume that

$$i_d \gg \frac{2eB^* E_b^N(\lambda)}{E(\lambda)}$$

and we can write

$$E_s^{min}(\lambda) \simeq \frac{hc}{\lambda \eta(\lambda)} \left\{ \frac{F_G i_d}{2eB\xi_e} \right\}^{1/2} (SNR)^{min} \qquad (89)$$

The typical value for the photocathode dark current of a photomultiplier is around $10^{-15}$ amp, which means that the inequality expressed above is not likely to be satisfied unless the photon arrival rate is less than about $10^3 \, s^{-1}$. On the other hand, for practical bandwidths, most infrared detectors tend to be dark current limited. Thus, we shall introduce the term "detector detectivity" $D^*$ (124) as

$$D^* \equiv \frac{S_d}{(2ei_d/A_d)^{1/2}} \tag{90}$$

where $S_d$, defined by Eq. (76), represents the detector sensitivity as seen by the relation

$$i_s \equiv S_d E_s(\lambda)/\tau_d$$

where $A_d$ represents the detector area. In this case, using this equation with Eqs. (67) and (90), we may express Eq. (89) in the form

$$E_s^{min}(\lambda) = \frac{1}{D^*} \left\{ \frac{F_G A_d}{4B\xi_e} \right\}^{1/2} (SNR)^{min} \tag{91}$$

For most solid-state detectors the factor $F_G/\xi_e$ can be replaced with unity, and representative detectivity $D^*$ values for a number of such detectors (89) are shown in Fig. 6.20. Solid-state devices have typical detector areas of a few square millimeters. Under these circumstances the threshold density is

$$[N_i]^{min} = \frac{R^2 \exp[\epsilon(R - R_0)](SNR)^{min}}{E_l T(R_0)\xi(R) U^*(\lambda)(\sigma_i/4\pi)} \qquad \text{\{dark current noise limited\}} \tag{92}$$

where we have introduced the modified system parameter

$$U^*(\lambda) \equiv D^* T_r(\lambda) A_r c \left\{ \frac{\tau_d}{2A_d} \right\}^{1/2} \tag{93}$$

In certain instances, namely, when studying fluorescence or Raman scattering, *backscattered laser radiation can limit the lasor's sensitivity.* This can arise in one of two ways. If insufficient spectral rejection is employed, some fraction of this laser return will arrive coincidental with the signal. On the other hand, if inadequate care is given the design of the lasor, it is possible for the laser backscattered radiation to induce fluorescence within some optical component (such as a lens) that is placed before the spectral analyzer. This situation is often avoided by the inclusion of a laser blocking filter ahead of all other vulnerable components. Nevertheless, considerable care must be given this problem.

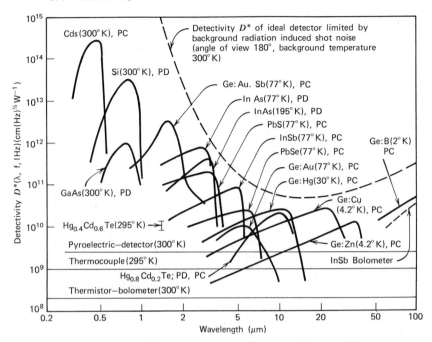

**Fig. 6.20.** Spectral dependence of detectivity $D^*$ for high-sensitivity photoconductors (PC) and photodiodes (PD). Representative values are given based on literature and manufacturer's data. From H. Melchior, "Demodulation and Photodetection Techniques," in *Laser Handbook*, Vol. 1, C 7, F. T. Arecchi and E. O. Schulz-Dubois, Eds., North-Holland, Amsterdam (1972). Reproduced by permission.

An estimate of the degree of spectral rejection necessary can be evaluated from Eq. (75). The criterion relevant to any given situation will of course depend on the predominate source of noise competing with the backscattered laser radiation. Under daylight operating conditions, the appropriate value of the spectral rejection ratio $T_r(\lambda_l)/T_r(\lambda)$ can be ascertained from Eqs. (37) and (60):

$$\frac{T_r(\lambda_l)}{T_r(\lambda)} \ll \frac{8\pi\Omega_r R^2 S_b(\lambda)\,\Delta\lambda}{\nu_\eta c\beta E_l} \exp\left\{2\int_0^R \epsilon(\lambda_l)\,dR\right\} \qquad (94)$$

where $\beta$ represents the atmospheric backscattering coefficient at $\lambda_l$. The range dependence of this criterion arises as a result of taking the backscattered laser radiation return that originates from the same location as the signal. In principle, any earlier (thereby larger) component can be discriminated against by temporal means.

If the value of the spectral rejection ratio is to be determined on the basis that the signal arising from the laser backscattered radiation should be small compared to the true signal, then

$$\frac{T_r(\lambda_l)}{T_r(\lambda)} \ll \frac{\xi(R)N_i\sigma_i}{\nu_n\beta} \exp\left\{\int_0^R [\epsilon(\lambda_l) - \epsilon(\lambda)]\, dR\right\} \qquad (95)$$

In situations where a comparison is made between two backscattered signals at two slightly different wavelengths, the magnitude of the difference between the signals originating from adjacent locations is instrumental in determining the threshold of detection. From Eqs. (67) and (71) we can express the difference in the output current attained at the two wavelengths of operation and arising from signals that originate from the same location as follows:

$$\Delta I_s(R) \equiv \frac{eG\xi_e\lambda_l\eta(\lambda_l)}{hc\tau_d}[E_s(\lambda_l + \delta\lambda, R) - E_s(\lambda_l, R)] \qquad (96)$$

where we assume the small difference in wavelength $\delta\lambda$ can be neglected with regard to the $\lambda_l\eta(\lambda_l)$ factor. Byer and Garbuny (54) have equated this difference to the minimum return signal, at the laser wavelength corresponding to the stronger absorption, in order to estimate the minimum laser energy required for detecting a given concentration at a specific range. In addition, they have indicated that the presence of a background concentration of the species of interest necessitates a degree of detuning from the peak absorption cross section if optimum operation is to be achieved.

Although such an approach can be used to indicate the requirements for ensuring a measurable difference between the "on" and "off" frequency laser returns from a given range, it does not address the question of making a meaningful measurement over a small range increment as would be the case for a pollution plume. Under such circumstances we would require that

$$\Delta I_s(R + \Delta R) - \Delta I_s(R) > I_s(\lambda_l, R)/SNR \qquad (97)$$

That is to say, the incremental change in the output signal current difference should exceed the output noise current associated with the smaller signal return for range $R$. Thus, the signal-to-noise ratio used in relation (97) refers to the return at wavelength $\lambda_l$ from range $R$ and

$$I_s(\lambda_l, R) = \frac{eG\xi_e\lambda_l\eta(\lambda_l)}{hc\tau_d}E_s(\lambda_l, R) \qquad (98)$$

In this case we can write

$$[E_s(\lambda_l + \delta\lambda, R + \Delta R) - E_s(\lambda_l, R + \Delta R)] - [E_s(\lambda_l + \delta\lambda, R) - E_s(\lambda_l, R)]$$
$$> \frac{E_s(\lambda_l, R)}{SNR} \qquad (99)$$

If the laser tuning is such that the "off" frequency return is relatively insensitive to the specie of interest, then it may be reasonable to assume that over the range increment $\Delta R$, $E_s(\lambda_l + \delta\lambda, R + \Delta R) \simeq E_s(\lambda_l + \delta\lambda, R)$. In this case if we introduce $x$ as the fractional change in the signal (at $\lambda_l$) received between $R$ and $R + \Delta R$, i.e.,

$$x \equiv \frac{E_s(\lambda_l, R) - E_s(\lambda_l, R + \Delta R)}{E_s(\lambda_l, R)} \tag{100}$$

equation (99) can then essentially be restated in the form

$$x > \frac{1}{SNR} \tag{101}$$

The appropriate lasor equation in this instance takes the form

$$E_s(\lambda_l, R) = E_l c\tau_d A_r T_r(\lambda)\xi \frac{\beta(\lambda_l)}{8\pi R^2} \exp\left\{-2\bar{\epsilon}(\lambda_l)R - 2\int_0^R \sigma_i^A(\lambda_l) N_i(R)\, dR\right\} \tag{102}$$

This is based on Eq. (37) except that we have introduced $\beta(\lambda_l)$ as the atmospheric volume backscattering coefficient and $\bar{\epsilon}(\lambda_l)$ as the mean attenuation coefficient, excluding the effects of the species of interest. If we use Eq. (102) in Eq. (100), then relation (101) takes the form

$$1 - \left(\frac{R}{R + \Delta R}\right)^2 \exp\left\{-2\bar{\epsilon}(\lambda_l)\,\Delta R - 2\int_R^{R+\Delta R} \sigma_i^A(\lambda_l) N_i(R)\, dR\right\} > \frac{1}{SNR}$$

If the first term within the braces is negligible compared to the term of interest, then we can write

$$2\int_R^{R+\Delta R} \sigma_i^A(\lambda_l) N_i(R)\, dR > \ln\left\{\frac{SNR}{(SNR - 1)(1 + \Delta R/R)^2}\right\} \tag{103}$$

from which it is clear that a conservative value for the threshold density that can be detected over the range increment $\Delta R$ (corresponding to a $SNR \gtrsim 1.5$) is given by

$$[N_i]^{min} \simeq \frac{1}{2\sigma_i^A(\lambda_l)\,\Delta R} \tag{104}$$

A more accurate estimate for the threshold density can be calculated by using Eq. (75) with Eq. (103).

To evaluate the minimum pulse of laser energy required under these circumstances, we set

$$E_s(\lambda_l, R) - E_s(\lambda_l, R + \Delta R) > MDE$$

The result will obviously depend on which form of noise limiting situation is applicable. For a dark current-limited detector,

$$E_l^{min} \simeq \left\{ \frac{A_d}{4BD^{*2}} \right\}^{1/2} \frac{8\pi R^2 (SNR)^{min} \exp\left\{ 2\bar{\epsilon}(\lambda_l)R + 2\int_0^R \sigma_i^A(\lambda_l)N_i(R)\,dR \right\}}{[1 - \exp\{-2\sigma_i^A(\lambda_l)N_i(R)\,\Delta R\}]c\tau_d A_r T_r(\lambda)\beta}$$

(105)

A detailed error analysis of the DAS technique has been undertaken by Schotland (144).

### 6.4.6   Special Problems Associated with Remote Sensing Using Lasers

As we have seen lasers may be used to probe the atmosphere from the ground and to undertake airborne surveillance operations. Certain problems arise that are common to both, while others are almost entirely associated with a particular task. The more general expression for the lasor equation can be expressed in the form

$$E(\lambda,R) = E_l T(\lambda,R) T(\lambda_l,R) \xi(R) \frac{A_r}{R^2} \int_{\Delta\lambda} C(\lambda',\lambda_l) T_r(\lambda')\,d\lambda' \quad (106)$$

where $C(\lambda',\lambda_l)$ represents the "target return efficiency" at $\lambda'$ and, as we have shown above, can take on many forms depending upon the kind of the interaction (fluorescence, scattering, etc.) and the optical properties of the target (optically thick or thin, relaxation times, etc.).

Most hydrographic targets that are of interest in airborne remote sensing (e.g., oil films or water effluent plumes) have a well-defined boundary and are either optically thick or can approximately be treated so because of the small penetration depth of the laser beam (bathymetric surveying is an obvious exception) in the background medium. Under these circumstances, $C(\lambda,\lambda_l)$ can depend upon many factors, including the nature of the air–liquid interface (wave structure), the optical (attenuation, fluorescent, scattering, and reflection) properties of the background medium, and the physical extent of the target (homogeneous or inhomogeneous dispersion/layer). It is clear that in this situation the amplitude of the return signal conveys only limited information about the target under surveillance. Indeed, its correct interpretation requires a thorough knowledge of the laser energy (due to shot-to-shot fluctuation), the atmospheric transmission losses (variable due to wind-induced spray), the altitude and inclination of the aircraft, the overlap between the laser beam and the area of observation, and the instrumental calibration, in addition to the properties of the target and its background medium.

Where the amplitude of the return signal is needed either to estimate the concentration of some constituent of the target or to quantify the target in some other way, the use of a reference signal such as the OH-stretch Raman return from water (73,154) may aid in overcoming the above difficulties. Excitation at several wavelengths is another approach that might enable the amplitude information to be more usefully interpreted (75). However, one of the major roles envisaged for airborne laser fluorosensors is the remote identification of oil spills. For this application, the spectral distribution of the return signal may be adequate. Unfortunately, where a high degree of spectral discrimination is necessary for this purpose, many spectral channels are required. This increases considerably the cost and complexity of the system and would invariably degrade the signal-to-noise ratio by reducing the energy available per channel. This could represent a formidable obstacle to aerial reconnaissance where the monitoring time may be strictly limited by the size and patchy nature of the slick. Measures et al. (15,73) have indicated that the specificity of the laser fluorosensor could be considerably enhanced by making full use of the "fluorescence decay spectrum" associated with the target.

The airborne use of lasors is also fraught with many technical constraints, not the least of which are the weight, size, and power restrictions. Such systems have to also satisfy fairly strict standards of fire prevention and must be designed to operate through high vibrational loading cycles, particularly in helicopter applications.

### 6.4.7 Eye Safety

A general limitation that will eventually encompass all operational lasors is that of safety. This may be more relevant to earth-oriented, airborne systems than to their sky-facing counterparts, but invariably both will have to comply with fairly strict regulations due to the high energy or peak power transmitted by their probing laser beams. Recently, the American National Standards Institute (ANSI) has recommended eye safety maximum permissible exposure (MPE) levels for various regions of the spectrum (155). However, these values have been contested as not being sufficiently stringent (156,157).

Studies have shown that there are at least three principal mechanisms of injury: thermal-mechanical (acoustic transient), thermal, and photochemical. Threshold retinal lesions from short-duration exposure may result from an acoustic transient which accompanies localized heating in the vicinity of the highly absorbing pigment granules (158). For longer exposure times the damage is more likely to arise from thermal denaturation of complex organic molecules (159) or photochemical activity. Of

**TABLE 6.6**
**Maximum Permissible Exposure (MPE) Based on the ANSI Standard, with Some Allowance for the Work of Zuclich and Connolly (157)**

| Wavelength | Exposure time, s | Maximum permissible exposure, J cm$^{-2}$ |
|---|---|---|
| 310 to 400 nm | — | $10^{-3}$ |
| 400 nm to 1.4 $\mu$m | $10^{-9}$ to $10^{-5}$ | $5 \times 10^{-7}$ |
|  | $10^{-5}$ to 10 | $1.8 \times 10^{-3} t^{3/4}$ |
| 1.4 $\mu$m to 13 $\mu$m | $10^{-9}$ to $10^{-7}$ | $10^{-2}$ |
|  | $10^{-7}$ to 10 | $0.56\, t^{1/4}$ |

particular relevance to laser environmental sensing has been the recent work of Zuclich and Connolly (157), who have found evidence that the threshold for damage in the near ultraviolet may, as a result of photo-chemical injury, be much lower than the previously accepted value. Table 6.6 provides a guide as to the maximum permissible exposure (MPE) for the eye and is based on the ANSI standards, making some allowance for the work of Zuclich and Connolly (157).

The severity of the constraint imposed on any lasor system by an eye safety standard will depend to a large extent upon its mode of operation. Lasors that operate at night and are directed skyward will be restricted far less than airborne surveillance systems that operate during the day. The atmospheric lasor, operating at night, can expand its laser beam to meet the eye safety standard, then expand its field of view to avoid loss of signal return without incurring any appreciable extra noise (there would of course be some loss of spatial resolution). For daytime operation any expansion of the field of view will invariably lead to an increase in the background radiation accepted by the optical system, and so this option is not as readily available. In the case of an airborne lasor the increased risk of exposure and the patchy nature of the target will impose counter-demands with regard to operational ceiling and beam expansion, particu-larly in the light of the new findings by Zuclich and Connolly (157).

## 6.5  ATMOSPHERIC LASOR APPLICATIONS

### 6.5.1  Atmospheric Studies

The development of remote laser sensing techniques holds the promise of substantially improving man's understanding of the environment in which he is immersed. Although the application of lasers to studying the atmosphere is still in its infancy, it is readily apparent that the increased scope and quantity of measurements made possible by this means represent an effective quantum jump in our knowledge of the nature and behavior of the atmosphere.

The *oxygen/nitrogen balance* of the atmosphere represents an important measurement in light of the potential reduction in the sources of free oxygen, namely, decline in marine life and vegetation, and the increase in the rate of its consumption due to population growth and corresponding increased fossil fuel combustion. Schweisow and Derr (27) have indicated that a precise measurement of the $O_2/N_2$ balance in the atmosphere should be possible using laser Raman scattering. They point out that with this technique a two order of magnitude improvement in precision over other techniques should be possible, leading to an accuracy of 0.3 ppm for the $O_2/N_2$ ratio and of 0.006 ppm for the $CO_2/N_2$ ratio, which is adequate to determine the magnitude of any long-term drifts in the oxygen/carbon dioxide/nitrogen balance of the atmosphere.

Leonard (24) was the first to use the nitrogen laser to observe Raman backscattering from nitrogen, at 365.9 nm, and oxygen, at 355.7 nm, at a range of around 1 km. Of particular significance at the time was the fact that the peak output of the laser was only 100 kW. Leonard pointed out that in order to achieve similar results with a ruby laser, the peak power would have to be in excess of 21 MW due to the strong wavelength dependence of both the Raman cross section, Eq. (10), and the quantum efficiency of the detector photocathode, $\eta(\lambda)$. However, care must be used in making such a comparison because of the increased attenuation (due primarily to elastic scattering) at the shorter wavelength. Although Boudreau (160) has studied this problem his conclusions may be limited in their usefulness as he neglects the Mie scattering contribution to the attenuation coefficient and his calculations do not account for the improved red response photocathodes currently available (see Fig. 6.6). Nevertheless, the nitrogen laser's high repetition rate represents another important advantage—commercial $N_2$ lasers are available that operate at 1000 pps.

The first Raman measurement of the gaseous density profile was undertaken in 1968 by Cooney (26) using a 25-MW $Q$-switched ruby laser. The Raman vibrational–rotational return from nitrogen was observed, at night, up to an altitude of 3 km. Cooney used a combination of a 694.3-nm blocking filter and a 150-nm bandpass interference filter to provide a net spectral rejection of $10^7$. This was more than adequate to overcome the 500 intensity factor between the elastic backscattered return at 694.3 nm and the nitrogen Raman return at 828.5 nm. Recently,

Garvey and Kent (161) have extended the range of Raman investigations of atmospheric nitrogen well into the stratosphere (to a height of at least 40 km) and have obtained good agreement between their observations, balloon-mounted radiosonde measurements, and the U.S. Standard Atmosphere.

As we have seen, Eq. (8), the relationship between the atmospheric attenuation coefficient and the visual range is, at best, approximate and, in strong scattering situations, fraught with ambiguities. Consequently, several researchers have endeavored to circumvent this difficulty by attempting to separate the laser return due to the gaseous constituents from that of the aeroticulates. Cooney, Orr, and Tomasetti (25) and Cooney (162) have attempted to determine the Mie backscattering coefficient using the following relation:

$$\beta^M(\lambda_l, R) = \frac{E(\lambda_l, R) - E^{RAY}(\lambda_l, R)}{E^{RAY}(\lambda_l, R)} \beta^{RAY}(\lambda_l, R) \tag{107}$$

where

$$E(\lambda_l, R) = E_l \frac{c\tau_d}{2} \frac{A_r}{R^2} \frac{T_r(\lambda_l)}{4\pi} [\beta^M(\lambda_l, R) + \beta^{RAY}(\lambda_l, R)]$$

$$\times \exp\left\{-2\int_0^R \epsilon(\lambda_l, R)\, dR\right\} \tag{108}$$

from Eq. (37), represents the total elastic return from range $R$ at the laser wavelength, where $E^{RAY}(\lambda_l, R)$ is the Rayleigh component of $E(\lambda_l, R)$, $\beta^M(\lambda_l, R)$ and $\beta^{RAY}(\lambda_l, R)$ are the respective Mie and Rayleigh backscattering coefficients, and $\xi(R)$ was set to unity in this instance. In general,

$$\beta^{RAY}(\lambda_l, R) = \sum_i N_i(R)\sigma_i^{RAY}(\pi, \lambda_l) \tag{109}$$

where $N_i(R)$ represents the appropriate number density and $\sigma_i^{RAY}(\pi, \lambda_l)$ is the Rayleigh backscattering cross section for the $i$th constituent of the atmosphere. In reality, there is no direct way of ascertaining the Rayleigh contribution to $E(\lambda_l, R)$. However, the above authors (25, 162) attempted to evaluate the Rayleigh component from the nitrogen Raman return as follows:

$$E^{RAM}(\lambda, R) = E_l \frac{c\tau_d}{2} \frac{A_r}{R^2} \frac{T_r(\lambda)}{4\pi} N_\alpha(R)\sigma_\alpha^{RAM}(\pi, \lambda)$$

$$\times \exp\left\{-\int_0^R [\epsilon(\lambda_l, R) + \epsilon(\lambda, R)]\, dR\right\} \tag{110}$$

where $N_\alpha(R)$ represents the nitrogen number density at range $R$ and $\sigma_\alpha^{RAM}(\pi,\lambda)$ is the corresponding Raman backscattering cross section. In this case we can write

$$\beta^M(\lambda_l,R) = \frac{E(\lambda_l,R) - Y(\lambda_l/\lambda)E^{RAM}(\lambda,R)}{Y(\lambda_l/\lambda)E^{RAM}(\lambda,R)} \beta^{RAY}(\lambda_l,R) \qquad (111)$$

and

$$Y\left(\frac{\lambda_l}{\lambda}\right) = \sum_i \left[\frac{N_i(R)\sigma_i^{RAY}(\pi,\lambda_l)}{N_\alpha(R)\sigma_\alpha^{RAM}(\pi,\lambda)}\right] \frac{T_r(\lambda_l)}{T_r(\lambda)} \exp\left\{-\int_0^R [\epsilon(\lambda_l,R) - \epsilon(\lambda,R)]\, dR\right\} \qquad (112)$$

and can be assumed to be independent of $R$ if

$$\epsilon(\lambda_l,R) \simeq \epsilon(\lambda,R) \qquad (113)$$

and the mixing ratio of the gaseous constituents is a constant, independent of altitude.

While Cooney, Orr, and Tomasetti (25) normalized the elastic return with the nitrogen vibrational–rotational Raman return, Cooney (162) employed the anti-Stokes wing of the nitrogen pure rotational Raman backscattered return at 691.2 nm (excitation 694.3 nm) for this purpose. A spectrum rejection of $10^5$ was found to be adequate for this work. The latter approach has two obvious advantages: (i) The rotational Raman shift is only $\simeq 50$ cm$^{-1}$, as opposed to 2330 cm$^{-1}$ for the vibrational case, so that the assumption represented by Eq. (113) is more likely to be justified. (ii) There should be a marked improvement in the magnitude of the inelastic signal due to both a larger cross section and a better photocathode response. However, it should be noted that care must be used in choosing the filter characteristics in order to avoid a measurement that is unduly temperature sensitive (162). It is clear that the beauty of this scheme lies in the way the Raman measurement eliminates the need for an absolute calibration. Indeed, Leonard and Caputo (163) have shown that it is possible to build a reliable monostatic atmospheric transmissometer based on the measurement of the nitrogen Raman backscattered signal.

The first remote laser measurement of the *vertical water vapor profiles* in the atmosphere was made in 1966 by Schotland using a thermally tuned ruby laser (143) and the DAS technique. The earliest laser Raman measurements to yield the spatial distribution of water vapor in the atmosphere were performed by Melfi, Lawrence, and McCormick (83) and Cooney (164). Each used a frequency-doubled, $Q$-switched ruby laser and normalized their water vapor return with the nitrogen vibrational Raman return. Melfi et al. (83) and Cooney (165) were also able to

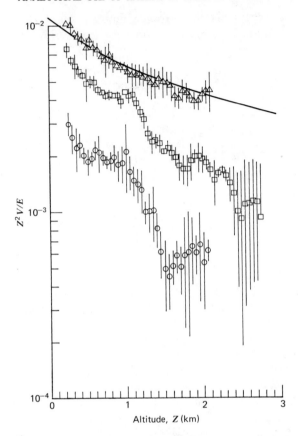

**Fig. 6.21.** Typical backscatter profiles normalized for range $Z$ and laser energy $E$ $(Z^2V/E)$ from the atmosphere: ($\Delta$) nitrogen, average of eight oscillograms, August 27, 1970; (O) water vapor, average of six oscillograms, August 27, 1970; ($\square$) aerosol, average of six oscillograms, with two No. 1 n.d. filters, August 27, 1970; (−) calculated return for nitrogen $(Z^2V)$; $V$ represents voltage signal from photomultiplier. From S. H. Melfi, *Appl. Opt.*, **11**, 1605 (1972). Reproduced by permission.

demonstrate good agreement between their lasor-evaluated profiles and measurements undertaken by radio sondes. A representative set of back-scattered profiles for nitrogen, water vapor, and aerosols is presented in Fig. 6.21, while an example of the close agreement obtained between the lasor water vapor ratio and the balloon sonde is shown in Fig. 6.22 taken from Melfi (166).

Although the general agreement obtained in the above work was excellent, the comparisons were made rather far apart in space and time. Strauch, Derr, and Cupp (31) avoided this problem by making a direct

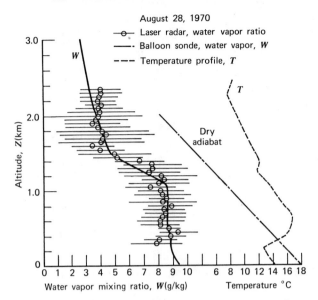

**Fig. 6.22.** Optical radar measurement of water vapor mixing ratio compared with standard balloon-sonde data. From S. H. Melfi, *Appl. Opt.*, **11**, 1605 (1972). Reproduced by permission.

comparison of the Raman lasor measurements with that of a standard humidity meter mounted on a tower some 30 m above the ground. Their results show an excellent correlation between the two measurements and indicate that their system should be capable of determining the water vapor profile to a range of about 4 km. An interesting observation made by these authors is that 365.4-nm radiation from mercury street lights falls very close to the $N_2$ Raman line at 365.8 nm (when excited with the nitrogen laser at 337.1 nm). They also point out that although allowance can be made for leakage of the $N_2$ Raman line into the $H_2O$ observation, fluorescence from the laser discharge and from aerosols, pollutant molecules, interference filters, and dust on the mirror surfaces can present serious sources of error and must be carefully avoided.

Recently, Murray, Hake, van der Laan, and Hawley (63) used a 1 J pulse$^{-1}$ $CO_2$ TEA laser to measure the water vapor content of the atmosphere over a range of about 1 km using the DAS technique. The differential absorption between the R(20), R(12) and R(18) rotational–vibrational transitions of the $CO_2$ laser is clearly seen in Fig. 6.23. The corresponding water vapor concentration profile, deduced from 100 paired pulses on the R(18) and R(20) lines, is shown in Fig. 6.24. It is apparent from these figures that, although the two return signals appear

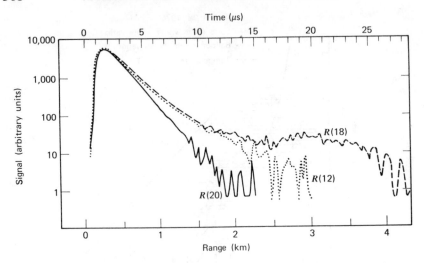

**Fig. 6.23.** Backscatter signal for R(12), R(18), and R(20) lines on 10μm band of $CO_2$. The R(20) is more strongly absorbed by ambient $H_2O$ vapor than is R(12) or R(18). From E. R. Murray, R. D. Hake, Jr., J. E. van der Laan, and J. G. Hawley, *Appl. Phys. Lett.*, **28**, 542 (1976). Reproduced by permission.

free of noise to about 1.5 km, the derived concentration becomes noisy around the 1-km range.

The idea of using a light beam to determine the atmospheric molecular density profile and thereby derive the *temperature profile* was first attempted by Elterman (167) in the early fifties. Employing a searchlight, he was able to evaluate the temperature from 10 to around 67 km by assuming

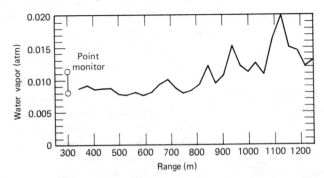

**Fig. 6.24.** Measured value of water vapor concentration as a function of range. A 60-m range cell was used in the data reduction. The point monitor was a calibrated recording hygrothermograph located near the laser line of sight. From E. R. Murray, R. D. Hake, Jr., J. E. van der Laan, and J. G. Hawley, *Appl. Phys. Lett.* **28**, 542 (1976). Reproduced by permission.

that beyond an altitude of 10 km the return was determined principally by Rayleigh scattering and that the ideal gas law could be combined with a hydrostatic relation to give the temperature in terms of the density change.

Strauch, Derr, and Cupp (34) have suggested that Raman backscattering from $N_2$ could be used to determine the molecular density profile and thereby the atmospheric temperature, given the ground level pressure. Although we have seen that Raman scattering can be used to evaluate the density profile, this is achieved by normalizing with respect to the nitrogen Raman return so that the effects of aeroticulate attenuation, $1/R^2$ corrections, $\xi(R)$ uncertainties, and $E_l$ fluctuations are all cancelled. An absolute measurement of the molecular density profile would be vulnerable to all these difficulties. Strauch et al. (34) were able to demonstrate a good correlation between the variations in the amplitude of the Raman return from nitrogen and the temperature fluctuations measured by a thermistor located in a tower.

Cooney (168) proposed that rotational Raman scattering from $N_2$ might serve as a more convenient vehicle for ascertaining the temperature profile of the atmosphere. He based his argument on the stronger scattering intensity and the temperature sensitivity of the rotational Raman spectrum.

Cooney (168) also suggested that a differential technique might be the best way of ensuring both adequate temperature sensitivity and cancelling out most of the extraneous factors, such as atmospheric transmission and the spectral response of the photodetector. Salzman (169) has demonstrated the feasibility of this approach by making temperature measurements from $-20°C$ to $30°C$ at a range of 100 m (indoors) with a resolution of 5 m and an accuracy of $±3°C$. He used two interference filters with circular apertures. These sampled the rotational anti-Stokes spectrum out in the wing and close to the exciting line. The ratio of the resulting intensities is solely dependent upon the gas temperature.

It should be noted, however, that if lines of equal quantum number $J$ on the Stokes and anti-Stokes wings are compared, their ratio is rather insensitive to the gas temperature. The high spectral rejection ratio $(10^5)$ required to eliminate interference from the elastic scattered return can be achieved by use of a multipass interferometer (170) or an iodine gaseous filter (171). Recently, Armstrong (32) has undertaken an analysis which indicates that the new interferometric technique of Barrett and Myers (172) could considerably improve the range and sensitivity of this approach by utilizing all of the rotational Raman lines simultaneously.

Mason (33) has suggested that the thermal distribution of rotational states within a molecule's absorption band could be probed by means of the DAS technique in order to derive the gas temperature. From Eq. (40) it is clear that the ratio of the population in two rotational states can be expressed in the form

$$\frac{N_\alpha^{J_1}(R)}{N_\alpha^{J_2}(R)} = \frac{\kappa_\alpha^{J_2}(\lambda_2,\lambda)}{\kappa_\alpha^{J_1}(\lambda_1,\lambda)} \left( \frac{\dfrac{d}{dR} \ln J(R,\lambda,\lambda_1)}{\dfrac{d}{dR} \ln J(R,\lambda,\lambda_2)} \right) \qquad (114)$$

where $N_\alpha^J(R)$ represents the number density of molecules in the rotational state with quantum number $J$; $\lambda_1$ and $\lambda_2$ represent the two laser wavelengths that are tuned to excite molecules originally in the $J_1$ and $J_2$ rotational states; $\lambda$ represents the laser wavelength chosen to give minimum absorption between the two transitions; $\kappa_\alpha$ represents the appropriate differential absorption cross section; and $J(R,\lambda,\lambda_1) \equiv E(\lambda,R)/E(\lambda_1,R)$ and $J(R,\lambda,\lambda_2) \equiv E(\lambda,R)/E(\lambda_2,R)$ are the respective "on–off" ratios for each of the transitions. For atmospheric conditions the rotational states may be assumed to be in equilibrium at the translational gas temperature $T$. In this case we may write

$$T = \theta[J_1(J_1+1) - J_2(J_2+1)][\ln\{\Lambda(J_1)/\Lambda(J_2)\}]^{-1} \qquad (115)$$

where

$$\Lambda(J_m) \equiv [g_{J_m}\kappa_\alpha^{J_m}(\lambda,\lambda_m)]^{-1} \frac{d}{dR}[\ln J(R,\lambda,\lambda_m)] \qquad (116)$$

$g_{J_m}$ being the degeneracy factor for the $J_m$ rotational state.

Mason (33) has shown, by way of a computer simulation, that temperature measurements might be possible to an altitude of 10 km using the magnetic dipole (A band) transition at 760 nm in $O_2$. However, an improvement in the accuracy might be achieved if a stronger transition, of some minor constituent of the atmosphere (such as $H_2O$ or $CO_2$), were to be employed.

### 6.5.2   Upper Atmospheric Probing

The development of high-energy, flash lamp-pumped, tunable dye lasers made it possible for Bowman, Gibson, and Sandford (173) to make the first ground-based measurement of the sodium atom concentration in the tenuous outer regions of the atmosphere using resonance fluorescence. Subsequently, several groups have studied the spatial and temporal variations of this layer of sodium atoms (7–11).

Hake et al. (7) observed a fourfold increase in the sodium layer content during the maximum of the Geminids meteor shower on the night of December 13–14, 1971. This observation, presented in Fig. 6.25, lends

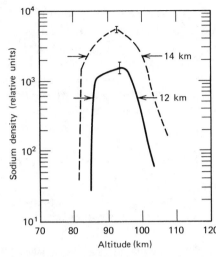

**Fig. 6.25.** Altitude profiles of free atomic sodium density obtained at 2315–2400 PST (solid line) and 0.255–0.310 PST (dashed line), before and slightly after transit of the radiant of the Geminids meteor shower on December 13–14, 1971, the peak of the shower. From R. D. Hake, Jr., D. E. Arnold, D. W. Jackson, W. E. Evans, R. A. Long, B. P. Ficklin, *J. Geophys. Res.*, **77**, 6839 (1972). Copyright by the American Geophysical Union, reproduced by permission.

support to the idea that meteor ablation represents an important source of this material. The laser used by Hake et al. (9) comprised a two-stage (oscillator–amplifier) rhodamine 6G dye laser with an output of 0.5 J in 300 ns at 589.0 nm. The spectral linewidth was less than 0.005 nm, and the final beam was collimated to better than 0.5 mrad. Additional confirmation of the meteor production theory was obtained by Aruga et al (10). Although most observations of this sodium layer were made at night, Gibson and Sandford (8) were able to modify their system sufficiently to map the spatial distribution of the sodium layer during the day and thereby dispel the notion of daytime enhancement. Felix et al. (11) have recently used this resonance fluorescence technique to study the potassium concentration of the upper atmosphere. In the future there is the distinct probability of using resonance fluorescence from high-flying aircraft, or even from the space shuttle, for attaining a more thorough knowledge of the species concentrations in the upper atmosphere.

### 6.5.3 Atmospheric Pollution Surveillance

The adverse effect of air pollution on human health is well established (174), and many government agencies have established maximum allowable levels of exposure for a wide range of atmospheric contaminants. Although there are many methods of monitoring the degree of atmospheric pollution, those based on lasers are in a class of their own. In the following section we look at the techniques that are capable of providing information on the general level of pollution, while in the subsequent section we focus our attention on those techniques that are better suited for effluent source monitoring.

The possibility of using multiple-line gas lasers to detect pollution over an extended path was first considered toward the end of the sixties by Hanst and Morreal (175). They showed that it was possible to detect gaseous pollutants such as CO, NO, $SO_2$, and $O_3$ down to concentrations of a few ppm over a net 1 km path using either a $CO_2$ or an $I_2$ laser and a retroreflector. To avoid absolute calibration of the system, a differential approach is usually employed. In this case, the mean level of pollution that can be detected over a range increment $\Delta R$ is given by an equation that is similar to Eq. (43):

$$\langle C \rangle \simeq \frac{1}{2\alpha(\lambda_l)\,\Delta R} \ln\left[\frac{E_s(\lambda_l + \delta\lambda)}{E_s(\lambda_l)}\right] \tag{117}$$

where $\langle C \rangle$ represents the mean concentration of the constituent of interest (ppm atm$^{-1}$) and $\alpha(\lambda_l)$ represents the corresponding absorption coefficient for 100% concentration (atm$^{-1}$ cm$^{-1}$). Note: $\alpha(\lambda) \equiv N_a\sigma^A(\lambda)$, where $N_a(\simeq 2.55 \times 10^{19}$ cm$^{-3}$ atm$^{-1}$) represents the molecular density at S.T.P. and $\sigma^A(\lambda)$ is the appropriate cross section (cm$^2$).

We have assumed that $\alpha(\lambda_l) \gg \alpha(\lambda_l + \delta\lambda)$; otherwise an appropriate correction factor would have to be introduced in Eq. (117). If we now introduce the fractional change in the received laser energy between the "on" and "off" laser wavelengths,

$$y \equiv \frac{E_s(\lambda_l + \delta\lambda) - E_s(\lambda_l)}{E_s(\lambda_l + \delta\lambda)} \tag{118}$$

we can then write

$$\langle C \rangle^{min} \simeq \frac{1}{2\alpha(\lambda_l)\,\Delta R} \ln\left\{\frac{1}{1 - y_{min}}\right\} \tag{119}$$

In Table 6.7 we have prepared a list of a large number of pollutants of interest and have indicated the minimum concentration (in ppm) that could be detected for a range increment of 100 m if $y_{min}$ is assumed to be 5%. Recently, Asai and Igarashi (42) have indeed measured 0.2 ppm of $O_3$ in a photochemical smog using differential absorption between the R(14) and R(16) $CO_2$ laser lines and a retroreflector situated approximately 80 m from the lasor. Henningsen, Garbuny, and Byer (52) have demonstrated that single-ended absorption measurements of CO (over a range of 107 m) are possible at 2.3 $\mu$m using topographical backscattering from foliage. Recently, Guagliardo, and Bundy (78) have developed an airborne differential lasor to measure the column concentration of $O_3$ between the aircraft and the earth, which acts as a topographical target. This system used two grating tuned TEA $CO_2$ lasers that were fired within 20 $\mu$s to minimize atmospheric scintillation and ensure that both beams strike the same spot on the ground.

## TABLE 6.7
### Pollution Monitoring by Absorption

| Pollutant | Band center, $\mu$m | Suitable laser | Absorption coefficient (per unit conc.), $atm^{-1} cm^{-1}$ | Refer-ence no. | Minimum concentration (for 100-m range increment and 5% signal difference), ppm |
|---|---|---|---|---|---|
| $O_3$ | 9.50 | $CO_2$ | 11.0 | 55 | 0.23 |
| | 9.50 | $CO_2$ | 14.0 | 56 | 0.18 |
| | 0.301 | frequency-doubled organic dye | 8.0 | 43 | 0.32 |
| $CH_4$ | 3.5 | He–Ne | 17.0 | 19 | 0.15 |
| $C_2H_4$ | 10.52 | $CO_2$ | 25.7 | 19 | 0.10 |
| | | | 33.0 | 55 | 0.08 |
| CO | 4.70 | $I_2$ | 5.13 | 54 | 0.50 |
| | 4.70 | $I_2$ | 0.26 | 19 | 9.86 |
| | 2.3 | parametric oscillator with Nd | 0.038 | 52 | 67.0 |
| $C_4H_{10}$ | 3.40 | $I_2$ | 10.3 | 19 | 0.25 |
| NO | 5.30 | $I_2$ | 1.0 | 19 | 2.56 |
| Peroxyacetyl nitrate (PAN) | 5.40 | $I_2$ | 10.3 | 19 | 0.25 |
| $NO_2$ | 0.40 | organic dye | 7.54 | 54 | 0.34 |
| | 0.448 | organic dye | 5.10 | 172 | 0.50 |
| $SO_2$ | 7.30 | semiconductor diode | 17.0 | 19 | 0.15 |
| | 0.29 | frequency-doubled organic dye | 9.16 | 54 | 0.28 |
| | 0.298 | frequency-doubled organic dye | 11.0 | 62b | 0.23 |
| $C_6H_6$ | 0.29 | frequency-doubled organic dye | 35.1 | 54 | 0.073 |
| $NH_3$ | 10.50 | $CO_2$ | 25.7 | 19 | 0.10 |
| | 9.22 | $CO_2$ | 120. | 55 | 0.02 |
| $CC_2F_2$ (Freon 12) | 10.84 | $CO_2$ | 92 | 55 | 0.028 |
| Na | 0.5896 | organic dye | $1.28 \times 10^7$ | 54 | $2 \times 10^{-7}$ |
| Hg | 0.2537 | frequency-doubled organic dye | $1.5 \times 10^6$ | 54 | $1.7 \times 10^{-6}$ |

In principle, much greater sensitivity can be obtained by either a multipath technique or positioning the retroreflector much further away. Indeed, the limit of sensitivity could easily be extended by a factor of 50, which would mean in some instances a sensitivity of better than a few ppb. Additional improvement can be achieved if heterodyne detection is employed (176). Furthermore, the very modest demand made of the laser energy for long-path absorption allows the use of a broad spectrum of lasers, including tunable infrared laser diodes and Raman spin flip lasers (23,115). Unfortunately, the range of situations amenable to installation of a retroreflector will invariably be restrictive. Topographical targets may be used to relax this constraint but at a cost of increased laser energy due to the return signal's range dependence, see Eq. (42).

As previously mentioned, Schotland (143) first proposed the combination of differential absorption and scattering (DAS, also known by the acronyms DASE and DIAL) as a means of studying the water content in the atmosphere. However, Measures (177) in a comparative study of laser techniques was the first to recognize the potential of this approach for pollution mapping. The superiority of DAS over fluorescence and Raman backscattering is clearly seen in Fig. 6.26. Unfortunately, it is also evident from Fig. 6.26 that high background concentrations of $SO_2$ can impose a premature range limitation due to strong attenuation of the laser beam. Slight detuning of the "on" wavelength will obviously relax this constraint. Equation (105) can be used to establish the minimum laser energy needed to measure a given concentration $N_p$ with a spatial resolution $\Delta R$ at a range $R$. In addition, it may be used to derive the cross section required to minimize the laser energy, all other things being equal. If we set $dE_l/d\sigma = 0$, then it is fairly easy to see that $E_l$ is a minimum for

$$\sigma_p^A(\lambda) = \frac{1}{2N_p(R)\,\Delta R} \ln\left(1 + \frac{\Delta R}{R}\right) \qquad (120)$$

which is well approximated by $[2N_p(R)R]^{-1}$ for $\Delta R < R/10$. Thus, in order to detect 1 ppm of *any species* at a range of 5 km with minimum laser energy, the "optimum cross section" is approximately $3.92 \times 10^{-20}$ cm$^2$, that is, $\alpha_p(\lambda) \simeq 1$ (atm$^{-1}$ cm$^{-1}$). Byer and Garbuny (54) have shown by a similar argument that where an excess density of some pollutant, $\Delta N_p$, is to be detected against a background density of the same constituent, $N_p$, the cross section appropriate to a minimum in the laser energy is given by

$$\sigma_p^A(\lambda) = \frac{1}{2L\,\Delta N_p} \ln\left(1 + \frac{L\,\Delta N_p}{RN_p}\right) \qquad (121)$$

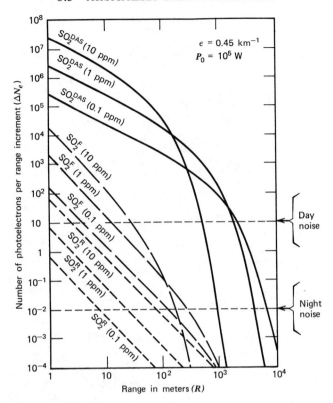

**Fig. 6.26.** Fluorescence, Raman, and DAS signals vs. range for sulphur dioxide. From R. M. Measures and G. Pilon, *Opto-Electron.*, **4**, 141 (1972). Reproduced by permission.

where $L$ is the extent of the pollutant plume and $R$ is its range from the lasor.

As we have seen from Table 6.7, most pollutants have absorption bands in the infrared. However, a few, notably $O_3$, $NO_2$, $SO_2$, $C_6H_6$, and many metals, have healthy absorption features that lie in the visible or near-ultraviolet part of the spectrum. Recently, several groups have undertaken studies of $SO_2$, $O_3$, and $NO_2$ using the DAS approach. Grant and Hake (43) and Grant et al. (57) used a 2.5-m sample cell at a range of 306 m to demonstrate that the kind of sensitivity indicated in Table 6.7 could in fact be achieved in practice for all three pollutants. Hoell, Wade, and Thompson (62) attained a measurement sensitivity of 10 ppb at a range of 0.8 km for $SO_2$ in actual field experiments. A frequency-doubled, flash lamp-pumped, tunable dye laser (Chromatrix Model CMX-4) having an output of $100 \mu J$ in a spectral bandwidth of less than 0.03 nm and a duration of $1.3 \mu s$ was used for these observations. Of

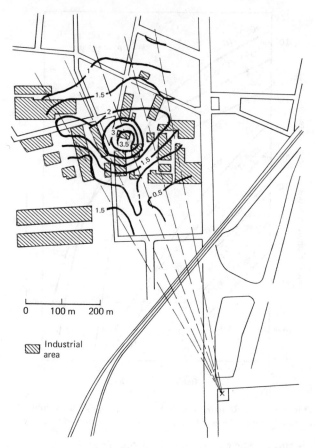

**Fig. 6.27.** $NO_2$ distribution over a chemical factory as derived from measurements performed in the indicated directions at an altitude of about 45 m. The concentrations are given in ppm. From K. W. Rothe, U. Brinkman, and H. Walther, *Appl. Phys.*, **4,** 181 (1974). Reproduced by permission.

equal significance has been the DAS studies undertaken by Rothe, Brinkman, and Walther (60,61) in Germany. They used a 1 mJ, 300 ns. flash lamp-pumped, tunable dye laser operating between 455 and 470 nm to demonstrate that $NO_2$ concentrations down to 0.2 ppm could be detected, at night, up to a range of 4 km over the city of Cologne (60). The most striking illustration of the DAS potential to date is represented by the $NO_2$ spatial distribution over a chemical factory, Fig. 6.27, as mapped by Rothe et al. (61) with a lasor situated 750 m from the chimney stack. By and large, the accuracy of most of these measurements has been limited by the delay incurred in switching the laser wavelength. Significant improvements can be expected when this time interval is reduced to the point where atmospheric fluctuations are negligible.

### 6.5.4  Pollution Source Monitoring

As we have seen, the differential absorption approach is unlikely to be rivaled for range and sensitivity. Nevertheless, the sophistication of the lasor and the complexity of signal interpretation associated with this technique provides considerable incentive to develop alternative approaches where the bounds on range and concentration are conducive. Such a situation is likely to be encountered where the spatial and temporal distributions of specific gaseous contaminants are to be monitored as they emerge from a pollution source.

An approach based upon laser-induced fluorescence may appear to be attractive for a number of constituents such as $SO_2$, $NO_2$, $I_2$, $O_3$, various hydrocarbon vapors, and aerosol pollutants. An analysis of the fluorescence return signal expected from a localized source of pollution was undertaken by Measures and Pilon (46). The results of this study revealed that above a certain peak concentration, a distortion of the returned signal could lead to a misinterpretation of both the range and concentration of the source. An example of the kind of distortion predicted for a Lorentzian distribution of $NO_2$ is presented in Fig. 6.28.

Unfortunately, in the case of atmospheric work, collision quenching and the broadband nature of the emission tend to restrict the remote sensing potential of this approach. In spite of this, fluorescence has been used by Tucker, Birnbaum, and Fincher (45) to measure the $NO_2$ concentration within a sampled volume of urban air with a sensitivity of less than 1 ppb, and calculations based on the experimental work of Gelbwachs and Birnbaum (178) have led Gelbwachs (179) to predict that on clear nights remote sensing of $NO_2$ laser-induced fluorescence could be undertaken with a sensitivity of about 50 ppb at a range of 1 km. Wang et al. (37) have recently used laser-induced fluorescence to make the first measurement of the natural abundance of the OH radical, a highly reactive trace component of the atmosphere that is believed responsible for controlling the worldwide conversion of CO to $CO_2$ (39).

Although the extraordinary small cross section associated with Raman scattering represents a considerable impediment to its use in remote sensing, it possesses several desirable characteristics which make it very attractive for pollution source monitoring:

1.   The spectral shift of the Raman backscattered radiation is specific to each molecule, see Fig. 6.2.

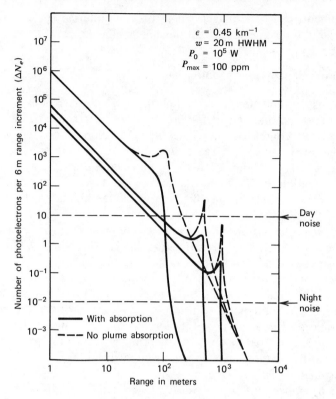

**Fig. 6.28.** Distortion of $NO_2$ fluorescence signal from a Lorentzian plume (due to absorption within plume) centered at one of three locations: $R_0 = 10^2$, $5 \times 10^2$ and $10^3$ m. From R. M. Measures and G. Pilon, *Opto-Electron.*, **4,** 141 (1972). Reproduced by permission.

2. The intensity of a given Raman signal is directly proportional to the density of the appropriate scattering molecule and independent of others. Consequently, a direct measurement of the concentration of a pollutant relative to nitrogen can be obtained without calibration problems.

3. The narrow spectral width and shift of the Raman return are conducive to spectral discrimination against both solar background radiation and elastic scattered laser radiation.

4. The inherent short duration of the Raman process can also be used to discriminate against solar background radiation when a small range interval is of interest.

5. Only a single, fixed-frequency laser is required to produce the simultaneous Raman spectra of all the pollutants within the region being probed. The multiplex advantage of this is obvious.

6. Good spatial and temporal resolution is possible since a backscattering process is involved.

There is one additional problem, alluded to earlier, that can arise in pollution monitoring. The Raman signal from a trace constituent of a plume could be masked by the O- or S-branch Raman signal from a major component. This can be appreciated by reference to Fig. 6.11. Inaba and Kobayasi, who were one of the first to consider Raman scattering for pollution monitoring (180), have addressed this problem and have found that in most instances spectral interference can be avoided by use of narrow spectral filters (22). An excellent example of the differences in the Raman backscattered spectra of ordinary air, oil smoke plume, and the exhaust gas of an automobile was presented by Inaba and Kobayasi (22) and is reproduced here in Fig. 6.29. The corresponding molecular concentrations relative to atmospheric $N_2$ are presented in Table 6.8; in the table the values inside the parentheses are deemed to be less reliable.

A foretaste of what is possible has been provided by Hirschfeld et al. (82) who designed and built one of the most powerful lasors to date. This system incorporates a frequency-doubled, $Q$-switched, 2-J (2 pps) ruby laser, a 36-in.-aperture $f/6.8$ Dall-Kirkham Cassegrain telescope, a polychromator, and an array of photomultipliers to provide multiplex detection. The high sensitivity achieved with this lasor is attested by the daylight Raman spectra (Fig. 6.30) obtained from a controlled plume of $SO_2$ and kerosene at a range of 200 m, with a 10-m range resolution. DeLong (181) has extended these measurements and has shown that sensitivities of the order of 100 ppm for a wide variety of constituents should be achieved even under unfavorable daytime weather conditions for a range of several hundred meters. Melfi, Brumfield, and Storey (40) have reported the detection of Raman scattering by $SO_2$ in the plume of a 200-MW coal-burning electrical generating plant at a slant range of 210 m. An example of the correlation observed between the output power of the electrical plant and the $SO_2$ return signal is shown in Fig. 6.31. An example of the kind of Raman lasor used is illustrated in Fig. 6.32.

## 6.6 HYDROGRAPHIC LASOR APPLICATIONS

### 6.6.1 Bathymetric Surveys, Turbidity Measurements, and Dye Studies

Despite the fact that lasers have been used to probe the atmosphere almost since their inception, it was a number of years before they were

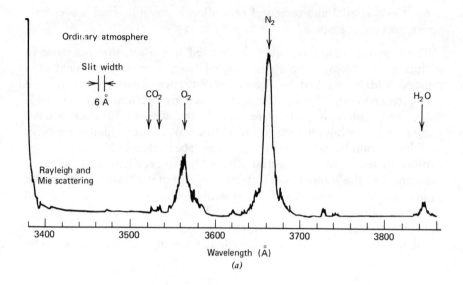

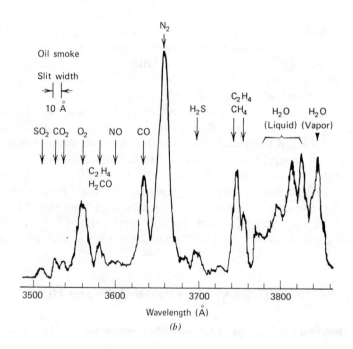

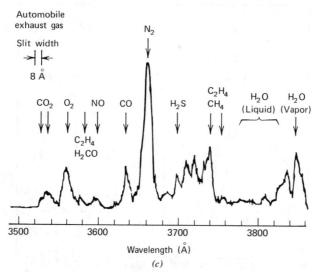

**Fig. 6.29.** (*a*) Measured Spectrum of Raman-shifted and unshifted backscatters from the ordinary atmosphere. (*b*) Spectral distribution of Raman-shifted components from a variety of molecular species in an oil smoke plume remotely analyzed by the laser Raman radar method. (*c*) Spectral distribution of Raman-shifted components from various molecules in an automobile exhaust gas detected by the laser Raman radar scheme. From H. Inaba and T. Kobayasi, *Opto-Electron.*, **4**, 101 (1972). Reproduced by permission.

**TABLE 6.8**
**Experimental Estimation of Molecular Concentrations Relative to That of Atmospheric $N_2$ Molecule from Raman Spectra Measured by Laser Raman Radar Techniques[a]**

| Observed spectrum | Molecular species | | | | | | | | | | | |
|---|---|---|---|---|---|---|---|---|---|---|---|---|
| | $SO_2$ | $CO_2$ | $O_2$ | $H_2CO$ + $C_2H_4$ | NO | CO | $N_2$ | $H_2S$ | $CH_4$ + $C_2H_4$ | $H_2O$ Liquid | Vapor |
| Ordinary atmosphere (Fig. 6.29*a*) | — | 0.02 | 0.26 | — | — | — | 1 | — | — | — | 0.016 |
| Oil smoke plume (Fig. 6.29*b*) | 0.01 | 0.06 | 0.24 | (0.12) | (0.03) | 0.43 | 1 | 0.015 | (0.069) | (0.11) | 0.10 |
| Automobile exhaust gas (Fig. 6.29*c*) | — | 0.11 | 0.21 | (0.04) | (0.09) | 0.28 | 1 | 0.024 | (0.057) | (0.05) | 0.071 |

[a] Values in parentheses are less accurate because of large experimental errors or uncertainties in Raman cross sections (22).

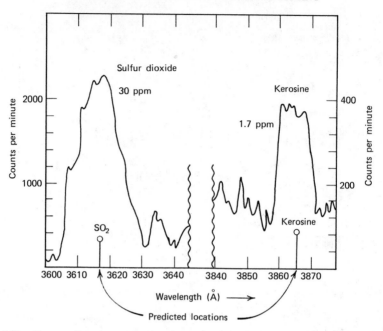

**Fig. 6.30.** Remote Raman returns for SO$_2$ and kerosene air pollution at 200 m. From T. Hirschfeld, E. R. Schildkraut, H. Tannenbaum, and D. Tannenbaum, *Appl. Phys. Lett.*, **22**, 38 (1973). Reproduced by permission.

mounted in aircraft and directed downward to investigate the hydrosphere. Toward the end of the sixties, Hickman and Hogg (64) demonstrated the feasibility of deploying an airborne blue-green laser for the purpose of mapping subsurface topography using the time interval between the surface and subsurface returns as a measure of the depth. A 60-$\mu$J pulsed neon laser was found adequate to record depths of close to 8 m from an altitude of 150 m over the shores of Lake Ontario at night. The 3-ns duration of their pulse laser enabled them to achieve a spatial resolution of about 0.34 m.

Coastal waters, which are likely to be prime targets for laser bathymetry, are often rich in microscopic marine life. Thus, not only is there likely to be enhanced scattering in such areas but also absorption due to phytoplankton. This is reflected in the different spectral profiles of the attenuation coefficient $\epsilon_w(\lambda)$, as seen in Fig. 6.33. The so-called "window on sea" can be seen to lie in the blue-green region of the spectrum. Extensive studies by Hickman (182) and Levis et al. (183) have revealed that the depth measuring capability of a bathymetric laser depends upon the ratio of the absorption coefficient to the scattering

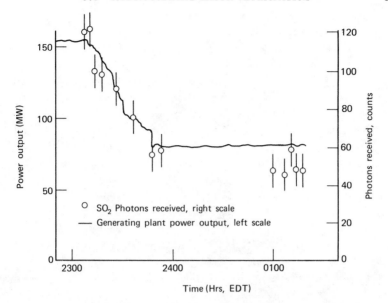

**Fig. 6.31.** Comparison of generating plant power output with the observed $SO_2$ Raman scattering from the plume. From S. H. Melfi, M. L. Brumfield, and R. M. Storey, Jr. *Appl. Phys. Lett.*, **22**, 402 (1973). Reproduced by permission.

coefficient ($\epsilon_W^A/\epsilon_W^S$) as well as on the reflectivity of the bottom and the overall attenuation coefficient ($\epsilon_W = \epsilon_W^A + \epsilon_W^S$).

For hydrographic applications several modifications to the lasor equation are required. The fraction of radiation emanating from a depth $R_W$ below the surface and accepted by the lasor is given by

$$\frac{\Omega_S}{4\pi} = \frac{A_S}{4\pi R_W^2} \tag{122}$$

where $A_S$ represents the surface area within the field of view of the receiver optics. However,

$$\frac{A_S}{(R_W/n)^2} = \frac{A_r}{(R_A + R_W/n)^2} \tag{123}$$

where $n$ is the refractive index of the water and $R_A$ is the altitude of the system. Therefore

$$\frac{\Omega_S}{4\pi} = \frac{A_r}{n^2(R_A + R_W/n)^2 4\pi}$$

and the energy received at wavelength $\lambda$ from depth $R_W$ and arising from

**Fig. 6.32.** 1972 NASA atmospheric lasor with a receiver aperture of 61 cm. From S. H. Melfi, M. L. Brumfield, and R. M. Storey, Jr., *Appl. Phys. Lett.*, **22**, 402 (1973). Reproduced by permission.

an interaction with a species having an appropriate cross section $\sigma_S(\lambda,\lambda_l)$ is

$$E_S(\lambda) = E_l T(R_A)\xi(R_A)T_r(\lambda)\frac{c\tau_d A_r \phi_W N_S \sigma_S(\lambda,\lambda_l)}{8\pi n^2(R_A + R_W/n)^2}$$

$$\times \exp\left[-\int_0^{R_W}\{\epsilon_W(\lambda_l)+\epsilon_W(\lambda)\}\,dR\right] \quad (124)$$

where $N_S$ represents the number density of the species at $R_W$, $T(R_A)$ is the two-way attenuation within the atmosphere, and $\phi_W$ is termed the two-way (air/water–water/air) "transmission factor" for the air/water interface. $\phi_W$ not only includes the transmission factors for a smooth interface but should also allow for the influence of surface roughness, by taking a sea-slope probability factor (183) into account. In the case of bathymetry, we may write for the surface return

$$E_1(\lambda_l) = E_l T_r(\lambda_l)\xi(R_A)\frac{A_r}{R_A^2}\frac{\rho_W(\lambda_l)}{\pi}\exp\{-2\epsilon_A(\lambda_l)R_A\} \quad (125)$$

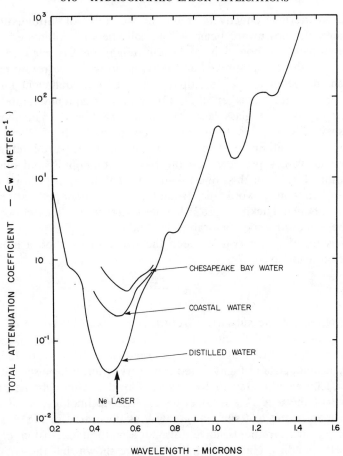

**Fig. 6.33.** Attenuation coefficient of water. Adapted from J. E. Tyler and R. W. Preisendorfer, in *The Sea*, M. N. Hill, Ed., Wiley-Interscience, NY (1962). Used by permission.

and for the subsurface return

$$E_2(\lambda_l) = E_1(\lambda_l) \frac{R_A^2 \phi_W}{(R_A + R_W/n)^2 n^2} \frac{\rho_B(\lambda_l)}{\rho w(\lambda_l)} \exp\{-[\epsilon_W^-(\lambda_l) + \epsilon_W^+(\lambda_l)]R_W\}$$

(126)

where we have assumed a common overlap factor and have introduced the water and subsurface scattering efficiencies $\rho_W(\lambda_l)$ and $\rho_B(\lambda_l)$, respectively.

The attenuation coefficient used in Eq. (126) for the downward $\epsilon_W^-(\lambda_l)$ and upward $\epsilon_W^+(\lambda_l)$ paths can be different due to beam spreading. Duntley (184) has shown that beyond a few exponential attenuation lengths, forward scattering effectively reduces the rate of attenuation. In bathymetry the downward beam will initially be well collimated, and so the "monopath" attenuation coefficient might be expected to apply. However, as the beam spreads and diffuses, forward scattering tends to replenish the beam and a "multipath" attenuation coefficient might be more appropriate. Levis et al. (183) using a frequency-doubled, $Q$-switched neodymium glass laser mounted in a tower 14 m above the water surface were able to detect targets to a depth of 26 m with a considerable sensitivity margin. In general, their measured returns fell between the values predicted on the basis of monopath and multipath attenuation. Based on this, one might expect that the maximum depth-measuring capability would lie around 8 to 10 monopath attenuation lengths, although Hickman (182) has indicated that values closer to 15 might be achieved under favorable conditions.

If $\Delta t$ is the time interval between the two returns, then a monopath assumption leads to

$$R_W = \frac{c \, \Delta t}{2n} \tag{127}$$

Obviously, where the subsurface return is observed by virtue of a multipath attenuation coefficient for part of the path, some error will be incurred in using this relation. Kim, Cervenka, and Lankford (66) have recently developed and flight tested a prototype airborne laser bathymeter over Chesapeake Bay in Key West, Florida. Although the primary objective of these tests was to evaluate ALB technology, a subsurface contour map of the test site was created, as illustrated in Fig. 6.34.

Turbidity measurements have formed a natural extension of these bathymetric studies. Hickman et al. (67) have shown that the volumetric backscattered signal can be directly related to the water turbidity for $\epsilon_W \leq 6 \, \text{m}^{-1}$ and that the greatest sensitivity was achieved at 440 nm. Although 10% accuracy may be feasible, use of several wavelengths would augment such measurements.

Fluorescent dyes have long been used to study water movement and dispersal rates for natural bodies of water. The airborne lasor represents a natural method of extending the range and capability of such measurements. Hickman et al. (182), in an extensive study of suitable dyes, have revealed that acridine red and rhodamine B have the highest product of optical density and quantum efficiency and are thereby likely to be the best candidates for detecting dye cloud movement and dispersion close to the surface. For studies that involve a greater penetration, their work indicates that 3,6-dichlorofluorescein (absorption maximum at 532 nm) might have some advantage.

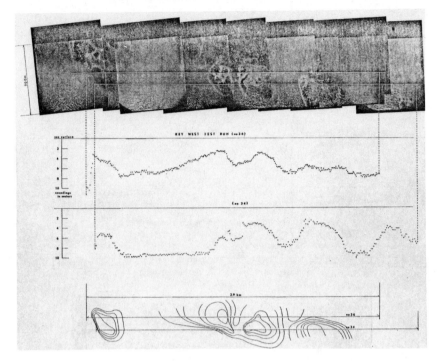

**Fig. 6.34.** Airborne lasor bathymetric survey results of Key West, Florida, undertaken by NASA. From Kim and Cervenka (66).

An interesting outgrowth of their investigation has been the suggestion that differences in the fluorescence sensitivity of dyes to their environment might be exploited to measure various parameters of importance. These include temperature, salinity, and pH. The approach to be adopted, in each instance, is to disperse a calibrated mixture of two appropriate dyes in the area of interest, then probe the resultant cloud with an airborne laser fluorosensor and evaluate the local condition by comparison of the fluorescence return from each of the dyes. The laboratory work of Hickman et al. (182) has suggested that temperature evaluation to an accuracy of ±0.5°C might be possible if the ratio of the fluorescent return from a mixture of rhodamine B with eosin Y is measured to within 2%. A very attractive feature of this technique is that almost all of the factors that would lead to uncertainty in an amplitude measurement cancel in such a ratio, provided the wavelength separation needed for spectral discrimination is not too large. Nevertheless, questions of differential dispersion and sensitivity to more than one parameter remain to be carefully investigated.

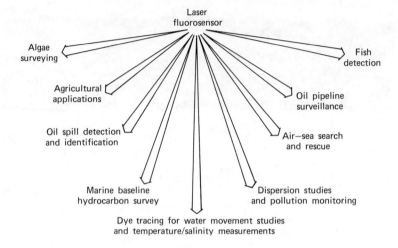

**Fig. 6.35.** Range of applications amenable to laser-induced fluorescence studies of the environment. From R. M. Measures, J. Garlick, W. R. Houston, and D. G. Stephenson, *Can. J. Remote Sensing*, **1**, 95 (Nov. 1975). Reproduced by permission.

### 6.6.2  Oil Spill Detection and Identification

Although the laser fluorosensor was originally conceived for airborne oil spill detection, Measures and Bristow (68) have indicated from the onset that this new form of active remote sensor was capable of undertaking a broad class of missions, some of which had never previously been considered to be within the realm of airborne surveillance. An appreciation for this spectrum of potential application can be obtained by reference to Fig. 6.35. The prototype laser fluorosensor developed by Measures and Bristow (68) is shown in Fig. 6.36, and its principle of operation is illustrated in Fig. 6.37. The airborne version is shown mounted within a DC-3 of the Canada Centre for Remote Sensing in Fig. 6.38.

In the case of oil pollution at sea, or on lakes and rivers, there are many different kinds of sensors that are capable of indicating the presence of some anomaly. These include multiband cameras, radar mappers, infrared scanners, and microwave radiometers. However, none of these can unequivocally recognize the presence of an oil patch on a 24-hour basis or offer any possibility of classifying the type of oil. Fantasia, Hard, and Ingrao (12) studied the fluorescence characteristics of some 29 samples of crude oil. Their results suggested that each sample could be uniquely characterized by a measurement of its peak emission wavelength, lifetime, and fluorescence efficiency, Eq. (58). Moreover, the magnitude of this fluorescence efficiency was sufficient to make airborne

**Fig. 6.36.** University of Toronto Institute for Aerospace Studies (UTIAS) laser fluorosensor at field test site, 1972.

measurements possible. This conclusion was also reached independently by Measures and Bristow (68) based on their program to develop an airborne laser fluorosensor and later confirmed in flight trials by several groups (13,70,71,72).

As a result of this work the role of the laser fluorosensor is likely to be complementary to that of other, high-altitude oil detection sensors. Once one of these sensors detects the presence of some anomaly, the reconnaissance aircraft would reduce altitude in order that the laser fluorosensor could probe the target to determine if the anomaly arose from an oil spill and, if so, characterize the oil. An advanced airborne version of the U.S. Coast Guard oil sensing lasor was mounted inside a helicopter and demonstrated the feasibility of this approach. This system employed an

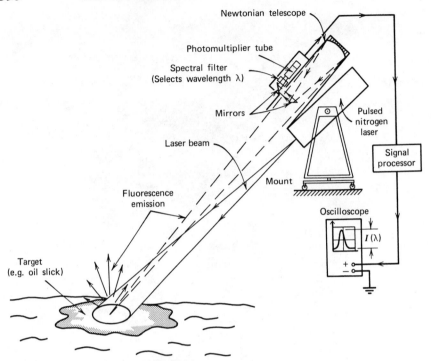

**Fig. 6.37.** Principle of operation of laser fluorosensor. From R. M. Measures, J. Garlick, W. R. Houston, and D. G. Stephenson, *Can. J. Remote Sensing*, **1**, 95 (Nov. 1975). Reproduced by permission.

image dissector and an optical multichannel analyzer (OMA) to record and process the complete fluorescence spectrum, and was arranged to operate in two modes. In the detection mode, two of the 35 OMA channels were calibrated for the ambient seawater fluorescence and were used to indicate the presence of an anomaly in the fluorescence return. Once such an anomaly was monitored, the operator was alerted and the system was switched to the classification mode.

In principle, the combination of lifetime measurement and the spectral distribution of the return signal may be adequate to specify the target in many instances. Unfortunately, in most situations the required degree of discrimination can only be achieved with fine spectral resolution. As seen with the U.S. Coast Guard system, this necessitates the use of many spectral channels, which adds considerably to the cost and complexity and may significantly degrade the signal-to-noise ratio by reducing the energy available per channel. This represents a formidable obstacle to aerial reconnaissance where the monitoring time may be strictly limited by the

**Fig. 6.38.** Canada Centre for Remote Sensing DC-3 mounted airborne version of UTIAS laser fluorosensor. Courtesy M. Bristow, private communication.

physical extent and the patchy nature of the target. One way to improve the specificity of the laser fluorosensor may be to excite at two or more wavelengths, as shown by Houston, Stephenson, and Measures (14).

### 6.6.3 Fluorescence Decay Spectroscopy

An alternative method of economically enhancing the identification capability of a lasor fluorosensor simply involves making better use of the temporal domain and stems from the recent discovery, by Measures, Houston, and Stephenson (14,15), of a new form of spectral signature—the *fluorescence decay time spectrum* (FDS). In the past, measurements of the temporal behavior of fluorescence usually involved the entire emission band and could, for pure substances, be attributed to the lifetime of the excited manifold of levels within a given molecule. In the case of a complex mixture of molecules, as is often of interest in remote sensing, a variation of the fluorescence decay time across the emission band has been observed (14,15) and leads to a new kind of spectral fingerprint. Figure 6.39 presents both the emission profiles and the decay time spectra for a selection of crude oils and refined petroleum products. In general, the crude oil fluorescence decay time is seen to decrease with decreasing API degrees, a trend first noted by Fantasia, Hard, and Ingrao (12), while for any given oil sample the decay time increases with increasing wavelength of emission.

What is of particular relevance in regard to remote sensing is the fact that decay time measurements can be undertaken for optically thick targets (such as oil spills) with relative freedom from the complexities of interpretation that plague amplitude measurements (see Sect. 6.4.6). To make an absolute measurement of the FDS, one has to make allowance for the finite duration of the laser pulse and the response time of the photodetector. Furthermore, the thickness of the target should be small to avoid depth effects interfering with the decay time measurements. This situation is virtually always true in the case of oil spills, as the optical thickness of such materials tends to be of the order $10^{-3}$ cm (12,68).

The recorded fluorescence temporal profile $f(t)$ can be expressed in the form of a convolution integral:

$$f(t) = \int_0^t i(t')\phi(t-t')\,dt'$$

where $i(t)$ represents the instrumental response function and

$$\phi(t) = \int_0^t P(t')F(t-t')\,dt'$$

represents the convolution of the finite laser excitation pulse $P(t)$, with the fluorescence medium possessing a fluorescence response function

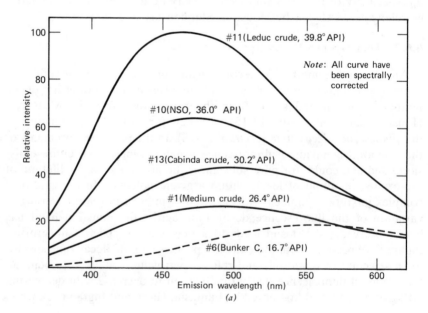

(a)

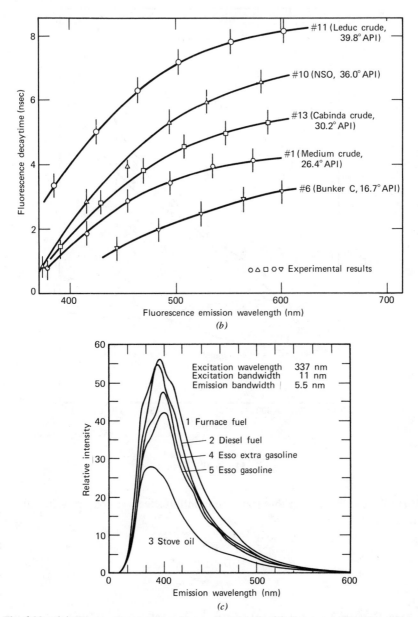

**Fig. 6.39.** (a) Fluorescence spectra of several crude oils. (b) Corresponding "fluorescence decay time spectra" of the same oils. (c) Fluorescence spectra of several refined petroleum products. (d) Corresponding "fluorescence decay time spectra" of four out of the five products. Adapted from R. M. Measures, W. R. Houston, and D. G. Stephenson, *Op. Eng.*, **13,** 494 (1974), by permission.

393

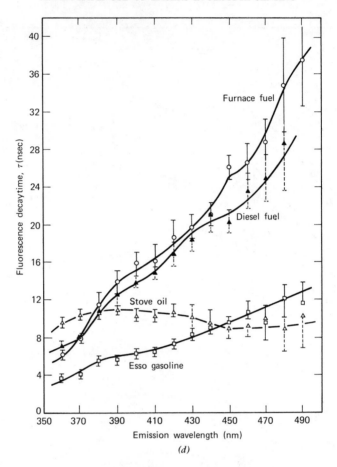

**Fig. 6.39.** *Continued.*

$F(t)$. Since convolution integrals are commutative, we may rewrite this in the form

$$f(t) = \int_0^t I(t')F(t-t')\,dt'$$

where $I(t)$ represents the modified instrumental response function, taking account of both the frequency response of the photodetector and the shape of the exciting pulse through the monitoring system. Unfortunately, there is no convenient analytical procedure for evaluating $F(t)$, given $I(t)$

and $f(t)$ as derived from the convolution integral. There are, nevertheless, relatively simple techniques for obtaining the required information (14).

In the event that the target, say, an oil slick, is simply to be characterized in order to match it to a sample taken from a suspected ship, deconvolution of the FDS may not even be necessary. However, before laser fluorosensor tagging of oil spills becomes accepted, much more will have to be known regarding the changes in the spectral characteristics of an oil slick due to weathering and the biodegradation effects of marine organisms.

### 6.6.4  Water Quality Monitoring and Chlorophyll Mapping

It has long been recognized that all water samples (except those of the very highest purity) can be made to fluoresce by exposure to ultraviolet radiation. This so-called "blue" water fluorescence has represented a considerable source of annoyance to those involved in fluorescent studies (95b). From the remote sensing point of view this fluorescence might constitute a useful parameter by which to study water quality. Indeed, the effluent from the pulp and paper industry has long been recognized as susceptible to measurement by fluorescent techniques (185). This has been corroborated in recent flight trials (71). An examination of the fluorescence of several different kinds of effluent was undertaken by Measures, Garlick, Houston, and Stephenson (73) and revealed that the spectral form of this fluorescence may not be as informative as its amplitude. In order to facilitate such amplitude measurements, Bristow, Houston, and Measures (70) suggested that the OH-stretch water Raman band could be employed as a built-in reference signal by which to normalize the fluorescent return. In this way many of the difficulties of amplitude measurement alluded to in Sect. 6.4.6 could be avoided. The wide variation observed in this fluorescence-to-Raman ratio $F/R$ for different water samples is illustrated in Fig. 6.40. The $F/R$ ratio used in these measurements was defined as the ratio of the peak amplitude of the fluorescence signal divided by the peak magnitude of the Raman component. The latter was obtained by subtracting the fluorescence component from the total signal at the Raman wavelength.

The results led Measures et al. (73) to speculate that the dispersion of an effluent plume might be rapidly mapped by a suitably positioned (tower-mounted or airborne) lasor that is equipped to monitor the $F/R$ ratio. To test this idea, samples of various kinds of effluent were diluted with water taken from an upstream location for each source, and measurements made of the corresponding change in the $F/R$ ratio. An example of the variation in this $F/R$ ratio as a function of dilution is presented as Fig. 6.41. The results of this work indicated that where the

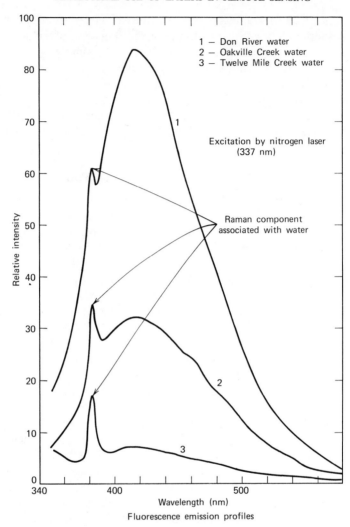

**Fig. 6.40.** Fluorescence emission profiles of three water samples, illustrating wide variation of peak fluorescence Raman ratio. From R. M. Measures, J. Garlick, W. R. Houston, and D. G. Stephenson, *Can. J. Remote Sensing,* **1,** 95 (Nov. 1975). Reproduced by permission.

effluent of an industry or a municipality enters a river of relatively good upstream quality, it might be possible to make fairly precise measurements of the dispersion of the effluent plume based upon the change in the $F/R$ ratio.

In addition, Measures et al. (73) found evidence of a correlation between the magnitude of this $F/R$ ratio and the total organic burden of

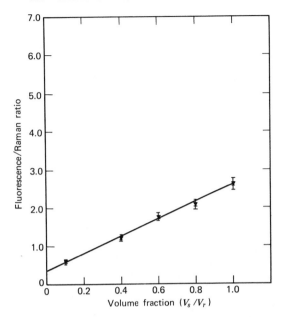

**Fig. 6.41.** Variation of the fluorescence/Raman ratio with dilution of effluent sample concentration in upstream Twelve Mile Creek water. From R. M. Measures, J. Garlick, W. R. Houston, and D. G. Stephenson, *Can. J. Remote Sensing,* **1,** 95 (Nov. 1975).

the water samples. This observation is in general agreement with other studies (186–188) and suggests the possibility of rapidly surveying the organic loading of natural bodies of water by an airborne laser fluorosensor.

Photosynthesis is the ultimate source of all food and the oxygen we breathe. It is the process by which all plants, including the aquatic varieties, convert solar energy into chemical energy. Chlorophyll *a* plays a leading role in this process, and so it is no surprise that Sorenzen (189) and El-Sayed (190) found a highly significant correlation between chlorophyll *a* concentration and the primary productivity in marine surface waters. Yentsch and Menzel (191) were able to show that a fluorescence technique could be used to determine the concentration of chlorophyll *a* within phytoplankton, while Sorenzen (192) developed an in vivo fluorescence approach to evaluating chlorophyll concentrations.

Hickman and Moore (74) first suggested, on the basis of laboratory studies, that chlorophyll measurements with a sensitivity of better than $10 \text{ mg m}^{-3}$ might be undertaken using an airborne, pulsed neon laser (output of about 20 kW) from an altitude of 100 m. An improved system, incorporating a 0.25-J, 300-ns dye laser, was shown by Kim (76) to be

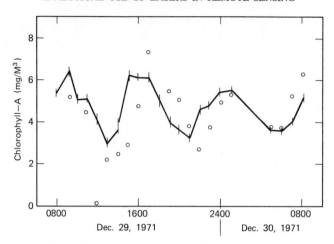

**Fig. 6.42.** Comparison of hourly averaged chlorophyll-$a$ readings as determined by laser chlorophyll analysis (straight line) and wet chemistry chlorophyll analysis (dots). From H. H. Kim, *Appl. Opt.*, **12**, 1454 (1973). Reproduced by permission.

capable of detecting chlorophyll $a$ concentrations down to a fraction of a mg m$^{-3}$ at an altitude of 30 m. Kim (76) has also drawn attention to the difference expected in the fluorescence profiles of algae and plant foliage. He also reports that excitation at 590 nm should yield the largest fluorescence signal and that in a pier-mounted test of a prototype chlorophyll laser fluorosensor fairly close agreement was obtained between laser measurements and those based on wet chemistry. These results are presented here in Fig. 6.42. Helicopter flight trials of this system revealed concentrations varying between 4 and 12 mg m$^{-3}$ across Lake Ontario.

Mumola, Jarrett, and Brown (75) have stated that accurate measurements of chlorophyll $a$ concentration can only be undertaken if allowance is made for the different color groups that may be present in any natural mixture of algae. Indeed, they show that additional fluorescence due to other pigments may often appear, as indicated in Fig. 6.43, even though the characteristic chlorophyll $a$ fluorescence peak at 685 nm is always present. Furthermore, the excitation spectrum is found to vary considerably from one phytoplankton color group to another, as seen in Fig. 6.43. In spite of this, the excitation and emission spectra for different species within one color group are observed, see Fig. 6.44, to be remarkably similar. Mumola, Jarrett, and Brown (75) have proposed a four-wavelength excitation scheme that exploits these features in order to evaluate the concentration of chlorophyll $a$ in an arbitrary mixture of phytoplankton color groups.

The optically thick form of the lasor Eq. (56) is appropriate to this

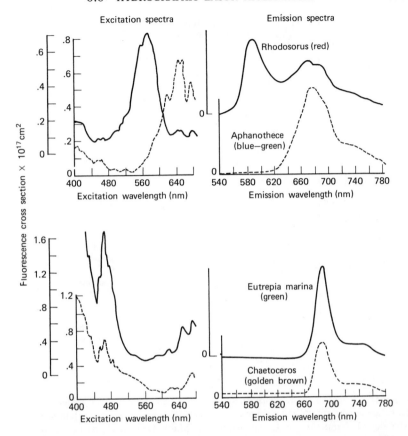

**Fig. 6.43.** Typical fluorescence cross-sections and emission spectra of red, blue-green, green, and golden-brown algae samples. From P. B. Mumola, O. Jarrett, Jr., and C. A. Brown, Jr. "Multiwavelength Lidar for remote sensing of chlorophyll *a* in Algae and Phytoplankton," NASA SP-375 (1973) pp. 137–145. Reproduced by permission.

situation, and so we can write the fluorescence return received in a band centered on the chlorophyll wavelength $\lambda_f$ (685 nm in all cases) when excited by a laser of wavelength $\lambda_i^i$ as follows:

$$E(\lambda_i^i) = E_l T_r(\lambda_f) \, \Delta\lambda_r T(R_A) \xi(R_A) \frac{A_r}{4\pi R_A^2} \frac{\phi_W \sum_j N_j \sigma_j^F(\lambda_i^i)}{[\epsilon(\lambda_i^i) + \epsilon(\lambda_f)]} \quad (128)$$

where $N_j$ represents the number density of chlorophyll *a* molecules in the phytoplankton color group *j*, $\sigma_j^F(\lambda_i^i)$ is the fluorescence cross section of the *j* color group molecules for emission into the 685-nm chlorophyll band when excited at $\lambda_l^i$, and $\phi_W$ is the two-way transmission factor for the air/water interface.

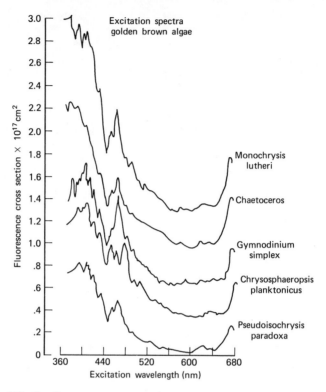

**Fig. 6.44.** Effective fluorescence cross sections of chlorophyll *a* in golden-brown algae. Data shown are normalized to single molecule values with a 5-nm resolution. From P. B. Mumola, O. Jarrett, Jr, and C. A. Brown, Jr., "Multiwavelength Lidar for Remote Sensing of Chlorophyll *a* in Algae and Phytoplankton," NASA SP-375 (1973), pp. 137–145. Reproduced by permission.

In principle, the color group densities can be evaluated by matrix inversion, given all of the other factors. Unfortunately, as seen from Eq. (128), the attenuation coefficients at each of the laser wavelengths must also be known and would have to be measured. As we have seen above, this could probably be achieved by simultaneously monitoring each of the laser backscattered signals. A prototype four-wavelength (454.4, 539.0, 598.7, and 617.8 nm) laser has been developed by Mumola, Jarrett, and Brown (75). This system operated with a minimum energy of 0.6 mJ pulse$^{-1}$ and was able to detect $<1$ mg m$^{-3}$ chlorophyll *a* from an altitude of 100 m during daytime operating conditions. It is also worth noting that at this altitude the system met the eye safety limits for a waterlevel observer. It may, however, be appropriate at this point to sound a warning. If $\epsilon(\lambda_i^j) \gg \epsilon(\lambda_f)$ due to strong absorption within the algae, then it might also be reasonable to assume that

$$\epsilon(\lambda_l) \simeq \Sigma_j N_j \sigma_j^A(\lambda_l^i) \tag{129}$$

In this case, if the quantum yield for each color group has a similar value, the return signal tends to become independent of the density [cf. Eq. (57)] and it will not be possible to measure the chlorophyll concentration. An additional issue raised by Yentsch (193) concerns the competitive absorption between algae and the organic material (often called Gelbstoff) carried to the sea by rivers. He suggests that over and above the chlorophyll measurements it may be necessary to excite at 350 nm while making fluorescence measurements at 500 nm in order to separate the influence of these two kinds of fluorescent material.

This section would be incomplete without some reference to the current interest in Raman studies of water and its contaminants. Slusher and Derr (194), Chang and Young (195), and Walrafen (196) have alluded to the possibility of making use of the OH-stretch water Raman band for the purpose of making temperature measurements, while Bradley (197) and Davis, Bristow, and Koningstein (198) have considered the possibility of using laser Raman spectroscopy for the detection and identification of molecular water pollutants. Lastly, Houghton (199), Dylis (200), and Walrafen (201) have viewed Raman spectroscopy as a means of measuring ionic concentrations in water. The remote sensing potential of any of these approaches seems at present to be indeed remote.

## 6.7 CONCLUSIONS

We have seen that the *lasor* constitutes an important new weapon in the arsenal of those who would study the environment. The range of this active probe extends from the tenuous outer regions of the atmosphere to below the surface of the hydrosphere. These instruments may be either mounted on the ground or in towers, or they may be operated from mobile platforms such as aircraft, helicopters, or possibly satellites under very special circumstances. The lasor may be used in a variety of roles, from simply evaluating the spatial distribution of a specific component within the environment to identifying the constituents of a complex target by essentially performing a spectral analysis at a distance.

By and large, we have concentrated on pulsed lasors as they possess the capability of making both spatial and temporal measurements and in certain situations are capable of greater sensitivity and specificity, via decay time measurements, than their continuous wave (cw) counterparts. Nevertheless, there are certain classes of applications where modulated cw systems might have some advantage regarding either sensitivity

(IR absorption) or cost and size. The latter situation can arise where a compact alarm sensor is required to simply indicate the threshold detection of some specific constituent at close range.

To speculate on the future of laser environmental sensing would be rather naive in the light of past dramatic advances. Nevertheless, it is clear that changes in three areas will play a major role in shaping the destiny of this subject. The first involves advances in laser technology, the second relates to improvements in infrared detector sensitivity, while the third pivots about the eye safety regulations finally adopted. One thing does, however, appear sure—laser environmental sensing is likely to blossom into one of the most important branches of remote sensing. Indeed, the "real time" analysis capability of lasors raises the fascinating specter of direct feedback control over sources of pollution emission.

## References

1.  G. B. Lubkin, *Physics Today* (Oct. 1975), pp. 34–39.
2.  A. L. Schawlow and C. H. Townes, *Phys. Rev.*, **112,** 1940 (1958).
3.  T. H. Maiman, *Nature*, **197,** 493 (1960).
4.  F. J. McClung and R. W. hellwarth, *J. Appl. Phys.*, **33,** 828 (1962).
5.  G. Fiocco and L. D. Smullin, *Nature*, **199,** 1275 (1963).
6.  M. G. H. Ligda, Proc. 1st Conf. Laser Technol., San Diego, California, (1963), pp. 63–72.
7.  R. D. Hake, Jr., D. E. Arnold, D. W. Jackson, W. E. Evans, R. A. Long, and B. P. Ficklin, *J. Geophys. Res.*, **77,** 6839 (1972).
8.  A. J. Gibson and M. C. W. Sandford, *Nature*, **239,** 509 (1972).
9.  A. J. Gibson and M. C. W. Sandford, *Atmos. Terrest. Phys.*, **33,** 1675 (1971).
10.  T. Aruga, H. Kamiyama, M. Jyumonji, T. Kobayasi, and H. Inaba, "Laser Radar Observation of the Sodium Layer in the Upper Atmosphere," *Rep. Ionosphere Space Res. Japan*, **28,** 65 (1974).
11.  F. Felix, W. Keenliside, G. Kent, and M. C. W. Standford, *Nature*, **246,** 345 (1973).
12.  J. F. Fantasia, T. M. Hard, and H. C. Ingrao, "An Investigation of Oil Fluorescence as a Technique for Remote Sensing of Oil Spills," Report #DOT-TSC-USCG-71-7, June 1971. Transportation Systems Centre, Dept. of Transport, Cambridge, Mass. (1971).
13.  J. F. Fantasia and H. C. Ingrao, "Development of an Experimental Airborne Laser Remote Sensing System for the Detection and Classification of Oil Spills," Proc. 9th Int. Symp. on Remote Sensing of the Environment, April 15–19, 1974, Paper 10700-1-X, pp. 1711–1745.

14. W. R. Houston, D. G. Stephenson, and R. M. Measures, "Lifes: Laser Induced Fluorescence and Environmental Sensing," NASA Conference on the Use of Lasers for Hydrographic Studies, NASA SP-375 (1973), pp. 153–169.

15. R. M. Measures, W. R. Houston, and D. G. Stephenson, *Op. Eng.*, **13**, 494 (1974).

16. Freeman F. Hall, Jr., "Laser Systems for Monitoring the Environment," in *Laser Applications*, Vol. II, M. Ross, Academic Press, New York (1974).

17. C. P. Wang, *Acta Astron.* **1**, 105 (1974).

18. V. E. Derry and C. G. Little, *Appl. Opt.*, **9**, 1976 (1970).

19. (a) J. A. Hodgeson, W. A. McClenny, and P. L. Hanst, *Science*, **182**, 248 (1973). (b) P. L. Hanst, *Appl. Spectrosc.*, **24**, 161 (1970).

20. R. L. Byer, *Opt. Quant. Electron.* **7**, 147 (1975).

21. G. W. Grams, *Atm. Technol.* 61 (Winter 1974–75).

22. H. Inaba and T. Kobaysi, *Opto-Electron.,* **4**, 101 (1972).

23. E. D. Hinkley, *Opto-Electron.*, **4**, 69 (1972).

24. D. A. Leonard, *Nature*, **216**, 142 (1967).

25. J. Cooney and J. Orr, *Nature*, **224**, 1098 (1969).

26. J. A. Cooney, *Appl. Phys. Lett.*, **12**, 40 (1968).

27. R. L. Schwiesow and V. E. Derr, *J. Geophys. Res.*, **75**, 1629 (1970).

28. M. C. W. Standford, *J. Atmos. Terreste. Phys.*, **29**, 1657 (1967).

29. J. Cooney and M. Pina, *Appl. Opt.*, **15**, 602 (1976).

30. G. B. Northam, J. M. Rosen, S. H. Melfi, T. J. Pepin, M. P. McCormick, D. J. Hofman, and W. H. Fuller, Jr., *Appl. Opt.*, **13**, 2416 (1974).

31. R. G. Strauch, V. E. Derr, and R. E. Cupp, *Remote Sensing Environ.*, **2**, 101 (1972).

32. R. L. Armstrong, *Appl. Opt.*, **14**, 383 (1975).

34. R. G. Strauch, V. E. Derr, and R. E. Cupp, *Appl. Opt.*, **10**, 2665 (1971).

35. G. Benedetti-Michelangeli, F. Congeduti, and G. Fiocco, *J. Atm. Sci.*, **29**, 906 (1972).

36. T. R. Lawrence, D. J. Wilson, C. E. Craven, I. P. Jones, R. M. Huffaker, and J. A. L. Thompson, *Rev. Sci. Instr.*, **43**, 512 (1972).

37. C. C. Wang and L. I. Davis, Jr., *Phys. Rev. Lett.*, **32**, 349 (1974).

38. E. L. Baardsen and R. W. Terhune, *Appl. Phys. Lett.*, **21**, 209 (1972).

39. C. C. Wang, L. I. Davis, Jr., C. H. Wu, S. Japer, N. Niki, and B. Weinstock, *Science*, **189**, 797 (1975).

40. S. H. Melfi, M. L. Brumfield, and R. M. Storey, Jr., *Appl. Phys. Lett.*, **22**, 402 (1973).

41. K. H. Becker, U. Schurath, and T. Tatarczyk, *Appl. Opt.*, **14**, 310 (1975).

42. K. Asai and T. Igarashi, *Opt. Quant. Electron.*, **7**, 211 (1975).

43. W. B. Grant and R. D. Hake, Jr., *Appl. Phys. Lett.*, **46**, 3019 (1975).

44. T. Kobayasi and H. Inaba, *Appl. Phys. Lett.*, **17,** 139 *August* (1970).

45. A. W. Tucker, M. Birnbaum, and C. L. Fincher, *Appl. Opt.*, **14,** 1418 *June* (1975).

46. R. M. Measures and G. Pilon, *Opto-Electron.* **4,** 141 (1972).

47. W. R. Fenner, H. A. Hyatt, J. M. Kellam, and S. P. S. Porto, *J. Opt. Soc. Amer.*, **63,** 73 (1973).

48. D. A. Stephenson, *J. Quant. Spectrosc. Radiat. Transfer*, **14,** 1291 (1974).

49. D. G. Fouche and R. K. Chang, *Appl. Phys. Lett.*, **18,** 579 (1971).

50. D. G. Fouche and R. K. Chang, *Appl. Phys. Lett.*, **20,** 256 (1972).

51. R. A. Smith, F. E. Jones, and R. P. Chasmar, *The Detection and Measurement of Infrared Radiation*, Oxford Press, London (1968).

52. T. Henningsen, N. Carbuny, and R. L. Byer, *Appl. Phys. Lett.*, **24,** 242 (1974).

53. E. D. Hinkley and P. L. Kelly, *Science*, **171,** 635 (1971).

54. R. L. Byer and M. Garbuny, *Appl. Opt.*, **12,** 1496 (1973).

55. W. Schnell and G. Fischer, *Appl. Opt.*, **14,** 2058 (1975).

56. J. Shewchun, B. K. Garside, E. A. Ballik, C. C. Y. Kwan, M. M. Elsherbiny, G. Hogenkamp, and A. Kazandjian, *Appl. Opt.*, **15,** 340 (1976).

57. W. B. Grant, R. D. Hake, Jr., E. M. Liston, R. C. Robbins, and E. K. Proctor, Jr., *Appl. Phys. Lett.*, **24,** 550 (1974).

58. D. C. O'Shea and L. G. Dodge, *Appl. Opt.*, **13,** 1481 (1974).

59. S. A. Ahmed, *Appl. Opt.*, **12,** 901 (1973).

60. K. W. Rothe, U. Brinkman, and H. Walther, *Appl. Phys.*, **3,** 115 (1974).

61. K. W. Rothe, U. Brinkman, and H. Walther, *Appl. Phys.*, **4,** 181 (1974).

62. (a) J. M. Hoell, Jr., W. R. Wade, and R. T. Thompson, Jr., Int. Conf. Environ. Sensing and Assessment, Las Vegas, Sept. 14–19 (1975); (b) R. T. Thompson, Jr., J. M. Hoell, Jr., and W. R. Wade, *J. Appl. Phys.*, **46,** 3040 (1975).

63. E. R. Murray, R. D. Hake, Jr., J. E. van der Laan, and J. G. Hawley, *Appl. Phys. Lett.*, **28,** 542 (1976).

64. D. Hickman and J. E. Hogg, *Remote Sensing Environ.*, **1,** 47 (1969).

65. C. E. Prettyman and M. D. Cermak, *IEEE Trans. Geosci. Electron.*, **GE-7** 235 (1969).

66. H. H. Kim and P. O. Cervenka, "Development of an Airborne Laser Bathymeter," NASA TN D-8079 (October 1975).

67. G. D. Hickman, A. H. Ghovanlou, E. J. Friedman, C. S. Gault, and J. E. Hogg, "A Feasibility Study for a Remote Laser Water Turbidity Meter," NASA CR-132376 (1974).

68. (a) R. M. Measures and M. Bristow, "The Development of a Laser Fluorosensor for Remote Environmental Probing," Joint Conference on Sensing of Environmental Pollutants, Palo Alto, Calif., Nov. 1971,

AIAAA Paper 71-112 (1971); (b) R. M. Measures and M. Bristow, *Can. Aeron. Space J.*, **17**, 421 (1971).

69. R. M. Measures, W. Houston, and M. Bristow, *Can. Aeron. Space J.*, **19**, 501 (1973).

70. M. P. F. Bristow, W. R. Houston, and R. M. Measures, "Development of a Laser Fluorosensor for Airborne Surveying of the Aquatic Environment," NASA Conference on the Use of Lasers for Hydrographic Studies, NASA SP-375 (1973), pp. 119–136.

71. R. A. O'Neill, A. R. Davis, H. G. Gross, and J. Kruus, "A Remote Sensing Laser Fluorometer," NASA Conference on the Use of Lasers for Hydrographic Studies, Wallops Island, Sept. 1973, NASA SP-375 (1973), pp. 173–195.

72. H. H. Kim and G. D. Hickman, "An Airborne Laser Fluorosensor for the Detection of Oil on Water", NASA Conference on the Use of Lasers for Hydrographic Studies, NASA SP-375 (1973), pp. 197–202.

73. R. M. Measures, J. Garlick, W. R. Houston, and D. G. Stephenson, *Can. J. Remote Sensing*, **1**, 95 (Nov. 1975).

74. G. D. Hickman and R. B. Moore, "Laser Induced Fluorescence in Rhodamine 3 and Algae," Proc. of 13th Conf. Great Lakes Res. 1970, Int. Assoc. Great Lakes Res. (1970), pp. 1–14.

75. P. B. Mumola, O. Jarrett, Jr., and C. A. Brown, Jr., "Multiwavelength Lidar for Remote Sensing of Chlorophyll *a* in Algae and Phytoplanktion," NASA Conference on the Use of Lasers for Hydrographic Studies, NASA SP-375 (1973), pp. 137–145.

76. H. H. Kim, *Appl. Opt.*, **12**, 1454 (1973).

77. F. Singer, *Appl. Opt.*, **7**, 1125 (1968).

78. (a) J. L. Guagliardo and D. H. Bundy, "Earth Reflected Differential Absorption Using TEA Lasers: A Remote Sensing Method for Ozone," 7th Int. Laser Radar Conf., Palo Alto, (1975); (b) "Differential Monitoring of Ozone in the Troposphere Using Earth Reflected Differential Absorption," Int. Telemetering Conference (Oct. 1974).

79. R. K. Seals, Jr., and C. H. Bair, *ISA, JSP*, 6675 (1973).

80. S. K. Poultney, *Space Res.*, **12**, 403 (1972).

81. G. W. Grams and C. M. Wyman, *J. Appl. Meteorol.* **11**, 1108 (1972).

82. T. Hirschfeld, E. R. Schildkraut, H. Tannenbaum, and D. Tannenbaum, *Appl. Phys. Lett.*, **22**, 38 (1973).

83. S. H. Melfi, J. D. D. Lawrence, Jr., and M. P. McCormick, *Appl. Phys. Lett.*, **15**, 295 (1969).

84. Lecroy Research Ltd., WD2000, and Tektronix, R7912.

85. E. E. Uthe and R. J. Allen, *Opt. Quantum Electron.*, **7**, 121 (1975).

86. C. L. frush, *Op. Quantum Electron.*, **7**, 179 (1975).

87. T. Hirschfeld, *Appl. Opt.*, **13**, 1435 (1974).

88. S. K. Poultney, "Single Photon Detection and Timing: Experiments and Techniques," *Advan. Electron. Electron Phys.*, **31,** 39 (1972).

89. H. Melchior, "Demodulation and Photodetection Techniques," in Laser Handbook, Vol. 1, C7, F. T. Arecchi and E. O. Schultz-Dubois, Eds., North-Holland, Amsterdam (1972).

90. M. Ross, *Laser Receivers (Devices, Techniques, Systems)* Wiley, New York (1966).

91. W. Wolber, *Res./Develop.*, **19,** 18 (1968).

92. J. G. Timothy and R. L. Bybee, *Rev. Sci. Instr.*, **46,** 1615 (1975).

93. R. J. Keyes and R. H. Kingston, *Phys. Today*, 48, (March 1972).

94. R. B. Emmons, S. R. Hawkins, and K. F. Cuff, *Opt. Eng.*, **14,** 21 (1975).

95. (*a*) D. A. Leonard, AVCO-Everett Research Report #362, Dec. 1970. "Feasibility Study of Remote Monitoring of Gas Pollutant Emissions by Raman Spectroscopy" (1970); (*b*) C. A. Parker, *Photoluminescence of Solutions*, Elsevier, New York (1968).

96. Monochromatic Interference Filters, Jenaer Glasswerk Schott & Gen., Mainz. (1965).

97. H. Walther and J. L. Hall, *Appl. Phys. Lett.*, **17,** 239 (1970).

98. M. Born and E. Wolf, *Principles of Optics*, 2nd ed., Pergamon Press, New York (1964).

99. P. J. Jacquinot, *J. Opt. Soc. Amer.*, **44,** 761 (1954).

100. A. Girard and P. Jacquinot, "Principles of Instrumental Methods in Spectroscopy," in *Advanced Optimal Techniques*, A. Van Heel, Ed., North-Holland, Amsterdam (1967), Chap. 3.

101. J. V. Ramsay, *Appl. Opt.*, **1,** 411 (1962).

102. *The Spex Speaker*, **XI,** No. 4, Fig. 4, (1966).

103. Jobin Yvon Optical Systems, "Diffraction Gratings" Handbook.

104. T. Hirschfeld and S. Klainer, *Optical Spectra*, pp. 63–66, (1970).

105. S. M. Klainer, T. Hirschfeld, and E. R. Schildkraut, "The Detection of Toxic Contaminants in the Atmosphere Using Single Ended Remote Raman Spectrometric Techniques," The Central States Section of the Combustion Institute, Houston, Texas, April 7–8 (1970).

106. R. T. H. Collis and E. E. Uthe, *Opto-Electronics*, **4,** 87 (1972).

107. H. Heard, *Nature*, 667 (1963).

108. D. A. Leonard, *Appl. Phys. Lett.*, **7,** (1965).

109. D. A. Leonard, in *Laser Raman Gas Diagnostics*, M. Lapp and C. M. Penney (Eds.), Plenum Press, New York (1974), pp. 45–61.

110. *Laser Focus*, 4 (Aug. 1975).

111. D. Basting, F. P. Schäfer, and B. Steyer, *Opto-Electron*, **4,** 43, (1972).

112. F. P. Schäfer, Ed., *Dye Lasers*, Springer Verlag, New York (1973).

113. T. W. Hänsch, *Appl. Opt.*, **11,** 895 (1972).

114. O. R. Wood, II, *Proc. IEEE*, **62**, 355 (1974).

115. K. W. Nill, *Opt. Eng.*, **13**, 516 (1974).

116. M. L. Bhaumik, R. S. Bradford, and A. R. Ault, *Appl. Phys. Lett.*, **28**, 23 (1976).

117. J. J. Ewing and C. A. Brau, *Appl. Phys. Lett.*, **27**, 350 (1975).

118. H. Rosen, P. Robrish, and O. Chamberlain, *Appl. Opt.*, **14**, 2703 (1975).

119. L. R. Kroller, *Ultraviolet Radiation*, 2nd ed., Wiley, New York.

120. A. E. S. Green, *The Middle Ultraviolet*, Wiley, New York (1966).

121. H. S. Gebbie, W. R. Harding, C. Hilsum, A. W. Pryce, and V. Roberts, *Proc. Roy. Soc. A*, **206**, 87 (1951).

122. M. Kerker, *The Scattering of Light*, Academic Press, New York (1969).

123. G. Mie, *Ann. Physik*, **25**, 377 (1908).

124. P. W. Kruse, L. D. McGlauchlin, and R. B. McQuistan, *Elements of Infrared Technology*, Wiley, New York (1963).

125. (*a*) S. Twomey and H. B. Howell, *Appl. Opt.*, **4**, 501 (1965); (*b*) R. W. Fenn, *Appl. Opt.*, **5**, 293 (1966).

126. B. G. Schuster, *Opt. Quant. Electron.*, **7**, 215 (1975).

127. G. Herzberg, *Molecular Spectra and Molecular Structure*, Vol. I, Van Nostrand, Toronto (1967).

128. G. Placzek, *Handbuch der Radiologie*, Vol. VI, Pt. 2, Akademische Verlag, Leipzig (1934), pp. 205–374 (UCRL-Trans-526L).

129. W. F. Murphy, W. Holzer, and H. J. Bernstein, *Appl. Spectrosc.*, **23**, 211 (1969).

130. R. L. Schiesow and N. L. Abshire, *J. Appl. Phys.*, **44**, 3808 (1973).

131. H. A. Hyatt, J. M. Cherlow, W. R. Fenner, and S. P. S. Porto, *J. Opt. Soc. Amer.*, **63**, 1604 (1973).

132. C. M. Penney, R. L. St. Peters, and M. Lapp, *J. Opt. Soc. Amer.*, **64**, 712 (1974).

133. J. J. Barrett, in *Laser Raman Gas Diagnostics*, M. Lapp and C. M. Penney, Eds., Plenum Press, New York (1974), 63–85.

134. W. M. H. Smith, *Opto-Electronics*, **4**, 161 (1972).

135. L. A. Woodward, in *Raman Spectroscopy*, H. A. Szymanski, Ed., Plenum Press, N. Y., 1967.

136. C. M. Penney, "Light Scattering and Fluorescence in the Approach to Resonance," in *Laser Raman Gas Diagnostics*, M. Lapp and C. M. Penney, Eds., Plenum Press, New York (1974), pp. 191–217.

137. D. G. Fouche, A. Herzenberg, and R. K. Chang, *J. Appl. Phys.*, **43**, 3846 (1972).

138. C. M. Penney, *J. Opt. Soc. Amer.*, **59**, 34 (1969).

139. D. G. Fouche and R. K. Chang, *Phys. Rev. Lett.*, **29**, 536 (1972).

140. R. L. St. Peters and S. D. Silverstein, *Opt. Commun.*, **7**, 193 (1973).

141.  R. L. St. Peters, S. D. Silverstein, M. Lapp, and C. M. Penney, *Phys. Rev. Lett.*, **30,** 191 (1973).

142.  P. F. Williams, D. L. Rousseau, and S. H. Dworetsky, *Phys. Rev. Lett.*, **32,** 196 (1973).

143.  R. M. Schotland, Proc. 4th Symposium on Remote Sensing of the Environment, April 12–14, 1966, Univ. of Michigan, Ann Arbor (1966), pp. 273–283.

144.  R. M. Schotland, *J. Appl. Meteorol.* **13,** 71 (Feb. 1974).

145.  W. L. Wolfe, Ed., *Handbook of Military Infrared Technology*, ONR Cat. No. 65–62266, U. S. Government Printing Office.

146.  H. Kildal and R. L. Byer, *Proc. IEEE*, 59, 1644 (Dec. 1971).

147.  R. M. Measures, "Lidar Equation Analysis," *Appl. Opt.*, **16,** 1092 (1977).

148.  W. K. Pratt, *Lidar Communication Systems*, Wiley, New York (1969).

149.  B. M. Oliver, *Proc. IEEE*, **53,** 436 (1965).

150.  R. Foord, R. Jones, C. Oliver, and E. Pike, *Appl. Opt.*, **8,** 1975 (1969).

151.  S. Nakahara, K. Ito, S. Ito, A. Fuke, S. Komatsu, H. Inaba, and T. Kobayasi, *Opto-Electronics*, **4,** 169 (1972).

152.  W. Schottky, *Ann. Phys. (Leipzig)*, **57,** 541 (1918).

153.  J. A. Topp, H. W. Schrotter, H. Hacker, and J. Bradmuller, *Sci. Instr.*, **40,** 1164 (1969).

154.  R. T. V. Kung and I. Itzkan, *Appl. Opt.*, **15,** 409 (1976).

155.  American National Standards Institute—Laser Standard Designated (Z 136.1 – 1973), ANSI, New York (1973).

156.  *Laser Focus*, 14 (Nov. 1972).

157.  J. A. Zuchlich and J. S. Connolly, *J. Opt. Soc. Amer.*, **66,** 79 (1976); *Laser Focus*, 41 (Jan. 1976).

158.  M. L. Wolbarsht and D. H. Sliney, in *Laser Applications in Medicine and Biology*, Vol. 2, M. L. wolbarsht, Ed., Plenum Press, New York (1974).

159.  A. Vassiliadis, in *Laser Applications in Medicine and Biology*, Vol. 2, M. L. Wolbarsht, Ed., Plenum Press, New York (1974).

160.  R. D. Bourdreau, *J. Appl. Meteoral.* 9, 316 (1970).

161.  M. J. Garvey and G. S. Kent, *Nature*, **248,** 124 (1974).

162.  J. Cooney, *Appl. Opt.*, **14,** 270 (1975).

163.  D. A. Leonard and B. Caputo, *Opt. Eng.*, **13,** 10 (1974).

164.  J. A. Cooney, *J. Appl. Meteorol.*, **9,** 182 (1970).

165.  J. A. Cooney, *J. Appl. Meteorol.*, **10,** 301 (1971).

166.  S. H. Melfi, *Appl. Opt.*, **11,** 1605 (1972).

167.  (*a*) G. Elterman, *J. Geophys.* **58,** 519, (1953); (*b*) *J. Geophys.*, **59,** 351 (1954).

168.  J. Cooney, *J. Appl. Meteorol.*, **11,** 108 (1972).

169. J. A. Salzman, "Low Temperature Measurements by Rotational Raman Scattering," in *Laser Raman Gas Diagnostics*, M. Lapp and C. M. Penney, Eds., Plenum Press, New York (1975), pp. 179–188.

170. J. R. Sandercock, *Opt. Commun.*, **2**, 73 (1970).

171. T. Kobayasi, H. Shimizu, and H. Inaba, "Laser Radar Techniques for Remote Measurement of Atmospheric Temperature," 1974 International Laser Radar Conf. Sept. 3–6, Sandai, Japan (1974), p. 49.

172. J. J. Barrett and S. A. Myers, *J. Opt. Soc. Amer.*, **61**, 1246 (1971).

173. M. R. Bowman, A. J. Gibson, and M. C. W. Sandford, *Nature*, **221**, 456 (1969).

174. S. J. Williamson, *Fundamentals of Air Pollution*, Addison-Wesley, Toronto.

175. P. L. Hanst and J. A. Morreal, *J. Air Pollut. Control Assoc.*, **18**, 754 (1968).

176. R. T. Menzies, *Opto-Electronics*, **4**, 179 (1972).

177. R. M. Measures, *Can. Aeronaut. Space J.*, **17**, 417 (1971).

178. J. Gelbwachs and M. Birnbaum, *Appl. Opt.*, **12**, 2442 (1973).

179. J. Gelbwachs, *Appl. Opt.*, **12**, 2812 (1973).

180. T. Kobayasi and H. Inaba, *Nature*, **224**, 170 (1969).

181. H. P. DeLong, *Opt. Eng.*, **13**, (1974).

182. G. D. Hickman, "Recent Advances in the Application of Pulsed Lasers to the Hydrosphere," NASA Conference on the Use of Lasers for Hydrograph Studies, NASA SP-375 (1973), pp. 81–88.

183. C. A. Levis, W. G. Swarner, C. Prettyman, and G. W. Reinhart, "An Optical Radar for Airborne Use Over Natural Waters" NASA Conference on the Use of Lasers for Hydrographic Studies, NASA SP-374, (1973), pp. 67–80.

184. S. Q. Duntley, *J. Opt. Soc. Amer.*, **58**, 214 (1963).

185. A. D. Thurston, Jr., *J. WPCF*, **42**, 1551 (1970).

186. K. Kalle, *Oceanogr. Mar. Biol. Ann. Rev.*, **4**, 91 (1966).

187. A. E. Sylvia, D. A. Bancroft, and J. D. Miller, "Detection and Measurement of Microorganics in Drinking Water by Fluorescence," Amer. Water Works Assoc. Tech. Conf. Proc. Dec. 2–3, 1974, Dallas, Texas (1974).

188. R. F. Christman and M. Ghasseni, *J. AWWA*, 723 (June 1966).

189. C. J. Sorenzen, *Limnol. Oceanogr.*, **15**, 479 (1970).

190. S. Z. El-Sayed, "Phytoplankton Production of the South Pacific and the Pacific Sector of the Antartic," Scientific Exploration of the South Pacific, National Academy of Sciences, Washington, D. C. (1970).

191. C. S. Yentsch and D. W. Menzel, *Deep Sea Res.*, **10**, 221 (1963).

192. C. J. Sorenzen, *Deep Dea Res.*, **13**, 223 (1966).

193. C. S. Yentsch, "The Fluorescence of Chlorophyll and Yellow Substance in Natural Waters: A Note on the Problems of Measurement and Their

Importance to Remote Sensing," NASA Conference on the Use of Lasers for Hydrographic Studies, NASA SP-375 (1973), pp. 147–151.

194. R. B. Slusher and V. E. Derr, *Appl. Opt.*, **14,** 2116 (1975).

195. C. H. Chang and L. A. Young, "Remote Measurement of Ocean Temperature from Depolarization in Raman Scattering," NASA Conference on the Use of Lasers for Hydrographic Studies, NASA SP-375, (1973), pp. 105–112.

196. G. E. Walrafen, *J. Chem.*, **47,** 114 (1967).

197. E. Bradley, *Water Res.*, **4,** 125 (1970).

198. A. Davis, M. Bristow, and J. Koningstein, "Raman Spectroscopy as a Water Quality Indicator," Remote Sensing and Water Resources Management Proc. 17, (1973), pp. 239–246.

199. W. M. Houghton, "Measurement of Raman Spectra of $H_2O$ and $SO_4^=$ in Sea Water," NASA Conference on the Use of Lasers for Hydrographic Studies Conference, NASA SP-375 (1973), PP. 113–118.

200. D. D. Dylis, *Opt. Eng.*, **13,** 502 (1974).

201. G. E. Walrafen, *J. Chem. Phys.*, **52,** 4176 (1970).

# AN INTRODUCTION TO
# SIGNALS, NOISE, AND MEASUREMENTS

## S. COVA and A. LONGONI

*Istituto di Fisica*
*Politecnico di Milano*
*Milano, Italy*

412   AN INTRODUCTION TO SIGNALS, NOISE, AND MEASUREMENTS

# 7.1   INTRODUCTION

Application of lasers in analytical spectroscopy requires detecting and measuring optical signals, that is, light coming either directly from a laser or from a scattering process or a fluorescent emission induced by a laser source. Optical instrumentation is not considered here. Instead, we deal only with the transduction from optical to electrical waveforms and with their subsequent processing. A schematic block diagram of detection and electronic processing is depicted in Fig. 7.1, showing the relevant interconnections between the various stages of the measurement. From this diagram one can see how random fluctuations and unwanted systematic contributions, usually called noise and background, respectively, arise in the early stages of the transduction, thereby influencing the quality of the measurement. The processed data can be in the form of analog (continuous) or digital (discrete) variables, and suitable means of conversion from one form to another are usually encountered in electronic systems, that is, analog to digital and digital to analog conversion (ADC and DAC). Several extensions of this scheme can of course be found in practice, such as elaborate systems with mixed analog and digital treatment. The overall picture is further complicated (and this should not be overlooked) by possible distortions (nonlinearities, etc.) in photodetectors, amplifiers, conversion processes, etc.

It is easy to realize that a thorough coverage in one chapter of all the aspects involved in the general scheme of Fig. 7.1 is impossible. Moreover, even a partial coverage requires a choice of arguments to be treated. In fact, the reader will find in the literature many excellent and detailed treatises on the design and use of electronic instruments, with emphasis on the practical aspects of measuring techniques. On the other hand, the *basis* for the understanding of the foundations of measurement techniques is found in books on information theory, statistical communications, systems, probability, and random functions, which are mostly devoted to extensive mathematical treatments of the matter. Analytical spectroscopy is certainly a field in which the former literature is warranted and indeed does exist, together with many review articles and monographs on special topics.

Instead of merely providing an additional review, we felt that it would be useful to cover the gap between these two basic aspects of the

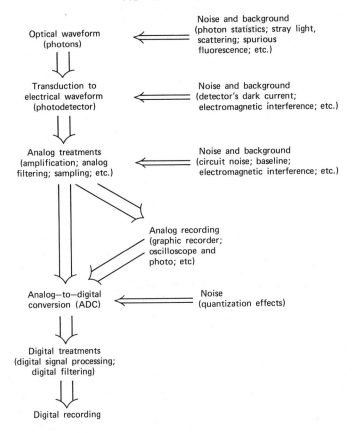

**Fig. 7.1.** Schematic block diagram of electronic processing of optical waveforms.

problem. Therefore, the aim of this chapter is to outline the theoretical foundations of the various measuring techniques and signal-processing methods to gain a better insight in their operation principles by individuating the essential parameters that characterize the quality of the data obtained. Rather lengthy demonstrations are omitted and physical intuition is preferred to mathematical rigor, whenever possible.

After an introduction on the representation of deterministic and random variables, signals, noise, and background are defined and characterized both in the time domain (Sect. 7.2.4.) and in the frequency domain (i.e., by means of Fourier transformation, Sect. 7.2.5.). The practical aspects of measurements based on these principles and their associated errors is emphasized. Linear filtering methods for the extraction of signals from noise are treated in Sect. 7.3, together with a brief discussion of the attainable limits, that is, of the optimum filtering pertaining to each particular physical situation. Sections 7.4 and 7.5 outline noise sources and photodetectors so as to frame the information content of waveforms and the type of noise that can obscure it. Finally, a brief treatment of analog and digital instrumentation, followed by a concise discussion of measurement types, associated problems, parameters, and relevant aspects, is presented.

## 7.2 REPRESENTATION OF SIGNALS AND NOISE

### 7.2.1 Premise

The matter treated in this section is covered in depth by texts on information theory, statistical communications, signals and systems, and probability and random functions (1–11). Rather than attempting to give a rigorous mathematical treatment, we want to stress the intuitive meaning and the physical significance of the various arguments. It is worth noting that any mathematical representation of the actual physical variables, in any system, is necessarily an approximation and that the degree of approximation is, in general, connected with the degree of mathematical complexity. So, the most suitable approximation for any specific variable can be different in different cases. For instance, a given voltage can be represented by a deterministic signal or by an irregularly fluctuating variable, depending on the resolution required in the measurement. It can be considered stationary or not, depending on the time interval over which it has been constant before the instant of the measurement and on the rate of response of the measurement system, and so on.

In order to clarify specific points, in the following we often refer to a typical example: a fluorescent emission excited by a laser source and viewed by one or more photomultiplier tubes (PMT), connected to electronic instrumentation (amplifiers, etc.). The physical variables observed in this example can be the optical radiation incident on the photodetector, the electrical current at its output, or electrical variables in the electronic circuits.

### 7.2.2 Random Processes, Random Variables, and Deterministic and Random Functions

In the example above, suppose let us use a repetitive, pulsed laser source and observe the detector current at time $t$ after the triggering of each laser flash. If the detected fluorescence intensity is high and everything is stable in the experiment, we observe fluorescence pulses with equal shape and equal peak amplitude, that is, equal area. We can write a precise mathematical expression that specifies the amplitude of each pulse at every instant $t$ after the laser flash. This means that the pulses can be considered a *deterministic function* of time. It will be useful to represent this function as a product $A f(t)$ of an intensity factor $A$ (for instance, the pulse's area or peak amplitude) and of a normalized shape function $f(t)$ (normalized to unit area or peak).

If the laser intensity varies irregularly and unpredictably from flash to flash, everything else remaining constant, the corresponding fluorescence will maintain the same shape, but its amplitude will fluctuate from pulse to pulse (see Fig. 7.2a). In this case, the amplitude of a pulse cannot be

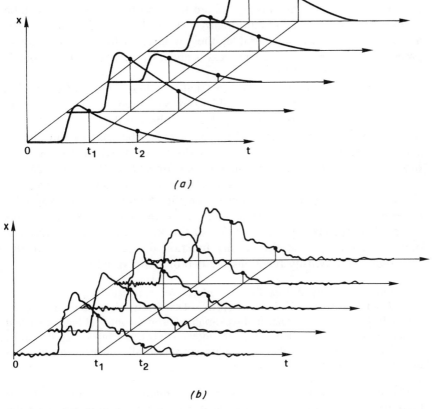

**Fig. 7.2.** (a) Representation of a random process whose sample functions are deterministic function of time. (b) Representation of a random process whose sample functions are random functions of time.

predicted from the knowledge of the preceding ones. The *ensemble* of all possible pulses that might have been observed in this case belongs to a *random process*; each function describing a particular pulse is a *sample function* of the ensemble. The value $x(t_1)$ of a sample function at some particular instant $t_1$ is a *random variable*, that is, its value fluctuates unpredictably from pulse to pulse. For such a variable we cannot specify a value, but we can specify the *marginal probability density function* $p(x)$, which is a normalized function ($\int_{-\infty}^{+\infty} p(x) \, dx = 1$) giving the probability $p(x) \, dx$ of finding the random variable lying in the interval between $x$ and $x + dx$. Note that, in this case, there is randomness from sample function to sample function, but there is no randomness at all from time instant to time instant within each sample function. In fact, each sample is deterministic. If we know its value at some particular time, we know its value at any other time: this means that all the random variables in the process (i.e., $x(t_1)$, $x(t_2)$, etc.) are strictly related to each other. The pulses can be represented as a product of a normalized-shape deterministic function $f(t)$ and of a single random variable, the intensity factor $A$. A complete *probabilistic description* of the process is given by specifying the probability density function $p(A)$.

If the fluorescence intensity is low, that is, a low number of photons is detected, even if everything is stable and our observation is accurately synchronized to the fluorescence pulse, we observe fluctuations not only from pulse to pulse but also within each pulse (because of the photon statistics, see Sect. 7.4.1). This is a random process where each sample function of the ensemble is a *random function of time* (see Fig. 7.2b). In order to describe this random process, we must consider more than one random variable and specify how much the different random variables (i.e., the amplitudes at different times $t$) are related to each other. This means that a probabilistic description of this kind of process specifies not only a marginal probability density function for each variable but also *joint probability* density functions for two or more variables.

The joint probability density function $p(x,y)$ for two variables $x$ and $y$ is a normalized function ($\int_{-\infty}^{+\infty} dx \int_{-\infty}^{+\infty} p(x,y) \, dy = 1$), which gives the probability $p(x,y) \, dx \, dy$ of the event that the first variable lies in the interval between $x$ and $x + dx$ *and* the second variable lies in the interval between $y$ and $y + dy$. If $x$ and $y$ are related to each other, that is, they are *not* statistically independent, $p(x,y)$ is *different* from the product $p(x)p(y)$.

It can be shown that in most cases it is sufficient to consider only two random variables, that is, the process at two instants of time, provided the time interval between these instants can be taken arbitrarily. It is worth noting that it may still be useful, even in this case, to represent a sample function $x(t)$ as a product of an intensity factor $A$ and a normalized-shape function $f(t)$. In this case, however, the $f(t)$ is a *random function* and the knowledge of the probability density function $p(A)$ alone does not give a complete probabilistic description of the process. For instance, if the laser intensity fluctuates, the fluorescence fluctuations arising from this cause may be separated by monitoring the intensity of each laser flash and renormalizing the measurement to it. Even with constant laser intensity, it may be required to classify the fluorescence pulse on the basis of a parameter $A$, for example, the pulse's area, in order to individuate different fluorescences, and so on.

### 7.2.3   Time Average and Ensemble Average; Signals, Noise, and Background

#### 7.2.3.1   Time Averages

The average value $[\langle x \rangle]_{2T}$ over a *finite time interval* $2T$ for a time function $x(t)$, deterministic or random, is defined by

$$[\langle x \rangle]_{2T} = \frac{1}{2T} \int_{-T}^{T} x(t)\, dt \qquad (1a)$$

The concept is to perform a weighted sum, where equal weight $dt/2T$ is given to all values $x(t)$ in the interval. The *average value* $\langle x \rangle$ of $x(t)$ is found from the above equation when an infinite time interval is considered:

$$\langle x \rangle = \lim_{T \to \infty} \frac{1}{2T} \int_{-T}^{T} x(t)\, dt \qquad (1b)$$

The definition is readily extended to functions of $x(t)$, for instance, to the square value $x^2(t)$, and to functions of $x(t)$ and of other time functions $y(t)$, for instance, a product $x(t)y(t)$. It is worth noting that in a random process a time-average value is defined for each sample function and that, depending on the particular process, these average values may be equal or different.

#### 7.2.3.2   Ensemble Averages

In the case of random processes a different and fundamental concept can be used. Let us consider a random variable $^k x(t_1)$, where $t_1$ defines the variable and $k$ individuates its values (i.e., the sample functions). The ensemble average $[\overline{x(t_1)}]_N$ *over a finite number $N$ of repetitions* for $x(t_1)$ is defined by

$$[\overline{x(t_1)}]_N = \frac{1}{N}\sum_{k=1}^{N} {}^k x(t_1) \tag{2}$$

The *ensemble average* $\overline{x(t_1)}$ (also called *mean* or *expected* value) for the random variable $x(t_1)$ is formed by extending Eq. (2) to an infinite number of repetitions. The concept is to perform a weighted sum, where the weight of each possible value $x_1$ is proportional to the number of events in the ensemble for which this variable lies in the range between $x_1$ and $x_1 + dx_1$; the normalized weight is therefore the probability $p(x_1)\,dx_1$:

$$\overline{x(t_1)} = \overline{x_1} = \int_{-\infty}^{+\infty} x_1 p(x_1)\,dx_1 \tag{3}$$

The definition is readily extended to functions of one random variable (e.g., mean square value $\overline{x_1^2}$) and, by using joint probability densities, to functions of two or more variables (products, etc.).

The concept of ensemble-averaging can <u>also</u> be used for random functions of time. The *average function of time* $\overline{x(t)}$ is such that its value at every instant $t_1$ is the mean value of the random variable $x(t_1)$.

### 7.2.3.3  *Signal, Noise, Background*

In measurements in the presence of random process, if the random fluctuations of the physical quantity $x(t)$ observed can be considered negligible, this quantity $x(t)$ is usually called a (pure) *signal*; otherwise, it is usual to say that the signal is masked or buried in *noise*. In accordance to this use, the *signal* can be defined as the ensemble-average function of time $\overline{x(t)}$:

$$s(t) = \overline{x(t)} \tag{4}$$

Correspondingly, the *noise* is defined as the random deviation of the sample function $x(t)$ from the average $\overline{x(t)}$:

$$n(t) = x(t) - \overline{x(t)} \tag{5}$$

It is worth remembering that, by definition, the noise has a *zero mean value* and is usually characterized by its root mean square value, $[\overline{n^2(t)}]^{1/2}$, which indicates the width of the distribution of the random function $x(t)$ around the average value $\overline{x(t)}$.

In order to evaluate the quality of a physical measurement, one usually refers to the concept of *signal-to-noise ratio*, which is defined as the ratio of the signal amplitude to the noise amplitude, measured by its root mean square value:

$$\frac{S}{N} = \frac{s(t)}{\sqrt{\overline{n^2(t)}}} = \frac{\overline{x(t)}}{\sqrt{\overline{[x(t) - \overline{x(t)}]^2}}} \tag{6}$$

Usually it is important to take into account that $\overline{x(t)}$ may be a sum of different contributions. In this respect, the true signal $s(t)$ will be defined as that part $\overline{x_s(t)}$ of the ensemble average $\overline{x(t)}$ due to the contributions object of the measurement; the remaining part $x_b(t)$ due to the undesired contributions will be defined as *background* $b(t)$:

$$\overline{x(t)} = \overline{x_s(t)} + \overline{x_b(t)} \tag{7a}$$

$$s(t) = \overline{x_s(t)} \tag{7b}$$

$$b(t) = \overline{x_b(t)} \tag{7c}$$

Obviously, any background must be excluded and only $\overline{x_s}$ must be used in Eq. (6). In the example considered above, the signal is the average current pulse due to the fluorescence to be measured; the background may receive contributions from spurious fluorescences, scattering, detected stray light, electromagnetic interference arising from the laser pulsing apparatus, dark current of the detector (see Sect. 7.5), leakage or bias currents in the circuits, etc.

It is worth noting that the characteristics of a random process depend on all the circumstances of the physical situations; in particular, a given physical source may cause background and/or noise, depending on the circumstances. In the example considered, an interference from the line voltage gives a background if the phase corresponding to the laser's triggering time is fixed; it gives only noise if this phase is a random variable with zero mean value. If the fluorescence pulses are superimposed on a baseline affected by drift (e.g., due to temperature dependence of some instrumental parameter in the circuits), the effects will depend on possible relations between the time behavior of the drift and the time of occurrence of the pulses.

The logical basis of the terminology here adopted is to associate the word "signal" with the quantity that is (or should be) the result of the measurement and the word "noise" with the random errors only. This terminology is consistent with the most commonly accepted meaning for such terms as "signal-to-noise ratio" in a measurement, etc. However, there is no general agreement on the terminology in the literature, and this probably reflects the variety of physical situations studied and the different interests of the various authors. Some authors denote by "noise" (see, for instance, noise current in PMTs) unwanted contributions with their fluctuations *and* their nonzero average value. For other authors, noise does not include the random fluctuations inherent in the nature of the quantity studied, which are considered separately, for instance, fluctuations due to the photon statistics in a fluorescence pulse. When the samples $x(t)$ are random functions, but may be represented (see Sect. 7.2.2.) as a product $x(t) = Af(t)$, other authors consider separately the fluctuations of the intensity $A$ and denote by noise only the deviations of $\underline{x(t)}$ from a function $A\overline{f(t)}$ having equal intensity $A$ and deterministic shape $\overline{f(t)}$. The latter definition reflects the intuitive concept derived from the observation of a single noisy waveform as recorded on a conventional oscillograph, while the definition used in this chapter may be visualized by referring to the trace of a fast sampling scope used in "equivalent time" mode. It is worth remembering, however, that $\overline{x(t)} = \overline{A}\,\overline{f(t)}$ *only if* fluctuations of the intensity $A$ and of the shape $f(t)$ are uncorrelated (see Sect. 7.2.4.).

### 7.2.3.4  *Averaging Measurement Techniques*

In principle, when speaking of ensemble averages, one should refer to the infinite number of all possible time functions $x(t)$ that might have occurred. In practice, it is usual to speak of *averaging measurement techniques* with reference to measurements of a finite number $N$ of samples $x(t)$. The *averaged value* $[\overline{x_1}]_N$ obtained for a random variable $x_1$ by means of taking the average over $N$ samples is still a random variable. In fact, it will fluctuate from one group of $N$ samples $x_1$ to another. However, this noise in the averaged value $[\overline{x_1}]_N$ will be in general progressively reduced as $N$ is increased, and $[\overline{x_1}]_N$ will tend to approximate the true ensemble average $\overline{x_1}$. For a given $N$, the reduction factor will depend on the possible relations between the fluctuations of the $N$ samples $x_1$ used in the averaging. When averaging is performed on a time series, that is, a series of events following each other in time (as in the example considered), one should not overlook contributions affected by variations that are slow with respect to the time interval between events (e.g., drift and low-frequency disturbances).

An approximation of an average function of time, $\overline{x(t)}$, can be obtained by means of a set of averaged values $[\overline{x(t)}]_N$. The matter will be treated in Sect. 7.2.4, but it is clear since now that the quality of the approximation will depend on the physical situation (type of noise, drift effects, etc.) and on the averaging procedure used (sequence of operations used for obtaining $[\overline{x(t)}]_N$, see Sect. 7.7.2).

### 7.2.4  Time Domain Representation—Correlation Functions

#### 7.2.4.1  *Representation of Waveforms*

Waveforms may be represented by elementary mathematical functions, and in some cases this is a good approximation. Thus, stationary values and simple periodical functions can be used although all real waveforms begin at some finite instant and have finite duration. Elementary *singularity functions* (4) are often useful, namely, the *unit step function* $1(t)$,

$$1(t) = 1 \quad \text{for } t \geq 0 \tag{8}$$
$$= 0 \quad \text{for } t < 0$$

and the *delta function* or *unit impulse function* $\delta(t)$, defined by its "sifting property,"

$$\int_{-\infty}^{\infty} f(t)\, \delta(t - t_0)\, dt = f(t_0) \tag{9}$$

which holds for any function $f(t)$ continuous at $t = t_0$. These functions are the idealizations of a real step (having unit amplitude and short, but finite, risetime $T_r$) and of its time derivative (having unit area and duration $T_r$), as shown in Fig. 7.3. Note that a factor multiplying a delta function indicates the impulse area; thus, if $A\delta(t - t_0)$ represents a current impulse located at time $t_0$, $A$ has the dimensions of a charge.

In many cases very convenient representations can be obtained by a linear combination of elementary functions. The mathematical theory of transforms and function spaces shows how a given $x(t)$ can be represented as a linear combination (weighted sum or series, weighted integral) of an ordered set of elementary functions, usually called *basis functions* (4).

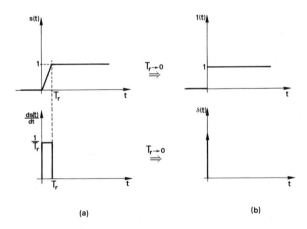

**Fig. 7.3.** (a) Real step function $s(t)$, with finite risetime $T_r$ and unit amplitude, and its time derivative function. (b) Idealization for $T_r \to 0$, with unit step function $1(t)$ and delta function or unit impulse function $\delta(t)$.

Various types of basis functions can be used: the well-known Fourier representation employs sinusoidal functions. For other functions (such as Walsh functions, etc.) and transforms (such as Laplace transforms, $Z$-transforms, etc.) the reader is referred to the literature (4).

#### 7.2.4.2   Energy and Power

Even though dimensionally incorrect, it is customary to refer to the integral of the square value of a waveform $x(t)$ for an infinite time interval as the "energy,"

$$E = \lim_{T \to \infty} \int_{-T}^{T} x^2(t)\, dt \tag{10}$$

and to the average square value as the "average power,"

$$P = \langle x^2 \rangle = \lim_{T \to \infty} \frac{1}{2T} \int_{-T}^{T} x^2(t)\, dt \tag{11}$$

This terminology originates from electrical circuit terminology, and if $x(t)$ is the voltage or the current in a unit resistance, then the names are indeed correct. Therefore, an *energy waveform* will have finite energy $E$ and zero average power and a *power waveform* will have infinite energy. Typical energy waveforms are rectangular pulses, decaying exponentials, etc., while power waveforms include all periodic waveforms. Some waveforms cannot be classified on this basis since both energy and power are infinite.

#### 7.2.4.3   Time Correlation Functions

It is often important to determine the degree of similarity between two measurements, the first at some instant on a waveform $x(t)$ and the second at a later time, after an interval $\tau$, on the same waveform $x(t)$ or on another waveform $y(t)$. For this purpose, the second waveform may be shifted in time by a delay $\tau$ with respect to the first one, and an *average* of a point-by-point comparison may be taken over the whole waveform, that is, *over all the time*. This is the concept of time correlation functions which give, as a function of $\tau$, a measure of this average similarity.

*Time autocorrelation* functions compare a waveform with itself, that is, $x(t) = y(t)$, and *time cross-correlation* functions compare two different waveforms, $x(t) \neq y(t)$. From our previous discussion, it follows that correlation functions should be defined as *energy-type* functions when at least one of the $x(t)$ and $y(t)$ is an energy waveform,

$$k_{xx}(\tau) = \lim_{T \to \infty} \int_{-T}^{T} x(t)x(t+\tau)\, dt \tag{12a}$$

$$k_{xy}(\tau) = \lim_{T \to \infty} \int_{-T}^{T} x(t)y(t+\tau)\, dt \tag{12b}$$

$$k_{yx}(\tau) = \lim_{T \to \infty} \int_{-T}^{T} y(t)x(t+\tau)\, dt \tag{12c}$$

and as *power-type* functions if both $x(t)$ and $y(t)$ are power waveforms:

$$K_{xx}(\tau) = \lim_{T \to \infty} \frac{1}{2T} \int_{-T}^{T} x(t)x(t+\tau)\, dt = \langle x(t)x(t+\tau) \rangle \tag{13a}$$

$$K_{xy}(\tau) = \lim_{T \to \infty} \frac{1}{2T} \int_{-T}^{T} x(t)y(t+\tau)\, dt = \langle x(t)y(t+\tau) \rangle \tag{13b}$$

$$K_{yx}(\tau) = \lim_{T \to \infty} \frac{1}{2T} \int_{-T}^{T} y(t)x(t+\tau)\, dt = \langle y(t)x(t+\tau) \rangle \tag{13c}$$

If a waveform is the sum of two or more components, its autocorrelation function will be the sum of *all* the autocorrelation functions plus the sum of *all* crosscorrelation functions of the components. It should be stressed that while a given waveform $x(t)$ leads to only one autocorrelation function $K_{xx}(\tau)$, the converse is not true. Indeed, one autocorrelation function can be associated with different waveforms.

Some general properties of time correlation functions are worth being recalled here. For *autocorrelation*, the function is an even function of $\tau$. Its value at $\tau = 0$ represents either the energy or the total power. The function can assume positive and negative values, but its absolute values are always lower than the one at $\tau = 0$.

If $x(t)$ is periodic, then $K_{xx}(t)$ is also periodic with equal period, while if $x(t)$ has zero average value and no periodic components, then $K_{xx}(\tau)$ has finite width. *Cross-correlations* are not even functions, but $K_{xy}(\tau) = K_{yx}(-\tau)$. Limits to their maximum values can be expressed in terms of the autocorrelation functions $|K_{xy}(\tau)| \leq (K_{xx}(0)K_{yy}(0))^{1/2}$ and $|K_{xy}(\tau)| \leq (K_{xx}(0) + K_{yy}(0))/2$.

### 7.2.4.4 Estimated Time Correlation Functions

In practice, correlation functions will be measured with a finite averaging time $2T$, and therefore the estimated function will be affected by some error. As Fourier representation can be used for many waveforms, some

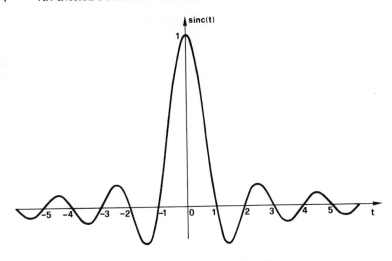

**Fig. 7.4.**   Plot of $\text{sinc}(t) = (\sin \pi t)/\pi t$.

general insight in this and other aspects of correlation analysis can be obtained by considering the elementary case of sinusoidal functions

$$x(t) = A \cos(\omega_1 t + \theta) \tag{14a}$$
$$y(t) = B \cos(\omega_2 t + \psi) \tag{14b}$$

The estimated correlation function (power type) will be

$$[K_{xy}(\tau)]_{2T} = \frac{1}{2T} \int_{-T}^{T} x(t)y(t+\tau)\,dt$$

$$= \frac{AB}{2}\{\text{sinc}[(\omega_1 - \omega_2)T]\cos(\omega_2\tau + \psi - \theta)$$

$$+ \text{sinc}[(\omega_1 + \omega_2)T]\cos(\omega_2\tau + \theta + \psi)\} \tag{15}$$

The function[1]

$$\text{sinc}(t) = \frac{\sin \pi t}{\pi t} \tag{16}$$

is plotted in Fig. 7.4.

Let us consider first the *true* correlation functions. They will be obtained with $T \rightarrow \infty$ or with finite $T$ values that are integer multiples of

---

[1] Although this is the commonly accepted definition of the sinc function, the reader is warned that throughout the chapter use is made of the definition $\text{sinc}(t) = (\sin t)/t$.

both the half-periods $\pi/\omega_1$ and $\pi/\omega_2$:

$$K_{xy}(\tau) = \lim_{T \to \infty} [K_{xy}(\tau)]_{2T} = \begin{cases} 0 & \text{if} \quad \omega_1 \neq \omega_2 \\ \dfrac{AB}{2} \cos(\omega\tau + \psi - \theta) & \text{if} \quad \omega_1 = \omega_2 = \omega \end{cases} \tag{17}$$

$K_{xy}(\tau)$ gives information about the *phase difference* $\psi - \theta$ between the two functions, but not about each phase $\theta$ and $\psi$, as it depends on the relative position of the two waveforms but not on their position with respect to the time origin $t = 0$. In particular, if the waveforms are in phase, that is, $\psi - \theta = 0$, then

$$K_{xy}(\tau) = \frac{AB}{2} \cos\omega\tau \tag{17a}$$

while if they are 90° out of phase, that is, $\psi - \theta = \pi/2$, then

$$K_{xy}(\tau) = -\frac{AB}{2} \sin \omega\tau \tag{17b}$$

For other values of $\psi - \theta$, the second sinusoid $y(t)$ may be expressed as a combination of two components, one in phase and the other $\pi/2$ out of phase with respect to $x(t)$. Correspondingly, $K_{xy}(\tau)$ as given by Eq. (17) may be expressed as a combination of two terms in quadrature. It is worth noting that at $\tau = 0$, the contribution of the in-phase component is maximum while that of the $\pi/2$ out-of-phase component is zero.

By setting $B = A$, we obtain the true autocorrelation function of a sinusoidal $x(t)$:

$$K_{xx}(\tau) = \frac{A^2}{2} \cos \omega\tau \tag{18}$$

It has to be emphasized that $K_{xx}(\tau)$ bears information about the amplitude $A$ and the angular frequency $\omega$ of $x(t)$ but does not depend at all on the phase $\theta$. For a function $x(t)$ which may be represented by a combination of sinusoidal functions (Fourier representation), it follows from Eq. (17) that the true autocorrelation function will be the sum of *all* and *only* the *auto*correlation functions of all components.

Let us now consider the estimated crosscorrelation function $[K_{xy}(\tau)]_{2T}$ by comparing Eqs. (15) and (17). If $\omega_1 = \omega_2 = \omega$, as $\text{sinc}(0) = 1$, Eq. (15) gives

$$[K_{xy}(\tau)]_{2T} = \frac{AB}{2} \{\cos(\omega\tau + \psi - \theta) + \text{sinc } 2\omega T \cos(\omega\tau + \theta + \psi)\} \tag{19}$$

This function is still sinusoidal with angular frequency $\omega$, but its amplitude and phase are altered by the second term in the brackets. The absolute magnitude of this error term is inversely proportional to the number of periods, $2\pi/\omega$, covered by the averaging time $2T$. On the other hand, if $\omega_1 \neq \omega_2$, the estimated correlation may be different from zero by an amount depending mainly on the values of $T$ *and* on the frequency difference $\Delta\omega = \omega_1 - \omega_2$. In fact, while the second term in Eq. (15) is negligible provided that $T$ covers a high number of periods of $x(t)$ and/or $y(t)$, the first term will be negligible only if

$$|\Delta\omega| \gg \frac{\pi}{T} \tag{20}$$

In other words, the estimated crosscorrelation function will not regard a frequency $\omega_1$ as different from $\omega_2$ if $\omega_1$ lies in a *bandwidth* of the order of $2\pi/T$ centered at $\omega_2$. Summarizing, we then have

$$[K_{xy}(\tau)]_{2T} \simeq K_{xy}(\tau) = \frac{AB}{2}\cos(\omega_2\tau + \psi - \theta) \qquad \text{if} \qquad |\Delta\omega| \ll \frac{\pi}{T} \tag{21a}$$

and

$$[K_{xy}(\tau)]_{2T} \simeq K_{xy}(\tau) = 0 \qquad \text{if} \qquad |\Delta\omega| \gg \frac{\pi}{T} \tag{21b}$$

The above discussion has been referred to pure sinusoidal functions but can be extended to periodic and nonperiodic waveforms as well. A periodic waveform $x(t)$ can be represented in terms of its Fourier components. If the period is $a$, then the fundamental frequency $\omega_f$ is equal to $2\pi/a$. In this case, the true autocorrelation function $K_{xx}(\tau)$ is also periodic with the same period $a$ and will be the sum of the autocorrelation functions of the components since the crosscorrelations are zero. It follows that the estimated autocorrelation will be approximately equal to the true function if

$$|\Delta\omega_{min}| = \omega_f = \frac{2\pi}{a} \gg \frac{\pi}{T} \tag{22a}$$

or

$$\frac{2T}{a} \gg 1 \tag{22b}$$

In other words, several periods $a$ of $K_{xx}(\tau)$ must be contained in the integration time $2T$. The same arguments apply for the crosscorrelation of two periodic waveforms whose fundamental frequencies are $\omega_f$ and $\omega'_f$. Here, $\Delta\omega_{min} = \omega_f - \omega'_f$.

For nonperiodic waveforms, $K_{xx}(\tau)$ will also be nonperiodic. In this case, it can be shown (1,3) that a $\Delta\omega_{min}$ must be considered, which is inversely proportional to the interval $|\tau| < \tau_{max}$ in which $K_{xx}(\tau)$ cannot be considered negligible. Thus, the estimated autocorrelation function will be approximately equal to the true one if

$$|\Delta\omega_{min}| \gg \frac{\pi}{T} \tag{23a}$$

or

$$\frac{T}{\tau_{max}} \gg 1 \tag{23b}$$

One can see that in both cases the degree of approximation is related to the ratio given by Eqs. (22b) and (23b). It can be evaluated (4) that in order to have in $[K_{xx}(\tau)]_{2T}$ a root mean square error less than $0.1 K_{xx}(0)$, these ratios must reach approximately 200.

### 7.2.4.5 *Representation of Random Processes*

Random processes (see Sect. 7.2.2) are *continuous* or *discrete* if their random variables $x(t)$ (i.e., amplitudes at the various times $t$) can respectively assume any or only certain specified values (possibly infinite in number) within their specified range. They are *deterministic* or *nondeterministic* if their sample functions $x(t)$ are respectively deterministic or random functions of time.

The probabilistic description of random processes generally requires the knowledge of the marginal probability density function (p.d.f.) $p(x)$ for every random variable $x$ and of the joint p.d.f., $p(x,y)$, for every couple of variables $x,y$. Sometimes, marginal p.d.f.'s may suffice, as in the deterministic example of Sect. 7.2.2. A very important case occurs when the random variables do not depend in any way upon each other: two variables $x,y$ are *statistically independent* if their joint p.d.f. can be factored into the two marginal p.d.f.'s (necessary and sufficient condition):

$$p(x,y) = p(x)\,p(y) \tag{24}$$

This description is quite complete: however, only in some cases can it be obtained for the output of known systems, such as amplifiers, filters, etc., from a known probabilistic description of the input. Moreover, it does not give direct indications of the time variation of the processed waveform. It is possible to derive a description that is less complete but more practical from the above standpoint. This *statistical description* is based on a set of average values of functions of random variables: mean, mean square, and variance of one variable $x$, and mean product and covariance of two variables $x$ and $y$. These are some of the *moments* and *central moments* of marginal and joint p.d.f.'s.

The moments of $p(x)$ and $p(x,y)$ are defined as

$$m_k = \overline{x^k} = \int_{-\infty}^{\infty} x^k p(x)\, dx \qquad\qquad k = 0, 1, 2, \ldots \qquad (25)$$

$$m_{kr} = \overline{x^k y^r} = \int_{-\infty}^{\infty} \int_{-\infty}^{\infty} x^k y^r p(x,y)\, dx\, dy \quad k, r = 0, 1, 2, \ldots \qquad (26)$$

The *order* of the moment is $n = k + r$. Zero-order moment is unity (normalized p.d.f.'s); first-order moments are the mean values $\bar{x}$ and $\bar{y}$; second-order moments are the mean squares $\overline{x^2}$ and $\overline{y^2}$ and the mean product $\overline{xy}$.

The *central moments* are defined as the moments of the deviations from the mean values:

$$\mu_k = \overline{(x - \bar{x})^k} = \int_{-\infty}^{\infty} (x - \bar{x})^k p(x)\, dx \qquad\qquad k = 0, 1, 2, \ldots \quad (27)$$

$$\mu_{kr} = \overline{(x - \bar{x})^k (y - \bar{y})^r} = \int_{-\infty}^{\infty} \int_{-\infty}^{\infty} (x - \bar{x})^k (y - \bar{y})^r p(x,y)\, dx\, dy$$

$$k, r = 0, 1, 2, \ldots \quad (28)$$

First-order central moments are obviously zero; second-order central moments are the *variance* $\sigma_x^2 = \overline{(x - \bar{x})^2}$ for one variable and the *covariance* $\mu_{11} = \overline{(x - \bar{x})(y - \bar{y})}$ for two variables. it can be easily shown that

$$\sigma_x^2 = \overline{(x - \bar{x})^2} = \overline{x^2} - \bar{x}^2 \qquad\qquad (29)$$

$$\mu_{11} = \overline{(x - \bar{x})(y - \bar{y})} = \overline{xy} - \bar{x} \cdot \bar{y} \qquad\qquad (30)$$

Two random variables $x, y$ are called *uncorrelated* if their covariance is zero, that is, if

$$\overline{xy} = \bar{x} \cdot \bar{y} \qquad\qquad (31)$$

The variance of the sum of uncorrelated variables is therefore the sum of the variances:

$$\sigma_{x+y}^2 = \overline{[(x + y) - (\overline{x + y})]^2} = \sigma_x^2 + \sigma_y^2 \qquad\qquad (32)$$

It is worth noting that two statistically independent variables are uncorrelated, see Eq. (24), but the converse is not always true. If the knowledge of the p.d.f.'s is sufficient to obtain a unique statistical description, the converse is not true. In fact, the statistical description takes into account only the most important moments of the p.d.f.'s, up to the second order. As far as these moments are concerned, different random variables may be equivalent, and therefore various *statistical models* can be associated with a given physical situation in which the necessary data for its statistical description are known.

The statistical description is more or less complete, depending on the p.d.f. of the process. In particular, in the case of *Gaussian variables*, it gives a *complete* description of the process. This special case is very important, both because it provides a good mathematical model for many (though not all) physical situations and because of its mathematical properties. For example, a linear combination of Gaussian variables is also a Gaussian variable; it is possible to carry out a complete statistical analysis of the modifications introduced in the process by both linear and nonlinear systems, etc. The normalized marginal p.d.f. for a Gaussian variable $x$ is

$$p(x) = \frac{1}{\sigma_x \sqrt{2\pi}} \exp\{-(x - \bar{x})^2/2\sigma_x^2\} \tag{33}$$

and is completely specified by knowing $\bar{x}$ and $\sigma_x^2$. The joint p.d.f. for two Gaussian variables is a generalization of Eq. (33) and is completely specified by knowing $\bar{x}$, $\bar{y}$, $\sigma_x^2$, $\sigma_y^2$, and the covariance of $x,y$ (3). In particular, it can be shown that if two jointly Gaussian random variables are uncorrelated, they are statistically independent.

### 7.2.4.6   *Random Processes and Correlation Functions; Ergodic Processes*

In a random process, the random variables are the amplitudes of the sample waveforms at the various times, that is, in Eqs. (24)–(33) we can set $x = x_1 = x(t_1)$ and $y = x_2 = x(t_2) = x(t_1 + \tau)$ for every instant $t_1$ and interval $\tau = t_2 - t_1$.

In *nonstationary* processes the marginal p.d.f. depends on $t_1$, the joint p.d.f. depends on $t_1$ and $\tau$:

$$p(x_1) = p(x_1; t_1) \tag{34}$$

$$p(x_1, x_2) = p(x_1, x_2; t_1, t_2) \tag{35}$$

The mean, mean square, and variance of one variable are functions of time $t_1$; the mean product of two variables is a function of $t_1$ and $\tau$ and is called the (*ensemble*) *autocorrelation function* of the process

$$R_{xx}(t_1,t_2) = R_{xx}(t_1,\tau) = \overline{x_1 x_2} = \overline{x(t_1)x(t_1+\tau)}$$

$$= \int_{-\infty}^{\infty} \int_{-\infty}^{\infty} x_1 x_2 p(x_1,x_2; t_1,t_2)\, dx_1\, dx_2 \quad (36)$$

Similarly, the *autocovariance* of the process is also a function of $t_1$ and $\tau$;

$$C_{xx}(t_1,t_2) = C_{xx}(t_1,\tau) = \overline{(x_1 - \bar{x}_1)(x_2 - \bar{x}_2)} = R_{xx}(t_1,\tau) - \overline{x(t_1)} \cdot \overline{x(t_1+\tau)}$$

$$(37)$$

For *stationary* processes, the marginal and joint p.d.f.'s do *not* depend on the time $t_1$; the joint p.d.f., however, depends on the interval $\tau$. The mean, mean square, and variance are constant; the autocorrelation and autocovariance are only functions of $\tau$.

If more than one random process is considered, joint random variables from two different processes $x(t)$ and $y(t)$ can be considered, that is, $x = x(t_1)$ and $y = y(t_2) = y(t_1 + \tau)$ in Eq. (24)–(33). Cross-correlation functions $R_{xy}$ and cross-covariance $C_{xy}$ can thus be similarly defined. If both processes are stationary, $R_{xy}$ and $C_{xy}$ are only functions of $\tau$.

Random processes may have a property of relevant conceptual and *practical* importance, namely, that their statistics can be determined by observing only one of their sample functions. This means that time averages equal ensemble averages; such a property is called *ergodicity*. Conditions for ergodicity are discussed in the literature (3). Stationariness is a necessary but not sufficient condition. In practice, it is difficult to prove ergodicity, but it is customary to assume it unless there are clear reasons for not doing so. For instance, many stationary processes having deterministic sample functions are not ergodic (4).

If a stationary process is ergodic, then the ensemble auto- and cross-correlation functions become equal to the power-type time auto- and cross-correlation functions:

$$R_{xx}(\tau) = K_{xx}(\tau) \quad (38)$$

and

$$R_{xy}(\tau) = K_{xy}(\tau) \quad (39)$$

Indeed, it can be shown that ensemble auto- and cross-correlation functions have characteristic properties which in stationary cases coincide with the ones previously reported for time correlation functions and which in nonstationary cases represent their natural extension (3).

### 7.2.4.7  Noise Types and Statistical Models

Noise has been defined here, Eq. (5), as a random process having a zero mean value. With other definitions, the considerations made here for mean square values and correlation functions are valid for variances and covariances.

If the noise $n(t)$ is a sum of various components $n_1(t) + n_2(t) + n_3(t) + \cdots$, its autocorrelation function will be the sum of all auto and cross-correlation functions of the components. In particular,

$$\overline{n^2(t)} = \overline{R_{nn}(t,t)} = \overline{n_1^2(t)} + \overline{n_2^2(t)} + \cdots + \overline{2n_1(t)n_2(t)} + \overline{2n_1(t)n_3(t)} + \cdots \tag{40}$$

If there is no correlation between different components, the mean square is just the sum of the mean squares:

$$\overline{n^2(t)} = \overline{n_1^2(t)} + \overline{n_2^2(t)} + \overline{n_3^2(t)} + \cdots = \sum_k \overline{n_k^2(t)} \tag{41}$$

In the presence of correlations, $\overline{n^2(t)}$ may be higher or lower than this value, depending on the type of correlation. In any case, from the properties of cross-correlation functions it follows that the root mean square is lower than the sum of the root mean squares:

$$\left[\overline{n^2(t)}\right]^{1/2} \leq \left[\overline{n_1^2(t)}\right]^{1/2} + \left[\overline{n_2^2(t)}\right]^{1/2} + \cdots = \sum_k \left[\overline{n_k^2(t)}\right]^{1/2} \tag{42}$$

From the statistical point of view, the *intensity* of the noise is measured by the mean square value, while the *type* of the noise is assessed by the shape of the autocorrelation function with respect to the delay $\tau$. A normalized shape is given by the autocorrelation coefficient, which is the autocorrelation function normalized to the mean square value:

$$\rho_{nn}(t,t+\tau) = \frac{R_{nn}(t,t+\tau)}{R_{nn}(t,t)} = \frac{R_{nn}(t,t+\tau)}{\overline{n^2(t)}} \tag{43}$$

In *nonstationary* cases, the autocorrelation shape in general depends on the time instant $t$, and the type of noise will be in general defined with reference to the time-average $\langle R_{nn}(t,t+\tau) \rangle$. There are cases, however, where the intensity varies with time, but the autocorrelation shape remains constant. A fundamental case is that of *white noise*, which is the noise that is *not* correlated with itself and is thus characterized in stationary and nonstationary conditions by a $\delta$-like autocorrelation

$$R_{nn}(t,t+\tau) = \lambda(t)\,\delta(\tau) \tag{44}$$

As previously seen, it is possible to use various models that are statistically equivalent to a given noise (i.e., as far as second-order moments are concerned). Two statistical models are particularly useful. The first one is a random process whose sample functions $x(t)$ are random superpositions of independent sinusoidal waveforms $A\cos(\omega t + \theta)$, where the phase $\theta$ is an independent random variable uniformly distributed in the $(0,2\pi)$ interval and $\omega$ is a random variable having (normalized) probability density $p(\omega)$, with $p(\omega) = p(-\omega)$. This process is stationary; and, due to the absence of correlation between different angular frequencies $\omega$, it has an autocorrelation

$$R_{xx}(\tau) = \overline{x(t)x(t+\tau)} = \frac{A^2}{2} \int_{-\infty}^{\infty} p(\omega) \cos \omega\tau \, d\omega \qquad (45)$$

If the Fourier transform $S_{xx}(\omega)$ of $R_{xx}(\tau)$ is considered, as $R_{xx}(\tau)$ is a real even function, $S_{xx}(\omega)$ is also real, and even

$$S_{xx}(\omega) = \mathcal{F}[R_{xx}(\tau)] = \int_{-\infty}^{\infty} R_{xx}(\tau)e^{-j\omega\tau} \, d\tau = \int_{-\infty}^{\infty} R_{xx}(\tau) \cos \omega\tau \, d\tau \quad (46)$$

$$R_{xx}(\tau) = \mathcal{F}^{-1}[S_{xx}(\omega)] = \frac{1}{2\pi} \int_{-\infty}^{\infty} S_{xx}(\omega)e^{j\omega\tau} \, d\omega = \frac{1}{2\pi} \int_{-\infty}^{\infty} S_{xx}(\omega) \cos \omega\tau \, d\omega$$

$$(47)$$

From Eqs. (45) and (47) it follows that

$$S_{xx}(\omega) = \pi A^2 p(\omega) \geq 0 \qquad (48)$$

As will be shown in Sect. 7.2.5, for any type of stationary noise $n(t)$, characterized by a given $R_{nn}(\tau) = \overline{n^2}\rho_{nn}(\tau)$, the Fourier transform $S_{nn}(\omega)$ of $R_{nn}(\tau)$ is called the *power spectrum of the noise*. Indeed, the mean square value can be expressed as the sum of contributions from the various frequency components, with distribution density given by $S_{nn}(\omega)$. In particular, in our model,

$$\overline{x^2} = R_{xx}(0) = \frac{A^2}{2} \int_{-\infty}^{\infty} p(\omega) \, d\omega = \int_{-\infty}^{\infty} S_{xx}(\omega) \frac{d\omega}{2\pi} = \frac{A^2}{2} \qquad (49)$$

The model process will therefore be statistically equivalent to $n(t)$ if $S_{xx}(\omega) = S_{nn}(\omega)$, that is,

$$\frac{A^2}{2} = \overline{n^2} \qquad (50a)$$

$$p(\omega) = \frac{S_{nn}(\omega)}{\pi A^2} = \frac{S_{nn}(\omega)}{2\pi \overline{n^2}} = \frac{1}{2\pi} \int_{-\infty}^{\infty} \rho_{nn}(\tau) \cos \omega\tau \, d\tau \qquad (50b)$$

For nonstationary noise types, this stationary model can be equivalent only as far as time-averaged autocorrelation is concerned, that is, it can have $R_{xx}(\tau) = \langle R_{nn}(t,t+\tau)\rangle$.

It is worth noting that a flat power spectrum, with constant amplitude at all frequencies, corresponds to the case of white noise. It is worth noting also that strictly white noise would have an infinite mean square value. A physical process, however, will only approximate white noise: its autocorrelation shape will have finite amplitude and a small, but nonzero width. The power spectrum will be flat up to a frequency of the order of the reciprocal of the time width of the autocorrelation and will decrease to zero at higher frequencies: the mean square value will be finite.

The second statistical model mentioned is a random process $x(t)$ consisting of a random sequence of pulses, of equal area $q$, equal shape described by a time function $h(t)$ (normalized to unit area), and Poisson distributed in time. This means that the probability that the starting point of one impulse falls in the interval $(t, t + dt)$ is $\lambda(t) \, dt$, where the density $\lambda(t)$ is finite; the probability of more than one point in the interval is of higher order in $dt$. These probabilities are independent of what happens outside the interval $(t, t + dt)$. The number $\lambda(t) \, dt$ is the mean (i.e., ensemble average) number of points in $(t, t + dt)$; $\lambda(t)$ is the mean repetition rate of pulses at time $t$. It can be shown (3) that the mean value of the process is given by the convolution integral of the mean rate $\lambda(t)$ and of the pulse $qh(t)$ as follows:

$$\overline{x(t)} = qh(t) * \lambda(t) = q \int_{-\infty}^{\infty} \lambda(t - \alpha) h(\alpha) \, d\alpha \tag{51}$$

and the autocovariance by

$$C_{xx}(t_1, t_1 + \tau) = R_{xx}(t_1, t_1 + \tau) - \overline{(x(t_1) \cdot x(t_1 + \tau))}$$
$$= q^2 \int_{-\infty}^{\infty} \lambda(t_1 - \alpha) h(\alpha) h(\alpha + \tau) \, d\alpha \tag{52}$$

If zero mean value is required, the process $[x(t) - \overline{x(t)}]$ can be considered. It is also possible to consider for this purpose a sequence where each pulse can be $+qh(t)$ or $-qh(t)$ at random, with equal probability independent of time. If $\lambda$ is constant, the process is stationary, and (Campbell's theorem)

$$\bar{x} = \lambda q \int_{-\infty}^{\infty} h(\alpha) \, d\alpha = \lambda q \tag{53}$$

$$\sigma_x^2 = \overline{x^2} - \bar{x}^2 = \lambda q^2 \int_{-\infty}^{\infty} h^2(\alpha) \, d\alpha \tag{54}$$

The stationary autocovariance is proportional to the time autocorrelation function $k_{hh}(\tau)$ of the energy signal $h(t)$:

$$k_{hh}(\tau) = \int_{-\infty}^{\infty} h(\alpha) h(\alpha + \tau) \, d\alpha \tag{55}$$

$$C_{xx}(\tau) = \lambda q^2 k_{hh}(\tau) = \lambda q^2 \int_{-\infty}^{\infty} h(\alpha) h(\alpha + \tau) \, d\alpha \tag{56}$$

The conditions of equivalence of the model to a stationary noise, characterized by $R_{nn}(\tau) = \overline{n^2} \rho_{nn}(\tau)$, are thus

$$\overline{n^2} = \lambda q^2 \int_{-\infty}^{\infty} h^2(\alpha)\, d\alpha = \lambda q^2 k_{hh}(0) \tag{57a}$$

$$\rho_{nn}(\tau) = \frac{\displaystyle\int_{-\infty}^{\infty} h(\alpha)h(\alpha+\tau)\, d\alpha}{\displaystyle\int_{-\infty}^{\infty} h^2(\alpha)\, d\alpha} = \frac{k_{hh}(\tau)}{k_{hh}(0)} \tag{57b}$$

It is worth remembering that there is some degree of freedom in the choice of $h(t)$, as it is not uniquely determined by the knowledge of its time autocorrelation function.

This statistical model is suitable also for nonstationary noise. From the mathematical standpoint, in general it is less simple than in stationary cases to determine the functions $qh(t)$ and $\lambda(t)$. However, in many cases (see Sect. 7.4) the real physical situation is a Poisson process itself and will suggest these functions.

### 7.2.4.8  Gaussian and Non-Gaussian Noise

A noise is said to be Gaussian or normal when all its probability density functions, marginal and joint, are Gaussian, see Eq. (33). In this case, the statistical description gives a complete description of the process: if second-order moments are known *and* the process is known to be Gaussian, higher-order moments can be determined. It is to be emphasized, however, that the knowledge that a noise is Gaussian does not give by itself any information about the second-order moments, and vice versa. A Gaussian noise can indeed be stationary or nonstationary; it may be "white" or have other autocorrelation behaviors. Likewise, the knowledge of the correlation behavior does not tell if a noise is Gaussian or not. Finally, it should be stressed that although in many physical situations the noise can be considered Gaussian, this is by no means universal, and other distributions are often met.

### 7.2.4.9  Autocorrelation of Signal Plus Noise

Let us consider a waveform $x(t)$ that contains a signal $s(t)$ (including possible background) and a noise $n(t)$. From the definitions of $s(t)$ and $n(t)$ (see Sect. 7.2.3) it follows that the (ensemble) cross-correlations

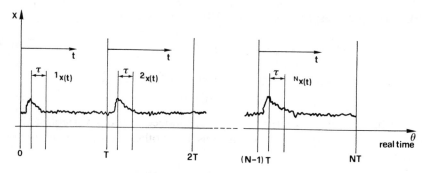

**Fig. 7.5.** Repetitive fluorescence waveform observed over a finite time interval after each laser triggering (see text).

between noise and signal are zero, and the autocovariance is due to noise only:

$$R_{xx}(t, t+\tau) = \overline{[s(t)+n(t)][s(t+\tau)+n(t+\tau)]} = s(t)s(t+\tau) + \overline{n(t)n(t+\tau)}$$

$$= s(t)s(t+\tau) + R_{nn}(t, t+\tau) \quad (58)$$

$$C_{xx}(t, t+\tau) = C_{nn}(t, t+\tau) = R_{nn}(t, t+\tau) \quad (59)$$

In the following, when considering a waveform $x(t)$, we shall consider separately signal and noise in it. Unless otherwise explicitly stated, we shall therefore call autocorrelation $R_{xx}$ at the point where $x$ is observed to be the autocorrelation of noise alone, although it would be more correct to speak of autocovariance $C_{xx}$.

In principle, correlation and covariances should be obtained from the observation of the infinite number of waveforms $x(t)$, which might happen in the physical situation considered. In practice, however, it is possible to observe a finite number $N$ of repetitions of $x(t)$, each repetition being observed over a finite time interval. Provided certain conditions are met, in this way an acceptable approximation to ensemble averaging may be obtained. Let us consider the example in Sect. 7.2.2 and observe the photodetector waveform $x(t)$ over a finite time interval after each triggering of the repetitive laser. In every repetition, the time $t$ will be measured with its origin at the triggering instant, so that $'x(t)$ will represent the $r$th repetition ($r = 1, 2, \ldots N$). For simplicity, although this is by no means necessary, we may consider a periodically repetitive laser with period $T$, as sketched in Fig. 7.5. In this case $'x(t) = x[t+(r-1)T]$. If the average over the finite set of $N$ repetitions is denoted by $\{\ \}_N$, the relationship corresponding to Eq. (58) can be written

$$\{x(t)x(t+\tau)\}_N = \frac{1}{N}\sum_{r=1}^{N} {}^r x(t)^r x(t+\tau)$$

$$= s(t)s(t+\tau) + \{n(t)n(t+\tau)\}_N$$

$$+ s(t)\{n(t+\tau)\}_N + s(t+\tau)\{n(t)\}_N \qquad (60)$$

These finite average values of noise are indeed random variables; they vary from one series of $N$ repetitions to another. If the mean square value,

$$\overline{\{n(t)\}_N^2} = \frac{1}{N^2}\left[\sum_{r=1}^{N} \overline{{}^r n^2(t)} + \sum_{r\neq k=1}^{N} \overline{{}^r n(t)^k n(t)}\right] \qquad (61)$$

is considered, it is seen that it depends on possible autocorrelation of the noise at time intervals equal to the ones between the various repetitions. If there is no correlation, $\overline{\{n(t)\}_N^2}$ decreases with $1/N$ as $N$ is increased and $\{n(t)\}_N$ approaches the true average $\overline{n(t)} = 0$. In the presence of such correlations, the decrease in $\overline{\{n(t)\}_N^2}$ and the approach of $\{n(t)\}_N$ to $\overline{n(t)}$ may be slower; for instance, this may happen with noise components that are slow (drift, low-frequency interference, etc.) with respect to the interval between repetitions. It may even happen that $\{n(t)\}_N$ does not at all approach $\overline{n(t)}$; for instance, if components correlated with the repetition rate are included in the noise.

It is worth noting that a nonperiodic repetition, with a statistical distribution of the time interval from one repetition to another, may be advantageous to avoid correlation with possible periodic components of the noise. With periodic repetition this may also be obtained if the repetition period is uncommensurate with respect to the one of the noise component.

It can be concluded, for the last two terms in Eq. (60), that an approximation to the ensemble average is obtained with high $N$ if the noise is not correlated from one repetition to another. The condition required in order to have $\{n(t)n(t+\tau)\}_N \rightarrow \overline{n(t)n(t+\tau)}$ at high $N$ is somewhat more restrictive, as it includes moments of order higher than the second, that is, higher-order autocorrelation functions, and will not be treated here. If the noise is statistically independent from one repetition to another, this condition is certainly met; we recall that if the noise is Gaussian, no correlation is equivalent to statistical independence.

Let us consider now the time autocorrelation function of $x(t)$

$$K_{xx}(\tau) = \lim_{A\to\infty} \frac{1}{A}\int_{-A/2}^{A/2} x(t)x(t+\tau)\,dt = K_{ss}(\tau) + K_{nn}(\tau) + K_{sn}(\tau) + K_{ns}(\tau)$$

$$(62)$$

or, in practice, the estimated function with a finite time interval $A$. In general, the cross-correlations $K_{sn}(\tau)$ and $K_{ns}(\tau)$ are not zero. However, if the ensemble average of Eq. (62) is considered, we see that these terms vanish, because it can be shown that the order of time and ensemble averaging can be interchanged in all cases of practical interest. The mathematical conditions for the validity of this interchange may be not satisfied by some idealized representation (e.g., with $x(t)$ not bounded), but they are always satisfied by physical waveforms $x(t)$. Therefore,

$$\overline{K_{xx}(\tau)} = \left\langle \overline{x(t)x(t+\tau)} \right\rangle = \left\langle \overline{x(t)x(t+\tau)} \right\rangle$$

$$= \langle s(t)s(t+\tau) \rangle + \left\langle \overline{n(t)n(t+\tau)} \right\rangle$$

$$= K_{ss}(\tau) + \langle R_{nn}(t,t+\tau) \rangle \qquad (63)$$

This relation is of practical importance because most of the available *correlators*, that is, the instruments measuring correlation functions, give in fact an approximation of such a time average of ensemble correlation functions. Indeed, in the case of *repetitive* (not necessarily periodic) signals plus noise, the time-autocorrelation function may be estimated by integrating over a finite time interval $A$, which is divided into a series of subintervals $B$, each one centered around a repetition. With reference to the example considered above, it is seen that if the noise is not correlated from one repetition to another, this is equivalent to averaging over a time interval $B$ an approximate ensemble correlation function. Therefore, the result will not contain contributions from signal–noise cross-correlations. As a matter of fact, in correlation computer instruments, the waveform $x(t)$ will be sampled at a set of suitably spaced time instants (see Sects. 7.2.5, 7.6, and 7.7) and the continuous time integral will be replaced by a discrete summation. This summation can be broken into a sum of subsummations each one corresponding to one interval $B$, and it is quite easy to verify the interchange of the approximate ensemble average with the approximate time averge:

$$[K_{xx}(\tau)]_{NT} = [K_{ss}(\tau)]_T + [\langle [R_{nn}(t,t+\tau)]_N \rangle]_T \qquad (64)$$

It is possible to extend the considerations in this subsection in order to consider cross-correlations, in particular, cross-correlation of two waveforms containing two repetitive signals with the same law of repetition.

### 7.2.5 Frequency Domain Representation—Power and Energy Spectra

### 7.2.5.1. *Representation of Waveforms*

As already outlined in Sect. 7.2.4, a time function $x(t)$ describing a waveform can be represented by means of the *Fourier transformation* as a linear combination of elementary sinusoidal functions.

The reader is referred to the literature for a complete treatment of the well-known Fourier transformation (1,2); only some aspects are here underlined. Aperiodic waveforms are described by a continuum of components having infinitesimal amplitudes. These amplitudes are written as a product $A(\omega)\,d\omega$, where the finite density $A(\omega)$ is called the *amplitude spectrum* of the waveform. Periodic waveforms (with period $T$) are described by components having $\omega$ integer multiples of $2\pi/T$ and finite amplitudes; they may be considered to have an amplitude spectrum consisting of equally spaced $\delta$ functions having various weights. The function $\theta(\omega)$ is called *phase spectrum* of the waveform. Great simplifications in computation are obtained by using a complex Fourier representation, that is, by considering complex elementary signals:

$$A \exp[j(\omega t + \theta)] = A \exp(j\theta) \exp(j\omega t) = X \exp(j\omega t) \qquad (65)$$

The complex Fourier transform $X = X(\omega) = A(\omega)e^{j\theta(\omega)}$ corresponds to an amplitude spectrum $|X(\omega)| = A(\omega)$, which is an even function of $\omega$, and to a phase spectrum $\arg X(\omega) = \theta(\omega)$, which is an odd function of $\omega$. It is therefore seen that complex components may be grouped in complex conjugate couples, equivalent to real sinusoids.

The amplitude spectrum is usually measured as a density with respect to the frequency $f = \omega/2\pi$. As $|X(\omega)| \, (d\omega/2\pi)$ has the dimensions of $x(t)$, it follows that the measurement unit of $|X(f)|$ is the unit of $x$ divided by the unit of $f$. For instance, if $x$ is a voltage, $|X(f)|$ is measured in $\mathrm{V\,Hz^{-1}}$.

The duration of a waveform in time and the width of its amplitude spectrum $A(\omega)$ must be defined in different ways for different kinds of waveforms; unfortunately, no definition is suitable for all cases. However, with any definition it turns out that time duration and frequency bandwidth are inversely proportional, and their product is of the order of $\pi$ (*uncertainty principle*).

A time function $x(t)$ representing a real waveform will be a *causal function*, that is, it will be identically zero prior to some specified time (for simplicity, it will be possible to set at this time the origin $t = 0$). It can be shown that $A(\omega)$ and $\theta(\omega)$ of causal functions are related to each other and $A(\omega)$ is subject to a mathematical condition (Paley-Wiener condition). Among other consequences, $A(\omega)$ cannot be identically zero over a finite interval of frequencies (i.e., in any "frequency band") and cannot decrease toward zero more rapidly than exponentially.

In practice, however, the Fourier transform is often negligibly small above a certain frequency $\omega_M$, and a bandwidth-limited spectrum is an acceptable approximation:

$$A(\omega) = 0 \qquad \text{for} \qquad |\omega| \geq \omega_M \qquad (66)$$

It can be shown (*sampling theorem*) that in this case $x(t)$ can be uniquely determined from its values at a sequence of equally spaced times separated by the interval $\Delta t = \pi/\omega_M$:

$$x_n = x\left(n\frac{\pi}{\omega_M}\right) \qquad n = 0, \pm 1, \pm 2, \ldots \tag{67}$$

that is, from *samples* taken periodically, with a frequency $f_s = (\omega_M/\pi) = 2f_M$. In fact, $x(t)$ can be determined from these samples by using $\mathrm{sinc}(\omega_M t)$ as interpolating function:

$$x(t) = \sum_{n=-\infty}^{\infty} x_n \, \mathrm{sinc}\left[\omega_M\left(t - n\frac{\pi}{\omega_M}\right)\right] \tag{68}$$

The frequency $f_s/2$ is referred to as the Nyquist frequency.

The sampling theorem has a counterpart in the frequency domain (*frequency sampling theorem*). If a waveform $x(t)$ is zero outside a finite interval, the time origin can be centered in this interval so that

$$x(t) = 0 \qquad \text{for} \qquad |t| \geq t_L \tag{69}$$

In this case, the Fourier transform $X(\omega)$ can be uniquely determined from its values at a sequence of equally spaced frequencies separated by $\pi/t_L$:

$$X_n = X\left(n\frac{\pi}{t_L}\right) \tag{70}$$

In fact, $X(\omega)$ is given by

$$X(\omega) = \sum_{n=-\infty}^{\infty} X_n \, \mathrm{sinc}\left[t_L\left(\omega - n\frac{\pi}{t_L}\right)\right] \tag{71}$$

### 7.2.5.2 Energy and Power Spectra

The quantity

$$X^*(\omega)X(\omega) = |X(\omega)|^2 = A^2(\omega) \tag{72}$$

is called the *energy spectrum* of the waveform $x(t) = \mathscr{F}^{-1}[X(\omega)]$. Parseval's theorem states indeed that the energy $E$ of *finite-energy signals* can be considered a sum of contributions from different frequencies $\omega$, with a density given by $|X(\omega)|^2$:

$$E = \int_{-\infty}^{\infty} x^2(t)\, dt = \frac{1}{2\pi}\int_{-\infty}^{\infty} |X(\omega)|^2\, d\omega \tag{73}$$

More generally, it can be shown that the (energy type) time correlation function and the energy spectrum are Fourier transforms:

$$k_{xx}(\tau) = \frac{1}{2\pi} \int_{-\infty}^{\infty} |X(\omega)|^2 \cos \omega t \, d\omega = \mathscr{F}^{-1}[|X(\omega)|^2] \qquad (74a)$$

$$|X(\omega)|^2 = \int_{-\infty}^{\infty} k_{xx}(\tau)\cos \omega\tau \, d\tau = \mathscr{F}[k_{xx}(\tau)] \qquad (74b)$$

The measurement unit of the energy spectral density $|X(\omega)|^2$ is, with respect to the frequency $f = \omega/2\pi$, the square of the unit of $x$ divided by the square of the frequency unit, for instance, $V^2 \, Hz^{-2}$.

In the case of *finite-power signals* $x(t)$, the energy is infinite and the integral of $|X(\omega)|^2$ diverges. Let us consider then the function $x_T(t)$, obtained by truncating $x(t)$ above $|t| = T$, and its Fourier transform:

$$X_T(\omega) = \mathscr{F}[x_T(t)] = \int_{-\infty}^{\infty} x_T(t)e^{-j\omega t} \, dt = \int_{-T}^{T} x(t)e^{-j\omega t} \, dt \qquad (75)$$

By using Parseval's theorem for $x_T(t)$, dividing both sides by $2T$ and taking the limit for $T \to \infty$ yields

$$P = \langle x^2 \rangle = \lim_{T \to \infty} \frac{1}{2T} \int_{-T}^{T} x^2(t) \, dt = \frac{1}{2\pi} \int_{-\infty}^{\infty} \left[ \lim_{T \to \infty} \frac{|X_T(\omega)|^2}{2T} \right] d\omega \qquad (76)$$

The real positive function

$$S(\omega) = \lim_{T \to \infty} S_T(\omega) = \lim_{T \to \infty} \frac{1}{2T} X_T(\omega)X_T^*(\omega) = \lim_{T \to \infty} \frac{|X_T(\omega)|^2}{2T} \geq 0 \qquad (77)$$

is called the *power spectrum* of $x(t)$. By writing

$$P = \langle x^2 \rangle = \lim_{T \to \infty} \frac{1}{2T} \int_{-T}^{T} x^2(t) \, dt = \frac{1}{2\pi} \int_{-\infty}^{\infty} S(\omega) \, d\omega \qquad (78)$$

it is seen that its interpretation is similar to that of the energy spectrum. As the time correlation function (power type) $K_{xx}(\tau)$ can be written as

$$K_{xx}(\tau) = \lim_{T \to \infty} \frac{1}{2T} \int_{-\infty}^{\infty} x_T(t)x_T(t+\tau) \, dt = \lim_{T \to \infty} \frac{k_{T,xx}(\tau)}{2T} \qquad (79)$$

by taking the Fourier transform, it follows that

$$\mathscr{F}[K_{xx}(\tau)] = S(\omega) \qquad (80a)$$

$$K_{xx}(\tau) = \mathscr{F}^{-1}[S(\omega)] \qquad (80b)$$

In fact, the power spectrum is often directly defined as the Fourier transform of the autocorrelation function.

Note that the measurement unit of the power spectral density is the *square* of the $x$ unit divided by the frequency unit, for instance, $V^2 Hz^{-1}$. Sometimes the square root $\sqrt{S(\omega)}$ is given, measured in units of $x$ divided by square root of frequency units, for instance, $V Hz^{-1/2}$.

Consider a case where $x(t) = x_1(t) + x_2(t)$, with $x_1$ and $x_2$ having non-zero correlation functions $K_{12}(\tau)$ and $K_{21}(\tau)$. It follows that $S(\omega)$ will be the sum of $S_1(\omega)$, $S_2(\omega)$, and the *cross-power spectra*

$$S_{12}(\omega) = \mathscr{F}[K_{12}(\tau)] = \lim_{T \to \infty} \frac{1}{2T} X_{1T}^*(\omega) X_{2T}(\omega) \tag{81}$$

$$S_{21}(\omega) = \mathscr{F}[K_{21}(\tau)] = \lim_{T \to \infty} \frac{1}{2T} X_{2T}^*(\omega) X_{1T}(\omega) \tag{82}$$

The extension to any number of components (and to energy spectra) is straightforward. The different properties of auto- and cross-correlation functions are reflected in the frequency domain; in fact, cross power spectra, instead of being real and positive, are complex functions of $\omega$.

It must be emphasized that while a unique power (or energy) spectrum corresponds to a given waveform, the converse is *not* true; these spectra do not depend on the phase $\theta(\omega)$ of the complex $X(\omega)$.

It is quite a common practice to indicate the square root of the spectral density $\sqrt{S(\omega)}$. It is worth remembering that, even if the *root* mean square $\sqrt{\langle x^2 \rangle}$ is sought, it is necessary, see Eq. (78), to take the *square* of $\sqrt{S}$ *before* integrating over the bandwidth concerned and to take the square root *after* integration.

### 7.2.5.3 Statistical Representation of Noise

Noise is a random process that is statistically characterized by its (ensemble) average values. The power spectrum can thus be defined by taking the ensemble average of $S_T(\omega)$ in Eq. (77) and then taking the limit for $T \to \infty$:

$$S(\omega) = \lim_{T \to \infty} \overline{S_T(\omega)} = \lim_{T \to \infty} \frac{\overline{X_T(\omega) X_T^*(\omega)}}{2T} \tag{83}$$

It can be shown that this is equivalent to defining the spectral density $S(\omega)$ as the Fourier transform of the time average of the (ensemble) autocorrelation function $R_{xx}(t, t + \tau)$:

$$S(\omega) = \mathcal{F}[\langle R_{xx}(t, t + \tau) \rangle] \tag{84}$$

If the noise is *stationary*, the autocorrelation function is independent of time; therefore

$$S(\omega) = \mathcal{F}[R_{xx}(\tau)] \tag{85}$$

If the process is ergodic, then Eq. (85) coincides with Eq. (80a). If the noise is a sum of cross-correlated components, the total spectral density contains also *cross-power density* components, which are the transforms of the cross-correlation functions, or their time average.

The results reported in this section may be used to obtain the power spectral density for the Poisson process considered in Sect. 7.2.4. If $H(\omega) = \mathcal{F}[h(t)]$ is the Fourier transform of the pulse shape, by taking into account that $\int_{-\infty}^{\infty} h(t)\, dt = H(0) = 1$, we obtain from Eqs. (56) and (53) (stationary noise)

$$S(\omega) = \mathcal{F}[R_{xx}(\tau)] = \mathcal{F}\left[ C_{xx}(\tau) + \left( \lambda q \int_{-\infty}^{\infty} h(t)\, dt \right)^2 \right]$$
$$= \lambda q^2 |H(\omega)|^2 + \lambda^2 q^2\, \delta(\omega) \tag{86}$$

The zero-frequency component $\lambda^2 q^2\, \delta(\omega)$ is associated to the mean value. The spectral density of the deviations from the mean is determined by the energy spectral density $|H(\omega)|^2$ of the pulse $h(t)$. The phase of $H(\omega)$ is not specified by $S(\omega)$ and is arbitrary; if $\sqrt{S(\omega)}$ satisfies the Paley-Wiener condition, the phase can be so chosen as to have a causal $h(t)$. In the case of a nonstationary process, the time-averaged rate $\langle \lambda(t) \rangle$ should be used in Eq. (86). This equation may be interpreted by taking into account that due to the independence of the starting times of the various pulses, the energy contributions will be uncorrelated for all frequency components, except the ones at $\omega = 0$.

Traditionally, the terminology and the treatment of noise have been based on the frequency-domain spectral representation. The spectral density of noise is often represented as a sum of components having frequency dependences of the type $|\omega|^\alpha$, with integer or fractional exponents $\alpha$, which are meant to correspond to particular noise types. As we have seen, components with $\alpha = 0$ are called "white noise." Components with $\alpha = -1$ (or close to $-1$) are met in a variety of physical situations and have therefore received various names: "flicker effect," "pink noise," "one over $f$ ($1/f$) noise," "excess noise," "contact noise," and "low frequency noise." Components with $\alpha = -2$ are sometimes named "random walk noise"; this name is related to their representation by the

Poisson model, which implies random positive and negative step pulses having $H(\omega) = \mathscr{F}[1(t)] = 1/j\omega$.

## 7.3  LINEAR FILTERING

### 7.3.1  Linear Systems: Definitions and General Properties

Physical systems (photodetectors, electronic circuits, etc.) are called *linear* when the *superposition principle* is valid for their input–output relationship. If an input variable $x_1$ results in an output variable $y_1$ and an input $x_2$ results in output $y_2$, then input $x_1 + x_2$ results in output $y_1 + y_2$. The response of a linear system may be unchanged under a shift of the time axis: if $x(t)$ produces an output $y(t)$, then $x(t + \tau)$ produces an output $y(t + \tau)$. If this is valid, the system is simply called a *linear* system (or constant-parameter linear system); if not, it is called a *time-variant linear* system (or time-dependent parameter linear system). Some examples may clarify the concepts. Figure 7.6a shows an elementary linear system, an approximate integrator with time constant $T = RC$, and its response to a step input. Figure 7.6b is a simplified scheme of an elementary time-variant linear system—a *boxcar*, or *gated switched integrator*. The switch $S$ is a series gate that can interrupt the connection to the voltage source and therefore the input signal path and the discharge path of $C$ through $R$. A step response, corresponding to a given sequence of gate opening and closing intervals, is also shown. This kind of system is sometimes called switching parameter (or piecewise constant parameter) linear system. An example of continuous time variation is shown in Fig. 7.7; the system is linear (i.e., the superposition principle holds for variables $x(t)$ applied at

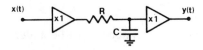

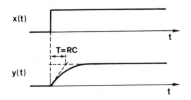

(a)

**Fig. 7.6a.**   Approximate $RC$ integrator (constant parameter linear system).

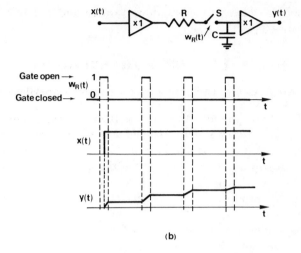

**Fig. 7.6b.** Boxcar integrator (switching parameter, time-variant linear system).

the same instant) because the reference time function $w_R(t)$ does not depend on the values of the input $x(t)$. Figure 7.7 is a simplified scheme of a wide class of time-variant filters called "*correlation filters.*"

It must be emphasized that for *any* type of linear system, constant-parameter or time-variant, the value of the output $y$ *at an instant t* can be regarded as being a *weighted integral* of the values of the input $x$ in the time $\tau$ preceding $t$. The system will thus be characterized by a weighting function $w(t,\tau)$ which in general will depend on both the $\tau$ and the $t$ considered:

$$y(t) = \int_{-\infty}^{t} x(\tau)w(t,\tau)\, d\tau \qquad (87)$$

Some properties of (constant-parameter) linear systems must be underlined. Linear transformations $x(t) \rightarrow y(t)$ are *reversible*: if a linear system

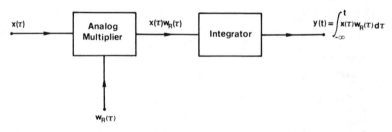

**Fig. 7.7.** Simplified scheme of a correlation filter (time-dependent parameter linear system).

performs a transformation $x(t) \rightarrow y(t)$, it is always possible to find another linear system that performs the reverse $y(t) \rightarrow x(t)$. A linear transformation consisting of a cascade of linear transformations is independent of the order in this sequence. This means that, as far as the overall input–output transformation is concerned, linear blocks in a cascade may be interchanged, provided the input–output transformation of each block is unaltered by the changing of connections (for instance, this may not be the case in a cascade of amplifiers, depending on input and output impedances). It also means that if the response $y_1(t)$ to a given $x_1(t)$ is known, the response to a waveform obtained from $x_1(t)$ by means of a linear operation (e.g., integration, differentiation, etc.) is obtained by applying this operation to $y_1(t)$. So, the response to a unit step pulse $1(t)$ is obtained from the response to a unit impulse $\delta(t)$ by integration, and conversely by differentiation.

The linear behavior of *physically realizable* systems is, of course, subject to limitations. The input–output relationship will be linear within finite ranges of values of the variables $x$ and $y$, called input and output *dynamic* or *linear range*. Moreover, the output $y$ will contain not only the response to the input $x$, but also additional noise generated in the linear system. In practice, reversibility can be achieved within the linear ranges provided the additional noise can be neglected; this will depend on the measurement type and on the type and level of signal and noise at the input. Similar considerations can be made for the exchange of order in a cascade.

### 7.3.2 Constant Parameter Filters

In the *time domain*, the *response $h(t)$ to $\delta(t)$ (unit impulse)* completely characterizes these systems. By taking into account that the response to an impulse $A\delta(t-\tau)$ of area $A$ at time $\tau$ is $Ah(t-\tau)$ and that any waveform $x$ can be broken up into elements of area $x(\tau) \, d\tau$, the value *at time $t$* of the output corresponding to $x$ can be expressed by the convolution integral

$$y(t) = \int_{-\infty}^{t} x(\tau)h(t-\tau) \, d\tau = x(t) * h(t) \tag{88}$$

By comparison with Eq. (87), it is seen that for these systems the *weighting function $w(t,\tau)$* is the *time-reversed impulse response displaced by time $t$*:

$$w(t,\tau) = h(t-\tau) \tag{89}$$

Figure 7.8 shows the $\delta$ response and the weighting function of the example in Fig. 7.6*a*.

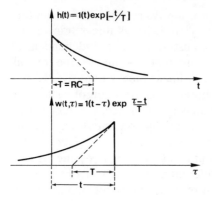

**Fig. 7.8.** Approximate $RC$ integrator response $h(t)$ to a unit impulse $\delta(t)$ and corresponding weighting function $w(t,\tau)$ for the output at time $t$.

The *response $u(t)$ to a unit step* $1(t)$ may be considered in order to specify the capability of response to fast waveforms. If $u(t)$ rises monotonically (or at least with small overshoots), the *risetime* $t_r$ is usually defined as the time taken from 10 and 90% of the flat top of $u(t)$ and gives a rough measure of the slowdown impressed by the system. For instance, in the example of Fig. 7.6a, $t_r = 2.2T$. It must be emphasized that this definition applied to $h(t)$ does not give a parameter having the same meaning. Figure 7.6 shows indeed a "slow" system having $h(t)$ with zero risetime. The response to fast signals depends on the *width* of $h(t)$, in accordance with the fact that $u(t) = \int_0^t h(\tau)\, d\tau$. This must be borne in mind, for instance, when evaluating the response of photomultiplier tubes, because some manufactures specify only the risetime of the response to a single photon (or to a very narrow light flash), which may have a quite longer falltime (see Sect. 7.5.1).

The modifications to ensemble correlation functions due to the passage through linear systems can be readily expressed in terms of $h(t)$, by taking into account that in Eq. (88) ensemble averaging acts only on $x$. The input–output cross-correlation is thus

$$R_{xy}(t_1,t_2) = \overline{x(t_1)y(t_2)} = R_{xx}(t_1,t_2) * h(t_2) \tag{90a}$$

$$R_{yx}(t_1,t_2) = R_{xx}(t_1,t_2) * h(t_1) \tag{90b}$$

and in stationary conditions

$$R_{xy}(\tau) = R_{xx}(\tau) * h(\tau) \tag{91a}$$

$$R_{yx}(\tau) = R_{xx}(\tau) * h(-\tau) \tag{91b}$$

The output autocorrelation is

$$R_{yy}(t_1,t_2) = \int_{-\infty}^{\infty} h(\alpha)R_{xy}(t_1 - \alpha,t_2)\, d\alpha = R_{xy}(t_1,t_2) * h(t_1)$$

$$= \int_{-\infty}^{\infty} h(\alpha)\, d\alpha \int_{-\infty}^{\infty} R_{xx}(t_1 - \alpha,t_2 - \beta)h(\beta)\, d\beta$$

$$= R_{xx}(t_1,t_2) * h(t_1) * h(t_2) \tag{92}$$

and in stationary conditions

$$R_{yy}(\tau) = \int_{-\infty}^{\infty} R_{xx}(\tau - \gamma)\, d\gamma \int_{-\infty}^{\infty} h(\alpha)h(\alpha + \gamma)\, d\alpha$$

$$= R_{xx}(\tau) * h(-\tau) * h(\tau) = R_{xx}(\tau) * k_{hh}(\tau) \tag{93}$$

Similar equations express the modifications to energy-type and power-type time correlation functions.

Many interesting conclusions can be drawn from the equations above. For instance, $h(t)$ can be determined from the knowledge of $R_{xx}$ and $R_{xy}$; in particular, if the input $x$ is stationary white noise, the input–output correlation $R_{xy}$ coincides with the $\delta$ response $h$. A type of stationary noise can be transformed in any other type by suitable selection of $h(t)$; practically any type can be obtained from white noise, and vice versa. These and other facts are readily interpreted in terms of the Poisson statistical model of noise (modifications to the shape of the random pulses). It is noteworthy that two $h(t)$ that have a different shape but equal autocorrelation function $k_{hh}(\tau)$ have equal effect as far as autocorrelation functions and therefore noise computations are concerned.

In the *frequency domain* (complex Fourier transform domain) the system is characterized by its (complex) transfer function $H(\omega)$, which is the Fourier transform of the $\delta$ response $h(t)$. As a convolution in the time domain corresponds to a multiplication in the frequency domain, and vice versa, Eq. (88) corresponds to

$$Y(\omega) = X(\omega)H(\omega) \tag{94}$$

where $Y(\omega)$ and $X(\omega)$ are the Fourier transforms of $y(t)$ and $x(t)$, respectively. Equation (94) is readily interpreted in terms of sinusoidal components. The response of a constant parameter linear system to a sinusoid at a frequency $\omega$ is a sinusoid having the same frequency and modified amplitude and phase. (A perfect sinusoid is considered, starting at $t = -\infty$; no initial transients are to be considered.) The module $|H(\omega)| = G(\omega)$ gives the output/input amplitude ratio, and the phase angle $\arg H(\omega) = \varphi(\omega)$ gives the output-input phase difference for the frequency $\omega$. $G(\omega)$ is usually called the *gain* or *amplification*, and $\varphi(\omega)$ is called the

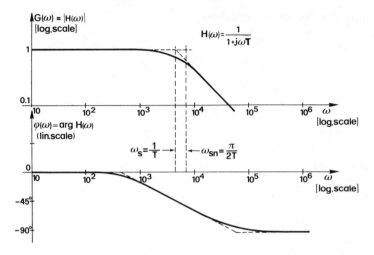

**Fig. 7.9.** Bode diagram for the approximate $RC$ integrator in Fig. 7.6a. Dashed lines correspond to the asymptotic diagram and to the reactangular approximations for calculations on signals $(\omega_s)$ and noise $(\omega_{sn})$, respectively.

*phase shift* or *phase response* of the system at $\omega$. *Bode diagrams* are often used for specifying $H(\omega)$: $\log_{10} G(\omega)$ and $\varphi(\omega)$ are plotted versus $\log_{10}\omega$.

Figure 7.9 shows the Bode diagrams for the approximate integrator of Fig. 7.6a. The gain $G$ is usually given in adimensional form, if necessary by dividing it by some reference value $G_{\text{ref}}$, and is measured in *decibels* (dB). This measure is by definition $20 \log_{10}[G(\omega)/G_{\text{ref}}]$. In simple cases, Bode diagrams can be approximated by piecewise straight-line diagrams, called asymptotic Bode diagrams. The "corners" in such diagrams correspond to characteristic parameters of the system. For instance, in Fig. 7.9 the "corner frequency" $\omega_s$ is related to the time constant $T = RC$ of the filter, namely, $\omega_s = 1/T$.

The modifications of power spectra due to the passage through the system are expressed by the Fourier transforms of the equations concerning correlation functions. Equation (93) thus corresponds to

$$S_{yy}(\omega) = S_{xx}(\omega) H^*(\omega) H(\omega) = S_{xx}(\omega) G^2(\omega) \qquad (95)$$

It must be noted that the phase response does not matter here; only the energy spectrum of the filter (see Sect. 7.2.5), that is, the square gain $|H(\omega)|^2 = G^2(\omega)$, enters into the output power spectrum, and therefore into computing noise contributions, see Eq. (78). This can be readily interpreted in terms of the sinusoidal model by remembering that the contribution given to the average power by each sinusoidal component depends only on its amplitude, not on its phase. If $S_{xx}(\omega)$ is the input (stationary) noise spectrum, the mean square noise at the output can be computed as

$$\overline{n_y^2} = \frac{1}{2\pi} \int_{-\infty}^{\infty} S_{xx}(\omega) G^2(\omega) \, d\omega$$

$$= \int_0^{\infty} 2S_{xx}\left(\frac{\omega}{2\pi}\right) G^2\left(\frac{\omega}{2\pi}\right) d\frac{\omega}{2\pi} = \int_0^{\infty} S'_{xx}(f) G^2(f) \, df \qquad (96)$$

where $S'_{xx}(f) = 2S_{xx}(\omega/2\pi)$ is the spectral density with respect to the frequency $f = \omega/2\pi$, which is usually considered only positive ($S_{xx}$ and $G^2$ are even functions of $\omega$). If we refer to the approximate integrator filter (Fig. 7.9), provided $S_{xx}(\omega)$ does not rise too steeply for $\omega > \omega_s$, an acceptable approximation could be to consider $G(\omega) = $ constant for $\omega < \omega_s$ and $G(\omega) = 0$ for $\omega > \omega_s$:

$$\overline{n_y^2} \simeq \frac{1}{2\pi} G^2(0) \int_{-\omega_s}^{\omega_s} S_{xx}(\omega) \, d\omega = G^2(0) \int_0^{f_s} S'_{xx}(f) \, df \qquad (97)$$

This filter is indeed called a *low-pass filter*: its *passband* extends from zero to approximately the upper frequency $f_s = \omega_s/2\pi = 1/2\pi T$, which is defined as *upper bandlimit*. Note that at $\omega = \omega_s$ the gain is $G(\omega_s) = G(0)/\sqrt{2} = 0.707 G(0)$, or $-3$ dB with respect to $G(0)$. The risetime of the filter's step response is related to the upper bandlimit:

$$t_r = 2.2T \simeq \frac{1}{3f_s} \qquad (98)$$

A wide variety of other low-pass filters is available, the upper bandlimit $f_s$ being usually defined by the 3-dB down point. It is still possible to use Eq. (98) for estimating the risetime, but in some cases this represents a quite crude approximation (the step response may also have remarkable "overshoot" and "ringing"). In a cascade connection of two or more filters, if suitable coupling is used (usually buffer amplifiers are required, as sketched in Fig. 7.6), the resulting transfer function is the product of each filter function, and Bode diagrams are simply *added*. For instance, in the gain diagram of $n$ equal integrators, the shape is $n$ times higher and the $-3$-dB point is lower by a factor of $(2^{1/n} - 1)^{1/2}$. It can be shown that in a cascade of filters having monotonically rising step response, the total risetime $t_r$ is the *quadratic* composition of those of the individual filters:

$$t_r = (t_{r_1}^2 + t_{r_2}^2 + \cdots + t_{r_n}^2)^{1/2} \qquad (99)$$

Thus, for $n$ equal filters, $t_r = \sqrt{n} t_{r_1}$.

A wide variety of *high-pass filters* is also available. For instance, by

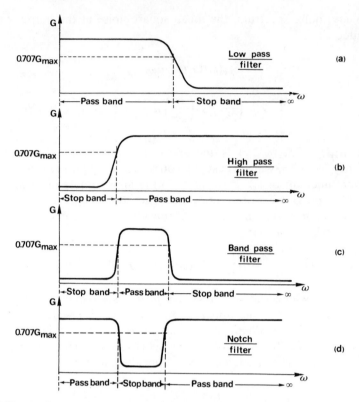

**Fig. 7.10.** Outline of gain diagrams, in linear scale, for the four basic filter types: (*a*) low-pass filter; (*b*) high-pass filter; (*c*) band pass filter; and (*d*) notch filter.

interchanging $R$ and $C$ in Fig. 7.6*a*, an *approximate differentiation* is obtained whose Bode diagrams can be obtained from the ones in Fig. 7.9—the gain diagram by a 180° rotation around a vertical axis at the corner frequency $1/RC = 1/T$ (left–right exchange) and the phase diagram by changing sign. The −3-dB point in high-pass filters is the *lower bandlimit* $f_l$ ($f_l = \omega_l/2\pi = 1/2\pi T$ in the approximate $RC$ differentiation). This limit is related to the duration of the step response of the filter in the time domain by an inverse relationship simiiar to Eq. (98).

By suitably combining high-pass and low-pass filters, one can realize a *bandpass filter*; −3-dB points define the passband limits and, for *wideband* types, are still related to the risetime, Eq. (98) and to the duration of the step response. The inverse can also be realized, namely, the band-stop or *notch filter*. The outline of the gain diagrams of the four basic types of filter is given in linear scale in Fig. 7.10. The generalization of Eq. (97) is a "rectangular" approximation of the energy spectrum of the filter (see Sect. 7.2.5): $G^2(\omega) = G_{ref}^2 = $ constant in the passbands, $G^2(\omega) = 0$ outside the passbands. The *area* of the spectrum thus obtained, by "truncating" the input $S_{xx}(\omega)$, must be compared to the *area* of the actual output spectrum $S_{yy}(\omega) = G^2(\omega)S_{xx}(\omega)$, in order to evaluate the degree of approximation.

The filter's bandwidth to be used in noise computations can therefore be different from the one defined by $-3$-dB points, and it may depend also on the shape of $S_{xx}(\omega)$. It is customary to make reference to "white noise" $S_{xx}(\omega) = $ constant and to use a rectangle having area and maximum amplitude equal to the ones of $G^2(\omega)$. For instance, an $RC$ approximate integrator has a noise equivalent bandwidth from $f = 0$ to $f_{sn} = 1/4RC$, to be compared to the $-3$-dB limit $f_s = 1/2\pi RC$. The gain function $G(\omega) = |H(\omega)|$ of realizable filters is subject to the Paley-Wiener condition (the $\delta$ response $h(t)$ is a causal function, see Sect. 7.2.5). Therefore, the above types of filters cannot be ideal; the gain can be zero in a set of discrete frequency points, not on a finite band, etc.

In practice, however, it is possible to obtain negligible gain over the stopbands. More generally, very sophisticated techniques have been developed in order to obtain acceptable approximations to almost any shape of $G(\omega)$ required in practice. Many different types of filter design known under a variety of names have been developed—Butterworth, Chebychev, Bessel, etc. Details can be found in texts devoted to the synthesis of filters (12). We simply note that (i) the gain is never strictly zero in the stopbands, (ii) it may be lower than unity in the passbands (an *insertion loss* can be correspondingly defined and measured in decibels); (iii) within the passband, and also in the stopband, the gain may not be flat in some filter types but be affected by some acceptable ripple; and (iv) the choice of a filter depends on the relative importance of many requisites, such as complexity of realization, sharp cutoff at the bandlimits, flat gain in the passband, etc.

### 7.3.3 Time-Variant Filters

#### 7.3.3.1 General Properties

The shape of the $\delta$ response of time-variant filters varies, in general, with the time position $\tau$ of the input $\delta$ pulse: $h = h(\tau, t - \tau)$. The weighting function $w(t, \tau)$, defined by Eq. (87), is used to characterize these filters. The Fourier transform of $w(t, \tau)$ with respect to $\tau$, denoted $W(t, \omega)$, will be thus defined as the *frequency domain weighting function*.

If the proper shape of $w(t, \tau)$ is considered (with its "truncations" $w(t, \tau) = 0$ for $\tau > t$ in causal systems), the upper limit of the integral in Eq. (87) can be set at $\infty$ and therefore, from Parseval's theorem in a more general form than Eq. (73), it follows that

$$y(t) = \int_{-\infty}^{\infty} x(\tau)w(t,\tau)\,d\tau = \frac{1}{2\pi} \int_{-\infty}^{\infty} X(-\omega)W(t,\omega)\,d\omega \qquad (100)$$

The modifications to correlation functions are easily expressed in terms of $w(t,\tau)$, as ensemble averaging does not concern $w(t,\tau)$. Thus,

$$R_{yy}(t_1,t_2) = R_{yy}(t_1,t_1+\theta) = \int_{-\infty}^{\infty} \int_{-\infty}^{\infty} R_{xx}(\alpha,\beta)w(t_1,\alpha)w(t_2,\beta)\,d\alpha\,d\beta$$

$$= \int_{-\infty}^{\infty} \int_{-\infty}^{\infty} R_{xx}(\alpha,\alpha+\gamma)w(t_1,\alpha)w(t_1+\theta,\alpha+\gamma)\,d\alpha\,d\gamma \qquad (101)$$

In particular, a *white noise input* $R_{xx}(\alpha,\beta) = \lambda(\alpha)\delta(\beta-\alpha)$ produces an output noise with mean square

$$\overline{n_y^2(t_1)} = R_{yy}(t_1,t_1) = \int_{-\infty}^{\infty} \lambda(\alpha)w^2(t_1,\alpha)\,d\alpha \qquad (102)$$

If the input noise is stationary, namely, $R_{xx}(\alpha,\alpha+\gamma) = R_{xx}(\gamma)$, then

$$R_{yy}(t_1,t_1+\theta) = \int_{-\infty}^{\infty} R_{xx}(\gamma)\,d\gamma \int_{-\infty}^{\infty} w(t_1,\alpha)w(t_2,\alpha+\gamma)\,d\alpha$$

$$= \int_{-\infty}^{\infty} R_{xx}(\gamma)k_{w12}(\gamma)\,d\gamma \qquad (103)$$

where $k_{w12}(y)$ is the (energy-type) time correlation function of $w(t_1,\tau)$ and $w(t_2,\tau)$. It follows for the mean square noise, from Parseval's theorem and Eq. (74$b$), that

$$\overline{n_y^2(t_1)} = \int_{-\infty}^{\infty} R_{xx}(\gamma)k_{w11}(\gamma)\,d\gamma = \frac{1}{2\pi} \int_{-\infty}^{\infty} S_{xx}(\omega)\,|W(t_1,\omega)|^2\,d\omega$$

$$(104)$$

Equations (100) and (104) show how the various input spectral components contribute to the output value and mean square value at any given time $t_1$ (frequency or time domain weighting).

It is worth noting that, in contrast to constant parameter filters, a constant input $x$ can result in a time-varying output $y(t)$, and a stationary input noise can produce a nonstationary output noise. The frequency domain representation of the output is considered below in connection with the various filter types.

### 7.3.3.2  Amplitude Modulation

This consists in the multiplication of the input $x(t)$ by a waveform $m(t)$, independent from $x(t): y(t) = x(t)m(t)$ [i.e., $w(t_1,\tau) = m(t)\delta(t-\tau)$]. For instance, *chopping a light beam* by opening and closing a mechanical shutter (rotating disk chopper, etc.) can be considered a modulation of the light intensity with an approximately square wave $m(t)$. In the frequency domain, the output $Y(\omega) = \mathscr{F}[y(t)]$ is the (complex) convolution of $X(\omega) = \mathscr{F}[x(t)]$ and $M(\omega) = \mathscr{F}[m(t)]$. Thus,

$$Y(\omega) = \frac{1}{2\pi} \int_{-\infty}^{\infty} X(\nu)M(\omega - \nu)\, d\nu = X(\omega) * M(\omega) \qquad (105)$$

It must be noted that this is an integral of a complex function; in general, $|Y(\omega)|$ is not equal to $|X(\omega)| * |M(\omega)|$ and the determination of $Y(\omega)$ is quite cumbersome. In many cases of practical interest, however, the matter is quite simple.

*a. Sinusoidal Modulation:* $m(t) = B\cos(\omega_m t + \varphi_m)$ and

$$M(\omega) = B\pi\{[\exp(j\varphi_m)]\delta(\omega - \omega_m) + [\exp(-i\varphi_m)]\delta(\omega + \omega_m)\}.$$

With a sinusoidal $x(t) = A\cos(\omega_x t + \varphi_x)$, the output spectrum $Y(\omega)$ is obtained from $X(\omega)$ by shifting it in frequency of amounts $\pm\omega_m$, by adding $\pm\varphi_m$ to the phase $\varphi_x$ and by multiplying $|X|$ by $B/2$. This result is readily extended to waveforms $x(t)$ having a low-frequency spectrum in the sense that $|X(\omega)| = 0$ for $|\omega| > \omega_m$, as sketched in Fig. 7.11. It is worth noting that in any case the $Y(0)$ (i.e., the output dc component) depends *only* on the component of the input spectrum $X(\omega)$ at $\omega = \omega_m$, more precisely on the part *in phase* with $m(t)$. Thus, only if $x(t)$ has a finite sinusoidal component $A\cos(\omega_m t + \varphi_x)$ has the output a finite dc component $Y(0) = (BA/2)\cos(\varphi_m - \varphi_x)$.

*b. Periodical Modulation:* $m(t) = m(t \pm nT_m)$, with fundamental frequency $\omega_m = 2\pi/T_m$. The $m(t)$ is a sum of harmonics having decreasing amplitudes $B_1, B_2, \ldots B_n$; the $M(\omega)$ is a series of lines spaced by $\omega_m$ having weights $\pi B_n$. If the input $x(t)$ has a low-frequency spectrum in the sense that $|X(\omega)| = 0$ for $|\omega| > \omega_m/2$, the $Y(\omega)$ can be found by superposition of the results corresponding to every harmonic of $m(t)$ (no overlap between the various spectra shifted by $n\omega_m$ and weighted by the $B_n/2$). In any case the *output dc component* $Y(0)$ will sum the input contributions at the various harmonics $n\omega_m$ weighted by the corresponding $B_n/2$.

Periodical modulation is often used to "shift" signals in frequency in order to bring them in a frequency range where the noise and the background added in the subsequent stages of the experimental apparatus (e.g., electrical noise in amplifiers, stray light, etc.) are lower. For instance, a slowly varying signal is shifted to higher frequencies by a rotating chopper disk. Modulation is then used (see later, correlation filters) also in the extraction of the modulated signal from the added

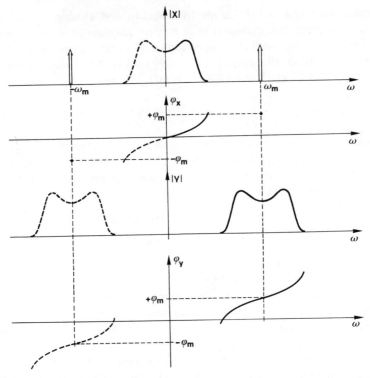

**Fig. 7.11.** Amplitude modulation, with sinusoidal $m(t)$ at $\omega = \omega_m$ and low-frequency input $x(t)$, i.e., $|X(\omega)| = 0$ for $|\omega| > \omega_m$. The output spectrum $Y(\omega)$ can be obtained by shifting the input spectrum $X(\omega)$.

background and noise. In many practical cases the waveform $m(t)$ can have a *nonzero time-average* value $\langle m(t) \rangle = B_0$. For instance, mechanical imperfections in a chopper wheel can produce unequal "opening" and "closing" time intervals and therefore an asymmetrical squarewave $m(t)$. The $m(t)$ then has a finite component $B_0$ at $\omega = 0$, $M(\omega)$ has a line $2\pi B_0 \delta(\omega)$, and the output $y(t)$ contains an unmodulated term $B_0 x(t)$. This can be a drawback—part of the input signal is not shifted in frequency. The output dc component $Y(0)$ contains a contribution coming from the input dc component $\langle x \rangle$ weighted by $B_0$.

With regard to noise, the output is nonstationary even with a stationary input: $\overline{n_y^2(t_1)} = R_{yy}(t_1, t_1) = R_{xx}(t_1, t_1) m^2(t_1) = \overline{n_x^2} m^2(t_1)$. The output power spectrum can be defined, Eq. (84), on a time-average basis. Thus, with a stationary input,

$$S_{yy}(\omega) = \mathscr{F}[R_{xx}(\theta)\langle m(t_1)m(t_1 + \theta)\rangle]$$
$$= \mathscr{F}[R_{xx}(\theta)K_{mm}(\theta)] = S_{xx}(\omega) * S_{mm}(\omega) \qquad (106)$$

The interpretation of this equation is simple because both $S_{xx}(\omega)$ and $S_{mm}(\omega)$ are real positive functions: $S_{yy}(\omega)$ is obtained by superposition of shifted and weighted replica of $S_{xx}(\omega)$. This can be readily interpreted in terms of the sinusoidal model of noise (Sect. 7.2.4). The $S_{mm}(\omega)$ of a periodical $m(t)$ is a series of lines spaced by $\omega_m$ having weights $(\pi/2)B_n^2$. The relative importance of higher harmonics therefore decreases more swiftly for the noise than for the signal. As an example, a symmetrical squarewave switching from $m = +B$ to $m = -B$ has $B_{2r+1} = (-1)^r(4B/\pi)[1/(2r+1)]$ and $B_{2r} = 0$. The relative weights of the harmonics for the signal are: $1, 0, -1/3, 0, 1/5, \ldots$; those for the noise are $1, 0, 1/9, 0, 1/25, \ldots$.

### 7.3.3.3 Gated Integration

The input waveform passes through a gate, which is open at time $t_0$, and then goes to a perfect integrator that has been reset to a definite initial state ($y = 0$) before starting the measurement. Practical examples of such filtering are found in many modern digital voltmeters. Mathematically, this can be expressed as

$$w(t,\tau) = \text{rect}(t_0,t) = 1(t - \tau) - 1(t_0 - \tau) \tag{107}$$

$$W(t,\omega) = (t - t_0) \text{ sinc } \omega \frac{t - t_0}{2} \exp[-j\omega(t + t_0)/2] \tag{108}$$

For convenience, we can take $t_0 = 0$. From Eq. (100) it follows that

$$y(t) = \int_0^t x(t)\, dt = t[\langle x \rangle]_t = t \int_{-\infty}^{\infty} X(-\omega) \text{ sinc } \frac{\omega t}{2} e^{-(j\omega t/2)} \frac{d\omega}{2\pi} \tag{109}$$

showing that as the interval of integration $(0 - t)$ is increased, the contribution to the output waveform is restricted to lower and lower input frequency components and that, if $\langle x(t) \rangle \neq 0$, the output value is divergent *as t*. As concerns noise, from Eq. (104) one can see that, with a stationary input,

$$\overline{n_y^2(t)} = R_{yy}(t,t) = t^2 \int_{-\infty}^{\infty} S_{xx}(\omega) \text{ sinc}^2 \frac{\omega t}{2} \frac{d\omega}{2\pi} \tag{110}$$

Gated integration performs a sort of low-pass filtering, and if $S_{xx}(0) \neq 0$, the mean *square* noise output is divergent *as t*. It is worth noting that the rectangular approximation to $\text{sinc}^2(\omega t/2)$, having equal area and maximum amplitude, defines a "bandwidth" from $\omega = 0$ to $\omega_{sn} = \pm\pi/t$ (i.e., $f_{sn} = \omega_{sn}/2\pi = 1/2t$). From Eqs. (101) and (103) it follows that the output correlation $R_{yy}(t_1, t_1 + \theta)$ is nonzero for any $\theta$, as intuitive. If the input signal has a long duration compared to the width of the input noise autocorrelation, the output signal-to-noise ratio is shown to increase with $\sqrt{t}$ by Eqs. (109) and (110) (approximately constant signal and approximately white noise).

It should be noted that there is a filtering type similar to gated integration but achievable with constant parameter filters, namely, integration from $t_0 = t - T$ to $t$, i.e., over an interval having a constant duration $T$ preceding the time $t$ of observation. This can be obtained, for instance, by delaying $x(t)$ by an amount $T$, subtracting it from the undelayed $x(t)$, and integrating the result. Such a constant parameter filter has a $\delta$ response $h(t) = \text{rect}(0,T)$ and a transfer function $H(\omega) = T \, \text{sinc}(\omega T/2) \exp(-j\omega T/2)$. It is a low-pass filter, with an upper frequency limit for noise calculations equal to $\omega_{sn} = \pi/T$ or $f_{sn} = 1/2T$ (see above). This filtering will be refered to as *running definite integration*. By comparing the respective weighting functions $w(t,\tau)$, it is seen that any constant parameter low-pass filter can be approximated by such an integration, and that the output can be considered as an approximate time average of the input over a suitable interval $T$, multiplied by $T$. This "rectangular" approximation of $w(t,\tau)$ is quite similar to the rectangular approximation of $|H(\omega)|$ in the frequency domain. In fact, as for the bandwidth, the interval $T$ to be considered depends on the output value to be computed. For instance, in the case of an $RC$ integrator, in computing the output signal corresponding to a constant input, $T = RC$; while in computing the output mean square noise corresponding to a stationary white input, $T = RC/2$, see Eq. (102).

### 7.3.3.4  Correlation Filters

As outlined in Fig. 7.7, they are the cascade of an amplitude modulation of the input $x(t)$ by a reference waveform $w_R(\tau)$ and of a low-pass filtering characterized by wighting function $w_{LF}(t,\tau)$ (that is, $h_{LF}(t-\tau)$ for constant parameter types, etc., see Sect. 7.3.2). The resulting weighting functions* for the time and frequency domain are

$$w(t,\tau) = w_R(\tau) w_{LF}(t,\tau) \tag{111}$$

$$W(t,\omega) = W_R(\omega) * W_{LF}(t,\omega) \tag{112}$$

A very common correlation filter is the *lock-in amplifier*, or phase-synchronous detector. It is often used to extract from superimposed

---

* In the general case of two subsequent linear filtering operations, characterized by the weighting functions $w_1(t,\tau)$ and $w_2(t,\tau)$, the resulting time-domain weighting function is

$$w(t,\tau) = \int_\tau^t w_1(\gamma,\tau) w_2(t,\gamma) \, d\gamma$$

unmodulated background and noise a low-frequency signal $a(t)$ "carried" by a high-frequency modulated waveform, corresponding to a high-frequency $m(t)$ (see modulation, example of the light chopper). The reference $w_R(\tau)$ is obtained from the $m(t)$ and is a power signal (Sect. 7.2.4), usually a periodic waveform. Very often the signal obtained from $m(t)$ is converted to a square wave by a circuit sensitive to the crossing of a reference level by $m(t)$ ("squarer" comparator). This simplifies the modulation circuitry, as multiplication by a square wave is simply obtained by switching the sign of an amplification in synchronism to the square wave. In order to reject the input dc component (see modulation), the $w_R(\tau)$ must have zero time-average value (e.g., symmetrical square wave). An adjustable shifting in time of $w_R(\tau)$ (i.e., phase shifting) is normally provided, because the reference must be properly synchronized (i.e., phase locked) to the modulated signal sought, which can be delayed with respect to $m(t)$ after passing through physical systems. A typical example is a fluorescent emission excited by a chopped source, with a chopping period not much greater than the emission lifetime and with reference obtained from the chopper waveform.

Another example are time-variant filters for pulse amplitude measurements (13). The reference $w_R(\tau)$ is an energy signal, namely, a pulse having shape equal to the one of the pulses to be measured. A reference generator is triggered by the input pulses so as to generate a properly synchronized $w_R(\tau)$. The low-pass filter is usually a gated integrator with suitable resetting after the measurement.

Accurate results can be computed from Eqs. (100) to (104) by using the actual $w(t,\tau)$ and $W(t,\omega)$ (the output noise spectrum is the tranform of $R_{yy}(\theta)$). Approximate results, sufficient in most practical cases, will be here quite intuitively obtained by using the time-domain rectangular approximation of $w_{LF}(t,\tau)$ (see above). Indeed, a simple integrator has been drawn in the simplified scheme of Fig. 7.7.

Correlation filters are so called because they can be considered to give as output $y(t)$, an estimation (Sect. 7.2.4), corresponding to the integration time interval $T$ characterizing the low-pass section, of the zero-shift value $k_{xw}(0)$ of the energy-type time cross-correlation function $k_{xw}(\tau)$ between the input $x(\tau)$ and the weight $w(t,\tau)$. If $x(\tau)$ contains a part equal in shape and time position to $w(t.\tau)$, this part will contribute with the highest efficiency to the output, as shown by the properties of auto- and cross-correlation functions. If $w_R(\tau)$ is a power signal, the output $y(t)$ can be thought of as the product of $T$ and of the estimated zero-shift value $K_{xR}(0)$ of the power-type cross-correlation between input $x(\tau)$ and reference $w_R(\tau)$. Thus, with a periodic $w_R(\tau)$ (as in the lock-in amplifier), only those components of $x(\tau)$ will be extracted that have, with respect to the significant harmonic components of $w_R(\tau)$, frequencies within bandwidths $\simeq 2\pi/T$ and *equal phases*. Indeed, the lock-in amplifier is known as phase-synchronous demodulator. In fact, the frequency domain weighting $W(t,\omega)$, Eq. (112), is $W_R(\omega)$ "smoothed" by the low-pass $W_{LF}(\omega)$ (bandwidth $\propto 1/T$).

The input–output transfer can also be analyzed completely in the frequency domain. The complex input spectrum $X(\omega)$ is weighted and shifted in frequency and phase by the $W_R(\omega)$ (see modulation). The resulting spectrum around $\omega = 0$ arises only from the input components with frequency and phase specified by the reference. The low-pass filter then extracts the frequencies lower than its upper bandlimit $\omega_s \propto 1/T$. The $W_R(\omega)$ must therefore be matched in amplitude and phase to the input signal to be extracted (which is equivalent to matching $w_R(\tau)$ in shape and time position). The effects of $\langle w_R \rangle \neq 0$ (i.e., of a dc component) and of possibly unnecessary higher harmonics in $w_R(\tau)$ can be evaluated. A constant parameter bandpass filter, centered at the fundamental frequency $\omega_m$, is often used before the lock-in amplifier in order to perform a prefiltering so as to avoid overloading of the input circuitry of the lock-in by intense wideband noise and eliminate unwanted harmonics that could contribute to the output background.

The output noise with stationary input noise and sinusoidal $w_R(\tau)$ (amplitude $B$, frequency $\omega_m$) can be easily computed with the usual approximation (remember that the $T$ values can be different for noise and signal calculations). As the square output value is concerned, two integrations over time are made, see Eq. (103). The first one implies taking the autocorrelation $K_{RR}$ of $w_R$, estimated over $T$ and multiplied by $T$. Thus, see Eq. (106), the input power spectrum $S_{xx}(\omega)$ is shifted in frequency and weighted by the power spectrum $S_{RR}(\omega)$ of $w_R$. The second integration then corresponds to a low-pass filtering. The resulting mean square output noise is $\overline{n_y^2} = B^2 T S_{xx}(\omega_m)/2$. Extension to periodic $w_R(\tau)$ is straightforward. It is worth noting that the signal-to-noise ratio increases as $\sqrt{T}$. Approximate results can also be obtained by using the reference $w_R(\tau)$ "truncated" at the ends of $T$, and its transform, in Eqs. (103) and (104). It can be seen that even with a white input noise, the output autocorrelation $R_{yy}(\theta)$ is wide and the output power spectrum $S_{yy}(\omega)$ is restricted to low frequencies (with a constant parameter low-pass section $R_{yy}(\theta) \neq 0$ for $\theta \leq T$ and $S_{yy}(\omega) \neq 0$ for $\omega < \pi/T$). The conclusions about noise can be readily interpreted by means of the sinusoidal model (Sect. 7.2.4).

Correlation filters can be compared to constant parameter filters having the same purpose. The former have $w(t,\tau)$ locked to the reference waveform $w_R(\tau)$, as the main role of $w_{LF}(t,\tau)$ is to determine the time interval $T$ where $w_R(\tau)$ is active. The latter have $w(t,\tau)$ with constant shape locked to, or "dragged" by, the observation time instant $t$. If a finite time interval is observed, the two types have unavoidably different $w(t,\tau)$ and therefore different output waveforms, autocorrelation functions, and power spectra. However, if only *one* instant $t$ is observed, the two types can be equivalent, as they can *at this time* have equal $w(t,\tau)$ and produce equal output value, variance, signal-to-noise ratio, etc. Phase discrimination on the input is accomplished in both cases by accurately synchronizing with the input signal sought the $w_R(\tau)$ or the $t$, respectively.

In principle, the performances of the two types are identical. Correlation filters, however, although generally more complicated and expensive, are more flexible (it is simpler to change the waveform $w_R(\tau)$ than a lot of component values in a filter) and have in practice superior performance, such as ease of obtaining very narrow bandwidths accurately and stably centered at the frequencies of interest. Furthermore, the slowly varying output waveform is often practical as it can be coupled to slow meters, graphic recorders, and so on.

### 7.3.3.5 *Boxcar Integrator*

This is a peculiar correlation filter, characterized by a synchronous, switching-parameter, low-pass filter. As shown in Figs. 7.6$b$ and 7.12, the modulating reference $w_R(\tau)$ is a binary asymmetrical square wave, gate open $w_R(\tau)=1$, gate closed $w_R(\tau)=0$. Within the aperture intervals $T_A$, the time constant $T$ of the approximate integrator is $T_F = RC$. In the closing intervals $T_C$, it is in principle $T_L \to \infty$, in practice $T_L = R_L C \gg T_F$, where $R_L$ is a finite leakage resistance. The exponential memory loss (i.e., discharging of $C$) should cease and the time should be virtually stopped during the $T_C$'s; in practice, the rate of loss is greatly reduced, and the time virtually slowed down. The corresponding weighting function $w_{LF}(t,\tau) = h_{LF}(\tau, t-\tau)$, shown in Fig. 7.12, can be compared to that of a constant parameter integrator in Fig. 7.8. In every $T_A$ and $T_C$, the relative loss factors are $f_A = 1 - \exp(-T_A/T_F)$ and $f_C = 1 - \exp(-T_C/T_L)$, respectively.

In spite of its remarkable practical importance, a fundamental difference between a boxcar integrator and a normal correlation filter that has the same reference $w_R(\tau)$ and a constant parameter low-pass filter is sometimes overlooked in the literature. The number of apertures $T_A$ weighted by the normal correlation filter is the one occurring in a fixed time interval, determined by its low-pass filter. The output value therefore varies as the repetition rate of $T_A$ is varied, even if the input amplitude is constant. On the other hand, the number of $T_A$'s weighted

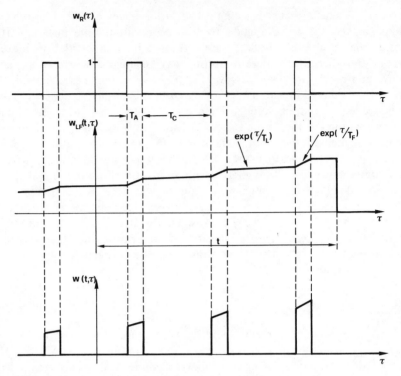

**Fig. 7.12.** Boxcar integrator: reference waveform $w_R(\tau)$ (gate open in $T_A$, closed in $T_C$); weighting function $w_{LF}(t,\tau)$ of the synchronous switching parameter filter; resulting weighting function $w(t,\tau) = w_R(\tau)w_{LF}(\tau)$. The $T_F/T_A$ shown corresponds to the "exponential averaging mode."

by the boxcar integrator depends only on the ratio $T_F/T_A$ and is independent of the repetition rate (periodic or not) as long as the losses during the closing times $T_C$ are negligible, that is, as long as $f_C \ll f_A$. As a matter of fact, the longest $T_C$ values usable in practice are limited not only by the unavoidable finite $R_L$ but also by slow drift, due mainly to spurious current generators in the circuitry connected to the memory capacitor $C$ (offset currents of electron devices, etc.). In order to circumvent these limits, digital storage is used in some instruments. After each $T_A$, the value stored in the analog memory $C$ is digitized and stored in a digital memory. This digital content is then converted to an analog level and the $C$ is held to this level until the next $T_A$ begins.

The output (signal, noise, autocorrelation, etc.) corresponding to any input condition can be accurately computed from Eqs. (100) to (104). A simple typical case will be discussed here. Let the input $x(t)$ be a signal having constant amplitude $s_x$ in the $T_A$'s (the value outside $T_A$ is irrelevant) plus white $[R_{xx}(\theta) = \lambda\delta(\theta)]$ or almost white noise $[R_{xx}(\theta) \simeq \overline{n_x^2}$ for $|\theta| \leq \theta_x < T_A$ and $R_{xx}(\theta) \simeq 0$ for $|\theta| > \theta_x]$. In the most frequently used operation mode, the ratio $T_F/T_A$ is moderately high and the output is a weighted integral of a number of $T_A$ repetitions of the order of $T_F/T_A$. This mode is usually called *exponential averaging*. Taking into account the actual $w(t,\tau)$ (see Fig. 7.12) in Eqs. (100) and (104) (with $w(t,\tau)$ normalized to unit area), one has

$$s_y = s_x \tag{113a}$$

$$\overline{n_y^2} = \lambda/2T_F = \overline{n_x^2}\theta_x/T_F \tag{113b}$$

and the signal-to-noise ratio is

$$(S/N)_y = s_x\sqrt{2T_F/\lambda} = (S/N)_x(T_F/\theta_x)^{1/2} \tag{113c}$$

For comparison, a measurement using a single repetition gives

$$s_y = s_x(T_A/T_F) \tag{114a}$$

$$\overline{n_y^2} = \lambda T_A/T_F^2 = \overline{n_x^2}2\theta_x T_A/T_F^2 \tag{114b}$$

$$(S/N)_y = s_x(T_A/\lambda)^{1/2} = (S/N)_x(T_A/2\theta_x) \tag{114c}$$

The exponential averaging thus enhances the $S/N$ ratio by a factor $\sqrt{2T_F/T_A}$. Provided that the correlation between different $T_A$ repetitions remains zero, this enhancement factor is valid also with noise having a wider $R_{xx}(\theta)$. In the presence of correlation between repetitions, the enhancement can be lower or higher, depending on the type of correlation (14).

The boxcar can also be operated with $T_F/T_A \to \infty$, that is, with a uniform integration. This mode, usually called *linear averaging*, is equivalent to a normal correlation filter using a perfect integrator as low-pass section. After each measurement, the memory $C$ is discharged by a reset device. The output is the integral of the whole number $N$ of apertures $T_A$ that occurred. In the conditions considered above,

$$s_y = s_x N T_A \tag{115a}$$

$$\overline{n_y^2} = \lambda N T_A = \overline{n_x^2}N\theta_x T_A \tag{115b}$$

$$(S/N)_y = s_x\sqrt{NT_A/\lambda} = (s_x/\sqrt{\overline{n_x^2}})\sqrt{NT_A/\theta_x} \tag{115c}$$

In the absence of correlation between repetitions, the enhancement factor is $\sqrt{N}$. Indeed, in the exponential averaging mode, $2T_F/T_A$ is often called the equivalent number of repetitions.

Circuits having the equivalent scheme of Fig. 7.6b but operating with $T_F/T_A \ll 1$ are called "*sample-and-hold*," not boxcar. After each $T_A$, they remember the input amplitude at the end of $T_A$. For very fast samplers, such as those of subnanosecond sampling oscilloscopes, the reference $w_R(\tau)$ is only roughly rectangular, and the operation corresponds to $T_F/T_A$ of the order of unity, so that a fraction of the input amplitude is memorized in $C$. This fraction is referred to as the sampling efficiency. Usually, oscilloscope circuits include devices to correct the stored value.

The boxcar is particularly valuable and versatile in measurements of pulsed signals with low duty cycles (i.e., pulse width/pulse spacing ratios) and/or variable or nonperiodic repetition rates. By using short $T_A$ and suitably delayed synchronization, the boxcar can sample and average the "instantaneous" amplitude at a given time position on the input waveform. It is worth noting that the accuracy and reproducibility of the parameters of $w(t,\tau)$, in particular of the gate aperture $T_A$, will in general determine the accuracy and stability in a series of measurements. Fairly accurate and stable $T_A$'s are now available down to $\sim 10$ ns. The range of measurements of instantaneous amplitudes in fast waveforms can be extended to the subnanosecond range by means of composite instruments, namely, fast sample-and-hold circuits followed by boxcar integrators.

Measurements free from contributions of slow background and noise components can be obtained with a "dual channel" boxcar operation (15). The modulating reference in the successive $T_A$'s is alternately $w_R = 1$ and $w_R = -1$, that is, the sign of an amplification is changed. Concurrently, the time position of $T_A$ is alternately synchronized at two different locations along the waveform, where ($a$) the signal amplitude (or its area) is to be measured and ($b$) the signal amplitude falls to zero but the slow background and noise are unchanged. A sort of lock-in amplifier having boxcar characteristics is thus obtained. The importance of having equal weight for the two positions (i.e., equal $T_A$ and $|w_R|$) is to be stressed (see lock-in amplifier, Sect. 7.3.3; input dc component).

It is worth remembering that the actual boxcar instruments include other features in addition to the essential ones, such as gate and switching integrator. Input stages (buffer amplifiers, etc.) have finite risetime, usually 2 to 10 ns, and therefore introduce some prefiltering. The $w_R(\tau)$ is not ideally rectangular but has transitions of a few nanoseconds (finite switching speed). Further low-pass filtering is usually provided by an output stage consisting of a constant parameter filter with a separately controllable bandpass limit.

### 7.3.4  Optimum Filtering

Let us consider a case that is well known in nuclear electronics (13,16–23), namely, the measurement of the amplitudes $A$ of pulse signals $s_x(\tau)$ that have a given shape $s_x(\tau)$ (normalized to unit amplitude) and a known time position and are accompanied by a noise $n_x(\tau)$ of given statistical properties. Because of the repetition rate of the pulses, a maximum limit $T_M$ is imposed on the interval available to each measurement. This is an example of the general problem of measuring a parameter of a signal whose detailed shape is known or not interesting, accompanied by a given noise, and with given constraints to the measurement. Among the quality factors of a measurement (see Sect. 7.7), the signal-to-noise ratio $(S/N)$ is essential. Any filtering method devised for maximizing it can only exploit the existing differences between signal and noise to separate as much as possible the signal from the noise. One should expect that corresponding to each situation a physical limit should exist, that is, a maximum (optimum) obtainable value $(S/N)_{opt}$. This problem is treated in the literature (19–22) and is only outlined here. A general analysis, taking into account both linear and nonlinear filtering methods, can be performed by relaxing the requisite that the values of the weighting function $w(t,\tau)$ be independent of the values of the waveform $x(\tau)$. It is found that an optimum filtering indeed exists, and it is a linear filtering (20). Some observations simplify the analysis, namely, (i) filtering by constant parameter linear filters is reversible (Sect. 7.3.2); (ii) for a given nonwhite noise, a *noise-whitening* constant parameter linear filter can be found that transforms it into white noise (this is quite intuitive in terms of the Poisson model of noise, Sect. 7.2.4); (iii) it is quite simple to compute the effect of filtering on white noise, see Eq. (102). Thus, the optimum filter can be divided in two parts. The first is a noise-whitening filter, which gives $R_{yy}(\tau,\tau+\theta) = \lambda(\tau)\delta(\theta)$, and also modifies the signal shape from $s_x(\tau)$ to $s_y(\tau)$, while the second part is characterized by a weighting function $w(t,\tau)$. If $t_M$ is the instant where the output waveform $u(t)$ is measured, then

$$(S/N)_u^2 = \frac{s_u^2(t_M)}{n_u^2(t_M)} = \frac{\left[\displaystyle\int_{-\infty}^{t_M} As_y(\tau)w(t_M,\tau)\,d\tau\right]^2}{\displaystyle\int_{-\infty}^{t_M} \lambda(\tau)w^2(t_M,\tau)\,d\tau} \tag{116}$$

More intuitive expressions can be written if $n_y$ is considered as only practically white, that is, as having an autocorrelation interval $\theta_y$ negligibly short with respect to the signal shape $s_y(\tau)$:

$$R_{yy}(\tau,\tau+\theta) \simeq \overline{n_y^2(\tau)} \quad \text{for} \quad \theta \leq \theta_y$$
$$R_{yy}(\tau,\tau+\theta) \simeq 0 \quad \text{for} \quad \theta > \theta_y$$

It can be shown  by using Schwartz's inequality (1), that with a weighting given by

$$w(t_M, \tau) \propto \frac{s_y(\tau)}{\lambda(\tau)} = \frac{1}{\theta_y} \frac{s_y(\tau)}{n_y^2(\tau)} \tag{117}$$

the optimum is reached as

$$(S/N)_{\text{opt}}^2 = \int_{-\infty}^{t_M} \frac{A^2 s_y^2(\tau)}{\lambda(\tau)} d\tau \simeq A^2 \int_{-\infty}^{t_M} \frac{s_y^2(\tau)}{n_y^2(\tau)} \frac{d\tau}{\theta_y} \tag{118}$$

which in case of stationary noise reduces to

$$(S/N)_{\text{opt}}^2 = \frac{A^2}{\lambda} \int_{-\infty}^{t_M} s_y^2(\tau) \, d\tau = \frac{A^2 \int_{-\infty}^{t_M} s_y^2(\tau) \, d\tau}{n_y^2 \theta_y} \tag{119}$$

Constraints to the measurement correspond to limitations to $w(t_M, \tau)$; for instance, truncation to an interval $T_M$ (see above). Equations (116)–(119) are quite intuitive and self-explanatory; it can be observed that in Eq. (119) $(S/N)^2$ is the ratio of the signal energy utilized by the whitened noise energy contained in the autocorrelation interval. Corresponding equations in the frequency domain can similarly be derived.

The filter that maximizes the $S/N$ in the presence of white noise is called a *matched* filter, because its weighting is matched to the form of its input signal. The reader is warned, however, that some texts use the word "matched" for the entire optimum filter in the presence of nonwhite noise, including the noise-whitening filter.

The matter of optimum filters is not only interesting from a theoretical point of view but has a remarkable practical value, as it gives criteria for the design of filters and the absolute evaluation of their performance. Optimum filters can also be considered for measurements of variables other than amplitude, for instance, in the timing of pulses (24).

### 7.3.5  Implementation of Linear Filtering

An electrical waveform can be represented and processed either as a variable in the *continuous* real time or as a set of sample values taken at *discrete* times. The information is preserved in sampling operations performed with suitable criteria (Sect. 7.2.5 and references 4, 15, and 25), and operations on sampled waveforms can be studied in the theoretical framework of the $Z$ transform (4,26,27). It is important to realize that, from the point of view of the obtainable results, there is no fundamental difference between the continuous and the discrete approach, and it is therefore possible to transfer concepts between the two areas (23). In fact, it has been proved that the theories of processing signals with linear time-invariant realizable filters are identical in the continuous-time and discrete-time cases (28).

A weighting function $w(t,\tau)$, representing a given approximation of optimum filtering, can therefore be implemented by various types of continuous-time or discrete-time linear filters. The choice is a matter of practical considerations (simplicity, reliability, economy, etc.) but *not* of $S/N$ performance, which depends only on the $w(t,\tau)$ adopted.

Continuous real-time filtering is performed by analog filters of the following classes:

a. *Passive* or longitudinal filters (23), that is, constant parameter linear networks, including only passive components such as resistances, capacitors, etc. An elementary example is the $RC$ integrator in Fig. 7.6a.

b. *Active* or recursive filters (23), that is, constant parameter linear circuits, including active devices (amplifers) and passive filters in feedback and input networks. The feedback design adds versatility, but stability criteria must be observed if oscillation is to be avoided.

c. *Time-variant* filters, as treated in Sect. 7.3.3.

Discrete-time filtering is now employed in the two following areas:

1. *Digital filtering.* The waveform is first converted into a representative set of numbers. Numerical processing is then performed by digital logic circuits (computers, microcomputers, and programmable logic circuits, hard-wired logic circuits). In this case, real-time operation is restricted to slower waveforms because of the limitations to the minimum sampling interval (Sect. 7.2.5) arising from the sampler and analog-to-digital converter (Sect. 7.6) and from the computation time. Digital filtering has been an area of intense activity in the past years (26,27).

2. *Transversal analog filters.* They represent an old idea that has found practical application only in recent years after the introduction of new electronic devices, principally the charge-transfer devices CTD (29,30) (for high-frequency applications, surface acoustic wave devices, SAW, are also used). The basic idea is illustrated in Fig. 7.13 and represents a discrete-time realization of the operation described by Eq. (87). An analog delay line has taps separated by delay intervals $\Delta\tau$. At time $t$, a sample of the waveform $x(\tau)$ is taken at each tap $k$ and multiplied by a weight $w_k$. All the weighted samples are then added to give the filter output

$$y(t) = \sum_k w_k x(t - k\,\Delta\tau) \tag{120}$$

The operation is repeated at times $t$ spaced by intervals $\Delta\tau$. The design of these filters is essentially a time domain, as it consists of the design of the

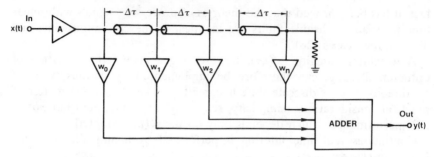

**Fig. 7.13.** Block diagram of discrete-time analog transversal filters.

set of $w_k$'s. The profile of this set versus the delay individuates the weighting function, that is, the time-reversed impulse response of the filter, see Eq. (89). These filters represent an exceedingly important hybrid of the analog and digital filtering approach. It is worth noting that the versatility of CTD filters is greatly enhanced by the fact that the delay $\Delta\tau$ (i.e., the time scale of the filtering) is controlled by external electrical signals. *Adaptive* filters can also be devised if provision is made to change the tap weights during operation. In principle, one can also use transversal filters in feedback configurations and obtain recursive transversal filters. The treatment of such filters, however, is much less intuitive, and again stability criteria must be observed if oscillation is to be avoided.

In some case the information, instead of being carried by the amplitude of an electrical waveform, is carried by the repetition rate of a sequence of pulses, periodical or random. For instance, this happens when single-photon counting is performed (Sects. 7.5.2 and 7.6.1) or when a voltage waveform is applied to a voltage-to-frequency converter circuit (Sect. 7.6.1). In such cases a digital variable (number of pulses) is to be processed, and it is quite simple in practice to implement accurate digital filterings characterized by weighting functions having square wave shapes, with both positive and negative values. In fact, integrating with constant weight corresponds to counting pulses, and changing the sign of the weight corresponds to reversing the sense of counting. Thus, real-time digital filtering operations equivalent to digital lock-in amplifiers (31) or boxcar integrators can be obtained by using gated counters that are reversible under the control of external electrical signals.

## 7.4   NOISE SOURCES

### 7.4.1   Physical Sources of Noise

Fundamental unavoidable causes of noise arise from (i) the thermal fluctuations of the matter, and (ii) the quantum nature of the matter (quantized charge, light, etc.). Additional causes will in practice always be present due to environmental disturbances, technological limitations in electronic devices, etc.

### 7.4.1.1   Shot Noise

In various electronic devices, the current is a superposition of random-time independent current pulses of charge $q$, corresponding to independent single electrons or holes. This is the case for a vacuum diode in the absence of space charge effects when the current is determined by the thermoelectronic emission. This can be recognized to be a Poisson process (Sect. 7.2.4). If the mean rate $\lambda$ is stationary, the mean current is $\bar{i} = \lambda q$. The time width of the noise autocorrelation is fixed by the width $T_h$ of the single pulses $h(t)$. In vacuum diodes, $T_h$ corresponds to the transit time, in the nanosecond range. The power spectrum of noise will be white from $\omega = 0$ to $\omega \simeq 1/T_h$, and within this range the density, referred to only positive frequencies (see Sects. 7.2.5 and 7.3.2) is

$$S_{ii}(\omega) = \frac{\overline{di_n^2}}{df} = 2q^2\lambda = 2q\bar{i} \tag{121}$$

In some cases the total current is a sum of more components, with possibly different sign. For instance, in $p-n$ junction diodes, carriers cross the barrier in both senses. While the total mean current can be zero, this cannot happen to the second-order statistical parameters of noise (Sect. 7.2.4). Thus, $p-n$ diodes at zero bias voltage carry two current components having an absolute mean value equal to the reverse saturation current $I_s$, opposite sign, and independent fluctuations. The total mean current is zero, and the total spectral density is (32)

$$S_{ii}(\omega) = 4qI_s \tag{122}$$

### 7.4.1.2   Photon Noise

Detailed treatments of photon statistics can be found in references 33–35. In practically all cases of interest for our purpose, the sequence of current pulses at the detector output, corresponding to the detection of single photons (SP), can be considered as a Poisson process, with mean rate $\lambda_p(t)$. Thus, in a time interval $T_M$ associated with a "photon signal," that is, with a mean number of photon $S_p = \int_{T_M} \lambda_p(\tau) \, d\tau$, there is inherently a "photon noise" $n_p$ having a mean square value equal to the signal $\overline{n_p^2} = S_p$. The photon signal-to-noise ratio is thus given by $\sqrt{S_p}$. If stray or scattered light is also detected, it causes both a background $S_B$ and an increase in noise, which becomes $\overline{n_{pB}^2} = S_p + S_B$. If the background light can be independently measured over an equal $T_M$ and subtracted, the resulting measurement will be affected by a noise $\overline{n_{pB}^2} + \overline{n_B^2}$ and will therefore have

$$S/N = [S_p^2/(S_p + 2S_B)]^{1/2}$$

It is worth stressing that a time-varying photon signal $S_p(t)$ will be inherently associated with a nonstationary noise component $\overline{n_p^2(t)} = S_p(t)$. This should not be overlooked, for instance, in the determination of optimum filtering (Sect. 7.3.4).

If the output current of the detector is processed by analog filters, photon noise corresponds to a shot noise current. It is worth noting, however, that this shot noise is often nonstationary and, if the detector (Sect. 7.5) has an internal gain $M$, it corresponds to pulses $h(t)$ having a charge $Mq$, where $q$ is the electron charge. The width of these single-photon response pulses $h(t)$ determines the width of the noise autocorrelation and the frequency range where it can be considered white.

### 7.4.1.3    Johnson or Resistance Noise

This noise is inherently associated with the resistance $R$ of every physical component. It corresponds to the superposition of the small electrical pulses induced by carriers in thermal motion. A pulse can be considered to correspond to the motion of each carrier in the very short time ($<10^{-12}$ s) between collisions with scattering centers. The situation can be considered a zero-average Poisson process, with positive and negative elementary pulses. The noise is white up to very high frequencies ($>10^{12}$ Hz). It can be characterized either as a noise current or as a noise voltage, with spectra given by $S_{ii}(\omega)$ and $S_{vv}(\omega)$, respectively (referred to positive frequencies). It can be shown that

$$S_{ii}(\omega) = \frac{\overline{di_n^2}}{df} = 4kTG = 4kT/R \qquad (123a)$$

$$S_{vv}(\omega) = \frac{\overline{dv_n^2}}{df} =: 4kTR \qquad (123b)$$

where $G = 1/R$ is the conductance, $k$ is the Bolzmann constant, and $T$ is the absolute temperature. One should note that Johnson noise does not vary when the resistor carries nonzero current. This is quite intuitive as the applied electrical field will perturb only very slightly the random electron motion. On the other hand, the small perturbations are strongly correlated so that the mean becomes different from zero.

### 7.4.1.4 One-over-f Noise

As mentioned in Sect. 7.2.5, noise spectra having slope $1/f^\alpha$, with $\alpha$ close to unity, have received various names, as they have been observed in several different cases (37), such as nuclear magnetic resonance NMR measurements; standards of time and frequency; biological systems; nuclear reactors; to mention a few. They are present in practically all amplifying electronic devices: in electron tubes, due to flicker effect in the thermoelectronic emission; in semiconductor devices, due to various fluctuations phenomena such as carrier trapping with distributed long-time constants, surface effects, etc.

The $1/f$ components are usually dominant in the noise spectra at low frequencies, say, below 100 Hz. They extend their range to very low frequencies; long-term drift can be considered as an example of very low frequency $1/f$ noise. They have a very wide autocorrelation and are represented by Poisson models with very long individual pulses. In fact, they are caused by fluctuation phenomena where the memory of individual independent perturbations is very long. These components have therefore a disastrous effect in measurements at low frequencies, that is, with weighting functions extending over long times. A selection of the type of filter to be used is relatively unimportant, because the output mean square value is not much sensitive to it (37). Indeed, if the filter bandwidth extends from $\omega_l$ to $\omega_s$, it is seen by using the rectangular approximation that the output mean square value is a logarithmic function of the ratio $\omega_s/\omega_l$, namely, $\overline{n_y^2} \propto \ln(\omega_s/\omega_l)$. Thus, the most important aspect is to individuate the source of $1/f$ noise in order to reduce it at the source, if possible. For instance, junction FET transistors are to be preferred to MOS types, which have higher $1/f$ noise. Furthermore, it is advisable to shift low-frequency signals to higher-frequency ranges, by modulating the signal (Sect. 7.3.3), *before* passing it through systems that contain $1/f$ noise sources, such as amplifiers etc.

### 7.4.1.5 Other Noise Sources

Semiconductors exhibit noise components due to different statistical phenomena. A major one is due to spontaneous fluctuations in the generation and recombination rate, which cause fluctuations in the free carrier concentration. Such a generation–recombination noise can be treated as a Poisson process with square pulses having random duration.

Noise caused by side effects in electronic components is sometimes called "excess noise." Some examples are the current-dependent noise in some type of resistors, besides Johnson noise, and fluctuations in amplification processes having a statistical nature, as in PMT (see Sect. 7.5).

Noise can also derive from environmental causes such as: vibrations, temperature fluctuations, among others. Particularly relevant is electromagnetic interference arising from power line frequency and its harmonics, radio stations, and motor and switch sparks, covering a wide frequency range. Various other sources of noise, many of them of quite prosaic nature, could be mentioned.

#### 7.4.1.6  Quantization Noise

In the quantization process of analog-to-digital conversion, the variance of a digitized variable turns out to be greater than that of the original analog variable (15,25). This fact (see Sect. 7.6) is referred to as quantization noise.

### 7.4.2  Noise in Electrical Circuits

When computing noise in electrical circuits, it is customary to treat electronic components as ideally noiseless and consider the noise to be produced by associated equivalent statistical generators. Usually, these are characterized by the power spectrum, as they represent stationary noise sources. For instance, associated with a resistor $R$ there will be either a noise voltage generator in series or a noise current generator in parallel, with $S_{vv}(\omega)$ or $S_{ii}(\omega)$ given by Eqs. (123$a$) and (123$b$).

As far as the output $U$ of a linear circuit is concerned, a noise generator $S_A(\omega)$ in a circuit position A can be equivalent to a generator $S_B(\omega)$ in a different location B. If $H_{AU}(\omega)$ and $H_{BU}(\omega)$ are the transfer function from A and B to $U$, the condition of equivalence is $S_A(\omega)|H_{AU}(\omega)|^2 = S_B(\omega)|H_{BU}(\omega)|^2$. Thus, equivalent noise sources reduced to the input of the circuit can be defined to compare the noise introduced by the circuit with the signal and noise received from preceding circuits or detectors. We note that in cascaded amplifier stages, the relative importance of a noise source is divided by the square of the gain $G^2(\omega)$ preceding it; usually only the first few stages need to be low-noise types. It is important to realize that at least two noise generators are necessary to give a complete equivalent representation at the input, as shown in Fig. 7.14. In fact, the parallel current generator does not produce noise at the output if the source resistence is very low, while the series voltage generator is ineffective with very high source resistance.

A measure of the low-noise performance of a circuit is given by the *noise figure NF*. It is defined as the ratio of the actual mean square noise at the output $\overline{n_{0a}^2}$ to the ideal mean square noise $\overline{n_{0i}^2}$, which would be obtained if the circuit were ideally noiseless and just processed the noise

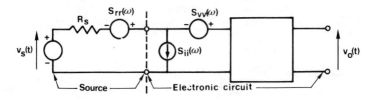

**Fig. 7.14.** Representation of noise of an electrical circuit by equivalent noise generators at the input. Johnson noise of the source is also represented.

received. The *NF* is usually measured in decibels (Sect. 7.3.2):

$$NF = 10 \log_{10}(\overline{n_{0a}^2}/\overline{n_{0i}^2}) = 20 \log_{10}(\overline{n_{0a}^2}/\overline{n_{0i}^2})^{1/2} \qquad (124)$$

The *NF* depends on the nature of the source and on the transfer function of the circuit, that is, on gain and bandwidth. It is usually specified by assuming that the source is at 20°C temperature and that the input is the Johnson noise of the source impedance. For amplifying devices, the *NF* corresponding to narrowband filtering is often shown by two-dimensional plots of its value as a function of the source resistance and of the center frequency.

## 7.5 PHOTODETECTORS

### 7.5.1 Basic Characteristics and Photodetector Types

The reader is referred to the literature (32,38,39,40) for descriptions of the various detector types and extensive treatment of their properties and operating conditions. This section is intended to underline the basic aspects and characteristics that determine the information content of the transduced signal and the internally generated noise. In our applications thermal detectors such as thermopiles, bolometers etc. are not considered. These detectors have their electrical output related to the total energy of the detected photons and are therefore true *energy rate detectors*. With photoelectric detectors, the detection of a photon in a photosensitive material corresponds to the release of a single carrier or of an electron-hole pair free to move in vacuum or in the material itself. The *quantum efficiency*, or conversion yield, $\eta_Q(\lambda)$, of the material is wavelength dependent; however, the electrical output generated by a detected photon is independent of $\lambda$. Thus, these detectors have electrical output related to the total *number* of photons detected and are therefore *photon rate detectors*. However, confusion often arises because specifications of a photodetector's performance are frequently given with incident optical radiation measured in watts.

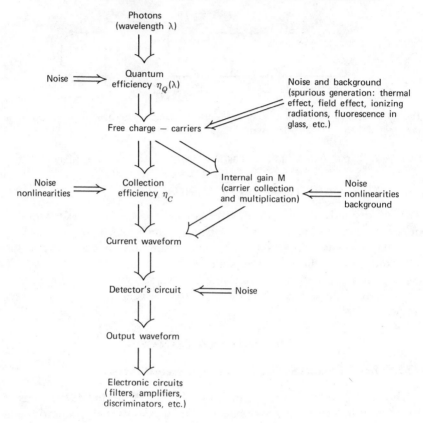

**Fig. 7.15.** Simplified block diagram of the various processes pertinent to photoelectric transducers.

One should stress that electrical signals do not carry any spectral information, and any processing involving $\lambda$ (e.g., separation of light at different $\lambda$ values) must be performed before photodetection.

A schematic diagram of the operation of these detectors is given in Fig. 7.15, and Table 7.1 lists the basic types, together with some indications of their performances. The basic characteristics to be considered are (*a*) photon detection efficiency $\eta_D(\lambda)$, (*b*) internal gain *M*, (*c*) speed of response, and (*d*) internal noise and background.

The reader is warned that different parameters are normally specified for the various detector types, depending on the typical or original application envisaged (e.g., photometry, communications, etc.) and on the test procedures traditionally used in the field. Also, the parameters specified often mix up different basic characteristics. For instance, the

**TABLE 7.1**
**Photoelectric Detectors: Basic Types and Range of Some Typical Characteristics**

| | Gain | Speed of response | Single-photon detection |
|---|---|---|---|
| *Solid-state ohmic devices:* | | | |
| Photoconductors | $10^5$ | 1 ms | No |
| *Solid-state junction devices:* | | | |
| Photodiodes | 1 | 10 ps–1 ns | No |
| Phototransistors | $10^2$ | 10 ns | No |
| Avalanche photodiodes (APD) | $10^2$–$10^4$ | 100 ps–2 ns | Not yet practical |
| *Vacuum tubes:* | | | |
| Photodiodes | 1 | 100 ps–1 ns | No |
| Photomultipliers (PMT) | $10^4$–$10^8$ | <1 ns–10 ns | Yes |
| Continuous channel multipliers (CCM) or channeltrons | $10^3$–$10^6$ | <1 ns–5 ns | Yes |
| Microchannel plate multipliers (MCP) | $10^3$–$10^6$ | <1 ns–5 ns | Yes |

anode radiant sensitivity of the PMT (38,39,40), measured in $mAW^{-1}$, mixes detection efficiency and internal gain. The noise-equivalent power *NEP* and the detectivity *D* of photoconductors (32,40) combine detection efficiency and internal noise spectrum.

### 7.5.1.1 Photon Detection Efficiency $\eta_D(\lambda)$

This parameter defines the probability of being detected for a photon incident on the photosensitive area. It is given by the product of the quantum efficiency $\eta_Q(\lambda)$ of the material and of the charge carrier collection efficiency in the detector, $\eta_c$; the latter may be independent or moderately dependent on $\lambda$. $\eta_D(\lambda)$ is of utmost importance: an incident photon not detected means that the information is lost and cannot be recovered in any way by amplification and processing of the electrical output. A variety of shapes of $\eta_D$ versus $\lambda$ are available covering the various spectral ranges (32,38,39,40). In recent years, the introduction of new photosensitive materials, in particular, negative-electron affinity materials, NEA, for photocathodes, has allowed efficiencies of 30–40% and even higher to be reached in the visible range.

It is worth remembering that the spectral sensitivity or responsivity $S_D(\lambda)$ usually specified for photocathodes in $mAW^{-1}$, although related to $\eta_D(\lambda)$, is a different function of $\lambda$, that is, $\eta_D(\lambda) = hcS_D(\lambda)/q\lambda$, where $h$ is the Planck constant, $c$ is the velocity of light, and $q$ is the electron charge.

### 7.5.1.2   *Internal Gain M*

This parameter is defined as the ratio of the charge available at the detector output to the charge of the original photogenerated free carriers; it can attain remarkably high values (see Table 7.1). This is a very useful characteristic inasmuch as it makes negligible, or at least less important, the noise sources in the subsequent circuits. However, it is less fundamental than $\eta_D$ because it can be at least in part substituted by external gain of amplifying devices. The gain mechanisms are random processes; the mean gain is considered, and some noise will be associated with it (see Sect. 7.5.1.4). Drift and fatigue effects on the gain should not be overlooked (38,39,40).

In photoconductors the gain is due to multiple recirculations of the photogenerated majority carriers, interrupted by carrier recombination. High gain implies long response time, as both are proportional to the carrier lifetime $\tau_c$. In fact, the internal amplification can be represented by a transfer function $H_I(\omega) = M/(1 + j\omega\tau_c)$, with $M = \tau_c/t_T$, where $t_T$ is the carrier transit time between electrodes.

Detectors using cascaded secondary electron emission, such as PMTs, CCMs, and MCPs, and detectors using avalanche multiplication, such as APDs, have very fast gain mechanism, in the nanosecond range. Increasing the applied voltage increases both $M$ and the internal background (see Sect. 7.5.1.4), the latter even faster in some voltage ranges (32, 38, 39, 40). In secondary emission types, the average signal current must be small compared to the bias current in the voltage divider in order to avoid changes in the voltage distribution and consequent nonlinear response. Nonlinearities can also arise from space charge effects. The *shape* of current pulses (41,42) can be already severely distorted in conditions where the pulse charge is not yet affected. This important point is often missed and causes great errors in measurements of waveforms using PMTs. Limitations to peak currents and suitable voltage dividers (increasing bias voltage in the last few stages) should be adopted in PMTs.

The matter of voltage divider distribution in a PMT is indeed quite complex, as conflicting requirements have to be taken into account and weighted differently from one application to another (38–40). For instance, uniform voltage dividers give maximum $M$; higher voltages in the first stages give lower statistical fluctuations in $M$ and in the transit time; higher voltages on the last stages give better linearity, and so on. In channeltrons and MCPs, the voltage divider is the internal resistive coating which gives secondary emission. The gain can intentionally be allowed to saturate (40,43) to reduce its fluctuations. In CCMs this can be used in a counting operation mode, in which amplitude information is discarded. On the other hand, in MCPs the amplitude information can be preserved as long as the probability of having more than one photoelectron per channel per transit time is negligible. Obviously, this therefore limits the maximum instantaneous mean rate of photons that can be processed linearly.

### 7.5.1.3 *Speed of Response*

The output current pulses due to single photons (SP) in some amplifying detectors (Table 7.1 and Sect. 7.5.2) can be individually processed and are usually called single-(photo)electron response (SER) (36,44). The SER pulses can be characterized by their time position (individuated by a characteristic point, e.g., barycenter) and by their shape. The delay from the photoelectron emission to the mean time position of the SER is called transit time. The variance of the time position is called time or delay jitter and depends mainly on the very first stages of the multiplication process (40,44). The shape of SER pulses, on the other hand, depends mainly on the fluctuations in the later multiplication stages. It is worth stressing that specification of the SER's risetime alone is *not* sufficient to characterize the speed of response. When using the detector as a linear system (Sect. 7.3.2) for processing multiphoton waveforms, one must consider the risetime of the response to a multiphoton step waveform $1(t)$, that is, the *width* of the response to a multiphoton $\delta$-like waveform. Such a width arises from the composition of the independent fluctuations of SER pulses, that is, it is the quadratic sum of the delay jitter and width of the SER. In practice, it will be almost coincident with the width of the SER, which ranges from 1 ns (or somewhat less) to 10 ns in the various PMTs, CCMs, and MCPs and is quite greater, sometimes 10 times greater, than the delay jitter.

For studying fast optical waveforms with techniques measuring the detector's electrical waveform, attention should also be given to the detector output circuit, because even low level "tails" and "ringing" of the SER pulse can severely affect the result. On the other hand, single photon techniques (36,45) are intended to measure the mean rate of photons as a function of time and are therefore concerned only with the time position of the SER pulses. In high-precision pulse timing techniques (36,45,16–19), the shape of the SER pulses has only minor effects on the precision of the measurement. The time resolution of SP techniques is therefore inherently superior, as it depends practically only on the delay jitter and not on the shape of SER pulses and factors affecting it (risetime of circuits, etc.)

The speed of response of the photodetector types not suitable for SP operation (32,40) can be limited by internal processes, such as slow gain mechanisms (see photoconductors) or transit time spread, or by the external circuit. It is normally specified in the same terms as are linear filters and amplifiers (risetime, bandwidth, etc.).

#### 7.5.1.4 *Background and Noise*

The output current arising from causes other than photon detection is usually called *dark current*. An internal background corresponds to its mean value and an internal noise to its fluctuations, which may include variations and drift in the value of the current averaged over comparatively short time intervals. Various causes produce dark current (40,45,46,47). In PMTs, CCMs, and MCPs some of these causes are after-pulsing, that is, current pulses correlated to previous ones.

In SP detectors, part of the dark current pulses has an amplitude distribution different from that of SER pulses. For instance, smaller pulses are due to emission from dynodes, while larger pulses are due to cosmic ray interaction, ion feedback, etc. Such differences can be exploited in SP techniques (36,40,45–47) to lower the background and noise by rejecting small and/or large pulses through the use of discriminators or single-channel pulse height analyzers (16–18).

Shot noise is due to the statistics of individual pulses in the detectors, arising from photon detection (Sect. 7.4.1) and dark current sources. Additional noise is generated by the statistics of the internal multiplication. If $M$ and $\sigma_M^2$ are the mean value and variance of gain, respectively, the effect can be evaluated (36,44) as an increase in the mean square shot noise by the factor $1 + \sigma_M^2/M^2$. The variance $\sigma_M^2$ has been reduced in modern PMTs by the introduction of GaP in the manufacturing of the first dynodes, which ensures secondary emission factors of 30–50, that is, 10 times larger than previously employed materials (40). In MCPs operation with saturated amplification gives $\sigma_M^2$ comparable to those of modern PMTs (43).

The current due to photon detection can be represented as a current generator in parallel with the output resistance $R_D$ of the detector. By representing Johnson noise of this resistance as a noise current, it is seen that its importance increases as $R_D$ and $M$ become lower. In addition to this, other causes of noise (32,40) are to be considered in semiconductor detectors. For instance, generation–recombination noise (Sect. 7.4.1) in photoconductors gives a shot noise "filtered" by the detector amplification (see Sect. 7.5.1.2), namely, with a "white spectral range" limited by the factor $1/(1 + \omega^2\tau_c^2)$.

#### 7.5.2 Single-Photon Detection

In order to process individual electrical pulses due to single photons (i.e., SER pulses), two basic conditions must be fulfilled by the detector and the circuits connected to its output (36,45–47):

*a.* The amplitude of SER pulses must be remarkably larger than the equivalent noise reduced to the input of the circuit. Obviously, the detector's shot and amplification noise is not to be included.

*b.* The time interval between the detector pulses must be larger than the resolving time of the pulse processing. This time is given either by the deadtime of standardizing and counting (or timing) circuits or by the pulse width in such circuits, in order to avoid pulse pile-up errors (16–18,36,45–47). Detector pulses to be considered are both SER pulses and dark current pulses of comparable amplitude.

Condition (*a*) demands detector requirements concerning gain, internal noise sources other than shot and multiplication noise, and duration of the SER pulses. Condition (*b*) sets a maximum limit to the allowable mean rate of SER plus dark current pulses, in order that the probability of having more than one pulse within a resolving time be much smaller than the probability to have just one pulse (36,45–47). Low-noise circuits are not fast, and conversely. In practice at least $M = 10^4$ is required for an acceptable limit rate. Table 7.1 summarizes the situation.

Single-photon detection leads quite naturally to the use of digital instrumentation such as counters, digital time sorters, etc. (16–18). Some general advantages pertaining to such instrumentation with respect to analog instruments (Sect. 7.6.2) have been sometimes considered as peculiar of SP techniques. This is not correct; techniques that measure the whole current waveform from the detector can also eventually use digital instruments by interposing suitable AD conversion stages (Sect. 7.6.1). A fair comparison between the techniques should therefore be made at the *input* of the digital processing stages. As specific advantages of SP techniques, shown in Sect. 7.5.1, one can consider (i) the possibility of using pulse discrimination techniques to reduce background and noise; (ii) the insensitivity (or tolerability, limited to second-order effects) to variations in *M*, and therefore to the multiplication noise, to slow drift, and to gain nonlinearities; (iii) the higher time resolution in the measurement of fast optical waveforms; (iv) the almost direct disposal of a digital variable (sequence of standard electrical pulses) and consequently (Sect. 7.3.5) the possibility of performing in real time accurate filtering operations with digital instruments. On the other hand, a specific disadvantage of SP techniques is the comparatively low limit to the maximum measurable rate of detected photons. It is possible, to some extent, to correct nonlinearities in pulse rates measured by counters (48) and time distribution of pulses measured by time sorters (36,45–47). In the last case, specially devised electronic circuits can be used to avoid distortion of the distribution shape (35,36). Anyway, higher photon rates can be managed with current waveform measuring techniques. A comparison of the suitability of the two methods should be made by taking into account the relevant aspects in each specific case. In general, SP techniques are advantageous for measuring *very low* light levels and *very fast* optical waveforms. In the latter case, however, they require averaging over a great number of repetitions of the waveform; and if the repetition rate is low, this may imply an unpractically long time (36). In such a case, if the available light level is higher than the maximum limit of the SP techniques, it would be better to use current waveform measurements. For instance, this is often the case when using pulsed lasers for exciting fluorescences.

## 7.6  ANALOG AND DIGITAL TECHNIQUES

### 7.6.1  From Analog to Digital Variables

A necessary premise to a comparison between analog and digital measurement instrumentation is to consider the aspects pertaining to the stages that are used to convert analog waveforms to digital variables. The reverse operation, DAC, has lower importance in measurement apparatus (15,49). It is sometimes used to couple single analog instruments to preceding digital instrumentation, for example, analog ratemeters to SP processing circuits. Digital variables suitable for further processing are usually obtained from (i) SP techniques using fast pulse standardizing and counting or timing circuits (ii) waveform sampling and AD conversion. Extensive information can be found in the literature, both on the former (15–18, 36, 45–48) and on the latter (15–18, 25, 49). The main aspects of SP techniques have been outlined in Sect. 7.5.2, and some aspects concerning waveform sampling have been described in Sects. 7.2.5 and 7.3.5.

Aliasing errors due to insufficient Nyquist frequency, that is, to excessive time interval between samples, can be evaluated with reference to the shape of the sampled waveform and the measurements to be performed on it (15,25). This can also be made for truncation errors due to insufficient extension of the time range covered by samples. The time scale of real-time sampling is thus limited by the deadtime of the sampler and ADC and by the data storage capability of the apparatus. An *aperture function* of the sampler can be defined as just the weighting function $w_A(t,\tau)$ of the used boxcar or sample-and-hold circuit (Sect. 7.3.3) having an *aperture time* $T_A$. A complete characterization of the waveform–sample transfer, however, must include the effect of other filters (amplifiers, etc.) preceding the sampler circuit (Sect. 7.3.3), characterized by a weighting function $w_F(t,\tau) = h_F(t - \tau)$ (assuming constant parameter filters). Thus, a *sampling function* $w_s(t,\tau)$ is defined as

$$w_s(t,\tau) = \int_{-\infty}^{t} w_A(t,\alpha) w_F(\alpha,\tau) \, d\alpha = \int_{0}^{\infty} w_A(t,\tau + \gamma) h_F(\gamma) \, d\gamma \quad (125)$$

Equation (125) may lead one to think that wide $w_A$ and narrow $w_F$ are equivalent to narrow $w_A$ and wide $w_F$. However, this is not completely true because of Johnson noise in the sampler circuit, which is transferred to the samples with a filtering given by $w_A(t,\tau)$ alone. The mean square value of the sample noise can be computed by Eq. (104); it will tend to diverge toward the one of the original white noise as the $w_A(t,\tau)$ becomes narrower. In fact, it is known that the noise of the sampler increases with the sampler's speed, that is, with $1/T_A$.

Provided the time intervals between samples are sufficiently short and the sampling function $w_s(t,\tau)$ is accurately known, it is possible to reconstruct fast waveforms from measurements performed with comparatively wide $w_s$ by "unfolding" the measured waveform from $w_s(t,\tau)$, namely, by reversing Eq. (100). In practice, however, the application is limited because it is necessary to know very precisely both the sampling function and the measured waveform down to very low levels.

Many different ADC methods have been devised and treated in the literature (15–18, 25, 49). The digitization process involves a weighting of the input analog waveform and a quantization of the samples. The two operations can be simultaneous in some ADC types, for instance, the ones using voltage-to-frequency converter and frequency digitizing by pulse counting. They are in cascade in types using separate sample-and-hold and quantizer circuits. Anyway, the relevant parameter for real-time operation is the overall digitizing time.

The weighted samples are classified by the ADC in a set of quantized amplitude *levels* usually spaced by a fixed *quantization interval* $\Delta x$. In nuclear electronics, these levels are called "channels." The analog amplitude $x$ of the samples has a continuous statistical distribution, described by its marginal p.d.f. $p(x)$. The digitized variable $x_q$ obtained from $x$ by means of the ADC has a related discrete statistical distribution. The relation between these distributions is similar, from a mathematical point of view, to the one between the continuous-time waveforms and their discrete-time representation obtained by sampling. Indeed, one can define the *quantization function* $v(k,x)$, also called *channel profile* in nuclear electronics, by analogy to the sampling function $w_s$; $v(k,x) \, dx$ is the probability for an amplitude in $x, x + dx$ of being classified in the $k$th level. In Fig. 7.16, two basic types of quantization functions are depicted in idealized form; real profiles will be somewhat more smoothed (18).

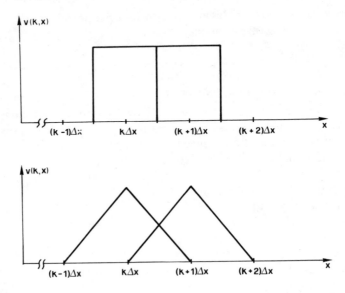

**Fig. 7.16.** Idealized quantization functions of ADCs: (*a*) rectangular channel profile; (*b*) triangular channel profile.

Rectangular channel profile is used in most instances. Triangular channel profile is obtained, for instance, when digital measurements of frequency and time are made without phase correlation between the clock oscillator and the waveform whose frequency or duration is to be measured. This is the case of some ADC using amplitude-to-time or amplitude-to-frequency conversion as a first stage.

If the detailed shape of the statistical distribution of $x$ has to be measured (Sect. 7.7.1), the necessary criteria about $\Delta x$ and $v(k,x)$ can be derived by transposing the analysis made for sampled time waveforms. If only the mean value is sought (Sect. 7.7), the quantized mean value $\bar{x}_q$ and variance $\sigma_{xq}^2$ must be compared to the original $\bar{x}$ and $\sigma_x^2$.

Rectangular channel profile classify $x$ in only one level when $\sigma_x^2 \ll (\Delta x)^2$, and this makes them widely accepted. If $\sigma_x^2$ is higher, the resulting $\sigma_{xq}^2$ becomes larger than $\sigma_x^2$ by a term called quantization noise (Sect. 7.4.1) which, for $\sigma_x^2$ comparable to $(\Delta x)^2$ or greater, is given by $(\Delta x)^2/12$ (15, 25, 51). On the other hand, rectangular profile produces a systematic deviation $\bar{x} - \bar{x}_q$, which can be as high as $\pm \Delta x/2$ for $\sigma_x^2 \ll (\Delta x)^2$, and decreases to a progressively smaller fraction of $\Delta x$ as $\sigma_x^2$ is increased beyond $(\Delta x)^2$ (52). With a given upper limit $x_{\max}$, a smaller $\Delta x$ means a larger number of levels $x_{\max}/\Delta x$, with associated practical complications (increased number of bits of ADC, increased digital memory size, etc.)

However, an accuracy better than $\Delta x$ and correspondingly a measurement dynamic range higher than $x_{max}/\Delta x$ can be obtained, and such complications can be avoided by increasing $\sigma_x^2$ to about $(\Delta x)^2$ with intentional addition of Gaussian noise to $x$ (15, 25, 52).

Triangular channel profiles always classify $x$ values in more than one level, even if $\sigma_x^2 \ll (\Delta x)^2$, and their quantization noise is $(\Delta x)^2/6$ for $\sigma_x^2$ comparable to $(\Delta x)^2$ or greater (51). However, they do *not* produce any deviation $\bar{x} - \bar{x}_q$ with any distribution of $x$. The accuracy and dynamic range are therefore *inherently* not limited by $\Delta x$. In our opinion, this is a very important feature for averaging measurements that is too often overlooked. In fact, it is currently adopted only in some measurement types, for instance, in time interval averaging by counters (53). Incidentally, similar quantization effects are met in the counting of random pulses when using a prescaler (54).

The most important parameter characterizing the linearity of ADC operation is the *differential* linearity (16–18). Differential nonlinearities are deviations of the probability of quantization in the various levels, that is, of the *areas* of the $v(k,x)$, from a constant value (in practice, deviations of the $\Delta x$ values from a constant). It is worth noting that even remarkable deviations of opposite sign may compensate and thus produce small deviations of the level positions from the nominal values $k \Delta x$, that is, small *integral* nonlinearities. Differential nonlinearities may arise both from the quantizer circuit and from the input circuit of the ADC, in particular, when pulse peak stretchers are used (16–18).

Possible drifts in the zero level and in the conversion ratio should not be overlooked.

## 7.6.2 Analog and Digital Electronic Instrumentation

It is not possible to conclude on an absolute basis that digital electronic instrumentation is superior to analog instrumentation, or vice versa. A comparison must be made in each specific case of measurement on the basis of its relevant aspects, and different conclusions can be drawn in different years because of the continuous evolution of the technological progress. Some general facts, however, are well established. The instruments that operate on digital variables, that is, on quantized variables coded by high-level standard electrical signals, are practically unaffected by various drawbacks and limitations of analog instruments (55), such as (*a*) internal sources of noise and background; (*b*) drift of the zero level and of the scale factor (gain); (*c*) inherently limited maximum value of the variable, necessity of various scales in measurements with wide dynamic range, inaccuracies in calibration ratios of the scales, internal noise related to full scale value, etc.; (*d*) nonlinearities; and (*e*) limited holding time of analog memories (memory leakage, see Sect. 7.3.3), in practice rarely exceeding 1 s.

Analog instruments are therefore inherently less suitable for implementing long-duration weighting function and averaging measurements such as are often required to recover low-level signals from noise (Sect. 7.3). Moreover, in constant parameter filters the holding time is tied to the time required for updating or discharging the memory; this is a serious drawback when performing a sequence of measurements, as it imposes a waiting time between successive measurements (see Sect. 7.7).

On the other hand, analog instruments can often perform real-time processing of waveforms on faster time scales than their digital counterparts (Sect. 7.3.5). As concerns factors such as compactness, simplicity, economy, etc., the comparison strongly depends on the operation considered and on the technological state of the art. Analog filters are considered to be more simple, compact, and economical in the implementation of simple filtering types, but the development of transversal filters shows promise of extending this consideration to a wider area. On the other hand, a steady improvement in large-scale integration circuits (LSI) is underway, and quite complex digital instruments become compact and economically feasible.

Last but not least, the data processed by digital instruments are inherently suitable for further processing by digital computers.

## 7.7   SOME COMMENTS ON INSTRUMENTATION AND MEASUREMENTS

### 7.7.1   Measurement of Single Parameters

In many cases the objective of the measurement is not the detailed shape of a signal but only a characteristic parameter (Sect. 7.3.4). For instance, in dc signals, the objective is the amplitude; in sinusoidal signals, the amplitude, the frequency, and the phase with respect to a reference signal; in pulse signals, the peak amplitude, the area, and the time position with respect to a reference signal. The parameter $A$ considered will have its own statistical distribution, individuated by a marginal p.d.f. $p(A)$. However, the actual output $A_m$ of a measurement system intended to measure $A$ will have a different distribution $p_m(A_m)$. This is due to the various phenomena that are the sources of noise, background, distortions, and drift. Noise will smooth and widen the distribution without modifying the mean value, which can be affected by the other causes.

*Single-shot measurements* give meaningful data if the $S/N$ is sufficiently high (and the original distribution $p(A)$ is not wide). In *repetitive measurements* it is possible to record either the average value only or the whole distribution of the measured values. The former are called *averaging measurements,* the latter *spectrum or distribution measurements.* The word "spectrum" is here intended in a statistical sense; it denotes the hystogram of the statistical relative frequency of occurrence of the various measured $A_m$ values. This spectrum will provide a quantized representation of the marginal p.d.f. $p_m(A_m)$ if a sufficient number of repetitions is recorded (16–19). Typical fields where both averaging and spectrum measurements are currently used are measurements of pulse amplitude and time intervals between pulses, as required by laser ranging applications, time-of-flight experiments, fluorescence experiments, and many others (16–19,24$b$,45,53,56,57). Quite obviously, in spectrum measurements it is practically necessary to use digital instrumentation such as multichannel pulse analyzers, etc.

In some instances two or more parameters, $A$, $B$, etc., have to be measured, either for obtaining the value of a quantity that is a function of these parameters, or because both individual values and their correlations have to be studied. Example of the former type are ratio measurements such as required to measure reflection, transmission, etc. In the latter cases both parameters must be simultaneously measured and classified, and a spectrum measurement is inherently required. In such cases the spectrum will be a two-dimensional or multidimensional hystogram of statistical relative frequency of occurrence as a function of the two or more measured parameters $A_m$, $B_m$, etc. With a high number of repetitions, the measured spectrum approaches the shape of the joint p.d.f. $p_m(A_m,B_m)$. Examples of this kind are found in lidar applications (see Chapter 6) where each backscattered pulse has to be classified in terms of amplitude and time of arrival. Other examples are met in biologic applications of lasers as automatic cell sorting methods, which use lasers to excite cell fluorescence and measure the fluorescence intensity of each cell in two different wavelength ranges (58). Such multidimensional spectrum measurements are sometimes called "correlation measurements," although this is not a strictly correct name.

In any measurement type, care must be taken to limit as much as possible alterations of the original distribution that modify the mean and mean square values. Distortions can be limited by accurately testing the linearity of all apparatus and by avoiding for all stages of the measurement that the dynamic range is exceeded by the variable processed, that is, by signal plus background plus noise. The dynamic range is usually defined in two ways: (i) the ratio of the maximum to the minimum value of the measurable signal, and (ii) the ratio of the noise intensity that produces overload to the signal corresponding to full-scale output. The latter definition is used when the $S/N$ at the input of the filtering apparatus is less than unity. In order to avoid the overloading of the input of high-performance apparatus such as lock-in amplifiers, prefilters of lower quality may be necessary so as to limit excessive input noise or background (Sect. 7.3.3).

As concerns drift, a distinction must be made between *zero* or *baseline drift* and *gain* or *scale factor drift.* The former can be considered as a very low frequency noise. The latter should be avoided as far as possible by careful choice of the apparatus and of its operating conditions (e.g., see Sect. 7.5). Automated stabilization systems have also been devised that use various methods (observation of some reference signal, etc.) for continuous monitoring of the output and correction by means of controlled gain stages (16,18).

As concerns noise and background, on the one hand the question is to minimize their sources, in particular $1/f$-like noise, which set an ultimate limit to the improvement in $S/N$ obtainable by using averaging measurements (37). On the other hand, one has to exploit all facts that make these disturbances different from the signal sought, to perform a selective treatment. The problem thus implies careful selection of apparatus and components and careful planning of the distribution of filtering stages, in particular, taking into account the position of all noise sources, including those associated to filters (e.g., see Sect. 7.6.1). It also implies producing signals that can be better distinguished from noise. For instance, in the presence of a dominant stationary noise not associated with the light to be measured, instead of using cw lasers it is better to employ pulsed lasers, even with lower time-average optical power. Modulation of signals before the stages adding low-frequency noise (in particular $1/f$ noise) is also very useful (Sect. 7.3.3).

Differences between signal and disturbances can be more easily individuated and exploited either in the time or in the frequency domain, depending on the circumstances. For instance, when using pulsed lasers, those unwanted scattering and spurious fluorescences that occupy different time intervals with respect to the signal can be rejected by suitable gating (time-resolved measurements). On the other hand, notch filters can efficiently reject electromagnetic interferences arising from powerline and radio stations. It is worth recalling that in some cases optical signals and disturbances have differences that are lost in the transduction to electrical waveforms. Thus, optical filtering may be the only way of discriminating a source scattering having equal behavior to the signal in both time and frequency domains.

Incidentally, one may note that in practice the distinction between noise and background may be somewhat arbitrary in some cases. For instance, in a series of measurements performed in a limited time interval, a very low frequency noise (e.g., a drift) can produce very correlated contributions in all the measurements. (See Sect. 7.2.4.9.)

One must stress that the type of optimum filtering for $S/N$ enhancement (Sect. 7.3.4) depends both on the signal and on the noise. For instance, in the measurement of a dc signal, uniform integration is the optimum only if the noise is white and stationary. In measurements of the amplitude of pulsed optical signals, if the statistics of the detected photons carrying the information is the dominant noise source, the optimum is a simple gated integration that just embraces the signal; the optimum depends instead on the signal shape if nonnegligible noise is added by stray light, dark current, etc. In the timing of pulses, quite different situations correspond to the cases of SP pulses, of multiphoton pulses from fast detectors having high gain, of multiphoton pulses from simple fast photodiodes, etc. Furthermore, the choice of filtering must take into account other random causes of possible deterioration of the $S/N$ obtained, such as pulse pile-up in the case of random time pulses (16–18, 22, 23) and systematic dependence of timing circuits on the amplitude of the pulses, producing random timing deviations because of random pulse amplitude distribution (16–19, 24, 44, 45).

### 7.7.2 Measurement of Waveforms (and Spectra)

The importance of this subject is wider than would appear at first sight, because many concepts in measurements of the shape of time waveforms can be extended to other measurements where parameters other than time are the relevant independent variable (e.g., parameters such as wavelength, polarization state, angle of incidence of light, frequency, etc.). This includes measuring optical spectra, angular distribution of scattering, power spectra, time correlation functions, etc. In cases where the independent parameter is scanned at a constant rate, this parameter is tied to the time by a constant scaling factor, and the function sought is indeed simply translated into a true time waveform. Extension of the concepts is possible also in other cases, as outlined below; for instance, when the independent parameter is scanned in steps, when measurements are performed in parallel (i.e., simultaneously for different values of the independent parameter), etc.

A set of measurements of instantaneous amplitudes corresponding to a set of time positions is required. The interval between these positions must not exceed a maximum value $\Delta t$ (time resolution) depending on the waveform shape (Sects. 7.2.5 and 7.6.1), and each amplitude measurement is subject to the considerations outlined in Sect. 7.7.1, together with further constraints.

Single-shot measurements are meaningful only if the $S/N$ is sufficient over the whole waveform. Repetitive measurements in practice are only of the averaging type, as the complexity of measurements of the statistical distribution of all the instantaneous amplitudes is such that they are performed only for very specific purposes (22). In principle, the result of averaging procedures is a measure of the cross-correlation function between the ensemble-average waveform and a $\delta$ pulse. In practice, because of the random errors $\theta$ in the synchronization, the result will be a cross-correlation with the p.d.f. $g_s(\theta)$ of the time jitter. In case of measurements other than time waveforms, this consideration is extended by taking into account random errors in the value of the independent parameter. It is worth noting that in the measurement of time autocorrelation functions $K_{xx}(\tau)$ of waveforms $x(t)$, a jitter of the reference signal with respect to the waveform may be unimportant, as it may shift rigidly the whole waveform, (i.e., it may affect the positions $t$ but not the interval $\tau$ considered). Therefore, an advantage of autocorrelation analysis lies in the fact that a reference signal accurately synchronized to the waveform is not required.

Measurements in continuous form can be obtained in the single-shot mode with analog instruments such as oscilloscope and camera, graphical recorder, etc. The possibility of unfolding the true waveform from filtering due to the apparatus is quite limited (Sect. 7.6.1). It is therefore necessary to have filtering characterized by a weighting function not wider than the time resolution $\Delta t$ required. This means that wideband noise can be effectively filtered only in the case of slow waveforms. In measurements with independent parameters other than time, the necessity of filtering noise thus imposes a limitation to the parameter scanning rate. For instance, if a wavelength resolution $\Delta\lambda$ is required for an optical spectrum and the width of the filtering adopted is $T_F$ (rectangular approximation in the time domain, Sect. 7.3.3), the wavelength scanning rate $r_\lambda = d\lambda/dt$ must not exceed $\Delta\lambda/T_F$.

Continuous measurements do not bring more information than measurements in sampled form with the proper $\Delta t$ value. Moreover, averaging in practice requires sampling (preferably not performed by hand on a graphical record!). The instruments that perform sampling can be divided in *single-sampler* and *multisampler* types on the basis of the number of physically separate *input* sampling channels, not from the point of view of the number of memories used to store and average the sampled values.

Multisamplers are more complicated and costly but can operate in real time on fast waveforms, beyond the limits of single-sampler instruments (Sect. 7.6.1). Their basic advantage in this field is that of collecting much more efficiently the available information in a single waveform. With moderate repetition rates, the time required for the averaging will be reduced by a factor equal to the number of sampled positions. With very low repetition rates and single events, measurements that would be impossible on single-samplers can be performed. On the other hand, for the different sampled positions, the instrumental parameters can be different, with consequent uncontrolled alterations in the shape of the measured waveforms. The various sampling channels can indeed have sampling functions (Sect. 7.6.1) with different weights (i.e., areas), different nonlinearities, and different drift and background. In the sampling of domains other than time, multisampler instruments can make simultaneous sampling at the various parameter positions, thus avoiding the necessity of scanning devices. For instance, multidetector systems coupled to suitable dispersive optical apparatus constitute multisamplers in the wavelength domain that can measure optical spectra even of single fast flashes.

Single-sampler types suffer from limits to real-time operation, but their behavior is independent of the sample position. Furthermore, with only one sampler to manage, greater care can be devoted to the instrumental features that determine the quality of the apparatus, and as linearity, long-term stability, etc. (59). When using single-samplers for time waveforms beyond their real-time limit, or for sampling parameters other than time, the time position (equivalent time) or the parameter concerned must be scanned, not necessarily at constant rate. By taking into account the corresponding time dependence of the parameter, the time domain sampling function is translated into a parameter sampling function, which must be accurately controlled to be suitable for the desired measure (see continuous measurement). A staircase scanning can be used (sequential sampling), with subsequent scans performed in an up–down or in up–reset–up fashion. Random scanning with uniform probability for all positions (random sampling) is also possible (60). Incidentally, within SP techniques, the so-called monophoton timing (36,45) has some analogy with random sampling. It is worth stressing that the effects of drift and low-frequency noise are different in the various scanning types. Dual-channel boxcar operation (Sect. 7.3.3) can be used in sequential scanning to filter out drift of the zero-level, low-frequency noise and slow background. Even a fast-varying background can be subtracted if it is possible to suppress at will only the signal without affecting the background. For instance, a mechanical chopper wheel can be used to this purpose in case of pulsed nitrogen lasers that irradiate electromagnetic interference synchronized to the light pulse. Also fast scanning rates, which minimize the time required to cover the range of the parameter (i.e., the single-scan time), and random sampling are effective in reducing the influence of slow-noise, zero-level drift and, as far as waveform shape only is concerned, also the gain drift (55). On the other hand, averaging over many fast scans will also provide an improvement in the $S/N$ equivalent to that offered by a single slow scanning of the parameter, where at each time position the weighting function either has a long duration or corresponds to averaging a high number of successive samples, as in the usual boxcar operation (Sect. 7.3.3).

Many other aspects concerning measurements of waveforms and spectra are beyond the scope of this introductory discussion. We shall just mention two very significant arguments. The first one is *derivative or modulation spectroscopy*, where an oscillation is superimposed on the scanning of the independent parameter (61–63). Lock-in or dual-channel boxcar techniques are then used to obtain direct measurements of the derivatives of the spectra with respect to the parameter, thus enhancing small features of the spectral shape. The second one is *time-resolved spectroscopy*, where a sort of two-dimensional waveform, with wavelength and time as two independent parameters, is measured. This kind of measurement can be used to great advantage in the study of fluorescent decays by enabling separation of the spectra of the emission from different levels by means of their different decay times (64,65). Application to other types of studies can be envisaged, and the technological progress in digital instrumentation can bring this type of measurement within the reach of many laboratories.

## References

1. A. Papoulis, *The Fourier Integral and Its Applications*, McGraw-Hill, New York (1962).

2. R. M. Bracewell, *The Fourier Transform and Its Applications*, McGraw-Hill, New York (1965).

3. A. Papoulis, *Probability, Random Variables and Stochastic Processes*, McGraw-Hill, New York (1965).

4. C. D. McGillem and G. R. Cooper, *Continuous and Discrete Signal and System Analysis*, Holt, Rinehart and Winston, New York (1974), also related titles in the Holt, Rinehart and Winston Series in Electrical Engineering, Electronics and Systems.

5. M. Schwartz, *Information Transmission Modulation and Noise*, McGraw-Hill, New York (1959).

6. Y. W. Lee, *Statistical Theory of Communications*, Wiley, New York (1960).

7. D. Middleton, *Introduction to Statistical Communication Theory*, McGraw-Hill, New York (1960).

8. W. D. Davenport and W. L. Root, *Random Signals and Noise*, McGraw-Hill, New York (1958).

9. J. S. Bendat, *Principles and Applications of Random Noise Theory*, Wiley, New York (1958).

10. L. A. Wainstein and V. D. Zubakov, *Extraction of Signals from Noise*, Prentice-Hall Int., Englewood Cliffs, N.J. (1962).

11. W. R. Bennet, *Electrical Noise*, McGraw-Hill, New York (1960).

12. J. K. Swirzynski, *Design Theory and Data for Electrical Filters*, Van Nostrand, London (1965).

13. V. Radeka, *Nucl. Instrum. Meth.*, **99**, 525 (1972).

14. R. R. Ernst, *Rev. Sci. Instrum.*, **36**, 1689 (1965).

15. H. V. Malmstadt, C. G. Enke, S. R. Crouch, and G. Horlick, *Electronic Measurements for Scientists* (in particular Modules 3 and 4), W. A. Benjamin, Reading, Penn. (1974).

16. E. Kowalski, *Nuclear Electronics*, Springer-Verlag, Berlin (1970).

17. J. J. Samueli, J. Pigneret, and A. Sarazin, *Instrumentation Electronique en Physique Nucléaire*, Masson, Paris (1968).

18. P. W. Nicholson, *Nuclear Electronics*, Wiley, New York (1974).

19. E. Baldinger and W. Franzen, *Advan. Electron. Electron Phys.*, **8**, 255 (1956).

20. M. Bertolaccini, C. Bussolati, and E. Gatti, *Nucl. Instr. Meth.*, **41**, 173 (1966).

21. V. Radeka and N. Karlovac, *Nucl. Instrum. Meth.*, **52**, 86 (1967).

22. M. Bertolaccini, C. Bussolati, S. Cova, I. De Lotto, and E. Gatti, *Nucl. Instrum. Meth.*, **61**, 84 (1968).

23. G. L. Miller and D. A. H. Robinson, *IEEE Trans. Nucl. Sci.*, **NS-22**, 2022 (1975).

24. (*a*) E. Gatti and V. Svelto, *Nucl. Instrum. Meth.*, **39**, 309 (1966); (*b*) E. Gatti and S. Donati, *Appl. Opt.*, **10**, 2466 (1971).

25. P. C. Kelly and G. Horlick, *Anal. Chem.*, **45**, 518 (1973).

26. A. V. Oppenheim and R. W. Schafer, *Digital Signal Processing*, Prentice-Hall, Englewood Cliffs, N.J. (1975).

27. L. R. Rabiner and B. Gold, *Theory and Application of Digital Signal Processing*, Prentice-Hall, Englewood Cliffs, N.J. (1975).

28. K. Steiglitz, *Inform. Control*, **8**, 455 (1965).

29. C. H. Sequin and M. F. Tompsett, "Charge Transfer Devices", *Advan. Electron. Electron Phys.*, *Suppl. No.* 8 (1975).

30. G. Horlick, *Anal. Chem.* **48**, 783A (1976).

31. (*a*) F. T. Arecchi, E. Gatti, and A. Sona, *Rev. Sci. Instrum.*, **37**, 942 (1966); (*b*) S. Cova, D. Dotti, and S. Tenconi, *IEEE Trans. Instrum. Meas.*, **IM-22**, 347 (1973).

32. S. M. Sze, *Physics of Semiconductor Devices*, Wiley, New York (1969).

33. R. J. Gambler, "Photon Statistics", in *Laser Handbook*, F. T. Arecchi and E. O. Schulz-Dubois, Eds., North-Holland, Amsterdam (1972), Part A1.

34. H. Haken, "The Theory of Coherence, Noise and Photon Statistics of Laser Light," in *Laser Handbook*, F. T. Arecchi and E. O. Schulz-Dubois, Eds., North-Holland, Amsterdam (1972), Part A1.

35. C. C. Davis and T. A. King, *Rev. Sci. Instrum.*, **41**, 407 (1970).

36. S. Cova, M. Bertolaccini, and C. Bussolati, *Phys. Stat. Solidi.* (a), **18**, 11 (1973).

37. V. Radeka, Proc. 1st Ispra Nuclear Electronics Symposium, Ispra, Italy, Euratom Publication EUR 4289e (1969).

38. RCA *Photomultiplier Manual*, Tech. Ser. PT-61, RCA, Harrison, N.J. (1970).

39. J. M. Schonkeren, *Photomultipliers*, Philips Publ. Dept., Eindhoven (1970).

40. D. H. Seib and L. W. Ankerman, *Advan. Electron. and Electron Phys.*, **34**, 95 (1973).

41. M. Bertolaccini, S. Cova, C. Bussolati, S. Donati, and V. Svelto, *Nucl. Instrum. Meth.*, **51**, 325 (1967).

42. G. Bianchetti and B. Righini, *Nucl. Instrum. Meth.*, **105**, 45 (1972).

43. G. Pietri, *IEEE Trans. Nucl. Sci.*, **22**, 2084 (1975).

44. S. Donati, E. Gatti, and V. Svelto, *Advan. Electron. Electron Phys.*, **26**, 251 (1969).

45. S. K. Poultney, *Advan. Electron. Electron Phys.*, **31**, 39 (1972).

46. G. A. Morton, *Appl. Opt.*, **7**, 1 (1968).

47. R. Foord, R. Jones, C. J. Oliver, and E. R. Pike, *Appl. Opt.*, **8**, 1975 (1969).

48. I. De Lotto, A. Ghirardi, and P. F. Manfredi, *Energia Nucl.*, **12**, 365 (1965).

49. D. F. Hoeschele, Jr., *Analog-to-Digital and Digital-to-Analog Conversion Techniques*, Wiley, New York (1968).

50. E. Gatti, *Nucl. Instrum. Meth.*, **2**, 26 (1958).

51. I. De Lotto and S. Osnaghi, *Nucl. Instrum. Meth.*, **56**, 157 (1967).

52. G. Amsel and R. Bosshard, *Rev. Sci. Instrum.*, **41**, 503 (1970).

53. (a) B. M. Oliver and J. M. Cage, Eds., *Electronic Measurements and Instrumentation*, McGraw-Hill, New York (1971); (b) Hewlett-Packard Application Note 162–1 "Time Interval Averaging."

54. S. Cova, G. Gambarini, and A. Longoni, *Rev. Sci Instrum.*, **46**, 1569 (1975).

55. S. Cova, G. Prenna, and G. Mazzini, *Histochem. J.*, **6**, 279 (1974).

56. S. Poultney, *IEEE Trans. Nucl. Sci.*, **NS-19**, 12 (1972), n. 3.

57. (a) C. A. Sacchi, O. Svelto, and G. Prenna, *Histochem. J.*, **6**, 251 (1974); (b) S. Cova, G. Prenna, C. A. Sacchi, and O. Svelto, in *Excited States of Biological Molecules*, J. B. Birks, Ed., Wiley, London (1976), p. 223.

58. (a) L. A. Kamentsky, in *Automated Cell Identification and Cell Sorting*, G. L. Wied and G. F. Bahr, Eds., Academic Press, New York (1970), p. 111; (b) L. A. Kamentsky and M. R. Melamed, *Proc. IEEE*, **57**, 2007 (1969).

59. G. Cottini, E. Gatti, and V. Svelto, *Electron. Lett.*, **3**, 569 (1967).

60. G. J. Frye and N. S. Nahman *IEEE Trans. Instrum. Meth.*, **IM-13**, 8 (1964).

61. G. Bonfiglioli and P. Brovetto, *Appl. Opt.*, **3**, 1417 (1964).

62. K. L. Shaklee and J. E. Rowe, *Appl. Opt.*, **9**, 627 (1970).

63. F. Grum, in *Physical Methods of Chemistry*, A. Weissberger and B. W. Rossiter, Eds., Wiley, New York (1972), Part IIIB, p. 207.

64. W. R. Ware, in *Creation and Detection of the Excited State*, Vol. 1, A. Lamola, Ed., Marcel Dekker, New York, (1971), Part A, p. 213.

65. W. E. Blumberg, R. E. Dale, J. Eisinger, and D. M. Zuckerman, *Biopolymers*, **13**, 1607 (1974).

# RECENT ADVANCES IN ANALYTICAL LASER SPECTROSCOPY[1]

## R. A. KELLER[2] and J. C. TRAVIS

*Analytical Chemistry Division*
*National Bureau of Standards*
*Washington, D.C.*

Several new, laser-based analytical techniques demonstrated in recent years show considerable promise but have not yet been developed into reliable analytical tools. Some of these techniques are described in this chapter, with the major emphasis on the use of tunable dye lasers, especially cw, in the visible and ultraviolet spectral regions, but with some

---

[1] Contribution of the National Bureau of Standards; not subject to copyright.

[2] Present address: CNC-2, Los Alamos Scientific Laboratory, Los Alamos, New Mexico.

inclusion of infrared and fixed-frequency results. The selection of topics reflects the authors' own interests and experience; consequently many important areas have been omitted. The text represents our understanding of the subject matter up to May of 1976. Section 8.1 treats laser properties as they relate to the general photoanalytical measurement. The remaining sections cover specific applications in the areas of absorption (Sect. 8.2), fluorescence (Sect. 8.3), microfluorescence (Sect. 8.4), and finally a few miscellaneous topics of more limited analytical value or at an earlier stage of development (Sects. 8.5 and 8.6). Previous work in this area has been discussed in several reviews (1–3).

## 8.1 RELEVANT LASER PROPERTIES

As a result of the large amount of laser development that has occurred in the last 10 years, coherent sources (fundamental and frequency-doubled laser emissions) now exist throughout the infrared, visible, and ultraviolet regions. A representative graph of the spectral regions covered by various sources is shown in Fig. 8.1. Included on this graph are both tunable lasers and some common fixed-frequency lasers. Several types of tunable lasers cover broad ($>10$ nm visible, $\geq 10^2$ cm$^{-1}$ infrared) spectral regions continuously, for example, dye (4,5), parametric oscillator (6), Raman spin flip (7,8), and diode (9,10). A more restricted continuous tuning range ($\sim 1$ GHz $\approx 0.03$ cm$^{-1}$) is available about each of the normal fixed-frequency CO and $CO_2$ laser lines for high-pressure waveguide lasers (11).

The spectral range from 200 to 400 nm is of particular interest to the analytical chemist. Tunable laser power in this wavelength region is presently available by frequency doubling the output of pulsed visible lasers (12–17). The current short wavelength limit (218 nm) is determined by the availability of appropriate nonlinear materials and has been significantly improved in the last several years (15–17). Further improvements are expected as new materials are developed. Although the spectral range from 220 to 1000 nm can be covered with pulsed systems, cw tunable lasers are presently limited to the range from 420 to 800 nm (18). The inaccessibility of the near UV represents a serious drawback of commercial cw lasers, but this limitation will soon be removed with the development of efficient doubling materials and techniques to give mW of cw tunable coherent radiation down to 220 nm. Tunable UV up to 50 mW has already been obtained in the range of 285–315 nm from a doubled cw dye laser (19).

For many applications the stability, narrow linewidth, and ease of data acquisition make the cw laser the system of preference. However, there

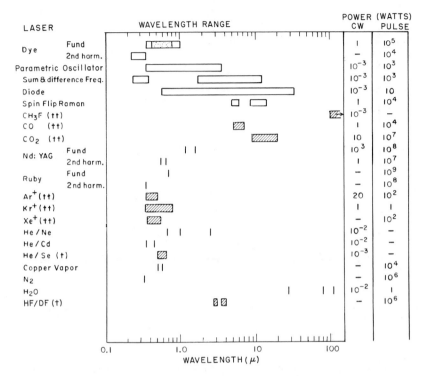

**Fig. 8.1.** Selected examples of currently available coherent light sources. Open bars indicate range of continuously tunable lasers; cross-hatched bars show range over which multiline, fixed-frequency lasers may be jump tuned. The dotted areas of the dye laser bars indicate the range of cw operation.

are certain applications for which the short pulse width and high peak powers of pulsed systems are very important. Pulse widths less than $10^{-11}$ s are now commonly available and have been used extensively for the study of the molecular dynamics of relaxation processes (20–22). Their use in analytical work to this time has been limited but they do possess potential for certain special applications. The high peak powers associated with pulsed systems have received considerable attention for analytical applications and discussed in detail in Chapter 4.

An interesting area of rapid growth is tunable laser emission in the infrared spectral region. To date, much of the infrared analytical work has been done with fixed-frequency lasers (23–25), but the development of the Raman spin flip laser (7) and the increased availability of steadily improving diode lasers (9) should promote new and interesting applications.

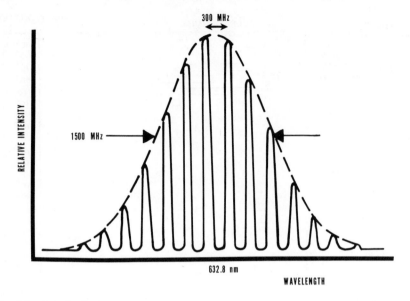

**Fig. 8.2.** Longitudinal mode structure within the gain curve of a 60-cm-long He–Ne laser. Many of these modes oscillate simultaneously, but an intracavity etalon can be used to confine the emission to a single longitudinal mode.

Although broadly tunable lasers are emphasized in this chapter, fixed-frequency lasers may be appropriate, or even preferred, for particular applications. *Actually, fixed frequency is a misnomer because even these lasers are continuously tunable to a limited degree.* All lasers possess longitudinal mode structure with mode spacing determined by the length of the optical cavity. For example, a $\frac{1}{2}$-m-long He–Ne laser will have several longitudinal modes oscillating under the Doppler profile of the emission line (see Fig. 8.2). Unless special precaution is taken, *all of these modes will be present in the laser output and the laser will not be single-frequency in any sense.* Insertion of an etalon or etalons into the optical cavity can suppress all but one of the longitudinal modes and result in single-frequency output. The optical length of the etalon determines which mode is selected. It is always possible to choose the mode corresponding to the peak of the Doppler curve ($\pm 0.0001$ nm), and therein lies the principal advantage of a fixed-frequency laser, namely, the ability to reproduce a single, known wavelength. Limited tuning of the selected mode can be accomplished by changing the cavity length. In fixed-frequency infrared lasers the spacing between longitudinal modes is usually so large that only one falls under the gain curve and single-frequency operation is the norm. However, the precise wavelength of the lasing mode within the gain curve (typically $0.01 \, cm^{-1}$) is still a function of the exact spacing of the laser mirrors. Either infrared or ultraviolet/visible lasers may thus be tuned in single-mode operation over their gain curve with appropriate electromechanical mirror/etalon controls. If a particular analytical need can be met by an accidental coincidence between a fixed-frequency laser and the absorption of interest, the fixed-frequency laser is probably the best for the task (23–25). Unfortunately, accidental coincidences are relatively rare and analytical applications were severely limited by this requirement before the development of tunable lasers.

Tunable lasers may also have longitudinal mode structure within their coarse tuning gain curve. Again, a single longitudinal mode can usually be selected by insertion of an etalon or etalons into the optical cavity (26). The spectral width of a single longitudinal mode is determined by the laser stability and typically is approximately 1–50 MHz ($10^{-6}$ nm at $\lambda = 500$ nm). Unlike fixed-frequency lasers, the wavelength of this emission may not easily be measured or reproduced with anything approaching the accuracy corresponding to the spectral width. For example, $10^{-6}$ nm represents 1 part in $10^{9}$ of a visible wavelength. The best spectrographs are capable of resolutions of 1 part in $10^{6}$ and 1 part in $10^{5}$ is more common. Complicated frequency measuring techniques are available for increasing the accuracy of the wavelength measurements, and they have been used in a few select cases (27), but no general solution to the wavelength measurement problem now exists.

All photoanalytical measurements, with the exception of emission spectroscopy, involve absorption of photons as the intial step. The subdivisions of such measurements concern themselves with various ways of detecting the fact that photons have been absorbed. There are several facets (e.g., resolution, absorption cross sections, and selectivity) of the absorption problem that are common to all of the subdivisions. It is necessary to examine how a laser affects each of these.

*Resolution* determines the minimum energy separation for which two absorptions can be separately excited. In many analytical cases the resolution of a particular analysis is determined by the spectrograph or monochromator and therefore is limited to 1 part in $10^{5}$ or $10^{6}$ depending upon the quality of the instrument. The spectral width of the laser gives an effective resolution in excess of $10^{8}$ and negates the need for a high-quality, expensive spectrograph. The next limit to resolution is often Doppler and collision broadening. Collision broadening can always be removed—often at the expense of concentration—by working at reduced pressure. Doppler-broadened absorption lines are typically 0.001 to 0.01 nm in width in the visible spectral region, and it is legitimate to wonder about the value of laser widths much less than this. Fortunately, it has been recently shown that it is possible to remove the effects of Doppler broadening and look at natural linewidths that are comparable to the spectral widths available from tunable lasers (see Sect. 8.3.3). These narrow linewidths permit analytical determinations that were inconceivable in Doppler-broadened situations (e.g., isotopic analysis where splittings are less than 0.001 nm).

In general, *absorption cross sections* are atomic or molecular properties and are not usually affected by laser irradiation.[3] However, the laser can make a large contribution to the detection of substances with small absorption cross sections because absorption and consequent emission of a very small fraction of the total power results in a relatively large signal due to the large incident laser power. It is important to realize that continual increase in incident laser power cannot indefinitely compensate for weak absorption because eventually all measurements are limited by background scattering and/or saturation effects.

Several factors contribute to *selectivity* in a particular analytical application. One of these, resolution, was previously discussed. Another important factor is modulation of either the excitation source or the sample. The narrow spectral width and the tunability of the laser allow relatively simple frequency modulation over spectral ranges comparable to Doppler absorption widths. The development of Doppler-free laser absorption techniques and the resultant narrow absorption widths permit perturbation modulation of the energy levels with small electric and magnetic fields in the visible spectral region as well as the infrared region. A third important contribution to selectivity is the ability to sample molecular or atomic absorption at many different wavelengths and thereby discriminate between several species that might have coincident absorption at one or more wavelengths. Tunable lasers in the visible and the ultraviolet have the capability of sampling many different wavelengths easily. A different approach to increased selectivity could exploit double-resonance techniques where two different laser wavelengths must match molecular transitions.

In the sections below we discuss analytical applications that are specific to the different categories of detection of absorption and illustrate important points with specific examples.

## 8.2   ABSORPTION MEASUREMENTS

It was stated above that many photoanalytical measurements involve absorption of photons as the initial step. In this chapter, "absorption

---

[3] An exception to this is *power broadening* of absorption lines which could be a problem in some cases.

measurements" means detection of the absorption process by measuring the ratio of the transmitted and incident light intensities. This is an intrinsically difficult procedure for detecting photon absorption because it involves measuring a small change in a large signal. In spite of this difficulty, absorption measurements are a widely used analytical tool because, in principle, the fractional power absorbed over a known pathlength may be related directly to the concentration of the absorbing species, without additional geometric and instrumental correction factors. *In a general sense, the high power of the laser does not lead to increased sensitivity in standard absorption measurements.* However, high directionality, coherence, narrow spectral widths, tunability, and the nonlinear properties of lasers have contributed to increased sensitivity and selectivity in several special areas. Contributions from high directionality and coherence are rather straightforward and have led to the use of lasers in long pathlength, multireflection cells, remote sensing devices (where the detector and emitter are separated by kilometers), and in situations where beam sizes as small as 10 $\mu$m in diameter are important (e.g., microabsorption).

### 8.2.1  Frequency Modulation

Modulation of one type or another has made a large contribution to absorption measurements. In fact, the field of microwave spectroscopy developed after the discovery of Stark modulation. If the wavelength of either the absorption resonance or the source emission can be rapidly and reproducibly cycled, the difficult problem of measuring a small change in a dc (or a chopped ac) signal becomes the relatively easy problem of detecting an ac signal, the amplitude of which is proportional to absorbance. Modulation techniques are simplified by the narrow spectral width and tunability of the laser and the narrow spectral lines of many absorbing species.

Modulation of the absorption of the species of interest is more selective than source modulation and has contributed significantly to measurements in the infrared spectral regions with "fixed frequency" gas lasers, for which accidental coincidences or near coincidences between molecular absorptions and laser emissions are required. In some cases these coincidences can be induced by the application of large electric or magnetic fields to shift the molecular energy levels. This technique, often called *perturbation spectroscopy* or *laser magnetic resonance* (for magnetic perturbation), has been routinely used in high-resolution spectroscopy but has just recently been applied to analytical problems (28–32). Large perturbation fields are used to induce coincidences with the laser emission, and secondary modulation fields, combined with synchronous

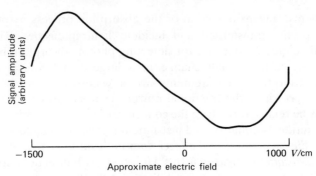

−1500         0       1000  V/cm<br>Approximate electric field

**Fig. 8.3.** Low electric field scan of a 9-ppm vinyl chloride in air sample corresponding to the absorption of the P(13) line of the 20–19 band of the CO laser. A high-frequency modulation imposed on the field and synchronous detection result in an approximate first derivative signal. Reprinted with permission from S. M. Freund and D. M. Sweger, *Anal. Chem.*, **47**, 930 (1975). Copyright by the American Chemical Society.

detection, give rise to easily observable absorption signals (see Fig. 8.3). Examples of the use of this method are analyses for vinyl chloride (28), OH (29), NO (29–31), and $NO_2$ (32). A working curve illustrating the detectability of low concentrations of $NO_2$ is shown in Fig. 8.4. This technique is selective since an interfering species must possess a similar molecular dipole moment as well as a coincident absorption. By the same token, the requirement of an adequate dipole moment reduces the general applicability.

In many cases (e.g., well-resolved atomic spectra) modulation of the laser frequency is just as satisfactory as modulation of the sample. Care must be taken to eliminate, or compensate for, amplitude modulation which often accompanies frequency modulation. The laser frequency can

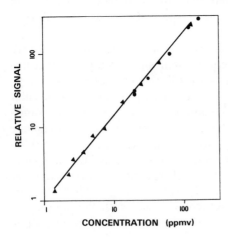

**Fig. 8.4.** Working curve for laser magnetic resonance determination of $NO_2$ in $N_2$ at 27-torr total pressure, using CO laser radiation at 1616 cm$^{-1}$. Reprinted with permission from S. M. Freund, D. M. Sweger, and J. C. Travis, *Anal. Chem.*, **48**, 1944 (1976). Copyright by the American Chemical Society.

be easily and rapidly modulated over an appreciable fraction of the absorption linewidth (0.01–0.001 nm) by using electromechanical devices to change the cavity length and effective thickness of internal etalons in a coordinated fashion. To our knowledge, the laser frequency modulation techniques described here have not yet been applied to analytical absorption determinations. Nevertheless, their potential for atomic flame absorption and isotopic ratio analysis appears promising.

### 8.2.2 Intracavity Absorption

It has been shown by several workers that the detection of small optical losses and gains can be enhanced several orders of magnitude by placing samples inside a laser cavity. A typical experimental arrangement is shown in Fig. 8.5. These measurements fall into two clearly different categories, which are illustrated in Fig. 8.6. In case (a), *the laser mode falls within the absorption linewidth* (inhomogeneously[4] broadened lasers with one mode falling within the absorption linewidth are also included). In case (b), *some or many modes of a homogeneously[4] broadened laser fall outside the absorption linewidth.* Insertion of absorbing samples into the laser cavity in case (a) causes an increased cavity loss and a consequential

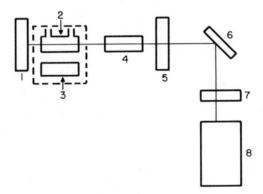

**Fig. 8.5.** Arrangement of apparatus for intracavity absorption: (1,5) mirrors which form the optical cavity; (2) dye cell; (3) excitation source; (4) sample cell; (6) mirror; (7) ground glass scatter plate; (8) spectrograph. Reprinted with permission from R. A. Keller, E. F. Zalewski, and N. C. Peterson, *J. Opt. Soc. Amer.,* **62,** 319 (1972).

[4] The terms homogeneous and inhomogeneous broadening refer to the coupling between laser modes. If each molecule in the laser gain medium can contribute to all modes, the laser emission is said to be homogeneously broadened. Conversely, if some molecules can contribute to only certain modes, then the laser is said to be inhomogeneously broadened (e.g., Doppler broadening, where different modes represent molecules with different velocity projections along the laser axis).

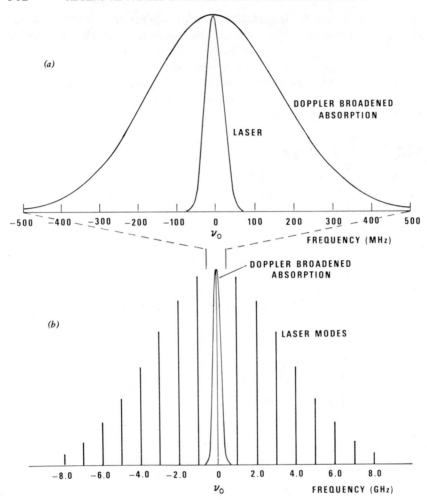

**Fig. 8.6.** Relation between absorption widths and mode structure for intracavity absorption in cases (*a*) and (*b*). Note the factor of 20 scale change between the two parts of the figure.

drop in laser power. In case (*b*), the absorption interacts with a relatively small number of laser modes and the power in these modes is shifted to other modes. In this case the total laser power is not an indicator of the presence of absorbing species, but instead a frequency-selective detector must be used to measure the power loss in the affected modes. For several reasons a theoretical analysis of case (*a*) is much easier than of case (*b*), and a description of the behavior of a single-mode laser in the presence of an overlapping absorption has recently been given (33). The essential conclusion of this treatment is that when the laser is well above threshold, the increased sensitivity is largely the result of the increased effective optical path through the sample. *When this is true, there is no advantage to placing the sample inside the laser cavity over the use of a multireflection cell of equivalent pathlength.* The sensitivity increases significantly near threshold, but at the same time the laser noise also increases and it is difficult to work in this region. With care, it appears possible to work close enough to threshold to attain an enhancement of 10–50 times that realized from the increased optical path.

In contrast to the single-mode case, case (*b*) in Fig. 8.6 involves a very subtle mode competition, and a rigorous theoretical description is difficult. Enhancement factors greatly in excess of the increased optical path are easily attained. A pictorial representation of the mode structure of a cw dye laser is shown in Fig. 8.7. The intermode spacing observed for a laser of length $L_1$ is much larger than that which would result from longitudinal modes ($c/2L_1$) and is determined by competition for gain molecules by the laser modes (spatial hole burning) (34,35). The effect of

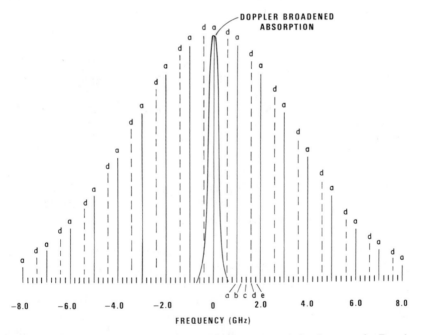

**Fig. 8.7.** Mode structure of a cw dye laser illustrating the relation between the Doppler-broadened absorption and the laser modes. Note that four fifths of the longitudinal modes are missing as the result of spatial hole burning. There are five equivalent sets of spatial hole burning modes (a, b, c, d, and e).

this competition is equivalent to the insertion of an etalon of spacing equal to twice the distance between the end laser mirror and the dye cell ($L_2$). The modes depicted in Fig. 8.7 are only one set of several possible sets. A displacement of all modes by a frequency of $nc/2L_1$ results in equivalent sets where $n = 1, 2, \ldots, L_1/2L_2$. In the presence of optical absorptions that overlap one or more of the oscillating modes, the mode structure of the laser shifts to best accommodate the absorption. Neither the theoretical nor experimental details of this complex process are known.

There have been several simplified theoretical descriptions of intracavity absorption which treat one or more of the concepts discussed above, and some of these descriptions are in relatively good agreement with experimental observations (36–44).

In spite of the difficulty in attaining a rigorous theoretical description, or even agreement on the best experimental parameters, many interesting results have been published. One fact which seems indisputable is that *cw laser systems are more sensitive to the presence of internal absorptions than are pulsed systems; and among pulsed systems, the longer the pulse length the more sensitive is the system.* In fact, there has been no data published for short-pulse (5 ns) $N_2$ laser-pumped dye laser systems, although some work has been reported for (50 ns) Nd laser-pumped systems (45). The decreased sensitivity of pulsed systems is probably due to surges in excited-state populations as the pump is turned on, to thermal distortions, and to the lack of sufficient time to establish good mode structure. Some selected experimental results are described below to illustrate several of these points.

Many workers have studied the enhancement of the detectability of trace absorptions by inserting samples into the cavities of single-mode or inhomogenously broadened lasers. The results obtained in this work are in general in agreement with theoretical predictions—the enhancement is equal to the gain in optical path length (caused by the multiple passes of the laser beam) and becomes larger as the lasers operate closer to threshold (46–48).

The apparatus for an interesting combination of perturbation spectroscopy, modulation techniques, and intracavity absorption (case ($a$) in Fig. 8.6) is shown in Fig. 8.8. This method has been used for the spectroscopic study of the radicals shown in Table 8.1. Concentrations of the OH radical as low as $2 \times 10^8$ cm$^{-3}$ were observable, and signals were found to be independent of laser power and proportional to the concentration of absorbing species over a concentration range of three decades (29). Although a more complete understanding of both the theoretical concepts and the experimental parameters must be realized, the high sensitivity,

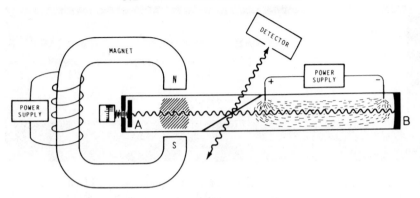

**Fig. 8.8.** Schematic diagram of laser magnetic resonance spectrometer. This apparatus combines the effect of intracavity absorption with magnetic modulation. The sample is contained in the compartment between the poles of the magnet. From Davies and Evenson (29), reproduced by permission.

selectivity, and linearity of the method demonstrates its potential as an analytical tool.

A comparison of the absorption spectrum of iodine inside and outside of the cavity of a flashlamp-pumped dye laser is shown in Fig. 8.9. Comparison of the top two traces in this figure illustrates the gain realized in the detectability of the absorption. Note that a gradual variation in laser power is discernible and that intracavity absorption is not an on–off phenomenon. A comparison of parts B and D in Fig. 8.9 shows that the enhancement factor is roughly 100. The observed enhancement factor of 100 exceeds that expected from a consideration of the increased optical path inside the laser cavity (reflectivity of output mirror ~60%). The enhancement factor was found to be independent of the excitation energy or the reflectivity of the resonator mirrors. Theoretical considerations show that the enhancement should be inversely proportional to the laser cavity length (36–41) and that increased sensitivity is not to be expected

**TABLE 8.1**
**Free Radicals Detected by Laser Magnetic Resonance**

| Radical | State | Radical | State |
|---------|-------|---------|-------|
| $O_2$ | $(X^3\Sigma^-)$ | PH | $(X^3\Sigma^-, a^1\Delta)$ |
| NO | $(X^2\Pi)$ | $NO_2$ | $(^2A_1)$ |
| OH | $(X^2\Pi)$ | HCO | $(^2A')$ |
| CH | $(X^2\Pi)$ | $HO_2$ | $(^2A'', {}^2A')$ |
| $^{14,15}$NH | $(X^3\Sigma^-)$ | $NH_2$ | $(^2B_1)$ |
| $^{14}$NH | $(X^3\Sigma^-, v=1)$ | $PH_2$ | $(^2B_1)$ |

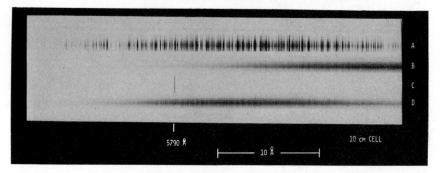

**Fig. 8.9.** Enhancement of the detectability of the absorption spectrum of iodine. Sample is placed inside the cavity of a *flash lamp*-pumped dye laser. (*A*) Sample cell inside cavity; (*B*) sample cell outside cavity; (*C*) Hg calibration; (*D*) sample cell inside cavity with vapor pressure reduced by factor of 100. From R. A. Keller, J. D. Simmons, and D. A. Jennings, *J. Opt. Soc. Amer.*, **63**, 1552 (1973). Reproduced by permission.

by going to longer sample and cavity lengths. The optimum configuration is to match the cavity length and the sample cell length as closely as possible. This prediction has not been tested experimentally.

An apparent optical density can be defined as the optical density measured with the sample inside the cavity. A plot of the apparent optical density versus the true optical density is shown in Fig. 8.10. These data were taken from photographic plates and exhibit a relatively large error. But within this error, the relation between the apparent optical density and true optical density appears linear (see also references 50 and 51). Substitution of diode arrays for photographic plates increases the accuracy of data acquisition at the price of a loss in resolution (52). Several other workers have obtained similar results with various flashlamp-pumped lasers, different dyes, and a variety of samples including solutions, flames, and transient radicals (53–59). The enhancement factors reported by these workers vary between 100 and 1000. An interesting application of this technique recently reported is its use in detecting weak Raman signals (60). In many cases the sensitivity and selectivity of the intracavity technique is limited by the resolution of the spectrograph or monochromator. The effective resolution of the system can be improved by inserting into the laser cavity an etalon of free spectral range greater than the resolution of the spectrograph. The laser output will then consist of a series of lines separated by the free spectral range of the etalon and of a width determined by the finesse of the etalon. One of these modes can be tuned to the center of the absorption and its intensity recorded as a function of absorber concentration (61).

The superior sensitivity of cw laser systems for trace analysis is shown

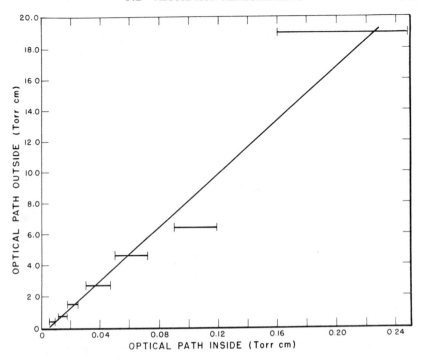

**Fig. 8.10.** Comparison of the apparent optical density of a sample inside the cavity of a flash lamp laser with the true optical density measured outside the laser cavity. From R. A. Keller, J. D. Simmons, and D. A. Jennings, *J. Opt. Soc. Amer.*, **63**, 1552 (1973). Reproduced by permission.

in Fig. 8.11 (62). In this case, the spectrograph was replaced by another sample chamber containing a reference sample of the material being analyzed, to increase the resolution of the detector. The decrease in laser power at wavelengths corresponding to sample absorptions was monitored by measuring the change in fluorescence intensity from the reference cell. An enhancement of $5 \times 10^3$ is indicated by the data. Enhancements as large as $10^5$ were obtained by using a balanced, $Z$-type, laser cavity configuration, with the dye cell in the center of the cavity (37). This configuration eliminates the effects of spatial hole burning and increases the number of modes present in the laser output ($L_2 = L_1/2$). This may have contributed to the excellent sensitivity observed. It is interesting to note that no commercial lasers use this cavity configuration. Another laser configuration that also eliminates spatial hole burning and increases the number of laser modes present is a traveling-wave ring laser (63). It would be interesting to investigate intracavity absorption in this laser configuration.

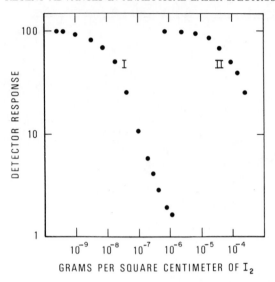

**Fig. 8.11.** Enhancement of the detectability of the absorption spectrum of iodine. The sample is placed inside the cavity of a cw dye laser. The concentration of iodine is correlated with the fluorescence intensity of iodine in a separate detector cell. (I) Sample cell inside laser cavity; (II) sample cell outside laser cavity. The displacement between the two curves indicates an enhancement of ~5000. From Keller and Travis, unpublished results.

Neodymium lasers represent an interesting combination of cases (a) and (b) shown in Fig. 8.6. Laser emission is from atoms distributed among different environments in the laser rod. The emission from each atom is homogeneously broadened, and the total emission is an inhomogeneous combination of the individual homogeneous atomic emissions. The relation of the absorption width to the homogeneous width of the laser emission determines whether the system behaves more like case (a) or case (b). Large enhancements in the detectability of trace absorptions have been observed by placing samples inside the cavity of a pulsed $Nd^{3+}$ glass laser (64–66). Particularly interesting was the observation of the rotational structure of vibrational overtone bands in gaseous molecules (Fig. 8.12).

An illustration of the effectiveness of intracavity absorption is shown in Fig. 8.13. The cell on the left contains a sample of $^{127}I_2$ and the cell on the right, a sample of $^{129}I_2$. The laser beam enters from the left with a spectral width of approximately 0.05 nm, which overlaps several transitions in each molecule. In the top photograph fluorescence is clearly visible from both samples. The bottom photograph illustrates the effect of placing a sample of $^{127}I_2$ inside the laser cavity. The laser emission is

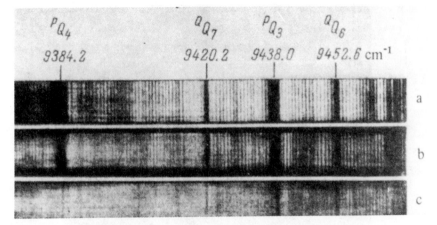

**Fig. 8.12.**   Overtone transitions in $HN_3$ detected by placing the sample inside the cavity of a $Nd^{3+}$ glass laser. Transitions at this frequency correspond to the second overtone. The enhancement factor was reported to be in excess of $10^6$. From T. P. Belikora, E. A. Sviridonkov, and A. F. Suchkov, *Opt. Spectrosc.* (USSR), **37,** 372 (1974). Reproduced by permission.

suppressed at mode wavelengths corresponding to $^{127}I_2$ absorptions but not those corresponding to $^{129}I_2$. This figure illustrates a potentially useful technique for isotope analysis or photochemical isotope enrichment (67). It is especially interesting because no sophisticated laser control is necessary.

Placing samples inside laser cavities is not only useful for detecting the presence of small optical losses, but the method is also very sensitive to the presence of small optical gains (68). The wavelength dependence of the output of a cw dye laser with a He–Se discharge inside the laser cavity shows gaps typical of the absorption processes discussed above and gaps with sharp lines in the center. These sharp lines correspond to optical transitions that exhibit optical gain—often not sufficient to overcome cavity losses and result in laser emission (see Fig. 8.14). Detection of these sharp lines is a very sensitive method for measuring small optical gain and promises to be useful in the search for new laser transitions.

## 8.3   FLUORESCENCE MEASUREMENTS

One of the most sensitive ways of measuring photon absorption is by detecting the reemitted light–fluorescence. All excited molecular and atomic levels fluoresce to some extent. Other processes such as radiationless decay, energy transfer, collisional quenching, and photochemical

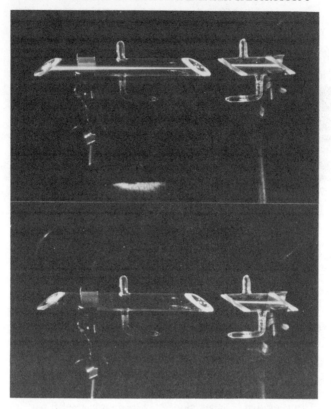

**Fig. 8.13.** Illustration of the use of intracavity samples for isotopic analysis. Top, laser beam exciting fluorescence from two cells, one containing $^{129}I_2$ and the other, $^{127}I_2$. Bottom, sample of $^{127}I_2$ placed inside laser cavity to extinguish laser emission at wavelengths corresponding to absorptions in $^{127}I_2$ with the result that no fluorescence is observable from the $^{127}I_2$ cell outside the cavity while fluorescence from $^{129}I_2$ is still clearly visible. Laser emission was from a cw dye laser with a bandwidth of 0.05 nm which was sufficient to excite several transitions in each molecule. From Keller and Travis, unpublished results.

reactions compete with normal fluorescence and in some cases make its observation difficult. With modern optical detection methods, the observation of 1–10 photons s$^{-1}$ is relatively straightforward. Geometric collection efficiencies as high as 10% are reasonable. A 1-W laser in the visible region emits approximately $10^{18}$ photons s$^{-1}$; so, in principle, the detection of fluorescence quantum yields of $10^{-15}$ is a possibility in bulk species. Alternatively, species with high quantum yields can be detected at very low concentrations. Unfortunately, other factors, principally scattering of the excitation source, determine the detectability limits actually

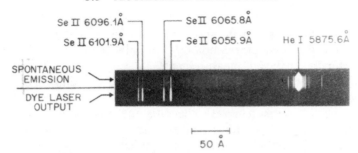

**Fig. 8.14.** Detection of small gains by placing samples inside the cavity of a cw dye laser. Shown are spontaneous emission from the discharge and emission from the dye laser locked to the selenium transitions. The 610.19 nm transition was not previously known to exhibit gain. From Klein, Shank, and Dienes (68), reproduced by permission.

observed. In spite of this limitation it was recently shown that concentrations of atomic species as small as $100$ atoms cm$^{-3}$ can be detected with some improvement still possible (see below).

There are ways to reduce the scattering problem. One of these techniques involves detection of nonresonance fluorescence (e.g., excitation of the transition A→B and observation of the transition B→C). There are several variations of this scheme, and some of them are discussed in Chapter 4. Although a laser is not necessary for the application of this method, often the high intensity of the laser permits the use of low cross section transitions for one or both steps. The narrow spectral width of the laser also contributes strongly to the reduction in scattered light because all of the incident power is effective for the excitation of the fluorescence transition; wavelengths corresponding to the wings of the absorption are not present.

Another process depends upon time resolution to remove the scattered light (69). The fluorescence lifetime of most atomic and molecular transitions is $10^{-9}$ s or longer. Recent advances in laser technology have resulted in tunable laser pulses with pulse lengths of approximately $10^{-12}$ s or less (20–22). Use of these pulses for fluorescence excitation, and appropriate gating of the detection electronics, should eliminate scattered light by time discrimination.

The availability of such short laser pulses, especially at high repetition rates (up to $10$ MHz), facilitates the measurement of another characteristic of the species, namely, the luminescence decay time. Experimental techniques and problems associated with these measurements are discussed in detail in Chapter 7. The improvement in selectivity gained by measuring this additional quantity that is characteristic of the species is especially important in condensed-phase fluorescence measurements, in which the spectral bands are broad ($\sim 10$ nm). In gas phase studies, comparison of measured fluorescence decay times with known fluorescence lifetimes yields information on collisional quenching, a very important factor for analytical measurements.

Modulation techniques discussed in Sect. 8.2.1 and removal of Doppler broadening by saturation and two-photon spectroscopy are also important to fluorescence measurements. The different contributions of lasers to fluorescence analysis are illustrated by discussing specific applications in the following paragraphs. Pulsed laser applications are covered in Chapter 4 and for the most part are omitted in this chapter.

### 8.3.1   Resonance Fluorescence Detection of Sodium in an Evacuated Cell (70–72)

When saturation effects are avoided, the fluorescence intensity at the end of a long sample cell illuminated axially by a laser beam is given by

$$I_f(c) = \alpha I_0 e^{-\varepsilon c L}(1 - e^{-\varepsilon c l}) \tag{1}$$

where $I_f$ (W cm$^{-2}$ sr$^{-1}$) is the fluorescence signal measured at the detector, $I_0$ (W cm$^{-2}$) is the intensity of the excitation laser, $\varepsilon$ ($l$. mole$^{-1}$ cm$^{-1}$) is the molar absorption coefficient, $c$ (mole $l.^{-1}$) is the concentration of fluorescing species, $L$ (cm) is the distance from the cell entrance to the observation region, $l$ (cm) is the length of the fluorescence observation region, and $\alpha$ (sr$^{-1}$) is the product of the geometric collection factor and the quantum efficiency. The fluorescence is detected at right angles to the direction of propagation of the laser beam. This function starts at zero, goes through a maximum at $c = 1/\varepsilon L$, and back to zero as $c \rightarrow \infty$. The absolute values of the concentration $c$ can be affixed to a graph of $I_f$ versus relative concentration by assigning to the maximum the absolute concentration $c = 1/\varepsilon L$. This requires a measurement of $\varepsilon$ that is independent of $c$ ($L$ can be measured very accurately in a long cell). Fortunately, for atomic systems there is a fundamental relationship between the radiative lifetime of the emission and $\varepsilon$. With modern techniques decay times can be measured with an accuracy of 1%. Measurements of $\varepsilon$ in this manner do not require a measurement of the concentration. Proper application of this technique gives absolute measurements of the concentration of the species independent of any other data (e.g., weighing samples, vapor pressures, etc.) and consequently can be used to determine vapor pressures as a function of temperature and related thermodynamic properties.

Vapor pressure data for atomic sodium obtained in this way (71) are shown in Fig. 8.15. The thermodynamic heats of vaporization derived from this curve match extremely well the heats of vaporization obtained by conventional methods in the region where the data overlap. Measurements were made down to a concentration of 100 atoms cm$^{-3}$. Actually, the active volume formed between the observation cone and the excitation beam was much less than 1 cm$^3$, and the total emission was from the equivalent of less than five sodium atoms, each absorbing and emitting many photons in 1 s. Scattered laser light is a problem at these low concentration levels, and its effect was minimized by frequency modulating the spectrally narrow laser emission and synchronously detecting the fluorescence. Laser powers of approximately 3 $\mu$W were sufficient for these measurements.

Trace analysis at these concentration levels is not only applicable to sodium. Most atomic species and many molecular species can be analyzed in this fashion at concentration levels comparable to those attained with sodium or, at the worst, with a loss in sensitivity of two to four orders of magnitude. The largest restriction which now exists is the availability of tunable cw lasers at wavelengths corresponding to strong absorption transitions in a wide variety of atoms and molecules. Since the power requirement for these measurements is quite modest, this restriction will soon be removed with the development of efficient, frequency-doubled cw dye laser emission (19).

Another application of this technique, which has not yet been explored, is to use the high sensitivity to study reaction kinetics and thermodynamic properties of materials that have very low vapor pressures at a reasonably attainable temperature (e.g., Na + OH $\rightleftharpoons$ NaOH at temperatures under 1000°C).

### 8.3.2 Sodium in Metal Oxide Semiconductor Processing Ovens (73)

In the fabrication of metal oxide semiconductors, a silicon wafer is placed in a refractory oven at 1000°C in an $O_2$ atmosphere to form an oxide layer. During this step sodium atoms become incorporated into the oxide layer and cause malfunctions of the semiconductor's switching process. It is postulated that sodium from the refractory material is transported through the oxygen atomosphere as free sodium atoms. The technique described in Sect. 8.3.1 can be adapted to the detection of trace sodium in the furnace atmosphere. The decrease in absorption cross section (due to Doppler and collisional broadening), the low fluorescence quantum yield at these pressures and temperatures, and the increased scattering from gases and dust particles in the furnace atmosphere all contribute to an increase in the detection limit. The fluorescence signal from a contaminated oven as the wavelength of the laser is swept through

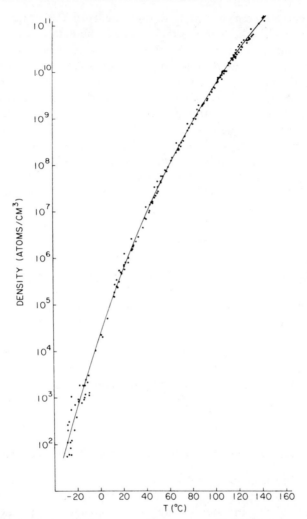

**Fig. 8.15.** Vapor density of sodium vs. temperature determined by resonance fluorescence. Note: Measurements were made down to a density of 100 atoms cm$^{-3}$. Solid line is a thermodynamically derived curve for $\Delta H_0^0 = 25{,}600$ cal mole$^{-1}$. From W. M. Fairbank, T. W. Hänsch, and A. L. Schawlow, *J. Opt. Soc. Amer.* **65,** 199 (1975). Reproduced by permission.

the sodium transition is shown in Fig. 8.16. Concentrations as low as $10^8$ atoms cm$^{-3}$ were detected with a signal-to-noise ratio of 100 to 1. Observation of such low concentrations of atomic sodium under these conditions was made possible by collisional mixing of the two excited $p$ states. Excitation was to one $p$ level and fluorescence was observed from the other. It

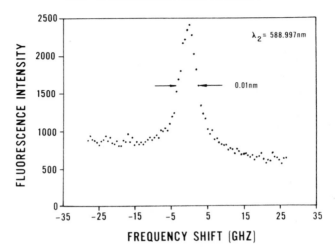

**Fig. 8.16.** Fluorescence emission from sodium atoms inside a semiconductor processing oven. The signal is from approximately $10^8$ sodium atoms $cm^{-3}$. From Mayo et al. (73), reproduced by permission.

remains to be shown whether free sodium does exist in the atmosphere of a clean processing oven and if this indeed is the source of the contamination. Preliminary measurements indicate that the free sodium atom concentration in a clean processing tube is less than $10^6$ atoms $cm^{-3}$.

### 8.3.3 Observation of Natural Linewidths

Doppler broadening results from the translational motion of the absorbing species. In a laboratory coordinate system all the photons can have a single frequency, but in a coordinate system moving with the absorbing species the photon frequency is determined by the projection of the velocity of the absorbing species onto the axis of propagation of the photon beam. A random distribution of velocities exists in a gaseous sample and each molecule, or atom, "sees" a different frequency. The resulting "Doppler" linewidth is a superposition of the narrow natural linewidths of all of the atoms (molecules) present in the system. Doppler-broadened linewidths are proportional to the square root of the temperature and inversely proportional to the square root of the mass. Typical Doppler widths are 0.005 to 0.0001 nm.

It is possible to remove Doppler broadening and observe natural linewidths by several techniques. In the most conventional of these methods, a molecular beam of the atomic or molecular species is prepared in which all of the components have their velocity vectors in one direction. Observation of the absorption spectrum perpendicular to this

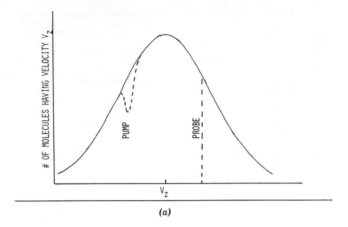

(a)

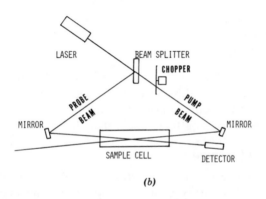

(b)

**Fig. 8.17.** Theory and apparatus for saturation spectroscopy: (a) distribution of molecular velocity projection along propagation direction of spectral source; (b) experimental apparatus for saturation spectroscopy.

direction results in natural linewidths at the "true" transition energy. This technique has received considerable attention in atomic and molecular spectroscopy, but since it need not involve lasers it is not considered further.

*Saturation spectroscopy* is a method of selecting from a randomly distributed set of velocity directions those molecules (atoms) that have a zero velocity component along the analysis beam (74). Consider the pictorial description shown in Fig. 8.17. For a random wavelength with a Doppler-broadened absorption, the two beams traveling in opposite directions will be at the transition wavelength ($\lambda_0 \pm$ Doppler component) for two different groups of molecules, and there will be no interaction

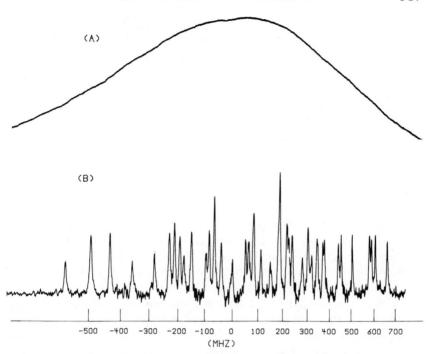

**Fig. 8.18.** Saturation fluorescence spectroscopy of iodine excited by the 514.5 nm Ar$^+$ laser line. The spectral range covered in this figure is approximately 0.001 nm. (A) Doppler-broadened line; (B) hyperfine structure which becomes visible upon removal of Doppler broadening. From Sorem and Schawlow (76), reproduced by permission.

between the beams. For the specific wavelength $\lambda_0$, the groups become identical, and there is an interaction between the two beams via a perturbation of the population of the ground state of the commonly selected group. When the laser pump beam is sufficiently intense to appreciably change the ratio of the population of the ground and excited states, modulation of the population by chopping the pump beam results in a corresponding modulation of the absorption of the probe beam. Synchronous detection of the intensity of the probe beam gives a signal only when the laser wavelength is within the natural linewidth of $\lambda_0$ (75). It is not necessary to detect the change in power of the probe beam (absorption measurement), since insertion of an additional chopper into the probe beam and synchronous observation of the "intermodulated fluorescence" at the sum frequency of the two choppers gives increased sensitivity concomitant with the usual fluorescence advantages (76,77) (see Fig. 8.18). Until recently, saturation spectroscopy was limited to

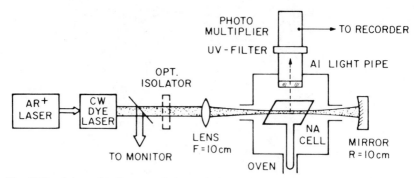

**Fig. 8.19.** Schematic diagram of the optical experimental setup for two-photon laser spectroscopy of Na vapor. From Hänsch et al. (79), reproduced by permission.

those cases where there was an accidental coincidence between a fixed frequency laser and an absorption of interest. The development of tunable lasers greatly expanded the applicability of this technique.

A serious disadvantage of saturation spectroscopy is that a very small fraction of the sample participates in the absorption (e.g., those atoms traveling perpendicular to the laser beams) and only a fraction of these participate in the saturation interaction. A recently developed *two-photon absorption* technique does not suffer from this deficiency (78–82). Although two-photon absorption probabilities are low, the high spectral power available from laser sources makes detection of this process relatively easy. If the two photons involved in the absorption process are at the same wavelength and traveling in opposite directions, the first-order Doppler broadening contributions cancel at each molecular (atomic) center, and natural linewidths are observed. In contrast to the case of single-photon saturation spectroscopy discussed above, *all* atoms or molecules in the laser beam may contribute to the absorption process, and signals can be quite strong. The apparatus for observing two-photon, Doppler-free transitions is rather simple and is illustrated in Fig. 8.19. Contributions from two photons going in the same direction are small because the cross sections are broadened by the Doppler effect. The very sharp natural linewidths ($\sim$50 MHz) observed in two-photon absorptions are useful for detailed signature analysis, even among isotopes (see below) or nuclear isomers, and can be selectively modulated with small fields (83) (see Fig. 8.20).

### 8.3.4 Isotope Analysis

The importance of the ability to do isotopic ratio analysis on both atomic and molecular systems should be stressed. Isotope shifts in spectral lines can be as large as several hundredths of a nanometer in atomic

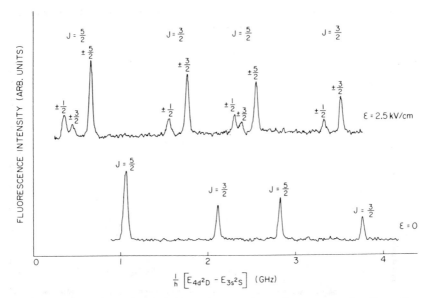

**Fig. 8.20.** Effect of an electric field on the two-photon fluorescence of sodium. From Harvey et al. (83), reproduced by permission.

systems and several tenths of a nanometer in molecular systems. It is well known that the method of standard additions (and/or internal standards) can be used for accurate analysis by eliminating errors that might result from sample handling. For example, a technique for analysis of lithium in blood samples is to add a known amount of $^6$Li to the sample before workup and then measure the resulting isotope ratio in an extensively prepared sample. This technique seems particularly applicable to laser analysis because the isotope splitting in lithium is large with respect to the Doppler width (Fig. 8.21) but well within the easily tunable range of the laser. Isotope shifts for other atoms and molecules can be much smaller and often fall within the Doppler width. In this case the ability to use saturation spectroscopy and two-photon absorption techniques to observe natural linewidths should permit isotope ratio analysis for many elements and expand a new area of analytical chemistry (85–86).

### 8.3.5 Flame Fluorescence

Atomic flame fluorescence is a well-known analytical technique that has achieved limited popular acceptance, partially due to the inadequacy of conventional excitation sources. The development of high-intensity tunable lasers in the last five years offered the possibility that substitution

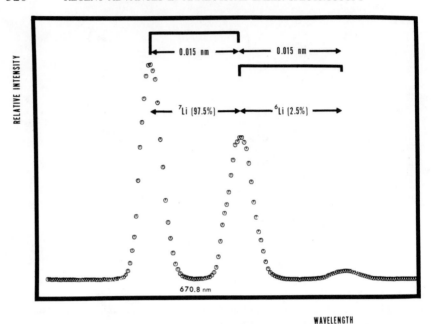

**Fig. 8.21.** Isotope shift and fine structure of lithium fluorescence at 670.7 nm. From Green, Keller, and Travis (84), reproduced by permission.

of lasers for conventional sources would result in a significant improvement in sensitivity and detection limits in atomic flame fluorescence. Extensive work with pulsed systems (see Chapter 4) has demonstrated that, indeed, laser excitation is a significant improvement over conventional excitation. However, the improvement is far less than that initially expected. In fact, in most cases the lowest detection limits are still obtained by flame emission and atomic absorption. Some of the problems associated with pulsed systems may be the cause of the less-than-anticipated results, such as rf noise, poor pulse-to-pulse repeatability, difficulty in attaining and maintaining a narrow and accurate wavelength, and data handling of low duty-cycle, pulsed outputs. It is true that careful attention to experimental details of complex laser technology and data acquisition may reduce these problems, but in practice they have contributed to the results reported in the literature.

In contrast to pulsed systems, cw dye lasers possess none of the characteristics described above. A series of traces of the fluorescence intensity from various concentrations of barium as a function of the wavelength of the laser is shown in Fig. 8.22. The laser output in this case consisted of two or three modes with a total width of approximately 0.003 nm—well within the Doppler and collision-broadened absorption linewidth (see Fig. 8.23). The scattered light level can be measured at

**BARIUM FLUORESCENCE**

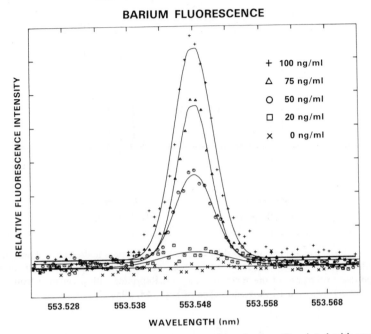

**Fig. 8.22.** Resonance atomic flame fluorescence from barium. Reprinted with permission from R. B. Green, J. C. Travis, and R. A. Keller, *Anal. Chem.*, **48**, 1954 (1976). Copyright by the American Chemical Society.

off-resonance wavelengths and suitable corrections made. A plot of the fluorescence intensity as a function of the concentration of barium is shown in Fig. 8.24. The detectability limit[5] derived from these data $(2 \text{ ng ml}^{-1})$ agrees well with that obtained in atomic flame emission analysis. In an $H_2$–$O_2$–Ar flame the laser-induced fluorescence signal is 3000 times as large as the flame emission signal from the active volume for a 100-mW, 2-mm-diameter laser beam. A comparison of the fluorescence signal with laser scatter from the blank and flame emission noise, for a low background flame, is shown in Fig. 8.25. The major contribution to the noise level is fluctuations in scattering of the laser light from the flame and aspirated solvent.

Detailed noise analysis indicates that an increase in laser power above 1–10 W will not increase signal-to-noise ratios for resonance fluorescence experiments (87), and this is probably the reason that pulsed laser systems with high peak powers do not significantly improve detection limits. The high peak powers available in the pulsed systems will be advantageous for cases where it is desirable to reduce quenching by saturating the absorption transition (88–89) and in the observation of nonresonance transitions

---

[5] Signal-to-rms noise level is equal to 2.

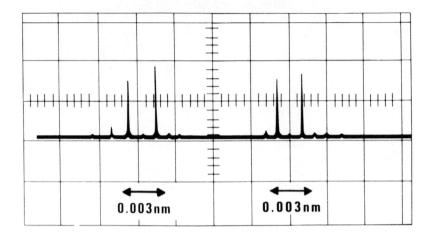

**Fig. 8.23.** Laser mode structure for the excitation of the barium fluorescence shown in Fig. 8.22. Mode spacing determined by spatial hole burning. The laser mode structure is repeated due to the scan of the spectrum analyzer. Reprinted with permission from R. B. Green, J. C. Travis, and R. A. Keller, *Anal. Chem.*, **48**, 1954 (1976).

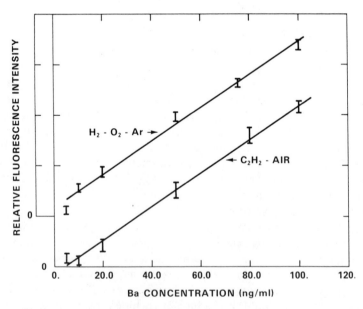

**Fig. 8.24.** Working curve for resonance atomic flame fluorescence from barium. Detection limit ~2 ng ml$^{-1}$. Reprinted with permission from R. B. Green, J. C. Travis, and R. A. Keller, *Anal. Chem.*, **48**, 1954 (1976). Copyright by the American Chemical Society.

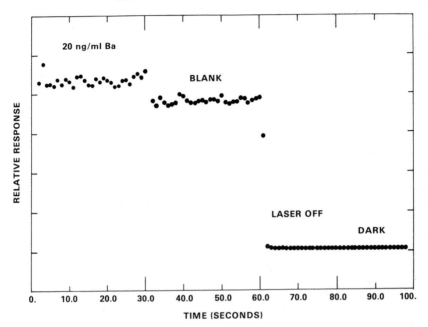

**Fig. 8.25.** Data for the determination of barium by atomic flame fluorescence to illustrate the sources of noise. Reprinted with permission from R. B. Green, J. C. Travis, and R. A. Keller, *Anal. Chem.*, **48**, 1954 (1976). Copyright by the American Chemical Society.

where weak oscillator strengths are involved. However, it is important to realize that the increased laser power is always accompanied by increased scattered light, and an optimum trade-off between these two effects must be established.

The fluctuations in the flame discussed above are also the major contributions to the detection limits observed in flame emission, and it is therefore reasonable that the detection limits for the two techniques are comparable. Detection limits for cw laser-excited atomic fluorescence analysis should significantly improve with the use of less noisy atomization sources, such as furnaces and sputtering chambers.

Another advantage of tunable laser excitation over fixed wavelength irradiation is shown for sodium in Fig. 8.26. The line profile and the introduction of a minimum at the line center at high concentrations is the combined result of self-absorption and self-reversal effects.* In fluorescence analysis there is often an ambiguity introduced because these

* *Editor's comment:* Recently (*Anal. Chem.*, **49**, 2111, 1977) Alkemade and Wijchers have pointed out that the dip observed in this profile can also be due to some spatial masking of the flame.

**SODIUM FLUORESCENCE**

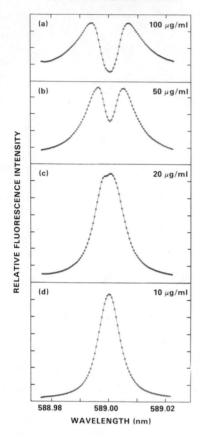

**Fig. 8.26.** Resonance atomic flame fluorescence of sodium illustrating the effects of self-reversal and self-absorption. Reprinted with permission from R. B. Green, J. C. Travis, and R. A. Keller, *Anal. Chem.*, **48**, 1954 (1976). Copyright by the American Chemical Society.

effects cause the working curve to be double valued. *Clearly, line profile analysis eliminates this ambiguity.*

The large fluorescence intensities and the spectral purity of the laser minimize the need for wavelength-selective detection of the fluorescence emission, especially for low-background atom reservoirs. This is an important concept because it shows that high effective resolution (in excess of 200,000) and good selectivity can be obtained without expensive spectrographs. If the laser is confined to a single longitudinal mode, the resulting linewidth (0.00001 nm) represents an effective resolution of 50 million. This resolution is in excess of that required for most analytical determinations but might prove useful in specialized applications, for example, isotope analysis.

At the present time, tunable cw laser excitation of fluorescence in atoms and molecules is restricted to those species that absorb between 420 and 800 nm. Fourteen elements have commonly used analytical lines in this wavelength range. The blue end of this restriction will be extended soon, as frequency doubling techniques improve (19) and milliwatts of power become available down to 220 nm.

### 8.3.6 Small Molecules and Radicals with Pulsed Lasers

A problem which is currently receiving much attention is the detection of trace contaminants in polluted atmospheres. It has recently been shown that laser-based fluorescence techniques work very well for several important contaminants, namely, $CH_2O$ (90), $SO_2$ (91), OH (92), and CN (93). The first three substances must be excited in the UV region which, at present, requires pulsed excitation and frequency doubling.

Formaldehyde has an estimated fluorescence quantum efficiency of $10^{-4}$ at atmospheric pressure. A frequency-doubled dye laser ($\lambda = 324$ nm) was used to initiate fluorescence from samples of formaldehyde in air, and the fluorescence intensity varied linearly with formaldehyde concentration at sample concentrations of 1–10 ppm (90). It was estimated that concentrations as low as 500 ppb could be easily and accurately detected. Interferences from $NO_2$, $SO_2$, and other aldehydes were shown to be insignificant at normal concentrations and could probably be recognized even at high concentrations by measuring excitation spectra.

The detection of very low levels of OH and CN radicals in ambient atmospheres has been a problem for pollution chemists and astrophysicists for several years. Recently both of these radicals have been detected in the number-density range of $10^6$–$10^8$ cm$^{-3}$ by a resonance fluorescence technique (92–93). The detection of OH required doubling the frequency of the output of a tunable dye laser to 280 nm for the excitation of the OH fluorescence. Interesting "spin-offs" from this research include a study of the two-photon dissociation of water to give OH radicals (94) and a study of the OH radical population distribution in flames (95).

Sulfur dioxide is also a constituent of polluted atmospheres. It is known that $SO_2$ fluoresces after excitation to two different electronic states. Fluorescence from the lowest electronic state can be induced by irradiation at wavelengths around 360 nm. However, the lifetime of emission from this state is long ($\sim 1$ ms), and the emission is easily quenched by molecular collisions at atmospheric pressure. The oscillator strength to the second electronic state around 220 nm is much larger, and the corresponding collision-free lifetime is 20 ns. Fluorescence from this state is readily observable, even at atmospheric pressure, and forms the basis of a recently developed $SO_2$ monitor (96). This monitor is based on the use of a Zn lamp for excitation, but recent work shows that a frequency-doubled dye laser at 218 nm works at least as well and has the added potential of ranging experiments (91). The current emphasis on fire research has prompted extensive studies into the basic principles of combustion processes. The ability to detect small amounts of some of the constituents of flames (OH, $C_2$, and CH) (95,97,98) and to measure their spatial distributions and the spatial distribution of the flame temperature from an energy level population analysis (95) has contributed to the understanding of these processes. Another important species ($N_2^+$) has been observed in an electrical discharge (99) but not yet in a flame.

### 8.3.7  Condensed Phase Fluorimetry

The detection of trace contaminants in ambient atmospheres is described in Sect. 8.3.6. A problem of equal importance and interest is the detection of trace materials in liquid solution. Condensed phase fluorescence techniques have played an important role in this area, and the application of laser excitation to the fluorescence analysis has greatly improved the sensitivity in certain applications. In most cases it is possible to excite at one wavelength and observe emission at some other, quite different wavelength. For these cases, if care is taken in the filter system chosen to isolate the emitted light from the excitation light, the effect of laser scattering is greatly reduced and large increases in sensitivity can be attained by using high-intensity laser excitation. The high degree of collimation in the laser beam also contributes to reduced scattering over conventional illumination methods.

The increase in sensitivity with laser excitation has recently been demonstrated by the detection of trace amounts of rhodamine 6G in water and ethyl alcohol by using a pulsed nitrogen laser with a peak power of approximately 80 kW for fluorescence excitation (100). Concentrations less than $10^{-13} M$ were readily detected, and a linear analytical curve was obtained between $10^{-13}$ and $10^{-6} M$.

An interesting application of this technique is the detection of fluorescent effluents of a chromatography column. A problem of great interest to grain producers and processers is the detection of the carcinogenic fungus aflatoxin. Standard chromatographic techniques are not sufficiently sensitive to detect this material at concentrations where it may prove harmful. Nitrogen laser-excited fluorescence has been found capable of detecting very small amounts of aflatoxin, both in thin-layer chromatography spot visualization (101) and in solutions similar to the effluent from a chromatography column (100). Concentrations as low as $8 \times 10^{-12} M$ have been detected.

## 8.3.8 Energy State Population Distribution

Another type of analytical determination which is not considered by most analytical chemists is energy state population distribution. Instead of measuring the total concentration of a particular species, it is sometimes of interest to study the population distribution among the various energy states (102–104). For example, the fluorescence excitation spectrum of BaO immediately after its formation by the collision of a barium atom and an oxygen molecule is shown in Figs. 8.27a and 8.27b. A low-resolution spectrum showing the intensity distribution among the various

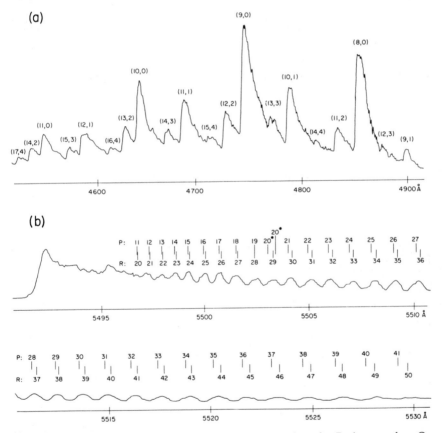

**Fig. 8.27.** Fluorescence from BaO formed by the intersection of a Ba beam and an $O_2$ beam. (a) Low-resolution vibrational spectrum; (b) high-resolution vibrational–rotational spectrum. An analysis of the relative intensities of the transitions in this figure enable the vibrational and rotational temperatures to be determined. From Schultz, Cruse, and Zare (102), reproduced by permission.

vibrational bands is shown in part (a). An analysis of the intensity variation combined with a knowledge of the Franck-Condon factors can be used to determine the vibrational temperature of the newly formed species. In part (b), a higher-resolution spectrum shows the intensity distribution of the rotational transitions inside a single vibrational transition. Analysis of these data leads to the rotational temperature of the BaO. Finally, if the resolution ($\Delta\lambda$) of the laser is sufficiently good, the width of the rotational transitions can be used to give information about the translational temperature of the BaO. In this manner, a complete energy distribution of the newly formed BaO can be determined. Information of this type is of considerable importance to kineticists interested in determining the detailed molecular dynamics of reacting systems. It is interesting to note that a population of $5 \times 10^4$ molecules in a particular vibrational–rotational level is adequate for detection.

A similar example is the use of the vibrational excitation spectrum of OH radicals to determine vibrational temperatures in different parts of flames for combustion research (95).

## 8.4 MICROFLUORESCENCE

Microfluorescence is an application in which it is clear that laser irradiation is much better than conventional irradiation. The high coherence and unidirectional properties of lasers enable extremely high irradiation densities to be attained. A laser beam can be focused into a spot of less than 1 $\mu$m in diameter through a ploem-type illuminator on a high-quality microscope with little loss in total power for fluorescence studies of single cells (105,106). A comparison of the irradiance attainable from different illumination sources is shown in Table 8.2. The laser illumination is much more intense. Microfluorescence analysis can benefit from the increased irradiance in several ways.

If care is taken to reduce scattered light and appropriate optics can be found to isolate observation wavelengths from excitation wavelengths, the

**TABLE 8.2**
**Comparison of Conventional and Laser Irradiation Sources**

| Source | Irradiance[a] |
| --- | --- |
| Xenon arc | 2.4 W cm$^{-2}$ |
| Argon ion laser (200 mW) | 200 W cm$^{-2}$ |
| Pulsed dye laser (0.3 MW) | 300 MW cm$^{-2}$ |

[a] All sources focused to a 300 $\mu$m diameter spot. See (107) and (108).

**Fig. 8.28.** Photomicrograph of fluorescent particles excited with flash lamp-pumped dye laser pulses. Particle diameters ranged from 5 to 40 $\mu$. From Berquist (107), reproduced by permission.

increased illumination can contribute significantly to decreased detection limits. The isolation of the excitation light from the emitted light is the heart of the problem. Even when extreme care is taken in the choice of nonfluorescent filters and the removal of scattered light, the signal-to-noise ratio is still limited by the excitation light that somehow reaches the detector. The high sensitivity has been used to detect very small particles in both fluorescent antibody analysis (107–109) and analysis of fluorescent contaminants adsorbed on dust particles. Fluorescence from single particles as small as 0.5 $\mu$m in diameter can be detected with cw laser illumination (110).

Perhaps an even more important aspect of the availability of high irradiance is the possibility of significant reductions in analysis time. A photomicrograph of fluorescence from particles 5–40 $\mu$m in diameter (107) is shown in Fig. 8.28. In this work the peak irradiance of 30 MW cm$^{-2}$ was estimated to be nearly two orders of magnitude above that necessary to saturate the transition. The effective exposure time for these photographs is 1–10 $\mu$s. Equivalent photographic densities require 1–10 s with conventional illumination. Although saturation effects limit the usable laser energy and the resulting fluorescence is insufficient to expose a photographic emulsion with a single pulse, high-sensitivity vidicon tubes (with near single-photon capability) are now available for single-shot analysis. This reduction in analysis time can be very important in analytical laboratories that must process thousands of samples daily. Recent work has demonstrated that laser-initiated microfluorescence is applicable to the analysis of antibodies associated with syphilis infections (109). A photomultiplier tube was substituted for the camera to increase sensitivity and make the analysis more quantitative. The detection of antibodies was rapid and at least as reliable as conventional techniques and showed no fading in repeated analysis of the same sample (see below).

It is important to measure the increase in signal-to-noise ratio attainable with this technique. If the five orders-of-magnitude increase observed in the fluorescence signal can be obtained without a significant increase in noise, fluorescence signals from single antibodies should be observable and would provide an aid in the early detection of many diseases (107).

Another illustration of the importance of high irradiance is in the use of "flying spot" analyzers to map the fluorescence intensity distribution from a microscopic sample. In this application, the excitation beam scans the sample, and fluorescence is recorded as a function of irradiation position. The dwell time in each sample segment is determined by the time necessary to collect a specified number of luminescence photons that give a signal clearly above dark current. This dwell time is related to irradiance and spot size (resolution). Power densities of approximately 3 kW cm$^{-2}$ are required to produce a signal-to-dark current ratio of 10 for a 6-$\mu$s dwell time (108). This power level is easily attainable with just milliwatts of total power from cw dye lasers focused to less than 1 $\mu$m spot sizes. The dwell time would have to be increased to 1 ms for conventional illumination. This increase in dwell time is intolerable for analytical applications that require resolutions of 0.25 $\mu$m diameter (160,000 resolution elements). Research is in progress to develop automated antibody analysis of the type described.

An added, unexpected benefit of pulsed laser analysis for fluorescent antibodies is that pulsed excitation does not lead to fading of the fluorescence emissions even though the sample receives a photon dose greater than 10 times that which produces fading in conventional illumination (105,107,109). The fading of the fluorescence during conventional analysis is a serious problem which can lead to errors in the results and prevent repeated analysis of the same sample to determine precision. The absence of fading after pulsed irradiation could be explained by a two-step photodissociation process where some photosensitive intermediate is formed in the first step. If the kinetics for the formation of the photosensitive intermediate are slow, an insignificant concentration will accumulate during the pulsed irradiation.

A very different type of microfluorescence experiment is laser cell sorting (111,112). A cw laser beam is focused on a high-velocity stream of suspended cells selectively tagged with an appropriate fluorophor. The resulting fluorescence signal from a single tagged cell is used to trigger a charging pulse and an electrical deflection sorting system. Modification of the technique to include correlated low-angle scattering from a second laser beam results in enrichment factors up to 500 at processing rates of several thousand cells per second. Relative cell populations as low as $10^{-5}$ can be identified with fluorophor concentrations of only a few thousand per cell.

Laser-initiated microfluorescence analysis is in the exploratory stage. If the initial promising results hold true with further research, this area can become very important in medical research and analysis.

## 8.5   COHERENT ANTI-STOKES RAMAN SPECTROSCOPY

A number of nonlinear optical interactions of high-intensity laser beams with matter are currently under investigation (113,114). Two of these, saturation spectroscopy and two-photon absorption, have already been discussed. A third technique, coherent anti-Stokes raman spectroscopy (CARS), a four-wave mixing technique, is attracting wide interest and has already seen limited analytical application. The unique attraction of CARS is that the emitted optical signal, resulting from the interaction of three photons from two incident laser beams in the sample, is itself a spatially and temporally coherent beam (the "fourth wave"). The technique thus shares the geometric advantage of laser absorption experiments of having no inverse-square losses in either the incident or emitted light from the sample. Thus, the spatial coherence of the emitted beam can be used to investigate inaccessible samples, for example, the inside of a jet engine (115).

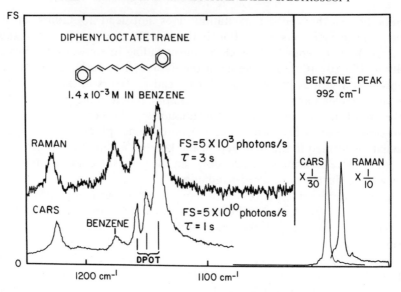

**Fig. 8.29.** Comparison of the CARS and Raman spectra of DPOT in benzene. From Chabay, Klauminzer, and Hudson (119), reproduced by permission.

The CARS spectral "signature," obtained by scanning the difference frequency between the two incident laser beams, resembles the corresponding Raman spectrum (119) (see Fig. 8.29). The resolution, however, is determined by the sample or laser system and not by the detection optics (as in conventional Raman spectroscopy). Indeed, little if any frequency-selective detection is required, especially in condensed phase samples.

The signal intensity for CARS is proportional to the square of the third-order susceptibility of the sample, composed of a Raman contribution and a background contribution. At high sample concentrations ($\geq 10\%$), signal intensities are proportional to the square of the sample concentration and may be typically $10^5$ times (in liquids and solids) to $10^{10}$ times (in gases) as strong as corresponding Raman signals. At lower concentrations, the cross product of resonant and nonresonant components yields a region of linear dependence of signal upon concentration. Finally, the analytical curve approaches zero slope in the low concentration limit as the background term becomes dominant (116). Detection limits of 10 ppm for $H_2$ in a flame (117,118) and $10^{-5} M$ for diphenyloctatetraene in benzene (119) have been observed or estimated. The limitation provided by small nonresonant susceptibility coefficients may be reduced or eliminated by techniques currently under study (120).

## 8.6 GALVANIC DETECTION OF OPTICAL ABSORPTIONS

Recent papers (121,122) have reported a method of obtaining electrical signals, corresponding to laser-induced optical absorptions, directly from the sample medium without conventional optical detection. The method relies upon the sample medium being sufficiently energetic to maintain a small fraction of the analyte population in the first ionic state. The ion fraction of the analyte species is appreciably altered when the energy level population distribution is perturbed by the absorption of laser light. The change in ion density may be readily sensed and quantified with common electrical measurement techniques. To date, the effect has been observed in discharges (121), analytical flames (122), and the effluent of a carbon rod atomizer (123). The sensitivity for sodium in the flame (<1 ppb) was comparable to that obtained in fluorescence with the same system (87).

A particularly useful application of the effect is the stabilization of cw dye lasers to atomic transition wavelengths (124).

## ACKNOWLEDGMENTS

The authors are grateful to Dr. Robert B. Green, Dr. Santos Mayo, and Dr. Rance Velapoldi of the National Bureau of Standards for informative discussions about the current status and future prospects for laser-based analytical techniques and to Dr. Green for his helpful comments on the manuscript.

### References

1. R. A. Keller, *Chemtech*, 626 (1973).
2. J. R. Allkins, *Anal. Chem.*, **47,** 752A (1975).
3. J. I. Steinfeld, *CRC Crit. Rev. Anal. Chem.*, **5,** 225 (1975).
4. J. Kuhl and W. Schmidt, *Appl. Phys.*, **3,** 251 (1974).
5. C. V. Shank, *Rev. Mod. Phys.*, **47,** 649 (1975).
6. R. L. Byer, in *Laser Spectroscopy*, R. G. Brewer and A. Mooradian, Eds., Plenum Press, New York (1974), p. 77.
7. C. K. N. Patel, in *Laser Spectroscopy*, R. G. Brewer and A. Mooradian, Eds., Plenum Press, New York (1974), p. 471.
8. J. S. Wells, F. R. Petersen, G. E. Streit, P. D. Goldan, and C. M. Sadowski, NBS Tech. Note 670, Washington, D.C. (Jan. 1976).
9. F. A. Blum and K. W. Nill, in *Laser Spectroscopy*, R. G. Brewer and A. Mooradian, Eds., Plenum Press, New York (1974). p. 493.
10. E. D. Hinkley, K. W. Nill, and F. A. Blum, in *Topics in Applied Physics*, Vol. 2, *Laser Spectroscopy of Atoms and Molecules*, H. Walther, Ed., Springer-Verlag, Berlin (1976), pp. 125–196.

11.  R. L. Abrams, *Appl. Phys. Lett.*, **25**, 304 (1974).

12.  J. Kuhl and H. Spitschan, *Opt. Commun.*, **5**, 382 (1972).

13.  R. W. Wallace, *Opt. Commun.*, **4**, 316 (1972).

14.  F. B. Dunning, E. D. Stokes, and R. F. Stebbings, *Opt. Commun.*, **6**, 63 (1972).

15.  F. B. Dunning, F. K. Tittel, and R. F. Stebbings, *Opt. Commun.*, **7**, 181 (1973).

16.  C. F. Dewey, W. R. Cook, Jr., R. T. Hodgson, and J. J. Wynne, *Appl. Phys. Lett.*, **26**, 714 (1975).

17.  H. J. Dewey, *IEEE J. Quant. Electron.*, **QE-12**, 303 (1976).

18.  J. M. Yarborough, *Appl. Phys. Lett.*, **24**, 629 (1974).

19.  D. Frölich, L. Stein, H. W. Schröder, and H. Welling, *Appl. Phys.*, **11**, 97 (1976).

20.  C. V. Shank and E. P. Ippen, in *Laser Spectroscopy: Proceedings of the Second International Conference, Megève, June 23–27, 1975*, S. Haroche, J. C. Pebay-Peyroula, T. W. Hänsch, and S. E. Harris, Eds., Springer-Verlag, Berlin (1975), p. 408.

21.  J. M. Harris, R. W. Chrisman, and F. E. Lytle, *Appl. Phys. Lett.*, **26**, 16 (1975).

22.  K. J. Kaufmann and P. M. Rentzepis, *Acc. Chem. Res.*, **8**, 407 (1975).

23.  J. Shewchun, B. K. Garside, E. A. Ballik, C. C. Y. Kwan, M. M. Elsherbiny, G. Hozenkamp, and A. Kazandjian, *Appl. Opt.*, **15**, 340 (1976).

24.  J. J. Ball and R. A. Keller, *J. Air Pollut. Control Assoc.*, **25**, 633 (1975).

25.  L. B. Kreuzer, N. D. Kenyun, and C. K. N. Patel, *Science*, **177**, 347 (1972).

26.  John P. Goldsborough, *Opt. Eng.*, **13**, 523 (1974).

27.  T. W. Hänsch, M. H. Nayfeh, S. A. Lee, S. M. Curry, and I. S. Shahin, *Phys. Rev. Lett.*, **32**, 1336 (1974).

28.  S. M. Freund and D. M. Sweger, *Anal. Chem.*, **47**, 930 (1975).

29.  P. B. Davies and K. M. Evenson, in *Laser Spectroscopy: Proceedings of the Second International Conference, Megève, June 23–27, 1975*, S. Haroche, J. C. Pebay-Peyroula, T. W. Hänsch, and S. E. Harris, Eds., Springer-Verlag, Berlin (1975), p. 132.

30.  A. Kaldor, W. B. Olson, and A. G. Maki, *Science*, **176**, 508 (1972).

31.  P. A. Bonczyk, *Rev. Sci. Instrum.*, **46**, 456 (1975).

32.  S. M. Freund, D. M. Sweger, and J. C. Travis, *Anal. Chem.*, **48**, 1944 (1976).

33.  H. K. Holt, *Phys. Rev.*, **A11**, 625 (1975).

34.  C. T. Pike, *Opt. Commun.*, **10**, 14 (1974).

35.  I. V. Hertel and A. S. Stamatov, *IEEE J. Quant. Electron.*, **11**, 210 (1975).

36.  R. A. Keller, E. F. Zalewski, N. C. Peterson, *J Opt. Soc. Amer.*, **62**, 319 (1972).

37. T. W. Hänsch, A. L. Schawlow, and P. E. Toschek, *IEEE J. Quant. Electron.*, **QE-8,** 802 (1972).

38. H. Von Weyssenhoff and U. Rehling, *Z. Naturforsch.*, **A29,** 256 (1974).

39. W. Brunner and H. Paul, *Opt. Commun.*, **12,** 252 (1974).

40. W. Brunner and H. Paul, *Ann. Phys.* (Leipzig), **32,** 366 (1975).

41. R. I. Stepanov and A. N. Rubinov, *Spectrosc. Lett.*, **8,** 621 (1975).

42. E. N. Antonov, V. G. Koloshnikov, V. R. Mironenko, *Opt. Commun.*, **15,** 99 (1975).

43. K. Tohma, *Opt. Commun.*, **15,** 17 (1975).

44. K. Tohma, *J. Appl. Phys.*, **47,** 1422 (1976).

45. B. Soep, C. Michel, A. Tramer, and C. Lindquist, *Chem. Phys.*, **2,** 293 (1973).

46. C. Chackerian, Jr., and M. F. Weisbach, *J. Opt. Soc. Amer.*, **63,** 342 (1973).

47. N. Djeu, *J. Chem. Phys.*, **60,** 4109 (1974).

48. N. Konjevic and M. Kokovic, *Spectrosc. Lett.*, **7,** 615 (1974).

49. R. A. Keller, J. D. Simmons, and D. A. Jennings, *J. Opt. Soc. Amer.*, **63,** 1552 (1973).

50. R. C. Spiker and J. S. Shirk, *Anal. Chem.*, **46,** 572 (1974).

51. H. Schroder, H. J. Neusser, and E. W. Schlag, *Opt. Commun.*, **14,** 395 (1975).

52. G. Horlick and E. G. Codding, *Anal. Chem.*, **46,** 133 (1974).

53. N. C. Peterson, M. J. Kurylo, W. Braun, A. M. Bass, and R. A. Keller, *J. Opt. Soc. Amer.*, **61,** 746 (1971).

54. R. J. Thrash, H. Von Weyssenhoff, and J. S. Shirk, *J. Chem. Phys.*, **55,** 4659 (1971).

55. H. Von Weyssenhoff, R. Thrash, and J. S. Shirk, *Ber. Bunsenges. Phys. Chem.*, **76,** 1114 (1972).

56. G. H. Atkinson, A. H. Laufer, and M. J. Kurylo, *J. Chem. Phys.*, **59,** 350 (1973).

57. H. W. Latz, H. F. Wyles, and R. B. Green, *Anal. Chem.*, **45,** 2405 (1973).

58. R. B. Green and H. W. Latz, *Spectrosc. Lett.*, **7,** 419 (1974).

59. W. J. Childs, M. S. Fred, and L. S. Goodman, *Appl. Opt.*, **13,** 2297 (1974).

60. W. Werncke, J. Klein, A. Lau, K. Lenz, and G. Hunsalz, *Opt. Commun.*, **11,** 159 (1974).

61. M. Maeda, F. Ishitsuk, and Y. Miyozoe, *Opt. Commun.*, **13,** 314 (1975).

62. R. A. Keller and J. C. Travis, unpublished results.

63. J. M. Green, J. P. Hohimer, and F. K. Tittel, *Opt. Commun.*, **7,** 349 (1973).

64. T. P. Belikova, E. A. Sviridenkov, A. F. Suchkov, L. V. Titova, and S. S. Churilov, *J. Exp. Theor. Phys.*, **62,** 2060 (1972).

65. T. P. Belikova, E. A. Sviridenkov, and A. F. Suchkov, *Quant. Electron.* (USSR), **1,** (1974).

66. T. P. Belikova, E. A. Sviridenkov, and A. F. Suchkov, *Opt. Spectrosc.* (USSR), **37,** 372 (1974).

67. S. Datta, R. W. Anderson, and R. N. Zare, *J. Chem. Phys.,* **63,** 5503 (1975).

68. M. B. Klein, C. V. Shank, and A. Dienes, *Opt. Commun.,* **7,** 178 (1973).

69. N. Omenetto, *Anal. Chem.,* **48,** 75A (1976).

70. D. A. Jennings and R. A. Keller, *J. Amer. Chem. Soc.,* **94,** 9249 (1972).

71. W. M. Fairbank, T. W. Hänsch, and A. L. Schawlow, *J. Opt. Soc. Amer.,* **65,** 199 (1975).

72. F. C. M. Coolan and H. L. Hagedoor, *J. Opt. Soc. Amer.,* **65,** 952 (1975).

73. S. Mayo, R. A. Keller, J. C. Travis, and R. B. Green, *J. Appl. Phys.,* **47,** 4012 (1976).

74. P. H. Lee, *IEEE Trans. Instrum. Meas.,* **IM-21,** 384 (1972).

75. T. W. Hänsch, I. S. Shahin, and A. L. Schawlow, *Phys. Rev. Lett.,* **27,** 707 (1971).

76. M. S. Sorem and A. L. Schawlow, *Opt. Commun.,* **5,** 148 (1972).

77. G. W. Hills, D. L. Philen, R. F. Curl, Jr., and F. K. Tittel, *Chem. Phys.,* **12,** 107 (1976).

78. F. Biraben, B. Cagnoc, and G. Grynberg, *Phys. Rev. Lett.,* **32,** 643 (1974).

79. T. W. Hänsch, K. C. Harvey, G. Meisel, and A. L. Schawlow, *Opt. Commun.,* **11,** 50 (1974).

80. M. D. Levenson and N. Blombergen, *Phys. Rev. Lett.,* **32,** 645 (1974).

81. D. Pritchard, J. Apt, and T. W. Ducas, *Phys. Rev. Lett.,* **32,** 641 (1974).

82. J. A. Gelbwachs, P. F. Jones, and J. E. Wessel, *Appl. Phys. Lett.,* **27,** 551 (1975).

83. K. C. Harvey, R. T. Hawkins, G. Meisel, and A. L. Schawlow, *Phys. Rev. Lett.,* **34,** 1073 (1975).

84. R. B. Green, R. A. Keller, and J. C. Travis, unpublished results.

85. D. E. Roberts and E. N. Fortson, *Opt. Commun.,* **14,** 332 (1975).

86. Y. Kato and B. P. Stoicheff, in *Laser Spectroscopy: Proceedings of the Second International Conference, Megève, June 23–27, 1975,* S. Haroche, J. C. Pebay-Peyroula, T. W. Hänsch, and S. E. Harris, Eds., Springer-Verlag, Berlin (1975), p. 452.

87. R. B. Green, J. C. Travis, and R. A. Keller, *Anal. Chem.,* **48,** 1954 (1976).

88. N. Omenetto, P. Benetti, L. P. Hart, J. D. Winefordner, and C. Th. J. Alkemade, *Spectrochim. Acta,* **28B,** 289 (1973).

89. N. Omenetto, L. P. Hart, P. Benetti, and J. D. Winefordner, *Spectrochim. Acta,* **28B,** 301 (1973).

90. K. H. Becker, U. Schurath, and T. Tatarczy, *Appl. Opt.,* **14,** 310 (1975).

91. H. J. Dewey and P. K. Schenck, National Bureau of Standards, unpublished results.

92. C. C. Wang and L. I. Davis, *Phys. Rev. Lett.*, **32,** 349 (1974).

93. W. M. Jackson, *J. Chem. Phys.*, **59,** 960 (1973).

94. C. C. Wang and L. I. Davis, Jr., *J. Chem. Phys.*, **62,** 53 (1975).

95. C. C. Wang and L. I. Davis, Jr., *Appl. Phys. Lett.*, **25,** 34 (1974).

96. F. P. Schwarz, H. Okabe, and J. K. Whittaker, *Anal. Chem.*, **46,** 1024 (1974).

97. K. H. Becker, D. Hoaks, and T. Tatarczy, *Z. Naturforsch.*, **A29,** 829 (1974).

98. R. H. Barnes, C. E. Moeller, J. F. Kircher, and C. M. Verber, *Appl. Opt.*, **12,** 2531 (1973).

99. P. C. Engelking and A. L. Smith, *Chem. Phys. Lett.*, **36,** 21 (1975).

100. A. B. Bradley and R. N. Zare, *J. Amer. Chem. Soc.*, **98,** 620 (1976).

101. M. R. Berman and R. N. Zare, *Anal. Chem.*, **47,** 1200 (1975).

102. A. Schultz, H. W. Cruse, and R. N. Zare, *J. Chem. Phys.*, **57,** 1354 (1972).

103. R. N. Zare and P. J. Dagdigian, *Science*, **185,** 739 (1974).

104. P. J. Dagdigian, H. W. Cruse, A. Schultz, and R. N. Zare, *Chem. Phys.*, **61,** 4450 (1974).

105. C. A. Sacchi, O. Svelto, and G. Prenna, *Histochem. J.*, **6,** 251 (1974).

106. S. Cova, G. Prenna, C. A. Sacchi, and O. Svelto, in *Excited States of Biological Molecules*, J. B. Birks, Ed., Wiley-Interscience, New York (1976), p. 223.

107. N. R. Berquist, *Scand. J. Immunol.*, **2,** 37 (1973).

108. G. I. Kaufman, J. F. Nestor, and D. E. Wasserman, *J. Histochem. Cytochem.*, **19,** 469 (1971).

109. S. S. Kasatiya, N. G. Lambert, and R. A. Lawrence, *Appl. Microbiol.*, **27,** 838 (1974).

110. R. A. Velapoldi, Analytical Chemistry Division, National Bureau of Standards, private communication.

111. W. A. Bonner, H. R. Hulett, R. G. Sweet, and L. A. Herzenberg, *Rev. Sci. Instrum.*, **43,** 404 (1972).

112. H. R. Hulett, W. A. Bonner, R. G. Sweet, and L. A. Herzenberg, *Clin. Chem.*, **19,** 813 (1973).

113. R. F. Begley, A. B. Harvey, L. R. Byer, and B. S. Hudson, *Amer. Lab.*, **6,** 11 (1974).

114. Y. R. Shen, *Rev. Mod. Phys.*, **48,** 1 (1976).

115. W. M. Roquemore and P. W. Schrieber, Air Force Aero-Propulsion Laboratory Colloquium, National Bureau of Standards (1976).

116. R. F. Begley, A. B. Harvey, and R. L. Byer, *Appl. Phys. Lett.*, **25,** 387 (1974).

117. P. R. Regnier and J.-P. E. Taran, *Appl. Phys. Lett.*, **23,** 240 (1973).

118. P. R. Regnier and J.-P. E. Taran, in *Laser Raman Gas Diagnostics*,

Marshall Lapp and C. M. Penney, Ed., Plenum Press, New York (1974), p. 87.

119. I. Chabay, G. K. Klauminzer, and B. S. Hudson, *Appl. Phys. Lett.*, **28,** 27 (1976).

120. H. Lotem, Department of Chemistry, Harvard University, N.B.S. Colloquium (1976).

121. R. B. Green, R. A. Keller, G. G. Luther, P. K. Schenck, and J. C. Travis, *Appl. Phys. Lett.*, **29,** 727 (1976).

122. R. B. Green, R. A. Keller, G. G. Luther, P. K. Schenck, and J. C. Travis, *J. Amer. Chem. Soc.*, **98,** 8517 (1976).

123. J. C. Travis, M. S. Epstein, P. K. Schenck, D. M. Sweger, and J. R. DeVoe, unpublished results.

124. R. B. Green, R. A. Keller, G. G. Luther, P. K. Schenck, and J. C. Travis, *IEEE J. Quant. Electron.*, **QE13,** 63 (1977).

# INDEX